D1521906

IEE POWER AND ENERGY SERIES 30

Series Editors: Professor A. T. Johns
Dr A. Ter-Gazarian
D. F. Warne

Flexible ac transmission systems (FACTS)

Other volumes in this series:

Volume 1 **Power circuits breaker theory and design** C. H. Flurscheim (Editor)
Volume 2 **Electric fuses** A. Wright and P. G. Newbery
Volume 3 **Z-transform electromagnetic transient analysis in high-voltage networks** W. Derek Humpage
Volume 4 **Industrial microwave heating** A. C. Metaxas and R. J. Meredith
Volume 5 **Power system economics** T. W. Berrie
Volume 6 **High voltage direct current transmission** J. Arrillaga
Volume 7 **Insulators for high voltages** J. S. T. Looms
Volume 8 **Variable frequency AC motor drive systems** D. Finney
Volume 9 **Electricity distribution network design** E. Lakervi and E. J. Holmes
Volume 10 **SF_6 switchgear** H. M. Ryan and G. R. Jones
Volume 11 **Conduction and induction heating** E. J. Davies
Volume 12 **Overvoltage protection of low-voltage systems** P. Hasse
Volume 13 **Statistical techniques for high-voltage engineering** W. Hauschild and W. Mosch
Volume 14 **Uninterruptible power supplies** J. D. St. Aubyn and J. Platts (Editors)
Volume 15 **Digital protection for power systems** A. T. Johns and S. K. Salman
Volume 16 **Electricity economics and planning** T. W. Berrie
Volume 17 **High voltage engineering and testing** H. M. Ryan (Editor)
Volume 18 **Vacuum switchgear** A. Greenwood
Volume 19 **Electrical safety: a guide to the causes and prevention of electrical hazards** J. Maxwell Adams
Volume 20 **Electric fuses, 2nd Edn.** A. Wright and P. G. Newbery
Volume 21 **Electricity distribution network design, 2nd Edn.** E. Lakervi and E. J. Holmes
Volume 22 **Artificial intelligence techniques in power systems** K. Warwick, A. Ekwue and R. Aggarwal (Editors)
Volume 23 **Financial and economic evaluation of projects in the electricity supply industry** H. Khatib
Volume 24 **Power system commissioning and maintenance practice** K. Harker
Volume 25 **Engineers' handbook of industrial microwave heating** R. J. Meredith
Volume 26 **Small electric motors** H. Moczala
Volume 27 **AC-DC power system analysis** J. Arrillaga and B. C. Smith
Volume 28 **Protection of electricity distribution networks** J. Gers and E. J. Holmes
Volume 29 **High voltage direct current transmission** J. Arrillaga

Flexible ac transmission systems (FACTS)

Edited by
Yong Hua Song
and Allan T Johns

The Institution of Electrical Engineers

Published by: The Institution of Electrical Engineers, London,
United Kingdom

The Institution of Electrical Engineers,
Michael Faraday House,
Six Hills Way, Stevenage,
Herts. SG1 2AY, United Kingdom

British Library Cataloguing in Publication Data

A CIP catalogue record for this book
is available from the British Library

ISBN 0 85296 771 3

Printed in England by TJ International Ltd., Padstow, Cornwall

Contents

Preface

The rapid development of power electronics technology provides exciting opportunities to develop new power system equipment for better utilization of existing systems. During the last decade, a number of control devices under the term "Flexible AC Transmission Systems" (FACTS) technology have been proposed and implemented. FACTS devices can be effectively used for power flow control, loop-flow control, load sharing among parallel corridors, voltage regulation, enhancement of transient stability and mitigation of system oscillations. A large number of papers and reports have been published on these subjects. In this respect, it is timely to edit a book with an aim to report on the state of the art development, internationally, in this area. By covering all the major aspects in research and development of FACTS technologies, the book intends to provide a comprehensive guide, which can serve as a reference text for a wide range of readers.

Chapter 1 focuses on the fundamentals of ac power transmission to provide a necessary technical background for understanding the problems of present power systems and the power electronics-based solutions the Flexible AC Transmission System (FACTS*)* offers. In Chapter 2 the principles of power electronic converters are introduced, covering the basics of power electronics systems as well as structures suitable for the design of high power converters for transmission level voltages and currents. Although the inclusion of high voltage dc transmission (HVdc) in this book seems to be a contradiction to some people, the boundaries between HVdc and FACTS will gradually become 'blurred'. For example, the back-to-back dc link may also be considered as a FACTS device. Thus an introduction to HVdc technology is given in Chapter 3.

The principles and applications of shunt, series, phase shifter and unified compensations are discussed in Chapters 4, 5, 6 and 7 respectively. Chapter 4 describes the principles, configuration and control of two major types of shunt static compensations – static var compensator (SVC) and Static Synchronous Compensator (STATCOM). Their practical applications are also reported, including recent relocatable SVC applications in the UK system. Chapter 5 examines the Thyristor-Controlled Series Capacitor (TCSC) and the Static Synchronous Series Compensator (SSSC) and their applications for damping of electromechnical oscillations and for mitigation of subsynchronous resonance. The main objectives of Chapter 6 are to describe the principles of operation, operational characteristics, technical merits and limitations and potential applications of phase shifters. More recently, one of the potentially most versatile class of FACTS device – the Unified Power Flow Controller (UPFC) was proposed. This device, with its unique combination of fast shunt and series compensation, offers a versatile device for the relief of transmission constraints. Chapter 7 contains an in-depth look at the basic operating principles, characteristics, control and dynamic performance of the UPFC. The first UPFC installation is also reported.

Chapters 8, 9, 10, 11 and 12 address the system aspects of FACTS applications. Various models, suitable for different studies including electromagnetic transient studies, steady-state and dynamic analysis, are presented. Effective control strategies for power flow and stability control, and novel protection schemes are proposed in these chapters. A review of FACTS development in Japan is reported in Chapter 13. Applications of power electronics in distribution systems are summarized in Chapter 14.

Finally, we, the editors, are very grateful to authors for their cooperation and patience. We wish to thank Sarah Daniels of the IEE for her help in the production of the book. We would also like to thank Xing Wang for re-setting the style of the whole book by overcoming incompatible word processing format.

Yong Hua Song, Brunel University
Allan T Johns, The University of Bath

October 1999

Contributors

L. Gyugyi and C.D. Schauder
FACTS & Power Quality Division
Siemens Power Transmission & Distribution
1310 Beulah Road
Pittsburgh, Pennsylvania 15235-5098, USA

G. Joos
Department of Electrical and Computer Engineering
Concordia University
1455 de Maisonneuve Blvd W.
Montreal, Que.
Canada H3G 1M8

J. Arrillaga
Department of Electrical and Electronic Engineering
University of Canterbury
Private Bag 4800, Christchurch, New Zealand

H. L. Thanawala, D. J. Young and M. H. Baker
Power Electronic Systems Ltd
Alstom, PO Box 27
Stafford S17 4LN, UK

M. Noroozian, L. Ängquist and G. Ingeström
Sweden ABB Power Systems AB
PSA Department
S-721 69 Västerås
Sweden

M.R. Iravani
Department of Electrical & Computer Engineering
University of Toronto
10 King's College Road
Toronto, Ontario, Canada

Y.H. Song and J.Y. Liu
Brunel Institute of Power Systems
Brunel University
Uxbridge UB8 3PH, UK

H.F. Wang
Department of Electronic and Electrical Engineering
The University of Bath
Bath BA2 7AY, UK

R. Mihalič and P. Žunko
Faculty of Electrical Engineering
University of Ljubljana
Trzaska 25, 1000 Ljubljana, Slovenia

D. Povh
Siemens AG
EV HA2, P.O. Box 3220
D-91050 Erlangen, Germany

Q.Y. Xuan
Panasonic
Matsushita Communication Industrial
UK Ltd
Daytona Drive
Thatacham RG19 4ZD, UK

A.T. Johns
Department of Electronic and
Electrical Engineering
The University of Bath
Bath BA2 7AY, UK

Y. Sekine and T. Hayashi
CRIEPI
2-11-1, Iwato-kita, Komae-shi
201-8511Tokyo, Japan

N. Jenkins
Department of Electrical Engineering
and Electronics
UMIST
PO Box 88
Manchester M60 1QD, UK

Chapter 1

Power transmission control: basic theory; problems and needs; FACTS solutions

Laszlo Gyugyi

1.1 Introduction

What we now refer to as the electric power industry began over 100 years ago, in the 1880s. Almost from the very beginning two competitive systems started to emerge: d*irect current (dc)* power generation and transmission strongly pursued by Thomas Edison, and *alternating current (ac)* power generation and transmission initiated in Europe and transformed into a practical scheme with Nikola Tesla's inventions. This scheme, implemented by industrialist George Westinghouse, decisively won the early competition in 1896 when the famous Niagara hydro power generation project convincingly demonstrated viable "long distance" ac power transmission over a 20 mile, 11 kV "high voltage" line from Niagara Falls to the city of Buffalo, NY. The success of the prestigious Niagara project fuelled the universal acceptance and rapid development of ac power systems. The key to this acceptance was the technical feasibility of stepping up the alternating generator voltage by highly efficient magnetic transformers for transmission to minimize losses, then stepping it down for the consumer to meet domestic and industrial load requirements. The Niagara and subsequent ac power systems first utilized the high voltage transmission capability for remote power generation and ultimately for intertying separate power systems into a large area power grid characterizing modern supply systems today. This is in contrast to Edison's concept of a dc power system, which, due to transmission limitations, envisioned a large number of distributed and independent *dc central* (generation) *stations*, each supplying no more than a few square miles of distribution network for local loads.

Edison's dc system was, from the theoretical viewpoint, simple to visualise. Only real quantities, voltage, current, and resistance were involved, and Ohm's law defined a simple relationship between them. However, the practical problems at that time for transmission were formidable. Since direct voltage could not be transformed up or down, the nominal voltage from generation to load had to be the same and, for safety reasons, had to be rather low. Consequently, the I^2R loss prevented the transmission of even a modest amount of power over more than a couple of miles.

The transformability of alternating voltage seemingly solved the long distance transmission problem. However, ac transmission involves both real and reactive circuit parameters and variables which jointly determine the transmittable real power and overall transmission losses encountered. As will be seen, the unavoidable reactive power flow in ac lines present difficult problems and ultimately imposes severe limitations on traditional ac power transmission, many of which were not fully appreciated at the outset.

It is an ironic fact of power transmission history that, after more than half a century of Edison's pioneering work, dc transmission was reinvented with modern *electronics* technology to solve the problems of *long distance* power transmission. It adds to the curiosity of this situation that, today, *power electronics,* which made high voltage dc transmission possible, is also being applied to solve the outstanding problems of ac transmission. Whereas some may look at these developments as a still ongoing competition between ac and dc transmission (if not between Thomas Edison and George Westinghouse), a more objective view probably recognizes the fact that the two types of system complement each other and together often provide the optimal system solution.

1.2 Fundamentals of ac power transmission

The main constituents of an ac power system are: *generators, transmission (subtransmission),* and *distribution lines,* and *loads,* with their related auxiliary support and protection equipment. The generators are rotating *synchronous* machines. The transmission, subtransmission, and distribution lines are essentially distributed parameter, dominantly *reactive* networks designed to operate at high, medium, and low, alternating voltages, respectively. The loads may be synchronous, non-synchronous, and passive, consuming in general both real and reactive power. This chapter focuses on the fundamentals of ac power transmission to provide a necessary technical background for understanding the problems of present power systems and the power electronics-based solutions the *Flexible AC Transmission System (FACTS)* offers.

The modern transmission system is a complex network of transmission lines interconnecting all the generator stations and all the major loading points in the power system. These lines carry large blocks of power which generally can be routed in any desired direction on the various links of the transmission system to achieve the desired economic and performance objectives. Separate ac systems may be synchronously intertied with ac transmission lines to form a *power pool* in which energy can be transported among and between the systems. In this arrangement, at a given time some systems may be importing and others exporting power, while some systems may be just providing the service of "wheeling" power through their transmission network to facilitate particular transactions. The main characteristic of today's transmission system is an overall *loop* structure, as illustrated with a simple power system schematic in Figure 1.1, which provides a number of path combinations to achieve the functional versatility desired. This is in contrast to early day transmission (and present day subtransmission and distribution systems), which were (are) mostly *radial,* supplying power from generator to a defined load.

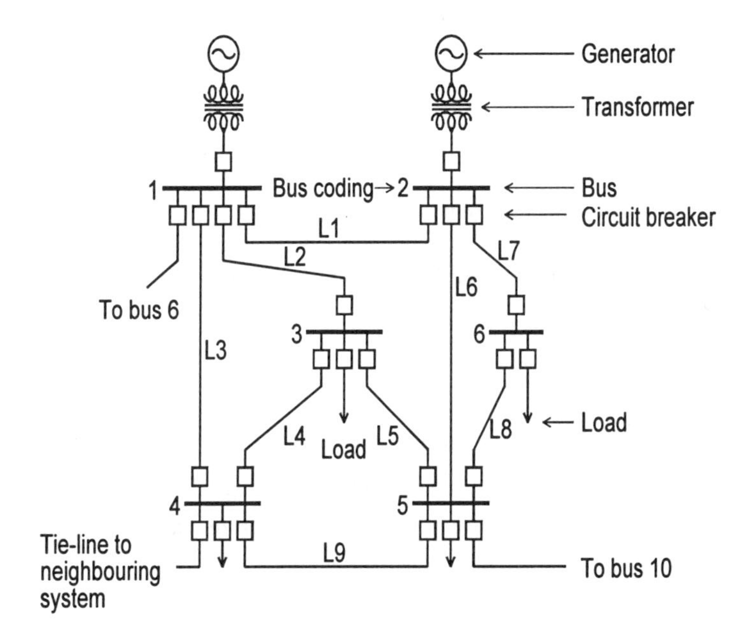

Figure 1.1 Typical power system structure

1.2.1 Basic relationships

In spite of the generally complex nature of an actual power system, the basic relationships of power transmission can be derived by a simple so-called two machine model, in which a sending-end generator is interconnected by a transmission line with a receiving-end generator (which is sometimes considered as an infinite power voltage bus). For the sake of generality, the sending-end and receiving-end generators in the model may also represent two independent ac systems, which are intertied by a transmission link for power exchange.

An ac transmission line is characterized by its distributed circuit parameters: the series *resistance* and *inductance*, and the shunt *conductance* and *capacitance.*

The characteristic behaviour of the line is primarily determined by the reactive circuit elements, the series inductance l and shunt capacitance c. With a customary lumped-element representation of the ac transmission line, the two machine transmission model is shown in Figure 1.2. (Bold-faced letters represent voltage and current phasors.)

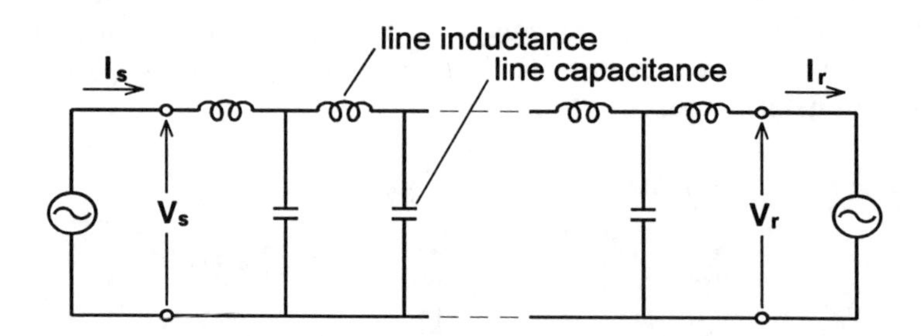

Figure 1.2 Lumped element representation of a lossless transmission line

The transmittable electric power of the system shown in Figure 1.2 is defined by the following equation:

$$P = \frac{V_s V_r}{Z_0 \sin\theta} \sin\delta \tag{1.1}$$

in which

V_s is the magnitude of the sending-end (generator) voltage, $\mathbf{V}_s$

V_r is the magnitude of the receiving-end (generator) voltage, $\mathbf{V}_r$

δ is the phase angle between V_s and V_r (*transmission* or *load* angle),

Z_0 is the *surge* or *characteristic* impedance given by

$$Z_0 = \sqrt{\frac{l}{c}} \tag{1.2}$$

θ is the *electrical length* of the line expressed in radians by

$$\theta = \frac{2\pi}{\lambda} a = \beta a \tag{1.3}$$

where λ is the wavelength and β is the number of complete waves per unit line length, i.e.,

$$\beta = \frac{2\pi}{\lambda} = \omega\sqrt{lc} = 2\pi f\sqrt{lc} \tag{1.4}$$

and a is the length of the line.

The lossless line considered exhibits an ideal power transmission characteristic at the *surge impedance* or *natural* loading, at which the transmitted power is:

$$P_0 = \frac{V_0^{\,2}}{Z_0} \tag{1.5}$$

where V_0 $(= V_s = V_r)$ *is* the nominal or rated voltage of the line. At natural loading the amplitude of the voltage remains constant and the voltage and current stay in phase with each other (but rotated together in phase) along the transmission line. Consequently, the transmission power, P_0, is *independent* of the length of the line. At surge impedance loading the reactive power exchange within the line is in perfect balance, and the line provides its own shunt *compensation*. That is, the reactive power demand of the series line reactance is precisely matched by the reactive power generation of the shunt line capacitance.

Unfortunately, economic considerations and system operation requirements rarely allow surge impedance loading. At lighter loads the transmission line is *over* compensated. The voltage increase across the series line reactance, due to the charging current of the shunt line capacitance, is greater than the voltage drop caused by the load current. As a result, the transmission line voltage increases along the line, reaching its maximum at the mid-point. This "surplus" charging current of course also flows through the sending-end and receiving end generators (or ac systems) forcing them to *absorb* the corresponding (capacitive) reactive power. At greater than surge impedance loading the transmission line is *under* compensated. That is, the voltage increase resulting from the shunt line capacitance is insufficient to cancel the voltage drop across the series line reactance due to the load current. Therefore, the voltage along the line decreases, reaching the minimum at the mid-point. In this case, the net reactive power demand of the line (inductive) must be supplied by the sending-end and receiving-end generators.

Equation (1.1) provides a generalized expression characterizing the power transmission over a *lossless,* but otherwise accurately represented line. For the explanation of the major transmission issues, and for the introduction of relevant FACTS concepts, it is convenient to use an approximate form of Equation (1.1) characterizing *electrically short* transmission lines, for which $\sin\theta \cong \theta = \beta a = \omega a\sqrt{lc}$. Then $Z_0\theta = \omega a \sqrt{lc}\sqrt{l/c} = \omega al = \omega L = X$, the series inductance of the line, and the transmitted power becomes:

$$P \cong \frac{V_s V_r}{X}\sin\delta \quad \text{or} \quad P \cong \frac{V^2}{X}\sin\delta \tag{1.6}$$

This simplified equation neglects the shunt capacitance of the line. The effect of shunt capacitance on the transmission for lines shorter than 100 miles is indeed negligibly small. Moreover, although the line capacitance, as explained above, can cause over-voltage problems for under loaded lines, since $\sin\theta \leq \theta$ and thus $Z_0 \sin\theta \leq X$, it *always* tends to *increase* the transmittable power. Thus, the neglect of shunt line capacitance represents a worse than actual case from the standpoint of maximizing the (steady-state or transient) transmittable power, but does not falsify the main considerations governing power flow control (which also determine the possible utilization of transmission assets and the application of FACTS technology). The voltage related problems of open circuited and underloaded lines are usually handled satisfactorily by permanently connected or switched shunt reactors in combination with the excitation control of generators.

Equation (1.6) can also be derived from the elementary techniques of ac circuit analysis, using complex *phasors*, an approach which allows a very effective treatment and illustration of different power flow control concepts employed in FACTS. Consider again the simple two machine model, but assume a purely inductive transmission line with zero shunt capacitance, as shown in Figure 1.3a with the corresponding phasor diagram in Figure 1.3b. If the sending- and receiving-end voltages are defined by

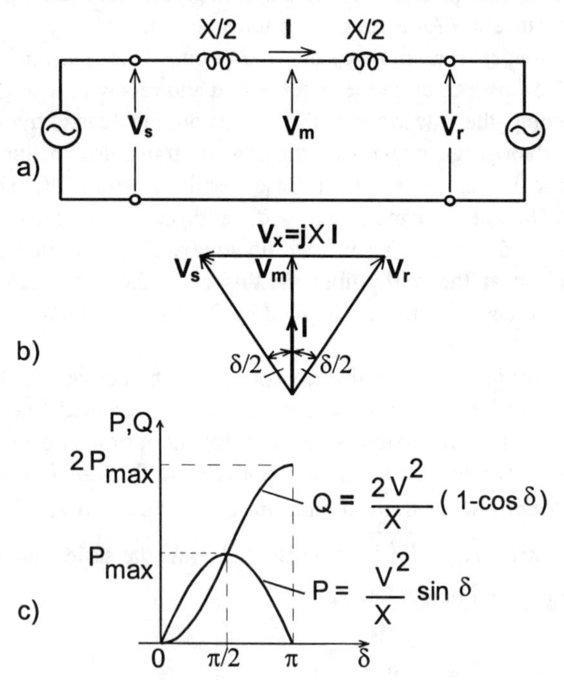

Figure 1.3 (a) Two machine power system with inductive line; (b) corresponding phasor diagram; and (c) power transmission vs. angle characteristic

$$V_s = Ve^{j\delta/2} = V(\cos\frac{\delta}{2} + j\sin\frac{\delta}{2}) \tag{1.7}$$

and

$$V_r = Ve^{-j\delta/2} = V(\cos\frac{\delta}{2} - j\sin\frac{\delta}{2}) \tag{1.8}$$

then the midpoint voltage is

$$\boldsymbol{V_m} = \frac{\boldsymbol{V_s} + \boldsymbol{V_r}}{2} = V_m e^{j0} = V\cos\frac{\delta}{2} \tag{1.9}$$

and the current through the line is given by

$$\boldsymbol{I} = \frac{\boldsymbol{V_s} - \boldsymbol{V_r}}{jX} = \frac{2V}{X}\sin\frac{\delta}{2} \tag{1.10}$$

In the case of the lossless line assumed, power is the same at both ends (and at the midpoint), i.e.,

$$P = V_m I = \frac{V^2}{X}\sin\delta \tag{1.11}$$

which is, of course, identical to that given by equation (1.6). The reactive power provided for the line at each end is

$$Q_s = -Q_r = VI\sin\frac{\delta}{2} = \frac{V^2}{X}(1-\cos\delta) \tag{1.12}$$

The relationships between real power P, reactive power Q, and angle δ are shown plotted in Figure 1.3c. As seen, at a constant voltage (V_s=V_r=V) and fixed transmission system (X=const) the transmitted power is exclusively controlled by angle δ. Note also that real power, P, cannot be controlled without also changing the reactive power demand on the sending- and receiving-ends.

A different illustration using voltage and current phasors to provide a physical explanation for the coupled variation of real power, reactive power, and of voltage along the transmission line, as a function of angle δ, is presented in Figures 1.4a and 1.4b. Figure 1.4a shows that, as angle δ is increased, the voltage phasor $\boldsymbol{V_x}$ *across* the series line reactance increases, which proportionally increases the line current phasor $\boldsymbol{I}$. At the same time, the (line to neutral) voltage along the line decreases from the ends towards the middle, reaching a minimum at the actual midpoint. It can also be observed that the relative angular position of the voltage along the line from the two ends is continuously changing (in the opposite direction) until at the midpoint it falls precisely in phase with the line current I (which is at a fixed 90° angle with respect to the voltage V_x). Thus, the product of the midpoint voltage and the line current, $V_m I$, yields the transmitted power P. It is evident from this relationship that the progressive increase of angle δ will not progressively increase the power P. This is because for $\delta > 90°$, the midpoint voltage will decrease more rapidly than the line current increases and, consequently, their product will decrease from its maximal value and ultimately reach zero at $\delta = 180°$.

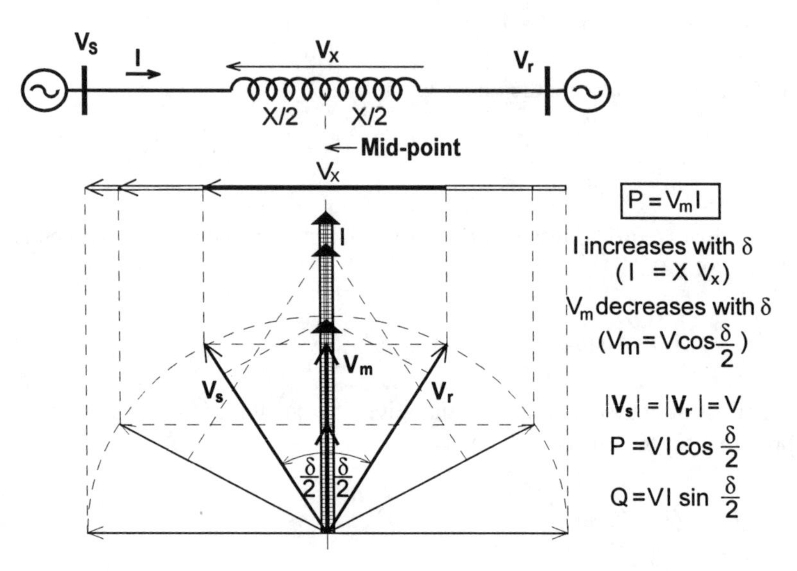

Figure 1.4a Variation of transmission line mid-point voltage, line current, and power with angle δ

Figure 1.4b illustrates the variation of the in-phase and quadrature components of the line current with respect to the (receiving-)end voltage, as angle δ is varied from zero to 180° (i.e., $\delta/2$ varied from zero to 90°). The quadrature component of the line current with respect to the sending-end, or the receiving-end, voltage progressively increases with δ until it becomes the total line current at $\delta = 180°$. By contrast, the in-phase component of the line current with respect to the end voltage increases with δ in the $0 < \delta < 90°$ interval, and then it decreases to zero as δ reaches 180°.

From the above observations it can readily be concluded that transmittable electric power at a given system voltage must be a function of the electrical length of the line, i.e., of the effective series line reactance X. Once the theoretical limit of steady-state power transmission is reached at $\delta = 90°$, the transmitted power would clearly decrease with increasing line length unless either the line voltage is increased or the effective line impedance is decreased.

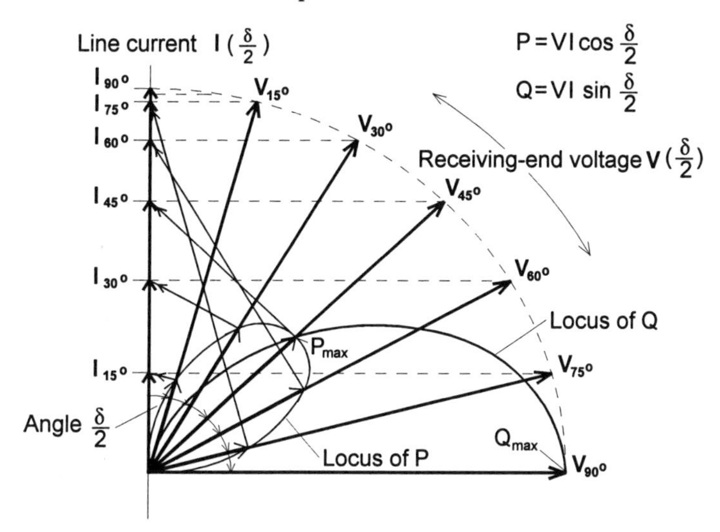

Figure 1.4b Variation of the transmitted real power P and the (receiving-end) reactive power Q with angle δ

1.2.2 Steady-state limits of power transmission

In the previous section it has been shown that the maximum power, $P_{max}=V^2/X$, transmittable over a *lossless* line at a given transmission voltage, is totally determined by the line reactance X and thus sets the *theoretical* limit for steady-state power transmission. A practical limit for an actual line with resistance R may be imposed by the I^2R loss that heats the conductor. At a certain temperature the physical characteristics of the conductor would irreversibly change (e.g., it could get deformed with a permanent sag). This sets a *thermal limit* for the maximum transmittable power. Generally, for long lines X, and for short lines R would provide the main transmission limitation. (It is mentioned here, without explanation, that sufficiently large R, at which the X/R ratio becomes relatively low $(X/R < 5)$, may impose practical limitation on the useful power transmission, before the thermal limit is reached, by increasing the reactive power flow in the line.)

AC loads are generally sensitive to the magnitude, and may be sensitive to the frequency of the applied alternating voltage. AC power systems are generally operated at a substantially constant (typically 50 or 60 Hz) frequency. The voltage levels in ac systems may moderately vary, but are not allowed to exceed well defined limits (typically +5 and −10%). This tight voltage tolerance may impose the primary transmission limitation for long *radial* lines (no generation at the receiving end) and for the so-called *tapped-lines,* which feed a number of (relatively small) loads along the transmission line.

Steady-state power transmission may also be limited by the so-called *parallel* and *loop* power flows. These flows often occur in a multi-line, interconnected power system, as a consequence of basic circuit laws which define current flows by the impedance rather than the current capacity of the lines. They can result in overloaded lines with thermal and voltage level problems.

1.2.3 Traditional transmission line compensation and power flow control

It has long been recognized that the steady-state transmittable power can be increased and the voltage profile along the line controlled by appropriate reactive compensation. The purpose of this reactive compensation is to change the natural electrical characteristics of the transmission line to make it more compatible with the prevailing load demand. Thus, shunt connected, fixed or mechanically switched reactors are applied to minimize line overvoltage under light load conditions, and shunt connected, fixed or mechanically switched capacitors are applied to maintain voltage levels under heavy load conditions. In the case of long transmission lines, *series capacitive* compensation is often employed to establish a *virtual short* line by reducing the inductive line impedance and thereby the electrical length, θ, of the line ($\theta = \beta a = \sqrt{X_l / X_c}$). In some multi-line system configurations, it can happen that the transmission angle imposed "naturally" on a particular line is inappropriate for the power transfer planned for that line. In this case, a *phase angle regulator* (or *phase shifter*) may be employed to control the angle of this line independent of the prevailing overall transmission angle.

In the following sections, basic approaches to increase the transmittable power by ideal *shunt-connected var compensation, series compensation,* and *phase angle regulation* will be reviewed. These basic approaches will provide the foundation for power electronics-based compensation and control techniques capable not only of increasing steady-state power flow but also of improving the stability and overall dynamic behaviour of the system.

1.2.3.1 Ideal shunt compensation

Consider again the simple two machine (two bus) transmission model in which an *ideal var compensator* is shunt connected at the midpoint of the transmission line as shown in Figure 1.5a. This compensator is represented by a sinusoidal ac voltage source (of the fundamental frequency), *in-phase* with the midpoint voltage, V_m, with an amplitude identical to that of the sending- and receiving-end voltages ($V_m = V_s = V_r = V$). The midpoint compensator in effect segments the transmission line into two independent parts: the first segment, with an impedance of *X/2,* carries power from the sending end to the midpoint, and the second segment, also with an impedance of *X/2*, carries power from the midpoint to the receiving end. Note that the mid-point var compensator exchanges only reactive power with the transmission line in this process. The relationship between voltages, V_s, V_r, V_m, (together with V_{sm}, V_{rm}), and line segment currents I_{sm} and I_{mr} is shown by the phasor diagram in Figure 1.5b.

For the lossless system assumed, the real power is the same at each terminal (sending-end, mid-point, and receiving-end) of the line, and it can be derived readily from the phasor diagram of Figure 1.5b using a similar computational process demonstrated in the previous section (see equations (1.7) through (1.11)).

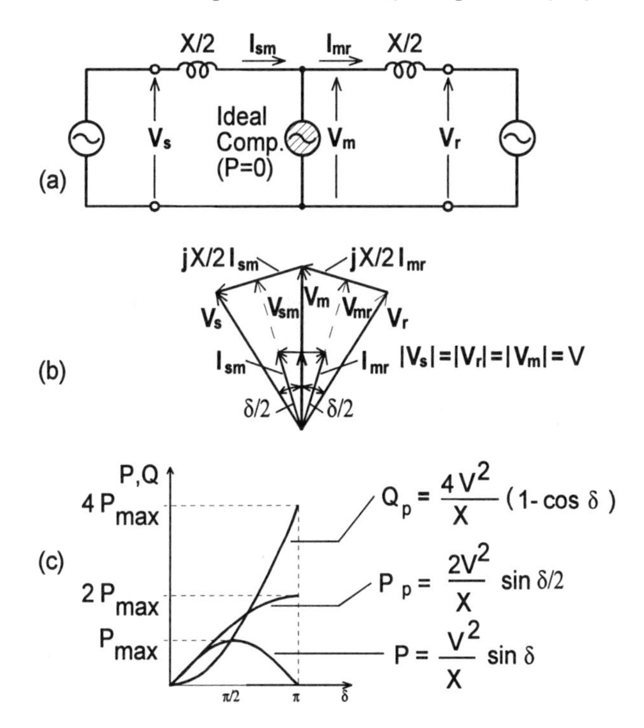

Figure 1.5 (a) Two machine power system with an ideal mid-point reactive compensator; (b) corresponding phasor diagram; (c) and power transmission vs. angle characteristic

With

$$V_{sm} = V_{mr} = V\cos\frac{\delta}{4};\ I_{sm} = I_{mr} = I = \frac{4V}{X}\sin\frac{\delta}{4} \tag{1.13}$$

the transmitted power is

$$P = V_{sm}I_{sm} = V_{mr}I_{mr} = V_m I_{sm}\cos\frac{\delta}{4} = VI\cos\frac{\delta}{4} \tag{1.14}$$

or

$$P = 2\frac{V^2}{X}\sin\frac{\delta}{2} \tag{1.15}$$

Similarly

$$Q = VI\sin\frac{\delta}{4} = \frac{2V^2}{X}(1-\cos\frac{\delta}{2}) \tag{1.16}$$

The relationship between real power P, reactive power Q, and angle δ for the case of ideal shunt compensation is shown plotted in Figure 1.5c. It can be observed that the mid-point shunt compensation can significantly increase the transmittable power (doubling its maximum value) at the expense of a rapidly increasing reactive power demand on the mid-point compensator (and also on the end-generators).

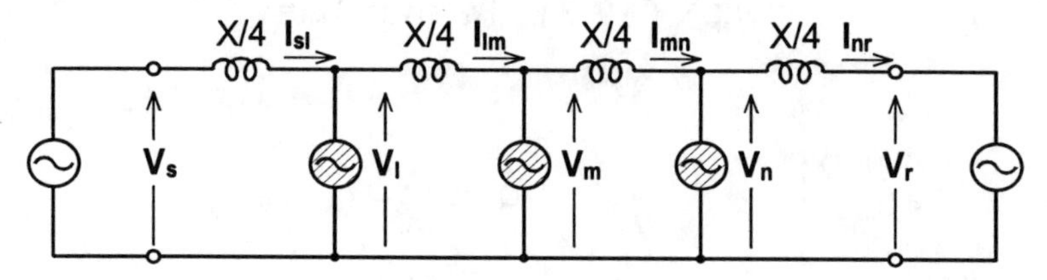

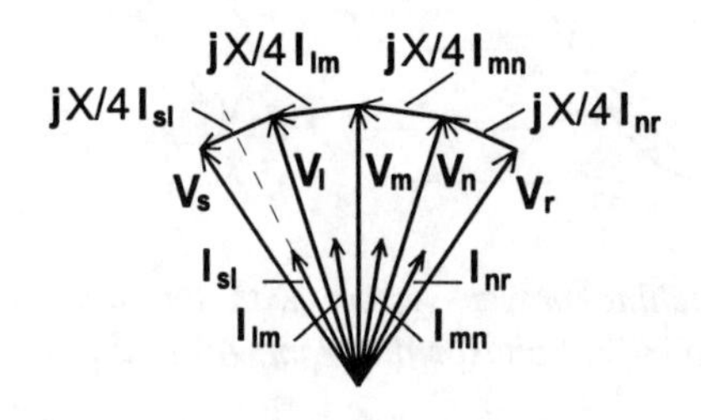

Figure 1.6 Two machine system with ideal reactive compensators providing multiple line segmentation, and associated phasor diagram

The concept of transmission line segmentation can be expanded to the use of multiple compensators, located at equal segments of the transmission line, as illustrated for four line segments in Figure 1.6. Theoretically, the transmittable power would double with each doubling of the segments for the same overall line length. Furthermore, with the increase in the number of segments, the voltage variation along the line would rapidly decrease, approaching the ideal case of constant voltage profile. Ultimately, with a sufficiently large number of line segments, an ideal distributed compensation system could theoretically be established, which would have the characteristics of conventional surge impedance loading, but would have *no power transmission limitations*, and would maintain a *flat voltage profile* at any load.

It will be appreciated that such a distributed compensation hinges on the instantaneous response and unlimited var generation and absorption capability of the shunt compensators employed, which would have to stay in synchronism with the prevailing phase of the segment voltages and maintain the predefined amplitude of the transmission voltage, independently of load variation. At the early conceptual stage of long distance ac transmission developments, rotating synchronous compensators (condensers) were visualized to provide the precise shunt compensation required. These are not likely to have the performance requirements under dynamic system conditions. Subsequently, saturating reactor type shunt compensators and more recently, thyristor-controlled var compensators were evaluated and, for limited voltage support functions, applied in practical systems. To visualize the operation and control coordination complexity of a generalized compensation scheme exhibiting ideal transmission characteristics, consider the lossless but otherwise correctly represented system of Figure 1.2. Assume that the line is provided with a sufficient number of shunt connected ideal var compensators. At no load (zero transmission), the voltages of all compensators would be in phase and they would be absorbing the capacitive vars generated by the distributed line capacitance. With increasing load (increasing δ), the relative phase angle between the voltages of adjacent compensators would increase, but their var absorption would continuously decrease up to the natural (surge impedance) loading, where it would become zero. With further increasing load, beyond the surge impedance loading, the compensators would have to generate increasing amount of capacitive vars to maintain the flat voltage profile. However, at sufficiently heavy loads, the relative phase angle between two adjacent compensators could become too large, resulting in a large voltage sag, at which the power transmission could not be maintained, regardless of the var generation capacity of the compensators, unless additional compensators would be employed to increase further the segmentation of the line.

From the above discussion it is evident that the controlled shunt compensation scheme approximating an ideal line, whose surge impedance is continuously variable so as to maintain a flat voltage profile over a load range stretching from zero to several times the actual surge impedance characterizing that line, would be too complex, and probably too expensive, to be practical, particularly if stability and reliability requirements under appropriate contingency conditions are also considered. However, the practicability of limited line segmentation, using thyristor-controlled static var compensators, has been demonstrated by the major, 600 mile long, 735 kV transmission line of the Hydro-Quebec power system built to transmit up to 1200 MW power from the James Bay hydro-complex to the City of Montreal and to neighboring US utilities. More importantly, the transmission benefits of voltage support by controlled shunt compensation at strategic locations of the transmission system have been demonstrated by numerous installations in the world.

Of course, there are many applications in which mechanically-switched capacitors are applied to control transmission line voltage where there are slow, daily and seasonal load variations. Although these provide economical solutions to steady-state transmission problems, their limited operating speed makes them largely ineffective under dynamic system conditions. Also, because of restrictions in the number of switching operations permitted, mechanically-switched capacitors often lack the flexibility of operation modern power systems may require.

Although, reactive shunt compensation has been discussed above in relation to a two machine transmission power system, this treatment can easily be extended to the more special case of radial transmission. Indeed, if a passive load, consuming power P at voltage V, is connected to the midpoint in place of the receiving-end part of the system (which comprises the receiving-end generator and transmission link $X/2$), the sending-end generator with the $X/2$ impedance and load would represent a simple radial system. Clearly, without compensation the voltage at the mid-point (which is now the receiving-end) would vary with the load (and load power factor). It is also evident that with controlled reactive compensation the voltage could be kept constant independent of the load. Shunt compensation in practical applications is often used to regulate the voltage at a given bus against load variations, or to provide voltage support for the load when, due to generation or line outages, the capacity of the sending-end system becomes impaired.

1.2.3.2 Series compensation

The basic idea behind series capacitive compensation is to decrease the overall effective series transmission impedance from the sending-end to the receiving-end (i.e., X in equation (1.11)). The conventional view is that the impedance of the series connected compensating capacitor cancels a portion of the actual line reactance and thereby the *effective* transmission impedance is reduced as if the line was physically shortened. An equally valid physical view, helpful to the understanding of power flow controllers, is that, in order to increase the current across the series impedance of a physical line (and thereby the transmitted power), the *voltage* across this impedance needs to be increased. This can be accomplished by a series connected *(passive* or *active)* circuit element that produces a voltage opposite to the prevailing voltage across the series line reactance. The simplest such element is a capacitor, but, as will be seen later, controlled *voltage sources* can accomplish this function in a much more generalized manner, ultimately facilitating full control of real and reactive power flow in the line.

Consider the previous simple two-machine model with a series capacitor compensated line, composed of two identical segments for the clarity of illustration, shown at Figure 1.7a. The corresponding voltage and current phasors are shown at Figure 1.7b. Note that the magnitude of the total voltage across the series line reactance, $V_x = 2V_{x/2}$, increased by the magnitude of the opposite voltage, V_C, developed across the series capacitor.

The *effective* transmission impedance X_{eff} with the series capacitive compensation is given by

$$X_{eff} = X - X_C \tag{1.17}$$

or

$$X_{eff} = (1-k)\,X \tag{1.18}$$

where k is the *degree of series compensation*, i.e.,

$$k = X_C/X \qquad 0 \leq k < 1 \tag{1.19}$$

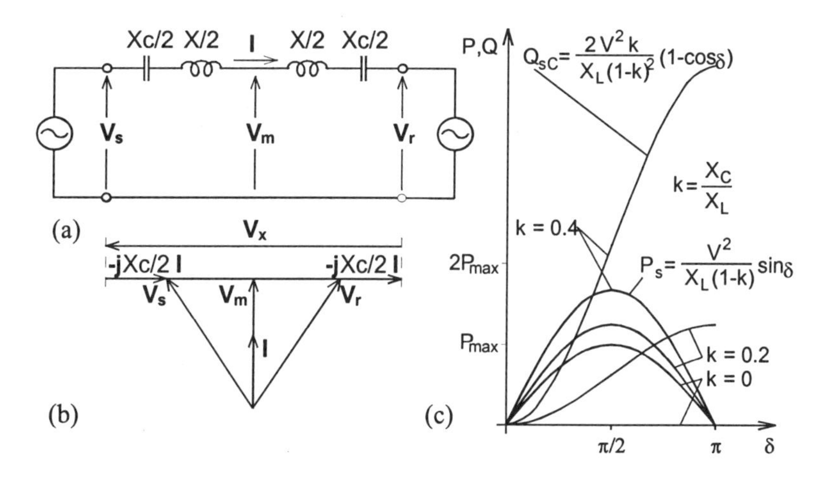

Figure 1.7 (a) Two machine power system with series capacitive compensation; (b) corresponding phasor diagram; (c) and power transmission vs. angle characteristic

The current in the compensated line and the real power transmitted, per equations (1.10) and (1.11) are

$$I = \frac{2V}{(1-k)X}\sin\frac{\delta}{2} \tag{1.20}$$

$$P = V_m I = \frac{V^2}{(1-k)X}\sin\delta \tag{1.21}$$

The reactive power supplied by the series capacitor can be expressed by

$$Q_C = I^2 X_C = \frac{2V^2}{X}\frac{k}{(1-k)^2}(1-\cos\delta) \tag{1.22}$$

The relationship between the real power P, series capacitor reactive power Q_C, and angle δ is shown plotted at various values of the degree of series compensation k in Figure 1.7c. It can be observed that, as expected, the transmittable power rapidly increases with the degree of series compensation k. Similarly, the reactive power supplied by the series capacitor also increases sharply with k and varies with angle δ in a similar manner to the line reactive power.

Series capacitors have been used extensively in the last 50 years throughout the world for the compensation of long transmission lines.

1.2.3.3. Phase angle control

In practical power systems it occasionally happens that the transmission angle required for the optimal use of a particular line would be incompatible with the proper operation of the overall transmission system. Such cases would occur, for example, when power between two buses is transmitted over parallel lines of different electrical length or when two buses are intertied whose prevailing angle difference is insufficient to establish the desired power flow. In these cases a *phase shifter* or *phase angle regulator* is frequently applied.

The basic concept is explained again in connection with the two machine model in which a phase shifter is inserted between the sending-end generator (bus) and the transmission line, as illustrated in Figure1.8a. The phase-shifter can be considered as a sinusoidal (fundamental frequency) ac voltage source with controllable amplitude and phase angle. In other words, the sending-end voltage V_s becomes the sum of the generator voltage V_g and the voltage V_σ provided by the phase shifter, as the phasor diagram shown in Figure 1.8b illustrates. The basic idea behind the phase shifter is to keep the transmitted power at the desired level, independent of the prevailing transmission angle δ, in a predetermined operating range. Thus, for example, the power can be kept at its peak value after angle δ exceeds $\pi/2$ (the peak power angle) by controlling the amplitude of quadrature voltage V_σ so that the effective phase angle $(\delta-\sigma)$ between the sending- and receiving-end voltages stays at π/2. In this way, the actual transmitted power may be increased significantly, even though the phase-shifter *per se* does not increase the steady-state power transmission limit.

With the above phase angle control arrangement the effective phase angle between the sending- and receiving-end voltages thus becomes $(\delta-\sigma)$, the transmitted power P can therefore be expressed as:

$$P = \frac{V^2}{X}\sin(\delta - \sigma) \tag{1.23}$$

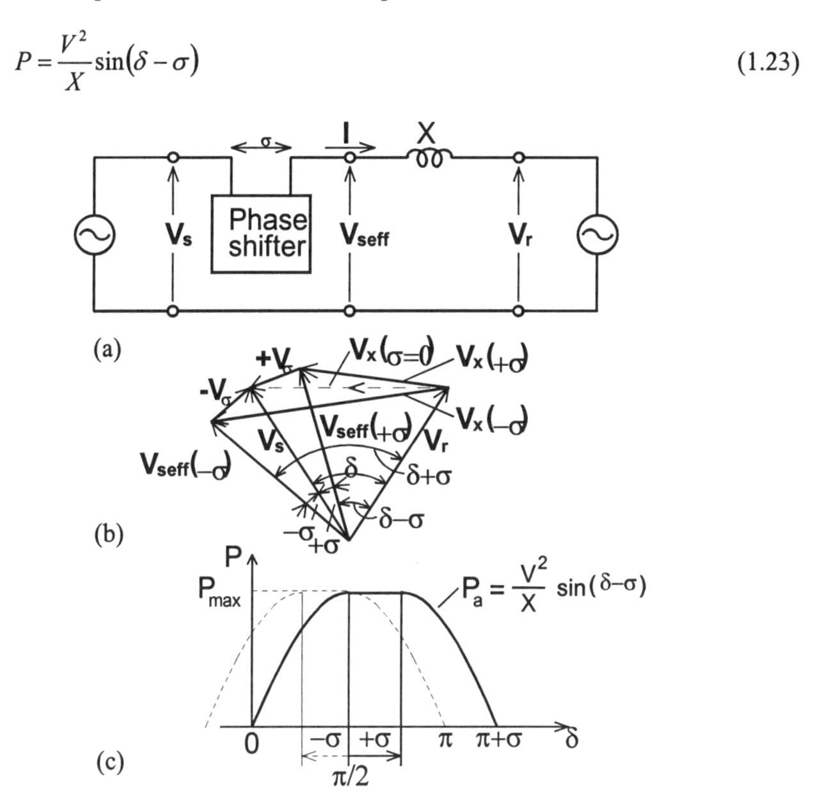

Figure 1.8 (a) Two machine power system with a phase shifter; (b) corresponding phasor diagram; (c) and power transmission vs. angle characteristic

The relationship between real power P and angles δ and σ is shown plotted in Figure 1.8c. It can be observed that, although the phase-shifter does not increase the transmittable power of the uncompensated line, theoretically it makes it possible to keep the power at its maximum value at any angle δ in the range $\pi/2 < \delta < \pi/2+\sigma$ by, in effect, shifting the P versus δ curve to the right. It should be noted that the P versus δ curve can also be shifted to the left by inserting the voltage of the phase shifter with an opposite polarity. In this way, the power transfer can be increased and the maximum power reached at a generator angle less than $\pi/2$ (that is, at $\delta = \pi/2 - \sigma$).

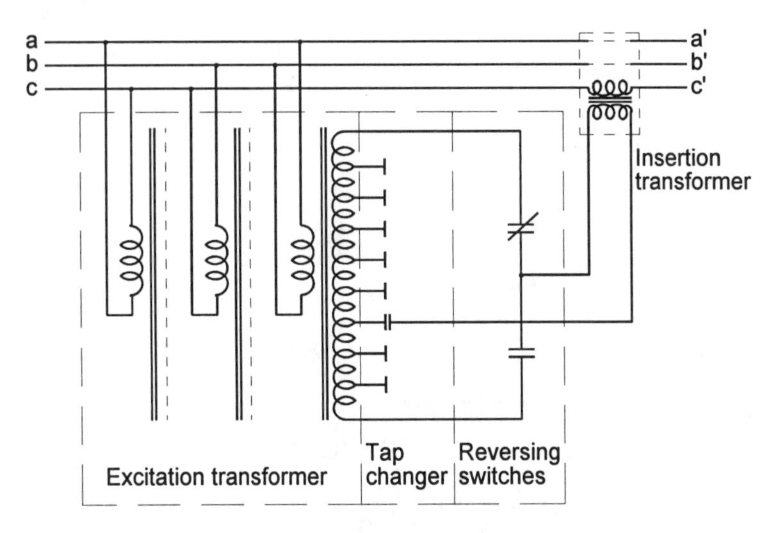

Figure 1.9 Conventional mechanically operated phase shifter

In contrast to the previously-investigated shunt and series compensation schemes, the phase-shifter generally has to handle *both* real and reactive powers. The *VA* throughput of the phase shifter (viewed as a voltage source) is

$$VA = |V_g - V_s|\,|I| = |V_\sigma|\,|I| = V_\sigma I \tag{1.24}$$

Assuming that the magnitude of the generator voltage and that of the sending-end voltage are equal, and using the expression given for the line current in equation (1.10), *VA* can be written in the following form:

$$VA = \frac{4V^2}{X}\sin\frac{\delta - \sigma}{2}\sin\frac{\sigma}{2} \tag{1.25}$$

In Equation (1.25), the multiplier $\sin[(\delta - \sigma)/2]$ defines the current, at a given system voltage and line impedance, flowing through the phase shifter, and the multiplier $\sin(\sigma/2)$ determines the magnitude of the voltage injected by the phase shifter.

Phase shifters employing a shunt connected excitation transformer with a mechanical tap-changer and a series connected insertion transformer (as illustrated schematically in Figure 1.9) to provide adjustable series voltage injection for phase angle control are often employed in transmission systems to control steady-state power flow and prevent undesired parallel and loop power flows.

1.2.4 Dynamic limitations of power transmission

AC power systems employ rotating *synchronous* machines for electric power generation. (They may also employ rotating synchronous compensators (condensers) for reactive power compensation.) It is a fundamental requirement of useful power exchange that all synchronous machines in the system operate *in synchronism* with each other maintaining a common system frequency. However, power systems are exposed to various dynamic disturbances (such as line faults, equipment failures, various switching operations), which may cause a sudden change in the real power balance of the system and consequent acceleration and deceleration of certain machines. The ability of the system to recover from disturbances and regain the steady-state synchronism under stipulated contingency conditions becomes a major design and operating criterion for transmission capacity. This ability is usually characterized by the *transient* and *dynamic stability* of the system. A transmission system is said to be *transiently stable* if it can recover normal operation following a specified *major* disturbance. Similarly, the system is said to be *dynamically stable* if it recovers normal operation following a *minor* disturbance. The dynamic stability indicates the damping characteristics of the system. A dynamic (or "oscillatory") instability means that a minor disturbance may lead to increasing power oscillation and the eventual loss of synchronism.

During and after major disturbances the transmission angle and transmitted power may significantly change from, and oscillate around their steady-state values. Consequently, a power system cannot be operated at, or even too close to its steady-state power transmission limit. An adequate margin is needed to accommodate the dynamic power "swings" while the disturbed machines regain their synchronism in the system.

The phenomenon of escalating decrease and eventual collapse of the terminal voltage as a result of an incremental load increase is referred to as *voltage instability*. Voltage collapse is the result of a complex interaction between induction motor type loads and certain voltage regulators, such as tap-changing transformers, which may take several seconds to minutes. The essence of this process is that decreasing terminal voltage results in increasing load current and poorer load power factor (induction motors) which tend to further decrease the terminal voltage. The voltage regulators (tap-changing transformers) are not able to change the character of this process and under sufficiently severe conditions (low system voltage and heavy loading) it degenerates (in a positive-feedback manner) into a voltage collapse. The *voltage stability limit* identifies for a given system the specific V and P condition at which the next increment of load causes a voltage collapse.

1.2.5 Dynamic compensation for stability enhancement

As seen in the previous sections, both shunt and series line compensation can significantly increase the maximum transmittable power. Thus, it is reasonable to expect that, with suitable and fast controls, these compensation techniques will be able to change the power flow in the system so as to increase the transient stability limit and provide effective power oscillation damping, as well as to prevent voltage collapse.

Similarly, the capability of the phase shifter to vary the transmitted power by transmission angle control can also be applied, with sufficiently fast controls, to the improvement of transient and dynamic system stability.

In the following two sections, the potential effectiveness of the three compensation approaches (shunt, series, and angle control) for transient stability improvement and power oscillation damping are explored and compared. In the subsequent third section, the use of shunt and series capacitive compensation for the increase of voltage instability limit for a radial transmission line is discussed.

1.2.5.1 Transient stability improvement

The potential effectiveness of shunt and series compensation and angle control on transient stability improvement can be conveniently evaluated by the *equal area criterion*. The meaning of the equal area criterion is explained with the aid of the simple two machine (the receiving-end is an infinite bus), two line system shown in Figure 1.10a and the corresponding P versus δ curves shown in Figure 1.10b. Assume that the complete system is characterized by the P versus δ curve "a" and is operating at angle δ_1 to transmit power P_1 when a fault occurs at line segment "1". During the fault the system is characterized by P versus δ curve "b" and thus, over this period, the transmitted electric power decreases significantly while mechanical input power to the sending-end generator remains substantially constant. As a result, the generator accelerates and the transmission angle increases from δ_1 to δ_2 at which the protective breakers disconnect the faulted line segment "1" and the sending-end generator absorbs *accelerating* energy, represented by area "A_1". After fault clearing, without line segment "1" the degraded system is characterized by P versus δ curve "c". At angle δ_2 on curve "c" the transmitted power exceeds the mechanical input power and the sending end generator starts to decelerate; however, angle δ further increases due to the kinetic energy stored in the machine. The maximum angle reached at δ_3, where decelerating energy, represented by area "A_2", becomes equal to the accelerating energy represented by area "A_1". The limit of transient stability is reached at $\delta_3 = \delta_{crit}$, beyond which the decelerating energy would not balance the accelerating energy and synchronism between the sending-end and receiving-end could not be restored. The area "A_{margin}", between δ_3 and δ_{crit}, represent the *transient stability margin* of the system.

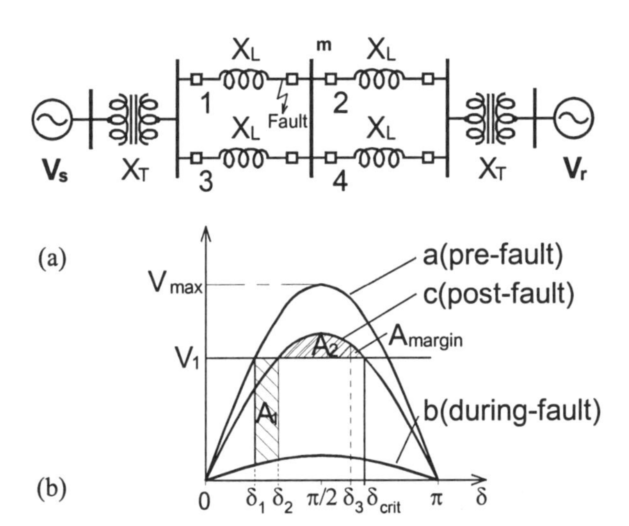

Figure 1.10 Illustration of the equal area criterion for transient stability

From the above general discussion it is evident that the transient stability, at a given power transmission level and fault clearing time, is determined by the *P* versus δ characteristic of the post fault system. Since, as previously shown, shunt and series compensation and angle control improve the natural transmission characteristic of the system, it can be expected that the judicious employment of these techniques would be highly effective in increasing the transmission capability of the post-fault system and thereby enhancing transient stability.

For comparison, consider the four basic two-machine (sending-end generator, receiving-end infinite bus) systems, with no compensation, mid-point shunt compensation, and phase angle control, as shown in Figures 1.4, 1.5, 1.7, and 1.8. For clarity, the above introduced equal-area criterion, is applied here in a greatly simplified manner, with the assumption that the original single line model represents both the pre-fault and post-fault systems. (The impracticality of the single line system and the questionable validity of this assumption has no effect on this qualitative comparison.) Suppose that in the uncompensated and all three compensated systems the steady-state power transmitted is the same. Assume that all four systems are subjected to the same fault for the same period of time. The dynamic behaviour of the four systems is illustrated in Figures 1.11a through 1.11d. Prior to the fault, each of the four systems transmits power P_m (subscript m stands for "mechanical") at angles δ_1, δ_{p1}, δ_{s1}, and δ_{a1}, respectively (Subscripts p, s, and a stand for "parallel", "series", and "angle".) During the fault, the transmitted electric power (of the single line system considered) becomes zero while the mechanical input power to the generators remains constant (P_m). Therefore, the sending-end generator accelerates from the steady-state angles δ_1, δ_{p1}, δ_{s1}, and δ_{a1} to angles δ_2, δ_{p2}, δ_{s2}, and δ_{a2} at which the faults clears. The accelerating energies in the four systems are represented by areas A_1, A_{p1}, A_{s1}, and A_{a1}. After fault clearing, the transmitted electric power exceeds the mechanical input power and the sending-end machine decelerates, but the accumulated kinetic energy further increases until a balance between the accelerating and decelerating energies, represented by areas A_1, A_{p1}, A_{s1}, and A_{a1} and A_2, A_{p2}, A_{s2}, and A_{a2}, respectively, is reached at δ_3, δ_{p3}, δ_{s3}, and δ_{a3}. The difference between angles δ_3, δ_{p3}, δ_{s3}, and δ_{a3}, representing the maximum angular swings, and the critical angles δ_{crit}, δ_{pcrit}, δ_{scrit}, and δ_{acrit} determines the margin of transient stability, that is, the "unused" and still available decelerating energy, represented by areas A_{margin}, $A_{pmargin}$, $A_{smargin}$, and $A_{amargin}$.

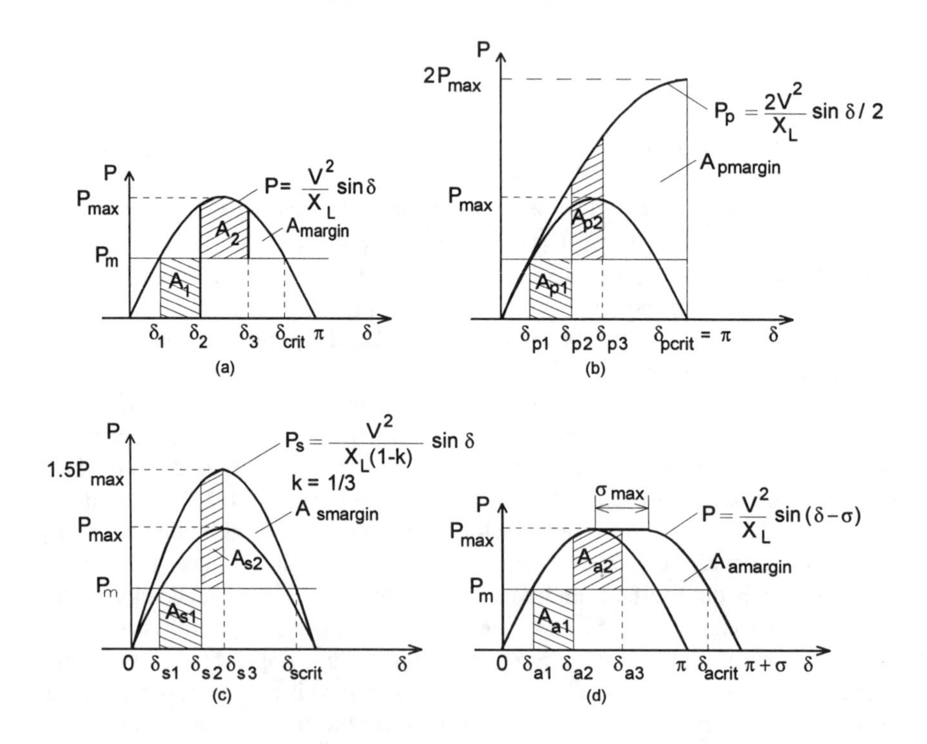

Figure 1.11 Equal area criterion to illustrate the transient stability margin for a simple two machine system: (a) without compensation; (b) with an ideal mid-point compensator; (c) with a series capacitor and (d) with a phase shifter

Comparison of Figures 1.11a through 1.11d clearly shows a substantial increase in the transient stability margin the three compensation approaches can provide through the control of different system parameters. The shunt-connected ("parallel') var compensation method provides the improvement by segmenting the transmission line and regulating the midpoint voltage. The series capacitive compensation approach reduces the effective transmission impedance and minimizes the transmission angle. Phase angle control keeps the transmission angle, $\delta-\sigma$, at $\pi/2$ for maintaining maximum power transmission while the generator angle δ swings beyond this value.

The use of any one of the three compensation approaches obviously can increase the transient stability margin significantly over that of the uncompensated system. Alternatively, if the uncompensated system has a sufficient transient stability margin, these compensation techniques can considerably increase the transmittable power without decreasing this margin.

In the explanation of the equal area criterion at the beginning of this section, a clear distinction was made between the "pre-fault" and "post-fault" power system. It is important to note that from the standpoint of transient stability, and thus of overall system security, the post-fault system is the one that counts. That is, power systems are normally designed to be transiently stable, with defined pre-fault contingency scenarios and post-fault system degradation, when subjected to a major disturbance (fault). Because of this (sound) design philosophy, the actual capacity of transmission systems is considerably higher than that at which they are normally used. Thus, it may seem technically plausible (and economically savvy) to employ fast acting compensation techniques, instead of overall network compensation, specifically to handle dynamic events and increase the transmission capability of the degraded system under the contingencies encountered

1.2.5.2 Power oscillation damping

In the case of an under-damped power system, any *minor* disturbance can cause the machine angle to oscillate around its steady state value at the natural frequency of the total electromechanical system. The angle oscillation, of course, results in a corresponding power oscillation around the steady-state power transmitted. The lack of sufficient damping can be a major problem in some power systems and, in some cases it may be the limiting factor for the transmittable power.

Until the late 1970s, the excitation control of rotating synchronous machines was the available active means for power oscillation damping. Later technological developments made it possible to vary rapidly reactive shunt and series compensation, as well as transmission angle, thereby facilitating highly effective power oscillation damping.

Since power oscillation is a sustained dynamic event, it is necessary to vary the applied compensation so as to achieve consistent and rapid damping. The control action required is essentially the same for the three compensation approaches. That is, when the rotationally oscillating generator accelerates and angle δ increases ($d\delta/dt > 0$), the electric power transmitted must be increased to compensate for the excess mechanical input power. Conversely, when the generator decelerates and angle δ decreases ($d\delta/dt < 0$), the electric power must be decreased to balance the insufficient mechanical input power. (The mechanical input power is assumed to be essentially constant in the time frame of an oscillation cycle.)

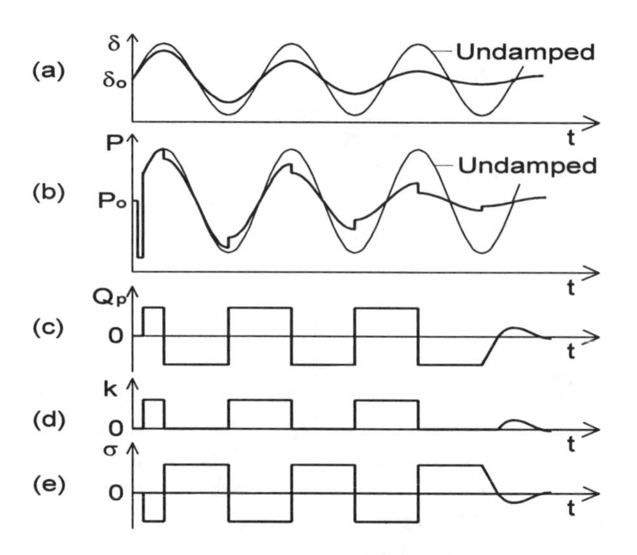

Figure 1.12 Waveforms illustrating power oscillation damping by shunt compensation, controllable series compensation, or phase angle control: (a) generator angle; (b) transmitted power; (c) var output of a shunt compensator; (d) degree of series compensation and (e) angle variation of a phase shifter

The requirements of output control, and the process of power oscillation damping, by the three compensation approaches, are illustrated in Figure 1.12a through 1.12e. Waveforms in (*a*) show the undamped and damped oscillations of angle δ around the steady-state value δ_0. Waveforms in (*b*) show the undamped and damped oscillations of the electric power P around the steady-state value P_0. (The momentary drop in power shown in the figure represents an assumed disturbance that initiated the oscillation.)

Waveform (*c*) shows the reactive power output Q_p of a shunt-connected var compensator. The capacitive (positive) output of the compensator increases the midpoint voltage and the transmitted power when $d\delta/dt > 0$, and it decreases those when $d\delta/dt < 0$.

Waveform (*d*) shows the required variation of $k=X_C/X$ for series capacitive compensation. When $d\delta/dt > 0$, k is increased and thus the line impedance is decreased. This results in the increase of the transmitted power. When $d\delta/dt < 0$, k is decreased (in the illustration it becomes zero) and the power transmitted is decreased to that of the uncompensated system.

Waveform (*e*) shows the variation of angle σ produced by the phase shifter. (For the illustration it is assumed that α has an operating range of $-\sigma_{max} \leq \sigma \leq \sigma_{max}$, and δ is in the range of $0 < \delta < \pi/2$.) Again, when $d\delta/dt > 0$, angle σ is negative making the power versus δ curve (refer to Figure 1.12c) shift to the left, which increases the angle between the end terminals of the line and, consequently, also the real power transmitted. When $d\delta/dt < 0$, angle σ is made positive, which shifts the power versus angle curve to the right and thus decreases the overall transmission angle and transmitted power.

As the illustrations show, a "bang-bang" type control (output is varied between minimum and maximum values) is assumed for all three compensation approaches. This type of control is generally considered the most effective, particularly if large oscillations are encountered. However, for damping relatively small power oscillations, a strategy that varies the controlled output of the compensator continuously, in sympathy with the generator angle or power, may be preferred.

1.2.5.3 Increase of voltage stability limit

Consider the simple radial system with feeder line reactance of X and load impedance $\boldsymbol{Z}$, shown in Figure 1.13a together with the normalized terminal voltage V_r versus power P plot at various load power factors, ranging from 0.8 lag and 0.9 lead. The "nose-point" at each plot given for a specific power factor represents the voltage instability corresponding to that system condition. It should be noted that the voltage stability limit decreases with inductive loads and increases with capacitive loads.

The inherent circuit characteristics of the simple radial structure, and the V_r versus P plots shown, clearly indicate that both shunt and series capacitive compensation can effectively increase the voltage stability limit. Shunt compensation does so by supplying the reactive load and regulating the terminal voltage ($V-V_r=0$) as illustrated in Figure 1.13b. Series capacitive compensation does so by canceling a portion of the line reactance X and thereby in effect providing a "stiff" voltage source for the load, as illustrated for a unity power factor load in Figure 1.13c.

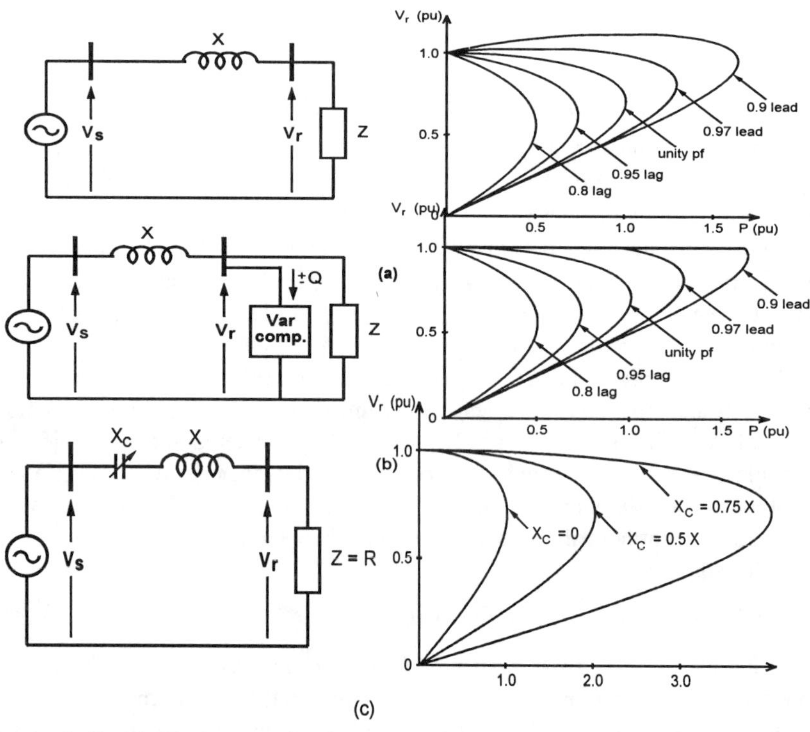

Figure 1.13 (a) Variation of voltage stability limit with load and load power factor; (b) extension of voltage stability limit by reactive shunt compensation and (c) extension of voltage stability limit by series capacitive compensation

1.3 Transmission problems and needs: the emergence of FACTS

The basic limitations of classic ac power transmission (distance, stability, and controllability of flow), which have necessitated the under-utilization of lines and other assets, and the potential of mitigating these limitations cost effectively by controlled compensation, provided the early incentives in the late 1970s to introduce *power electronics*-based control for reactive compensation. This normal evolutionary process has been greatly accelerated by more recent developments in the utility industry, which have aggravated the early problems and highlighted the structural limitations of power systems in a greatly changed socioeconomical environment. The desire to find solutions to these problems and limitations led to focused technological developments under the *Flexible AC*

Transmission System (FACTS) initiative of the *Electric Power Research Institute (EPRI)* in the United States with the ultimate objective to provide *power electronics*-based, *real time* control for transmission systems.

Apart from the obvious orientation of FACTS to the needs of US utilities, the structure of the American electric power system is, from the standpoint of the subject considered, also of particular interest because, first, it has the characteristics of that needed to supply a highly developed industrial society and, second, due to the physical size and geographic composition of the country, it presents similar system problems faced by other geographically large countries in the world.

1.3.1 Historical background

Historically, the US electric power industry has been one of the twentieth century's most phenomenal growth industries. From the very beginning, the unparalleled expansion of industrial activity, rapidly growing population, and available energy sources contributed to this growth. The number of utilities reached 3620 by 1902, and was more than 6000 by the early 1920s. Privately owned utilities dominated the production of electric power, but municipal organizations also participated in retail power distribution. Originally the utilities typically were small and locally owned, with very limited geographical service areas. However, by the late 1920s, as few as 15 large utility holding companies controlled over 80% of the industry's power generation capacity.

The largely unregulated holding company era, characterized by financial improprieties and abuses, was brought to an end by public indignation, which resulted in the passage of the Public Utility Holding Company Act of 1935, and the creation of the Federal Power Commission. The industry concentration was significantly decreased (15 largest utilities, out of some 3400, controlled less than 35 % of generation capacity) and, at the same time, regulation of interstate electric rates and other legislative measures were introduced. Federal intervention was not limited to regulatory measures. The economic depression of the 1930s prompted the government to participate in the production of electric power for regional economic development. This led to the establishment of a number of public utility entities and cooperatives, most of which were primarily distribution companies. Still by 1950, the total power supplied by the private sector decreased to about 80%. This structure of the utility industry has substantially prevailed until now.

1.3.2 Recent developments and problems

The unprecedented technological developments after the Second World War with rapid industrial growth resulted in a dramatic increase in the demand for electric power and the industry capacity expanded nearly tenfold from the 1950s to the early 1970s. This huge increase in power demand was answered by major expansion of generation and transmission facilities, and by the formation of regional power pools and increasing interconnection of individual power systems.

The socioeconomic conditions had unexpectedly begun to change during the 1970s with the utility industry facing a set of difficult economic, environmental, and social problems. The oil embargo in the mid 1970s, public opposition to nuclear power, and social focus on clean air and other environmental issues led to considerable increases in operating cost and governmental intervention. The national energy legislation, various environmental initiatives, and other restrictive regulations went into effect. Alternate energy development plans for solar, geothermal, oil shale, and others were initiated. At the same time the US manufacturing industry went through major restructuring: large, concentrated manufacturing facilities were closed down and production was distributed to smaller facilities at different geographic locations. This, combined with pronounced demographic changes (people moved from cold to warm climates), resulted in a considerable geographical shift in power demand.

All these would have required the relocation or construction of new generation facilities and transmission lines relatively quickly to match the geographically different power demand profile and accommodate a volatile fuel cost structure. Neither the strong economic base, nor the previous freedom of action existed for utilities to adopt these conventional solutions. Indeed, the increasing public concern about environment and health, and the cost and regulatory difficulties in securing the necessary "rights-of-way" for new projects, have often prevented or excessively delayed the construction of many generation facilities and transmission lines needed by the utilities.

The problems imposed by the new socioeconomic conditions fuelled the further growth of interconnection among neighbouring utility systems to share power with other regional pools and be part of a growing national grid. The underlying reason for this integration has been to take advantage of the diversity of loads, changes in peak demand due to weather and time differences, the availability of different generation reserves in various geographic regions, shifts in fuel prices, regulatory changes, and other factors which may manifest themselves differently in other time and geographic zones.

The US power system, evolving from originally isolated utility suppliers to regional power pool groups, did not have a flexible enough transmission grid to cope with the rapidly changing requirements under rapid economic and environmental changes. In the interconnected system "contracted" power was to be delivered sometimes from a distant generation site, often by "wheeling" it through the transmission systems of several utilities, to the designated load area. These arrangements inevitably led to uncontracted and undesired parallel- and loop-flows of power (since part of the line current from the sending-end flowed through each available parallel path in proportion to its admittance), which often overloaded some lines causing thermal and voltage variation problems. The receiving-end was also exposed to difficulties caused by the contingency loss of imported power and the consequent heavy overload condition on the local system, leading to severe voltage depression with the danger of possible voltage collapse. These problems greatly accelerated the use of capacitive compensation. Still, with the growing interconnected system it was increasingly difficult to maintain the traditional (conservative) stability margins without sufficient transmission reinforcement.

The voltage support and transient stability requirements of the expanding interconnected network, and the prevailing restrictions for new line construction, as well as economic considerations, also led to the increasing applications of controllable var compensators in transmission systems beginning in the late 1970s. To date, there are about 20,000 controllable Mvars installed in the USA, Canada, and Mexico, and about 70,000 Mvars worldwide.

1.3.3 Challenges of deregulation

The newest challenge the US utility industry faces is deregulation. In 1995 the Federal Energy Regulatory Commission (FERC) issued its Notice of Proposed Ruling (NOPR) for a more competitive electric industry. The main objective of the new rules is to facilitate the development of a competitive market by ensuring that wholesale buyers and sellers can reach each other through non-discriminatory, *open access transmission* services. The implication of "open access" is that power generation and transmission must be *functionally* "unbundled". The structural implementation of this ruling is still unclear. The control of the power transmission grid could be given to an Independent Grid Operation Company, or the utility transmission assets could be sold to one or more Transmission Companies that would own and operate the transmission grid. (This latter arrangement would be similar to that established earlier in the United Kingdom with the formation of the National Grid Company.) State-level regulations addressing the retail market are presently being worked out. The California Public Utilities Commission has already established its schedule for planned deregulation, starting with direct access for transmission level consumers and progressively reaching the ultimate objective of *direct access for all consumers* by year 2002.

The implementation of deregulation poses formidable challenges to the US utility industry. Some of these challenges in the area of power transmission – relevant to the present discussion – are readily foreseeable. The main economic emphasis of deregulation is to reduce the cost of electricity, that is, to minimize the cost of electric power generation by free (non-discriminatory) competition. This, as an opportunity, will considerably increase the number of independent power producers. Also, it will undoubtedly keep moving the geographical locations of lowest cost power generation, and vary the local generation levels, according to the relative cost of different fuels and other changing factors affecting the cost of energy production (e.g., environmental protection). The compulsory accommodation of the contracted (usually the least expensive) power by the transmission network will likely aggravate the parallel- and loop-flow problems, causing unpredictable line loading (thermal limits), voltage variation, and the potential decrease of transient stability. The unbundling of power generation from transmission will likely worsen these problems by eliminating the incentive for equipping the generators with – often costly – functional capabilities, such as the effective control of reactive power generation and absorption (high-response excitation systems, power system stabilizers, etc.) to aid power transmission. The consequent decrease of coordinated reactive power support provided traditionally by the generators will further aggravate the transmission problems in the areas of voltage variation, transient and dynamic stability. The potential effect of all these on the reliability and security of the overall power system, without effective counter measures, could be devastating.

The traditional solution to the above problems would be a massive reinforcement of the transmission network with new lines to re-establish, by the method of "brute force", the conventional voltage limit, stability, and thermal margins under greatly expanded contingency scenarios. Apart from the cost, such a major undertaking under the present environmental and regulatory constraints would not be possible. This leaves the solution to technological approaches. One such approach, the Flexible AC Transmission System (FACTS), the main subject of this book, relies on the large scale application of power electronics-based, and real time computer-controlled, compensators and controllers to provide cost effective, "high tech" solutions to the problems with the objective being the *full utilization* of transmission assets.

1.3.4 The objectives of FACTS

The basic transmission challenge of the evolving deregulated power system, whatever final form it may take, is to provide a network capable of delivering contracted power from any supplier to any consumer over a large geographic area under market forces-controlled, and thus continuously varying, patterns of contractual arrangements. The aggravating constraint to any potential solution is that, due to cost, right-of-way, and environmental problems, the network must substantially be based on the existing physical line structure.

The Electric Power Research Institute, after years of supporting the development of high power electronics for such applications as High Voltage DC (HVDC) Transmission and reactive compensation of ac lines, in the late 1980s formalized the broad concept of Flexible AC Transmission System (FACTS). The acronym *FACTS* identifies alternating current transmission systems incorporating power electronics-based controllers to enhance the controllability and increase power transfer capability. The FACTS initiative was originally launched to solve the emerging system problems in the late 1980s due to restrictions on transmission line construction, and to facilitate the growing power export/import and wheeling transactions among utilities, with two main objectives:

(1) To increase the power transfer capability of transmission systems, and
(2) to keep power flow over designated routes.

The first objective implies that power flow in a given line should be able to be increased up to the thermal limit by forcing the necessary current through the series line impedance if, at the same time, stability of the system is maintained via appropriate real-time control of power flow during and following system faults. This objective of course does *not* mean to say that the lines would normally be operated at their thermal limit loading (the transmission losses would be unacceptable), but this option would be available, if needed, to handle severe system contingencies. However, by providing the necessary rotational and voltage stability via FACTS controllers, instead of large steady-state margins, the normal power transfer over the transmission lines is estimated to increase significantly (about 50%, according to some studies conducted).

The second objective implies that, by being able to control the current in a line (by, for example, changing the effective line impedance), the power flow can be restricted to selected (contracted) transmission corridors while parallel and loop-flows can be mitigated. It is also implicit in this objective that the primary power flow path must be rapidly changeable to an available secondary path under contingency conditions to maintain the desired overall power transmission in the system.

It is easy to see that the achievement of the two basic objectives would significantly increase the utilization of existing (and new) transmission assets, and could play a major role in facilitating deregulation with minimal requirements for new transmission lines.

The implementation of the above two basic objectives requires the development of high power compensators and controllers. The technology needed for this is high power (multi-hundred MVA) electronics with its real-time operating control. However, once a sufficiently large number of these fast compensators and controllers are deployed over the system, the coordination and overall control to provide maximum system benefits and prevent undesirable interactions with different system configurations and objectives, under normal and contingency conditions, present a different technological challenge. This challenge is to develop appropriate *system optimization* control strategies, communication links, and security protocols. The realization of such an overall *system optimization control* can be considered as the third objective of the FACTS initiative.

1.4 FACTS controllers

The development of FACTS controllers has followed two distinctly different technical approaches, both resulting in a comprehensive group of controllers able to address targeted transmission problems. The first group employs reactive *impedances* or a tap-changing transformer with thyristor switches as controlled-elements; the second group uses self-commutated static converters as controlled *voltage sources*.

1.4.1 Thyristor controlled FACTS *controllers*

The first group of controllers, the *static var compensator (SVC), thyristor-controlled series capacitor (TCSC)* and *phase-shifter*, employ *conventional thyristors* (i.e., those having no intrinsic turn-off ability) in circuit arrangements which are similar to breaker-switched capacitors and reactors and conventional (mechanical) tap-changing transformers, but have much faster response and are operated by sophisticated controls. Each of these controllers can act on one of the three parameters determining power transmission, voltage (SVC), transmission impedance (TCSC) and transmission angle (phase-shifter), as illustrated in Figure 1.14.

Except for the thyristor-controlled phase shifter, all of these have a common characteristic in that the necessary reactive power required for the compensation is generated or absorbed by traditional capacitor or reactor banks, and the thyristor switches are used only for the control of the combined reactive impedance these banks present to the ac system. (Although the phase shifter does not inherently need a capacitor or reactor, neither is it able to supply or absorb the reactive power it exchanges with the ac system.) Consequently, conventional thyristor-controlled compensators present a variable reactive *admittance* to the transmission network and therefore generally change the character of the system impedance. Typically, capacitive shunt compensation coupled with the inductive system impedance results in a network resonance somewhere above the fundamental frequency (50 or 60 Hz) that may be at, or close to, the dominant harmonic frequencies (3rd, 5th, 7th) of the SVC (and of the ac system). The series capacitive compensation results in an electrical resonance below the fundamental that can interact with the mechanical resonances of the turbine-generators supplying the line and, in this way, may cause an overall system *sub-synchronous* resonance (SSR).

The network resonances above and below the fundamental can cause significant problems if they occur at frequencies at which sustained excitation is possible. For this reason, tuned LC filters are usually employed in the SVC to produce impedance zeros and thus prevent parallel resonances at the dominant harmonic frequencies. The mitigation of subharmonic resonance produced by series compensation may require "active" damping via the controls of the thyristor valves.

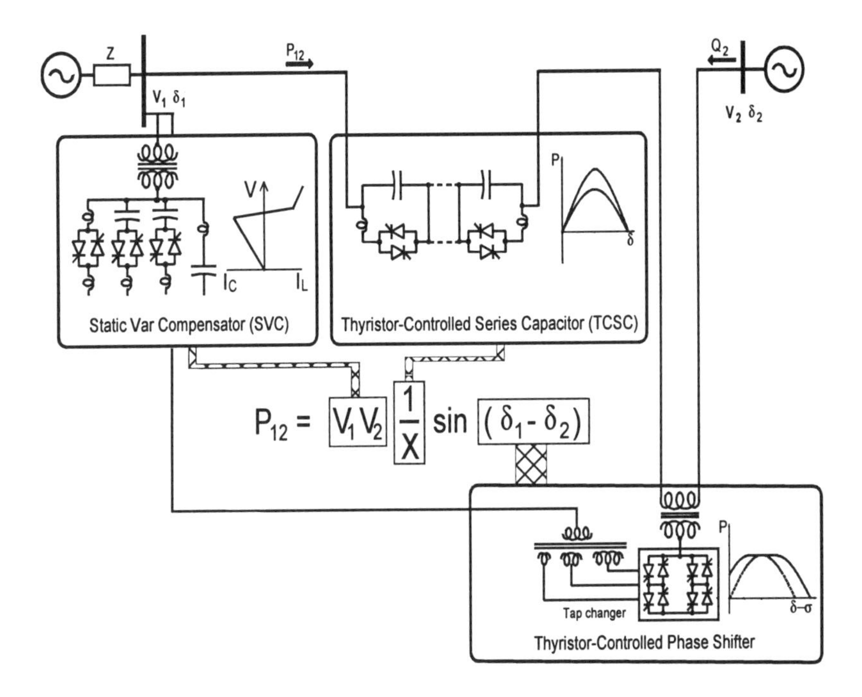

Figure 1.14 Conventional thyristor-based FACTS controllers

From the standpoint of functional operation, the SVC and TCSC act *indirectly* on the transmission network. For example, the TCSC is inserted in series with the line for the purpose of developing a compensating *voltage* to increase the voltage across the series impedance of the given physical line that ultimately determines the line current and power transmitted. Thus, the actual series compensation provided is inherently a function of the line current. Similarly, the SVC is applied as a shunt impedance to produce the required compensating current. Thus, the shunt compensation provided is a function of the prevailing line voltage. This dependence on the line variables (voltage and current) is detrimental to the compensation when large disturbances force the TCSC and SVC to operate outside of their normal control range.

The basic operating principles and characteristics of the conventional thyristor-controlled FACTS controllers are summarized below.

1.4.1.1 Static var compensator

Thyristor-controlled static var compensators are the forerunners of today's FACTS controllers. Developed in the early 1970s for arc furnace compensation, they were later adapted for transmission applications. A typical shunt-connected static var compensator, composed of thyristor-switched capacitors (TSCs) and thyristor-controlled reactors (TCRs), is shown in Figure 1.15. With proper coordination of the capacitor switching and reactor control, the var output can be varied continuously between the capacitive and inductive ratings of the equipment.

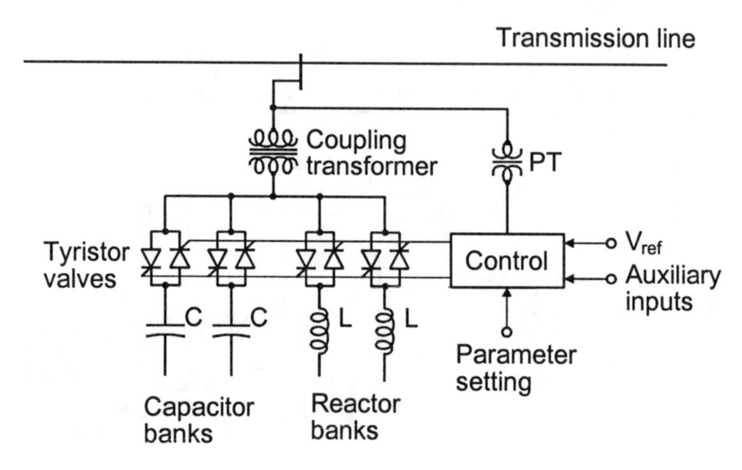

Figure 1.15 Static var compensator employing thyristor-switched capacitors and thyristor-controlled reactors

The compensator is normally operated to *regulate the voltage* of the transmission system at a selected terminal. The V–I characteristic of the SVC, shown in Figure 1.16, indicates that regulation with a given *slope* around the nominal voltage can be achieved in the normal operating range defined by the maximum capacitive and inductive currents of the SVC. However, the maximum obtainable capacitive current *decreases* linearly (and the generated reactive power in quadrature) with the system voltage since the SVC becomes a fixed capacitor when the maximum capacitive output is reached. Therefore, the voltage support capability of the conventional thyristor-controlled static var compensator rapidly deteriorates with decreasing system voltage.

In addition to voltage support, SVCs are also employed for *transient* (first swing) and *dynamic* stability (damping) improvements. The effectiveness of the SVC for the increase of transmittable power is illustrated in Figure 1.17, where the transmitted power P is shown against the transmission angle δ for a simple two-machine model at various capacitive ratings defined by the maximum capacitive admittance B_{Cmax}. It can be observed that the SVC behaves like an ideal mid-point compensator with a P versus δ relationship, as given by equation (1.15), until the maximum capacitive admittance B_{Cmax} is reached.

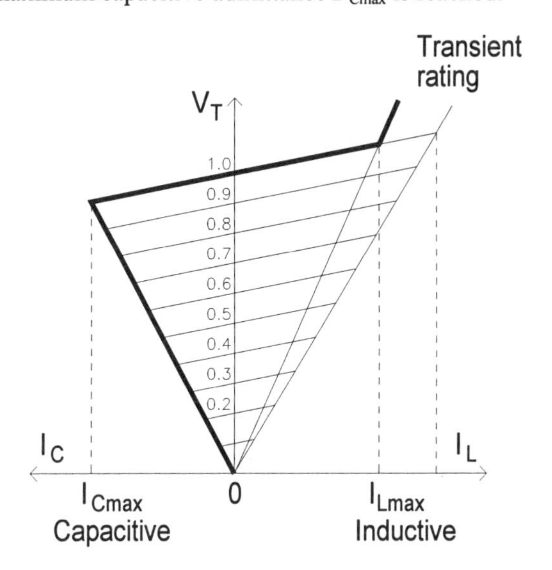

Figure 1.16 V–I characteristic of the Static Var Compensator

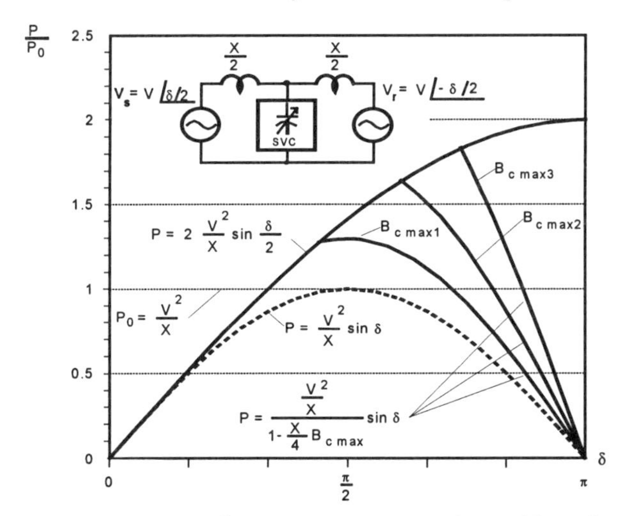

Figure 1.17 Power transfer capacity increase obtainable with a mid-point SVC

From this point on, the power transmission curve becomes identical to that obtained with a fixed, mid-point shunt capacitor whose admittance is B_{Cmax}. The first swing stability improvement is proportional to the $\int P d\delta$ area between the compensated and uncompensated P versus δ curves obtained after fault clearing. As can be seen, this area for relatively large δ swings ($\delta > \pi/2$) sharply decreases unless the rating of the SVC is very large.

The *dynamic stability* improvement (power oscillation damping) can be obtained by alternating the output of the SVC between appropriate capacitive and inductive values so as to oppose the angular acceleration and deceleration of the machines involved. The idea is to increase the transmitted electrical power by increasing the transmission line voltage (via capacitive vars) when the machines accelerate and to decrease it by decreasing the voltage (via inductive vars) when the machines decelerate. The effectiveness of the SVC in power oscillation damping is, of course, a function of the voltage variation it is able, or allowed, to produce.

Static var compensators are extensively employed in electric utility systems with over 750 world wide installations.

1.4.1.2 Thyristor-controlled series capacitor

The two basic schemes of thyristor-controlled series capacitors, using thyristor-switched capacitors and a fixed capacitor in parallel with a thyristor-controlled reactor, are shown schematically in Figures 1.18a and 1.18b.

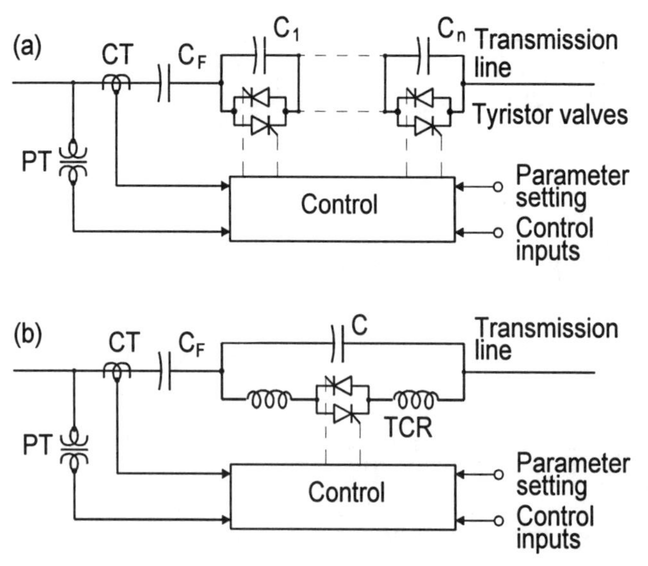

Figure 1.18 (a) Controllable series compensator scheme using thyristor-switched capacitors and (b) a thyristor-controlled reactor in parallel with a series capacitor

In the thyristor-switched capacitor scheme of Figure 1.18a, the degree of series compensation is controlled by increasing or decreasing the number of capacitor banks in series. To accomplish this, each capacitor bank is inserted or bypassed by a thyristor valve (switch). To minimize switching transients and utilize "natural" commutation, the operation of the thyristor valves is coordinated with voltage and current zero crossings. Since, the voltage across the series capacitor is a direct function of the line current, the prevention of damaging overvoltage during faults and other surge current conditions usually necessitates the use of a ZnO type voltage clamping device or other by-pass arrangement in parallel with the thyristor-switched capacitor banks.

In the fixed-capacitors, thyristor-controlled reactor scheme of Figure 1.18b, the degree of series compensation in the capacitive operating region (the admittance of the TCR is kept below that of the parallel connected capacitor) is increased (or decreased) by *increasing* (or *decreasing*) the thyristor conduction period, and thereby the current in the TCR. *Minimum* series compensation is reached when the TCR is *off*. The TCR may be designed to have the capability to limit the voltage across the capacitor during faults and other system contingencies of similar effect.

The two schemes may be combined by connecting a number of TCRs plus fixed capacitors in series in order to achieve greater control range and flexibility.

The P versus δ characteristic of the variable series capacitive compensation (Figure 1.7c) indicates that, apart from steady-state control of power flow, the TCSC can be effective in transient stability improvement, power oscillations damping and balancing flow in parallel lines.

The first three TCSC installations were completed in the USA in the early 1990s. A few other installations are on order in other countries.

1.4.1.3 Phase-shifter

Although there is no high power, non-mechanical phase-shifter in service, the principles for using a phase-shifting transformer with a thyristor tap-changer are well established. Most conventional phase-shifters with a mechanical tap-changer, like their thyristor-controlled counterparts also provide *quadrature voltage* injection.

A thyristor-controlled phase-shifting transformer arrangement is shown in Figure 1.19. It consists of a shunt-connected excitation transformer with appropriate taps, a series insertion transformer and a thyristor switch arrangement connecting a selected combination of tap voltages to the secondary of the insertion transformer. The excitation transformer has three non-identical secondary windings, in proportions of 1:3:9. It can produce a total of 27 steps using only 12 thyristor switches (of three different voltage ratings) per phase with a switching arrangement that can bypass a winding or reverse its polarity.

The phase angle requirements for power flow control can be determined from angle measurements, if available, or from power measurements. The thyristor-controlled phase-shifting transformer could be applied to *regulate the transmission angle* to maintain *balanced* power flow in multiple transmission paths, or to *control* it so as to increase the transient and dynamic stability of the power system.

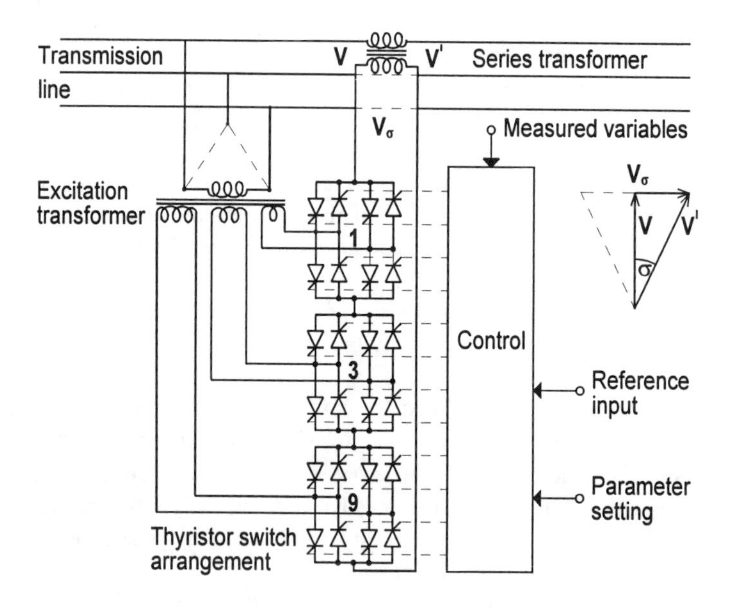

Figure 1.19 Thyristor-controlled phase shifting transformer scheme for transmission angle control

It is to be noted that the phase angle between the voltage injected by the phase-shifter (which is, by design, in quadrature with the line to neutral terminal voltage) and the line *current* is arbitrary, determined by the pertinent parameters of the overall power system. This means that, in general, the phase-shifter must exchange, via the series insertion transformer, *both* real and reactive power with the ac system. Since the tap-changing transformer type phase-shifter *cannot* generate, or absorb, either real *or* reactive power, it follows that *both* the real and reactive power this type of phase-shifter supplies to, or absorbs from, the line when it injects quadrature voltage *must* be absorbed from it, or supplied to it, by the ac system. As a consequence, the Mva ratings of the excitation and insertion transformer are substantially the same.

The fact that the tap-changing transformer type phase-shifter cannot generate or absorb reactive power can be a significant disadvantage in practical applications. If the reactive power, exchanged as a result of the quadrature voltage injection, has to be transmitted through the line, the corresponding voltage drop may be substantial. In order to avoid large voltage drops across the line (and their adverse effects on power transmission), the tap-changing type phase-shifter must be either complemented with a controllable reactive shunt compensator to supply the necessary reactive power locally, or must be located close to the power generator. This inherent operating characteristic (together with the relatively high cost) is probably the main reason why this type of phase-shifter has not yet been developed for high power applications.

1.4.2 Converter-based FACTS controllers

1.4.2.1 Basic concepts

The FACTS controller group discussed in this section employs self-commutated, *voltage-sourced switching converters* to realize rapidly controllable, static, synchronous ac voltage or current sources. This approach, when compared to conventional compensation methods employing thyristor-switched capacitors and thyristor-controlled reactors, generally provides superior performance characteristics and uniform applicability for transmission voltage, effective line impedance, and angle control. It also offers the unique potential to exchange *real power* directly with the ac system, in addition to providing the independently controllable reactive power compensation, thereby giving a powerful new option for flow control and the counteraction of dynamic disturbances.

The synchronous voltage source (SVS) considered here is analogous to an ideal, rotating synchronous machine which generates a balanced set of (three) sinusoidal voltages at the fundamental frequency, with controllable amplitude and phase angle. This ideal machine has no inertia, its response is practically instantaneous, it does not significantly alter the existing system impedance, and it can internally generate *reactive* (both capacitive and inductive) power. Furthermore, it can exchange *real power* with the ac system if it is coupled to an appropriate energy source that can supply or absorb the power it supplies to, or absorbs from, the ac system.

A functional model of the switching-converter-based synchronous voltage source is shown schematically in Figure 1.20. Reference signals Q_{ref} and P_{ref} define the amplitude V and phase angle ψ of the generated (sinusoidal) output voltage and thereby the reactive and real power exchange between the voltage source and the ac system. If the function of real power exchange is not required ($P_{\mathrm{ref}} = 0$), the SVS becomes a self-sufficient reactive power source, like an ideal synchronous compensator (condenser), and the external energy source or energy storage device can be eliminated.

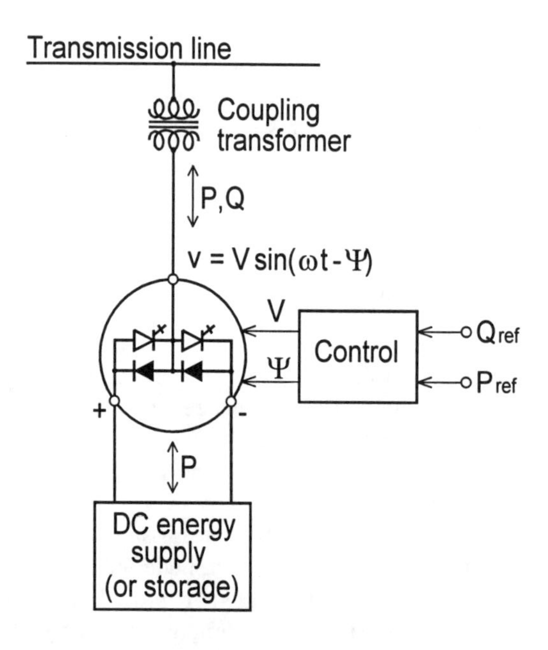

Figure 1.20 Switching converter-based synchronous voltage source shown in shunt connection

The synchronous voltage source facilitates a *forcing-function* approach to transmission line compensation and power flow control. That is, the SVS can apply a defined voltage to force the desired line current (or a defined current to force the desired terminal voltage). In contrast to the controlled impedance approach, the applied compensation provided by an SVS remains largely independent of the network variables (line current, voltage, or angle) and thus it can be maintained during major system disturbances (e.g., large voltage depressions, current and angle swings).

The SVS is an alternating voltage source (with a substantially sinusoidal output) which, with fixed control inputs, would operate only at the fundamental frequency. Its output impedance at other frequencies would theoretically be zero. (In practice, it has a small inductive impedance provided by the leakage inductance of the coupling transformer.) Consequently, the SVS, in contrast to impedance type compensators, is unable to form a classical shunt or series resonant circuit with the ac transmission network.

The SVS also has the inherent capability of executing a bi-directional *real* (active) power flow between its ac and dc terminals. Thus, it becomes possible to couple the dc terminals of two or more SVSs and thereby establish paths for real power transfer between selected buses and/or lines of the transmission network. With the appropriate combinations of SVSs unique FACTS controller arrangements able to control independently real and reactive power flow in individual lines, and balance real and reactive flows among lines, can be devised.

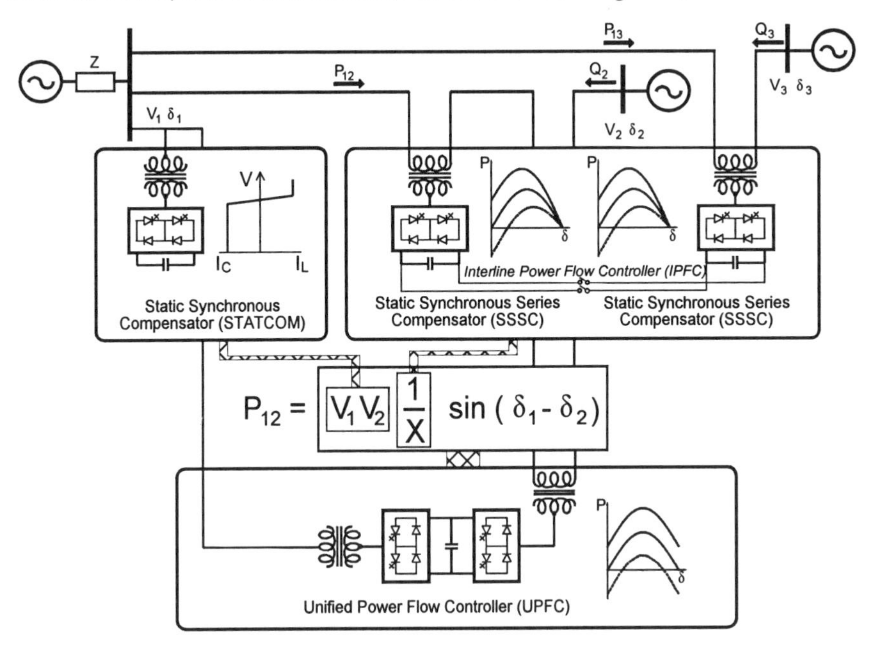

Figure 1.21 Family of switching converter-based FACTS controllers

The group of FACTS controllers employing switching converter-based synchronous voltage sources, the *STATic synchronous COMpensator (STATCOM),* the *Static Synchronous Series Compensator (SSSC),* the *Unified Power Flow Controller (UPFC)* and, the latest addition to the group, the *Interline Power Flow Controller (IPFC)* are shown with their functional control objective in Figure 1.21. The STATCOM, like its conventional counterpart, the SVC, controls transmission voltage by reactive shunt compensation. The SSSC provides series compensation by directly controlling the voltage across the series impedance of the transmission line thereby controlling the *effective* transmission impedance. The UPFC can control, individually or in combination, all three effective transmission parameters (voltage, impedance, and angle) or directly, the real and reactive power flow in the line. The IPFC is able to transfer real power between lines, in addition to providing reactive series compensation, and thereby can facilitate a comprehensive overall real and reactive power management for a multi-line transmission system. The basic operating principles and characteristics of the general switching converter-based SVS and the related four FACTS controllers, STATCOM, SSSC, UPFC, and IPFC are summarized below.

1.4.2.2 Switching converter-based synchronous voltage source

The static synchronous voltage source can be implemented by various *switching power converters*. These converters are discussed in detail in Chapter 2. Here, for the readers convenience, a brief summary is given for the multi-pulse, *voltage-sourced converter* presently favored for FACTS applications.

An elementary, *six-pulse*, voltage-sourced converter is shown in Figure 1.22a. It consists of six self-commutated semiconductor (usually *gate-turn-off thyristor*) switches, each of which is shunted by a reverse-parallel connected diode. (It should be noted that in a high power converter each switching device may consist of a string of series connected semiconductors to yield the required voltage rating.) With a dc voltage source (which may be a charged capacitor), the converter can produce a balanced set of three quasi-square voltage waveforms of a given frequency, as illustrated in Figure 1.22b, by connecting the dc source sequentially to the three output terminals via the appropriate converter switches.

Several of the elementary converters can be operated from the same dc source, each producing a set of three quasi-square voltage waveforms. By generating these voltage waveforms with appropriate successive phase displacement and summing the thus obtained component waveforms, usually with the use of an appropriate magnetic circuit (transformer) arrangement, a multi-pulse output voltage waveform can be obtained. With a sufficient number of elementary converters the output waveform can be made to approximate a sine-wave as closely as desired. A possible multi-pulse converter structure is shown schematically in Figure 1.23a, and the output voltage and current waveforms for a 48-pulse arrangement employing eight elementary six-pulse converters are shown in Figure 1.23b. (The current waveform is shown for 12% coupling transformer reactance when the converter generates capacitive vars.) As can be observed, at this pulse number (which would be a practical choice for high power applications) the output current is essentially a sine-wave and thus the converter can be considered for all practical purposes as a sinusoidal voltage source.

The reactive power exchange between the converter and the ac system (refer to Figure 1.20) can be controlled by varying the amplitude of the three-phase output voltage produced. That is, if the amplitude of the output voltage is increased above that of the ac system voltage, then the current flows through the reactance from the converter to the ac system, and the converter generates reactive (capacitive) power for the ac system. If the amplitude of the output voltage is decreased below that of the ac system, then the reactive current flows from the ac system to the converter and the converter absorbs reactive (inductive) power. If the output voltage is equal to the ac system voltage, the reactive power exchange is zero.

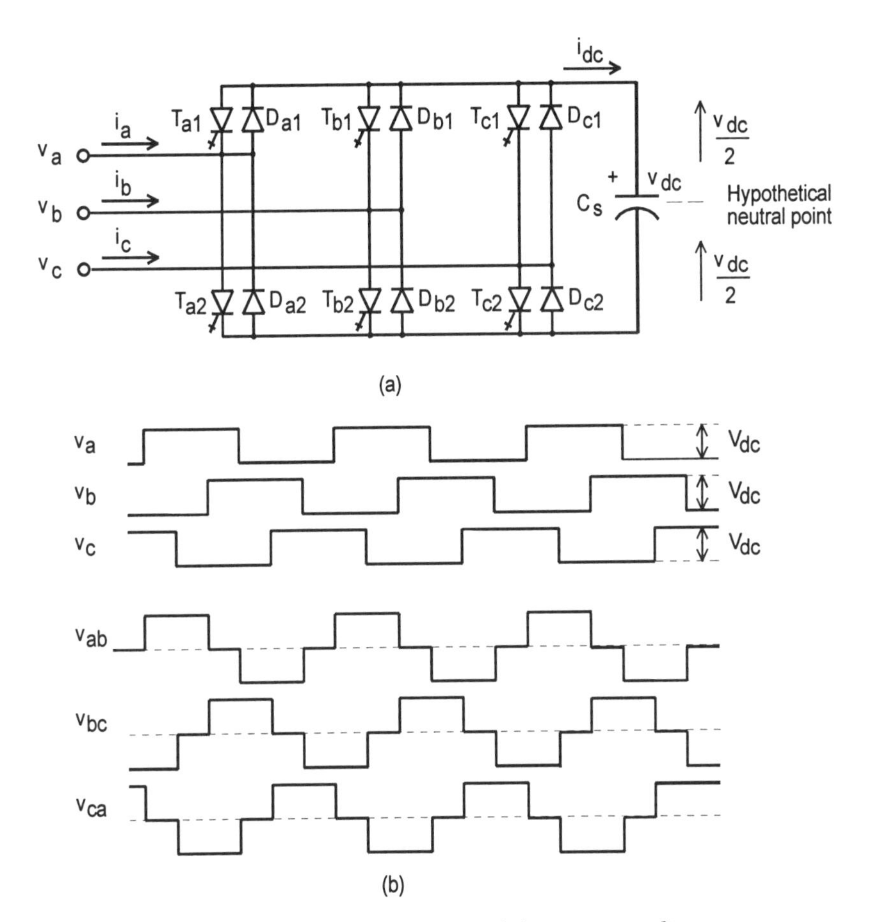

Figure 1.22 (a) Elementary six-pulse converter and (b) corresponding output voltage waveforms

Similarly, the real power exchange between the converter and the ac system can be controlled by phase-shifting the converter output voltage with respect to the ac system voltage. That is, the converter supplies real power from its dc energy storage to the ac system if the converter output voltage is made to lead the corresponding ac system voltage. (This is because this phase advancement results in a real component of current through the tie reactance that is in phase opposition with the ac system voltage.) By the same token, the converter absorbs real power from the ac system for dc energy storage, if the converter output voltage is made to lag the ac system voltage. (The real component of current flowing through the tie reactor is now in-phase with the ac system voltage.)

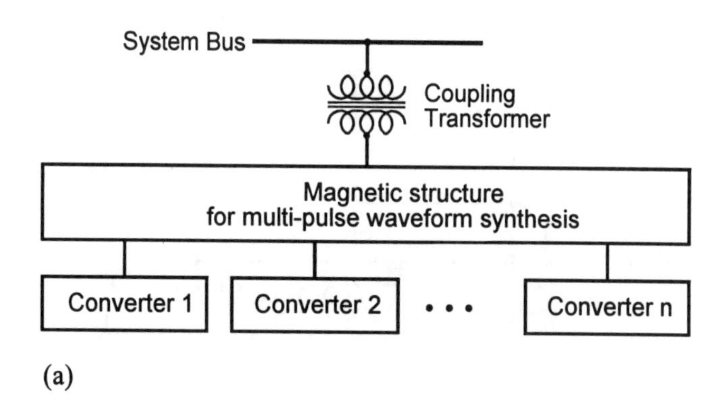

(a)

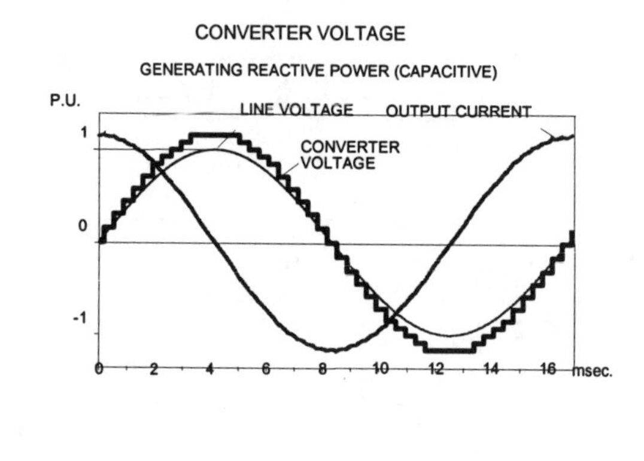

(b)

Figure 1.23 (a) Generalized multi-pulse converter structure and (b) 48-pulse (n=8) output waveforms

The mechanism by which the converter internally generates reactive power can be explained, without considering the detailed operation of the switch array(s) of the converter, simply by considering the relationship between the output and input powers of the converter. The key to this explanation resides in the physical fact that the process of energy transfer through the converter (consisting of nothing but arrays of switches) is absolutely direct, and thus it is inherent that the net instantaneous power at the ac output terminals must always be equal to the net instantaneous power at the dc input terminals (neglecting the losses of the semiconductor switches).

For the case where the converter is operated to supply only reactive output power, the real input power provided by the dc source has to be zero. Furthermore, since reactive power at zero frequency by definition is zero, the dc source supplies no input power and therefore it clearly plays no part in the generation of the reactive output power. In other words, the converter simply interconnects the three output terminals in such a way that the reactive output currents can flow freely between them. Viewing this from the terminals of the ac system, one could say that the converter establishes a circulating power exchange among the phases.

Although reactive power is internally generated by the action of switch operation, it is still necessary to have a relatively small dc capacitor connected across the input terminals of the converter. The need for the dc capacitor is primarily to satisfy the above-stipulated equality of the instantaneous output and input powers. The output voltage waveform of the converter is not a perfect sine-wave. (As shown in Figure 1.23b, it is a staircase approximation of a sine-wave.) However, the multi-pulse converter draws a smooth, almost sinusoidal current from the ac system through the tie reactance. As a result, the net three-phase *instantaneous* power (VA) at the output terminals of the converter fluctuates slightly. Thus, in order not to violate the equality of the instantaneous output and input powers, the converter must draw a fluctuating ("ripple") current from the dc storage capacitor that provides a constant terminal voltage at the input.

1.4.2.3 Static synchronous compensator (STATCOM)

If the SVS is used strictly for *reactive* shunt compensation, like a conventional static var compensator, the dc energy source (see Figure 1.20) can be replaced by a relatively small dc capacitor, as shown in Figure1.24a. (The size of the capacitor, as indicated above, is primarily determined by the "ripple" input current encountered with the particular converter design.) In this case, the steady-state power exchange between the SVS and the ac system can only be *reactive*, as illustrated in Figure 1.24b.

When the SVS is used for reactive power generation, the converter itself can keep the capacitor charged to the required voltage level. This is accomplished by making the output voltages of the converter lag the system voltages by a small angle. In this way the converter absorbs a small amount of real power from the ac system to replenish its internal losses and keep the capacitor voltage at the desired level. The same control mechanism can be used to increase or decrease the capacitor voltage, and thereby the amplitude of the output voltage of the converter, for the purpose of controlling the var generation or absorption. The dc capacitor also has the function of establishing an energy balance between the input and output during the dynamic changes of the var output.

The SVS, operated as a reactive shunt compensator, exhibits operating and performance characteristics similar to those of an *ideal* rotating synchronous compensator and for this reason this arrangement is called Static Synchronous

Compensator. (The term *advanced static var compensator* or *ASVC* has also been used in the older literature.) The characteristics of the STATCOM are superior to those attainable with the conventional thyristor-controlled static var compensator (SVC).

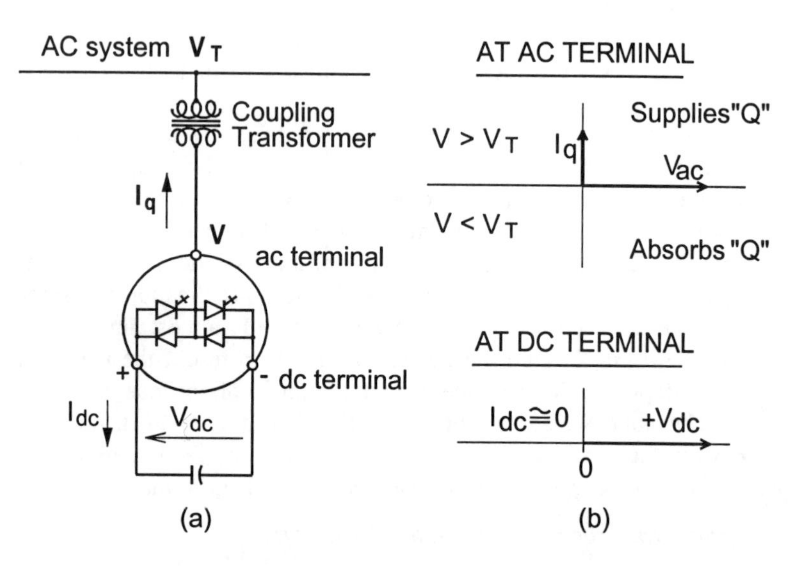

Figure 1.24 (a) Synchronous voltage source operated as a STATCOM and (b) corresponding steady-state power exchange diagram

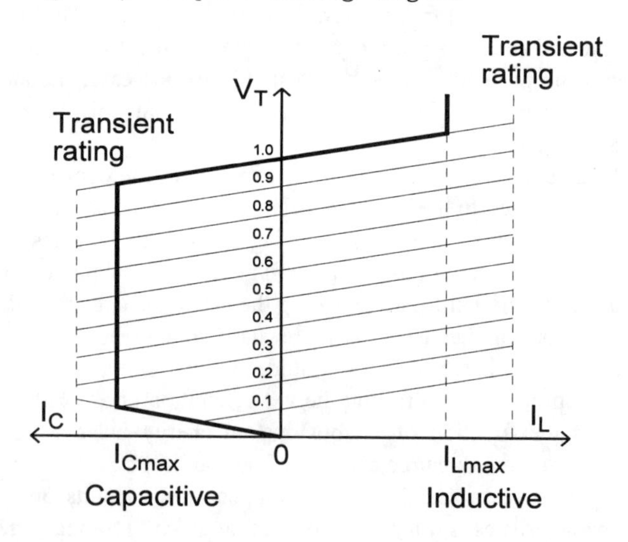

Figure 1.25 V–I characteristic of the STATCOM

The V–I characteristic of the STATCOM is shown in Figure 1.25. As illustrated, the STATCOM can provide both capacitive and inductive compensation and is able to control its output current *over the rated maximum capacitive or inductive range* independently of the ac system voltage. That is, the STATCOM can provide full capacitive output current at any system voltage, practically down to zero. By contrast, the SVC, being composed of (thyristor-switched) capacitors and reactors, can supply only diminishing output current with decreasing system voltage as determined by its maximum equivalent capacitive admittance.

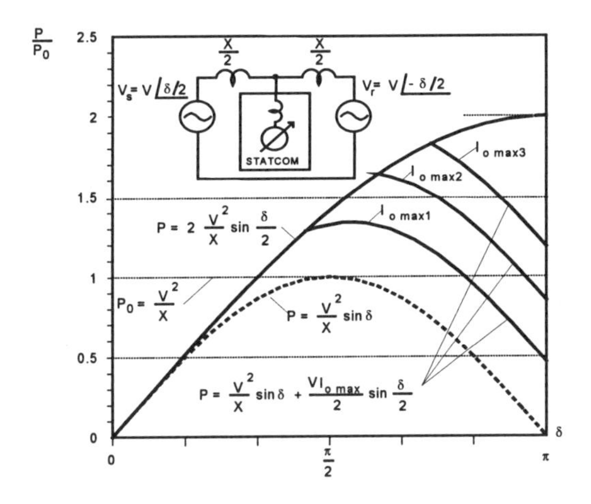

Figure 1.26 Power transfer capacity increase obtainable with a mid-point STATCOM

As Figure 1.25 illustrates, the STATCOM may have an increased transient rating in both the inductive and capacitive operating regions. (The conventional SVC has no means to increase transiently the var generation since the maximum capacitive current it can draw is strictly determined by the size of the capacitor and the magnitude of the system voltage.) The available transient rating of the STATCOM is dependent on the characteristics of the power semiconductors used and the junction temperature at which the devices are operated

The ability of the STATCOM to produce full capacitive output current at low system voltage also makes it highly effective in improving the *transient* (first swing) stability. The effectiveness of the STATCOM for the increase of transmittable power is illustrated in Figure 1.26, where the transmitted power P is shown against the transmission angle δ for the usual two-machine model at various capacitive ratings defined by the maximum capacitive output current I_{Cmax}.

(This figure is comparable to Figure 1.17, where the P versus δ plots are shown for an SVC of different capacitive ratings.) It can be observed that, as expected, the STATCOM, just like the SVC, behaves like an ideal mid-point shunt compensator with P versus δ relationship as defined by equation (1.15) until the maximum capacitive output current I_{Cmax} is reached. From this point on, the STATCOM continues to provide this maximum capacitive output *current* (instead of a fixed capacitive admittance like the SVC), independent of the further increasing δ and the consequent variation of the mid-point voltage. As a result, the sharp decrease of transmitted power P in the $\pi/2 < \delta < \pi$ region, characterizing the power transmission of an SVC supported system, is avoided and the obtainable $\int P d\delta$ area representing the improvement in stability margin is significantly increased.

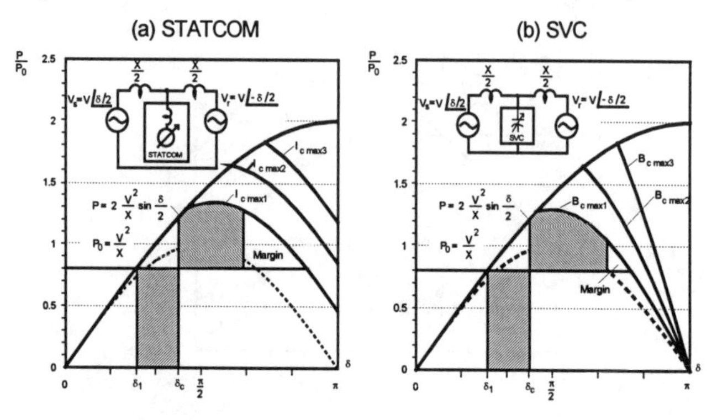

Figure 1.27 (a) Transient stability improvement provided by a mid-point STATCOM compared with (b) a mid-point SVC of the same rating

The increase in stability margin obtainable with a STATCOM over a conventional thyristor-controlled SVC of identical rating is clearly illustrated with the use of the *equal-area* criterion in Figure 1.27a and 1.27b. The same simple two-machine model considered previously (refer to Figures 1.5a) is compensated at the mid-point by a STATCOM and an SVC, respectively, of the same var rating. The illustrations in Figures 1.27a and 1.27b show the attainable increase in the margin of transient stability by the STATCOM and SVC, each with the same limited var rating, are similar to those shown in Figure 1.11 and are self explanatory. The meaning of the figures is that the transmittable power can be increased if the shunt compensation is provided by a STATCOM rather than by an SVC, or, for the same stability margin, the rating of the STATCOM can be decreased below that of the SVC.

Although the basic concepts of switching converter-based shunt compensators were established in the early 1970s, the development of practical hardware did not start until the mid 1980s due to the lack of suitable high power semiconductors.

The emergence of the FACTS initiative in the USA, and other incentives in Japan, accelerated the developments considerably in the 1990s. Presently there are three STATCOM installations both in the USA and Japan for transmission line compensation. Several others for arc furnace and other power quality applications are also in service. Additional installations are on order or in a planning stage.

1.4.2.4 Static synchronous series compensator (SSSC)

The concept of using the solid-state synchronous voltage source for series *reactive* compensation is based on the fact that the impedance versus frequency characteristic of the conventionally employed series capacitor, in contrast to filter applications, plays no part in accomplishing the desired line compensation. The function of the series capacitor is simply to produce an appropriate voltage at the *fundamental* ac system frequency to increase the voltage across the inductive line impedance, as illustrated in Figure 1.28, and, thereby, the fundamental line current and the transmitted power. (This of course has the same electrical effect as if the series line inductance was reduced to that of a shorter line.) Therefore, if an ac voltage source of fundamental frequency, which is locked with a quadrature (lagging) relationship to the line current and *whose amplitude is made proportional to that of the line current,* is injected in series with the line, a series compensation equivalent to that provided by a series capacitor at the fundamental frequency is obtained. Mathematically, this voltage source can be defined as follows:

$$V_c = -jkXI \tag{1.26}$$

where, as before, V_C is the injected compensating voltage phasor, I is the line current phasor, X is the series reactive line impedance, k is the degree of series compensation, and $j = \sqrt{-1}$. Thus, $kX = X_C$ represents a virtual series capacitor generating the same compensating voltage as its real counterpart. However, in contrast to the real series capacitor, the SVS is able to maintain a constant compensating voltage in face of variable line current, or control the amplitude of the injected compensating voltage independent of the amplitude of the line current.

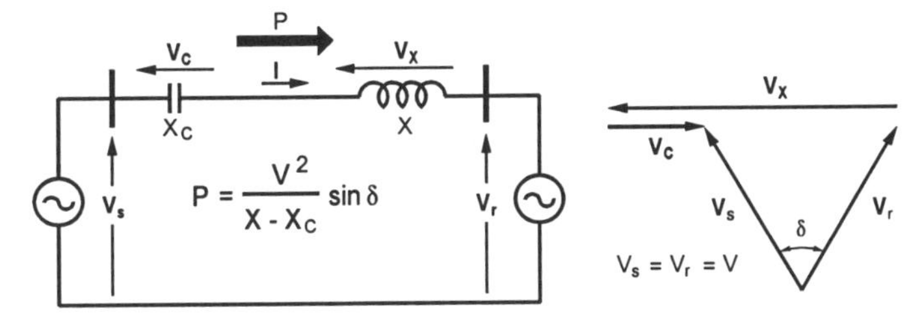

Figure 1.28 Traditional line compensation by series capacitor

For normal capacitive compensation, the output voltage lags the line current by 90 degrees. However, the output voltage of the SVS can be reversed by simple control action to make it lead the line current by 90 degrees. In this case, the injected voltage decreases the voltage across the inductive line impedance and thus the series compensation has the same effect as if the reactive line impedance was increased.

With the above observations, a generalized expression for the injected voltage, V_q, can simply be written:

$$V_q = \pm jV_q(\varsigma)\frac{I}{I} \tag{1.27}$$

where $V_q(\zeta)$ is the magnitude of the injected compensating voltage $(0 \le V_q(\zeta) \le V_{qmax})$ and ζ is a chosen control parameter. The series reactive compensation scheme, based on equation (1.27), is illustrated in Figure 1.29. Such series compensation scheme is termed the *Static Synchronous Series Compensator (SSSC)*.

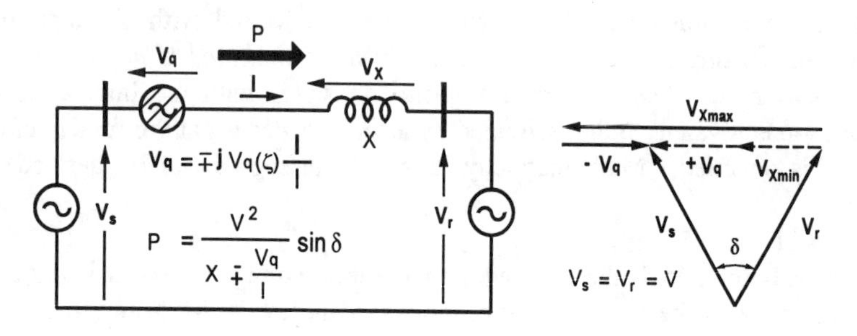

Figure 1.29 A synchronous voltage source operated as a Static Synchronous Series Compensator

The transmitted power P versus transmission angle δ, characterizing an SSSC compensated two machine system, can be expressed with injected voltage V_q as follows:

$$P = \frac{V^2}{X}\sin\delta + \frac{V}{X}V_q\cos\frac{\delta}{2} \tag{1.28}$$

The normalized power P versus angle δ plots as a parametric function of V_q are shown in Figure 1.30. Comparison of these plots to those shown in Figure 1.7c for the series capacitor (of comparable compensation range) clearly shows that the series capacitor increases the transmitted power by a fixed *percentage* of that transmitted by the uncompensated line at a given δ and, by contrast, the SSSC increases it by a fixed *fraction of the maximum power* transmittable by the uncompensated line, independent of δ, in the important operating range of

$0 \le \delta \le \pi/2$. From the standpoint of practical applications, steady-state flow control or stability improvements, the SSSC clearly has considerably wider control range than the controlled series capacitor of the same MVA rating.

A prototype SSSC installation of ±160 Mvar is completed in the USA.

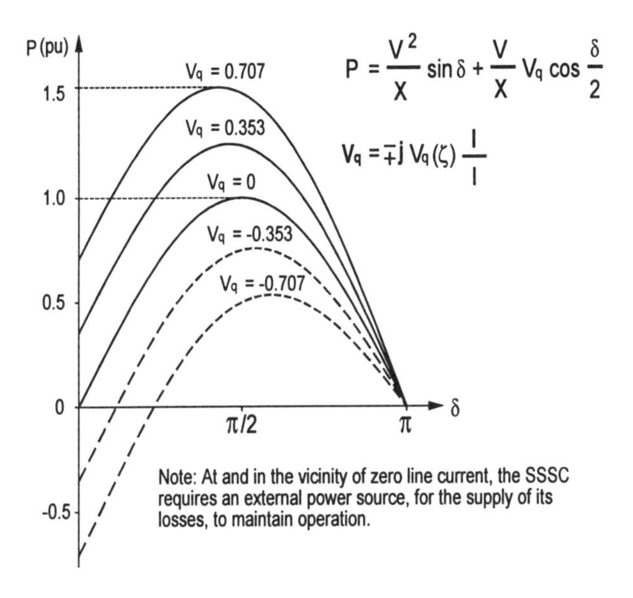

Figure 1.30 Transmitted power P versus transmission angle δ as a parametric function of the series compensating voltage V_q

1.4.2.5 Unified power flow controller (UPFC)

It has been shown in the previous section that, from the conventional viewpoint of reactive compensation, the SSSC can be considered as a FACTS controller acting, like a controlled series capacitor, on the effective transmission *impedance*. An equally valid viewpoint is that the SSSC does not alter the transmission impedance but changes the effective sending-end (or receiving-end) *voltage*. This duality of viewpoints can be extended to the Unified Power Flow Controller that theoretically can be considered as a generalized SSSC operated without the constraints of the quadrature relationship stipulated for the injected voltage with respect to the line current equations (1.26) and (1.27).

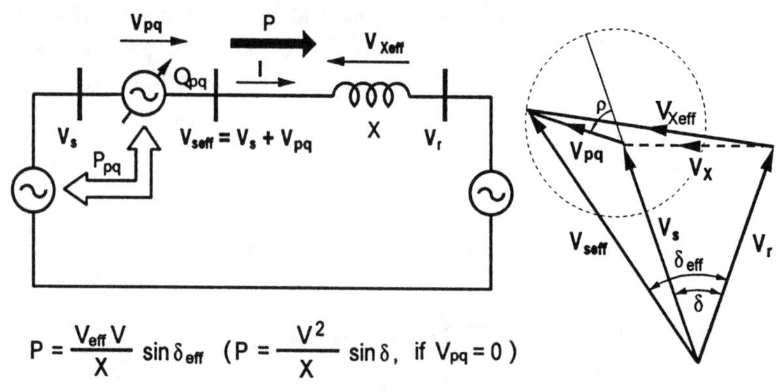

Figure 1.31 Basic concept of the Unified Power Flow Controller

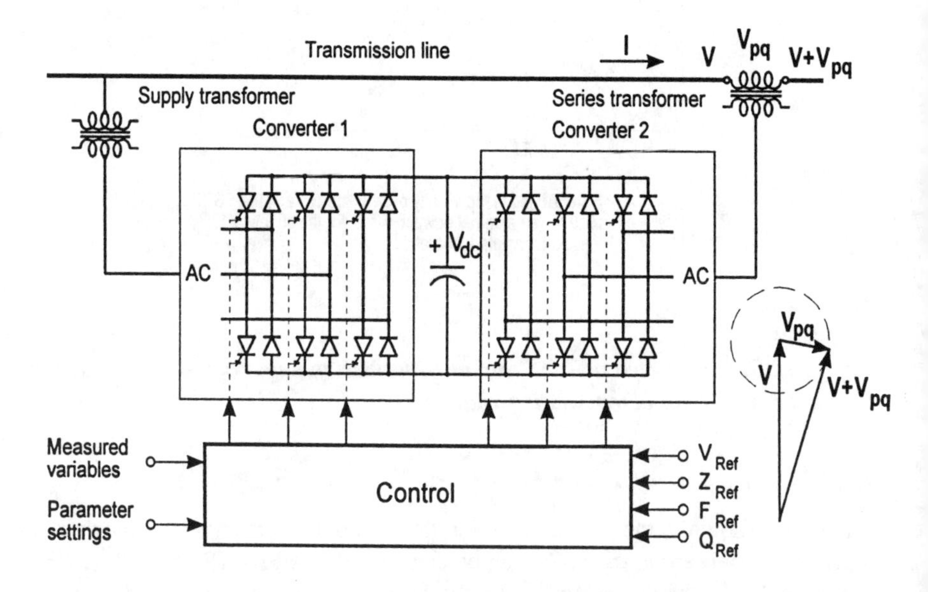

Figure 1.32 Implementation of the UPFC by two back-to-back voltage-sourced converters

Consider a generalized series compensation scheme shown in Figure 1.31. Assume that the voltage (V_{pq}) injected by the SVS in series with the line can be controlled *without* restrictions. That is, the phase angle of phasor V_{pq} can be chosen *independently* of the line current between 0 and 2π, and its magnitude is variable between zero and a defined maximum value, V_{pqmax}. This implies that voltage source V_{pq} generally exchanges *both* real and reactive power with the transmission line. As established in Section 1.4.2.2, a converter-based SVS can internally generate or absorb the reactive power, but the real power it exchanges with the line must be supplied to, or absorbed from its dc terminals. The available source of power, by definition, is the sending-end generator. Thus, it is reasonable to stipulate that the real power the SVS exchanges with the transmission line to accomplish the desired flow of power must be provided by the sending-end generator. This stipulation is indicated symbolically in Figure 1.31 by showing a bidirectional coupling of real power flow between the sending-end generator and the SVS. A possible practical implementation of this coupling is a back-to-back converter arrangement, as shown in Figure 1.32, in which the shunt-connected converter provides the real power (from the sending-end bus) the series-connected converter exchanges with the line. The self-sufficient power flow controller arrangement for unrestricted series voltage injection, conceptualized in Figures 1.31 and 1.32 is, is called Unified Power Flow Controller (UPFC).

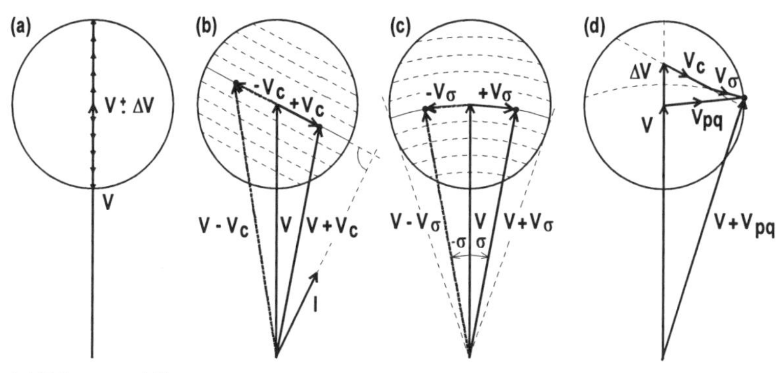

(a) Voltage regulation
(b) Line impedance compensation
(c) Phase shifting
(d) Simultaneous control of voltage, impedance and angle

Figure 1.33 Functional capabilities of the UPFC

From the conventional viewpoint of power transmission, the UPFC can provide multiple power flow control functions by adding the injected voltage phasor V_{pq}, with appropriate magnitude, V_{pq}, and phase angle, ρ, to the sending-end voltage phasor V_s. As illustrated in Figure 1.33, by the appropriate choice (control) of phasor V_{pq}, the three customary power flow control functions, and combinations thereof, can be accomplished:

Terminal voltage regulation or control, similar to that obtainable with a transformer tap-changer having infinitely small steps, is shown in (*a*) where

$V_{pq}=\pm\Delta V_o$ is injected in phase (or anti-phase) with V_s.

Series reactive compensation is shown in (*b*) where $V_{pq}=V_q$ is injected in quadrature with the line current I.

Phase-shift (transmission angle regulation) is shown in (*c*) where $V_{pq}=V_\sigma$ is injected with an angular relationship with respect to V_s that achieves the desired σ phase shift (advance or retard) without any change in magnitude.

Multi-function power flow control, executed by simultaneous terminal voltage regulation, series reactive line compensation, and phase shifting angle regulation, is illustrated in (*d*) where $V_{pq}=\Delta V_o+V_q+V_\sigma$.

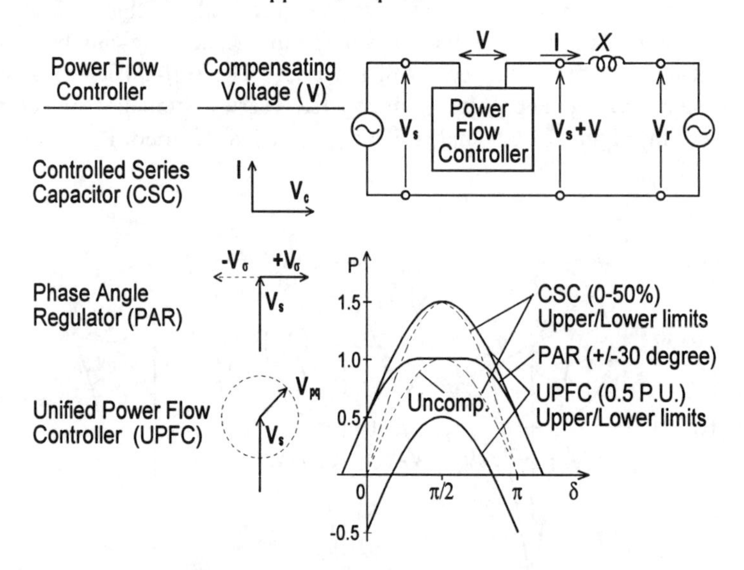

Figure 1.34 P versus δ control range for a UPFC controlled transmission line

The above functional capability is illustrated by a set of plots in Figure 1.34, where the control range for the transmitted power *P* versus angle δ is shown, as the UPFC (of 0.5 p.u. Mva rating) is operated to imitate series capacitor/reactor ($V_{pq}=V_q$), a phase shifter ($V_{pq}=V_\sigma$), and an unrestricted multi-function controller (V_{pq}) maximizing the control range for *P*. This set of plots also gives a comparison of relative effectiveness for flow control between the fully utilized UPFC and its conventional power flow controller counterparts, the controlled series capacitor and the phase shifter.

Applying the *forcing function viewpoint* of flow control, the functional capabilities of the UPFC can be put into a different, and from the standpoint of practical applications, more meaningful perspective. That is, the inherent two-dimensional control capability (manifested by the independent magnitude and angle control of the injected compensating voltage), implies that the UPFC is able to control directly *both* the real and reactive power flow in the line. From this viewpoint, the conventional terms of series compensation and phase shifting become irrelevant: the UPFC simply controls the magnitude and phase angle of the injected voltage so as to *force* the magnitude and angle of the line current, with respect to a selected voltage (e.g., the receiving-end), to such values which establish the desired real and reactive power flow in the transmission line.

The real and reactive control capability of the UPFC can be conveniently illustrated in the *{Q,P}* plane where the normalized reactive power *Q* (p.u.) is plotted against the normalized real power *P* (p.u.) of the simple two-machine model shown in Figure 1.31. For the uncompensated system, with $V^2/X=1$ (p.u.) stipulation, the function *Q=f(P)* can be readily written from equations (1.11) and (1.12):

$$Q = -1 - \sqrt{1 - P^2} \tag{1.29}$$

or

$$(Q+1)^2 + P^2 = 1 \tag{1.30}$$

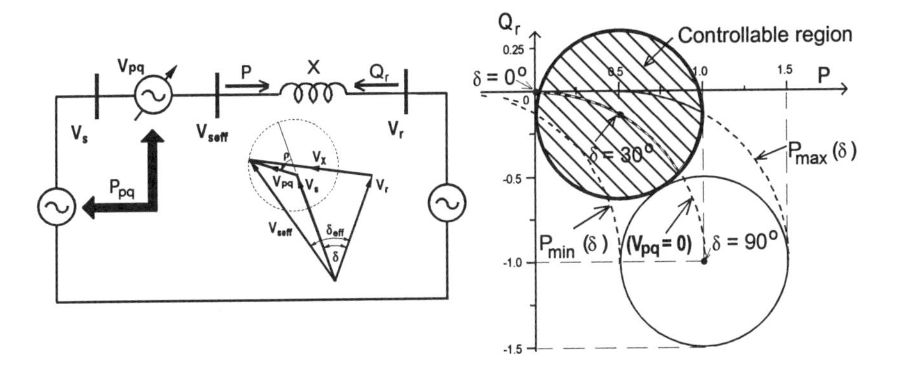

Figure 1.35 Q_r versus P control region for a UPFC controlled transmission line

Equation (1.30) shows that the *Q=f(P)* function for the uncompensated system describes a circular arc drawn with a radius of 1.0 around the center defined by coordinates $P_o = 0$, $Q_o = -1.0$, shown by dashed-line in Figure 1.35. Each point defined by its *P* and *Q* coordinates on this arc corresponds to a particular transmission angle δ; e.g., coordinates $P = 0$, $Q = 0$ corresponds to $\delta = 0$; $P = 1.0$, $Q = -1.0$ corresponds to $\delta = 90°$, etc., which identifies the starting compensation point of the UPFC. Assume that, as illustrated, the prevailing transmission angle, δ, is 30°. Then, the full (360°) rotation of the compensating voltage phasor V_{pq}, with its maximum magnitude V_{pq}(=0.5 p.u.), will describe a circle in the *{Q,P}* plane with a radius of 0.5 around its centre defined by coordinates $P_{30°} = 0.5$ and $Q_{30°} = -0.134$, which characterize the uncompensated system at $\delta = 30°$. The area within this circle defines all P and Q values obtainable by controlling the magnitude and angle of compensating voltage phasor V_{pq}. In other words, this circle in the $\{Q,P\}$ plane defines all P and Q transmission values attainable with the given rating of the UPFC for the power system considered. With changing transmission angle, the control area with the same circular boundary would simply move up or down along the arc characterizing the uncompensated system.

The concept of unrestricted series voltage injection opens up new possibilities for power flow control. This approach allows not only the combined application of phase angle control with controllable series reactive compensations and voltage regulation, but also the *real-time transition* from one selected compensation mode into another to match existing system structures and handle particular system contingencies more effectively. (For example, series reactive compensation could be replaced by phase-angle control or vice versa.) Moreover, the unrestricted voltage injection makes it possible to control directly real and reactive power flow in the transmission line.

The independent P and Q control capability of the UPFC can be very effectively applied to counteract dynamic system disturbances and balance both real and reactive power flow in parallel lines.

The first UPFC with a combined shunt and series converter rating of ±320 Mva was commissioned in 1998 in the USA.

1.4.2.6 Interline power flow controller (IPFC)

The Interline Power Flow Controller represents a novel concept with the objective of providing a flexible power flow control scheme for a multi-line power system, in which two (or more) lines employ an SSSC for series compensation. The IPFC scheme provides, together with the independent controllable reactive compensation of each line, a capability to transfer *real* power between the compensated lines. This capability makes it possible to equalize both real and reactive power flow between the lines, to transfer power demand from overloaded to underloaded lines, to compensate against resistive line voltage drops and the corresponding reactive line power, and to increase the effectiveness of the compensating system for dynamic disturbances (transient stability and power oscillation damping). In general, the IPFC provides a highly effective scheme for power transmission management at a multi-line substation.

An elementary Interline Power Flow Controller consisting of two converter-based SSSCs, connected back-to-back for real power transfer, is shown in Figure 1.36. Each SSSC is coupled to a different transmission line via its own series insertion transformer and is able to provide independent series reactive compensation to its own line.

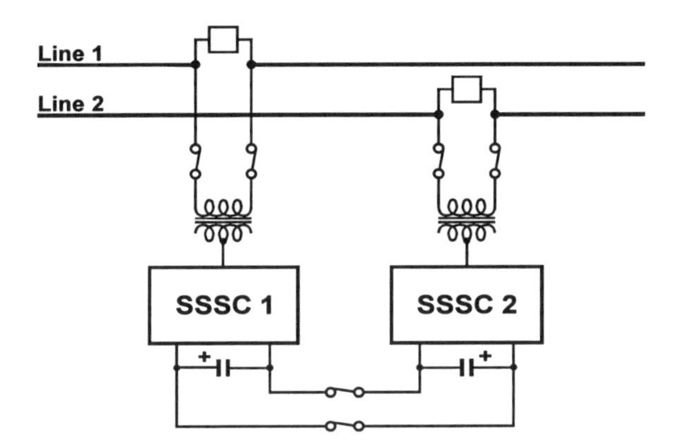

Figure 1.36 Elementary Interline Power Flow Controller consisting of two SSSCs operated with a common dc link

The converter of each SSSC produces a controllable ac output voltage at the fundamental frequency, which is synchronized to the voltage of the transmission line which that converter controls. The phase angle and magnitude of the two output voltages are controlled with respect to a selected bus (e.g., sending-end) voltage and the current of their own line. The injected voltages will generally have one component that is in quadrature and another that is in phase with the relevant line current. The quadrature components provide series reactive compensation for the lines and the in-phase components define the real power absorbed from one line and generated for the other. Since each converter is self-sufficient in generating or absorbing reactive power, the two quadrature voltage components can be independently controlled (within the converter rating) according to the reactive compensation requirements of the corresponding lines. However, since the real power exchanged by a converter at its ac terminals has to be supplied to, or absorbed from its dc terminals, the in-phase output voltage component of each of the two converters must be controlled so as to ensure a net zero real power balance at their common dc terminals. In other words, the real power compensation demand of one line must be fully supplied (or absorbed) by the other line.

The operation of the IPFC is illustrated in Figure 1.37 in connection with an elementary two line system. For this discussion assume that Line 1 is the *prime* line which is to be optimized for power transmission by means of independently controllable real and reactive power flow. Line 2 is assumed to have capacity to provide the real power needed for the optimization of the power flow in Line 1. (For clarity of the illustration, it is assumed that Line 1 and Line 2 are identical, although in practice they would usually be different.)

Figure 1.37a shows the single line Power System 1, with sending-end bus providing power for Line 1 (represented by line inductance X_1), the receiving-end bus, and the controllable ac voltage source representing the output of Converter 1 of the IPFC. In order to achieve the desired power flow in the primary Line 1, Converter 1 in the IPFC configuration injects an appropriate compensating voltage $\boldsymbol{V}_{1pq}$, (with controllable magnitude V_{1pq} and angle ρ) to change the magnitude and angle of the line current $\boldsymbol{I}_1$ so as to force the flow of the desired real power P_1 and reactive power Q_1. The corresponding phasor diagram at this figure illustrates the mechanism of line current control: voltage *phasor*, $\boldsymbol{V}_{1pq}$ is added to the existing sending-end bus voltage phasor, $\boldsymbol{V}_{1s}$, to produce the effective sending-end voltage phasor, $\boldsymbol{V}_{1seff} = \boldsymbol{V}_{1s} + \boldsymbol{V}_{1pq}$. The difference, $\boldsymbol{V}_{1seff} - \boldsymbol{V}_{1r}$, provides the voltage phasor, $\boldsymbol{V}_{1X}$, across the line impedance (X_1) needed to force the necessary line current phasor, $\boldsymbol{I}_1$, and thereby to establish the desired real and reactive power flow in the line. It is seen from this phasor diagram that, from the standpoint of the primarily controlled Line 1, the IPFC provides a two-dimensional compensation capability for the independent control of real line power P_1 and reactive line power Q_1, similar to that obtainable with the UPFC.

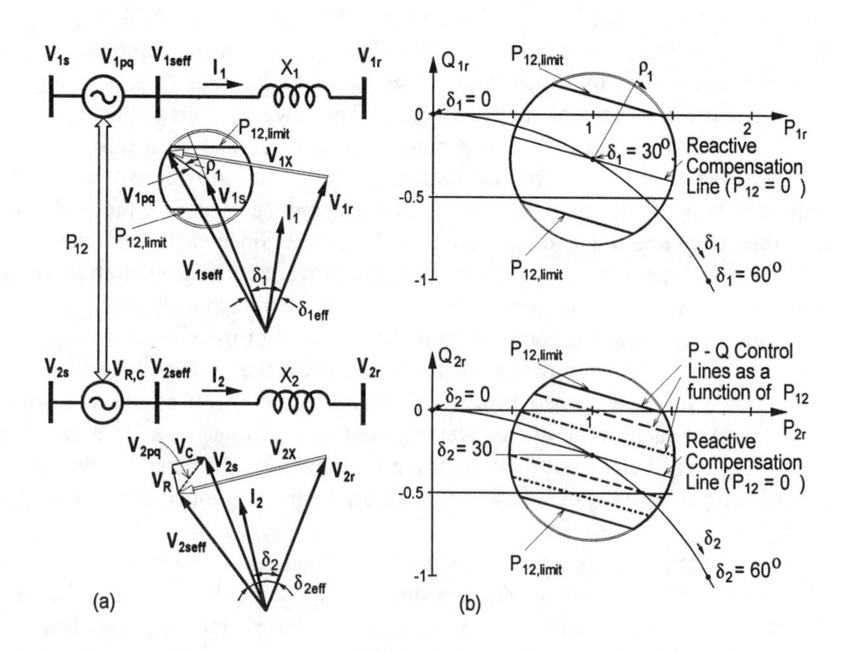

Figure 1.37 Phasor and Q_r versus P diagrams illustrating the operation of the elementary IPFC shown in Figure 1.35

The two-dimensional voltage injection in series with Line 1 generally results in the exchange of reactive power Q_{1pq} and of real power P_{1pq} between Line 1 and Converter 1 of the IPFC. The reactive power Q_{1pq} exchanged is provided by Converter 1 itself, however, the real power P_{1pq} appears as a real power demand at its dc terminals. To satisfy this power demand, the IPFC control makes Converter 2 supply real power $P_{2pq} = -P_{1pq} = (\pm)\, P_{12}$ from Line 2 to Line 1. In other words, Converter 2 is controlled to regulate the common dc bus of the IPFC and thereby continuously satisfy the relationship $V_{2pq}I_2\cos\psi_{2pq} = -\, V_{1pq}I_1\cos\psi_{1pq}$ (where ψ_{1pq} and ψ_{pq2} are the angles between the injected voltages and the corresponding line currents, that is, between phasors $\boldsymbol{V}_{pq1}$ and $\boldsymbol{I}_1$, and between $\boldsymbol{V}_{pq2}$ and $\boldsymbol{I}_2$, respectively). This constraint of course means that the real power exchange for Line 2 is *pre-defined* and therefore only its series reactive compensation can be freely varied to control the power flow in this line, in a manner done by controllable series reactive compensators such as the SSSC and TCSC. In other words, a two-converter IPFC configuration can provide a UPFC-type, two-dimensional series compensation for Line 1 and an SSSC-type series reactive compensation for Line 2.

The operation of Converter 2 is illustrated by the phasor diagram characterizing the simple, single-line power system of Line 2 in Figure 1.37a. The injected voltage of Converter 2 is controlled with respect to the prevailing current in Line 2 to meet the real power demand of Line 1 and to provide the desired reactive series compensation for Line 2. Accordingly, a voltage phasor component $\boldsymbol{V}_R$, *in-phase* (or anti-phase) with the line current phasor $\boldsymbol{I}_2$, is injected with a magnitude to satisfy the $P_{12} = V_R\, I_2$ $(= V_{2pq}\, I_2 \cos \psi_{2pq})$ real power demand of Line 1. Independent of voltage phasor component $\boldsymbol{V}_R$, another freely controllable voltage phasor component $\boldsymbol{V}_q$ is injected *in-quadrature* with $\boldsymbol{I}_2$ to provide series reactive compensation. The sum of the injected in-phase and quadrature phasor components determine the injected voltage phasor of Converter 2, that is $\boldsymbol{V}_{2pq} = \boldsymbol{V}_R + \boldsymbol{V}_q$ which must be within the rating of Converter 2.

The operation of the IPFC can be characterized by two, two-dimensional *P–Q* plots (similar to that introduced for the UPFC in Figure 1.35), showing the achievable control range for the reactive and real power in the two lines, as illustrated in Figure 1.37b. For the illustration the rating of both converters were assumed to be 0.25 p.u. Mva, the sending-end and receiving-end voltages 1.0 p.u., both line impedances 0.5 p.u., both transmission angles 30°, so that both systems transmit 1.0 p.u power *without* compensation ($V_{1pq} = V_{2pq} = 0$). The real power transfer between the two converters may be limited to a value, for example, to that just enough to make the power transmission at unity power factor possible for Line 1 throughout its compensation range, as illustrated in the figure.

For the primarily controlled Line 1, the *P–Q* plot, from the viewpoint of series compensation, is identical to that which characterizes the UPFC. Observe that with the power transfer limit stipulated, Line 1 can transmit real power at unity power factor, i.e., *without any reactive power flow*, over the total control range of the IPFC. However, independent control of reactive power flow in Line 2 is not possible due to the constraint imposed on the injected voltage by the real power demand of the primarily controlled Line 1. The straight, *reactive compensation control lines* in the control region of Line 2 represent the possible control range for real power P_{2R} at a *fixed* P_{12} supplied for (or absorbed from) Line 1. It can be seen that, with reasonable limit on the real power P_{12}, the range of real power control corresponding to the series reactive compensation of Line 2 is not significantly affected by the real power exchanged with Line 1. Therefore, the real power flow in Line 2 is fully controllable practically the same way as if Converter 2 was used separately for series reactive compensation (i.e., operated as an SSSC). However, for significant real power compensation of Line 1, Line 2 must have sufficient capacity to carry the additional reactive power resulting from the real power transfer to Line 1.

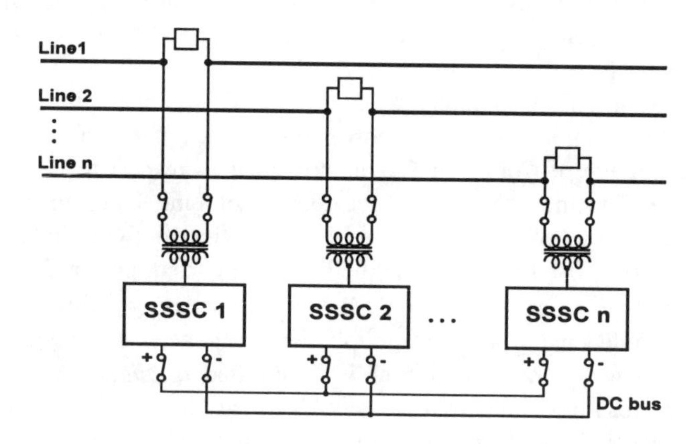

Figure 1.38 Multi-line IPFC consisting of n SSSCs with a common dc link

The plots in Figure 1.13b clearly show the *added flexibility* the IPFC configuration provides for series compensation: it is able to equalize not only the real power flow in a two- or multi-line transmission system, but *also* can equalize or control the reactive power flow in the lines. The IPFC provides an excellent tool to solve economically power flow problems in a multi-line transmission system in which the actual power flows are not proportional to the capacities of the corresponding lines or to the desired power transmissions in the individual lines, or in which the desired real power transmission in some lines is hindered by relatively high reactive power flow.

It is evident from the above discussion that the IPFC concept, characterized so far for two lines, can be extended to multiple (n) lines as illustrated in Figure 1.38.

The underlying idea of this generalized IPFC approach is that the strong or under-loaded lines are forced to help the weaker or over-loaded lines in order to optimize the utilization of the whole transmission system. The operation of a multi-line IPFC requires that the sum of the real power exchanged by the total number of converters must be zero. However, the distribution of the positive and negative power exchanged with the individual lines by the individual converters within this overall constraint is arbitrary (as long as the Mva rating of the individual converters is not exceeded). This arrangement would provide a UPFC-type two-dimensional compensation for some lines and an SSSC-type reactive compensation for the others.

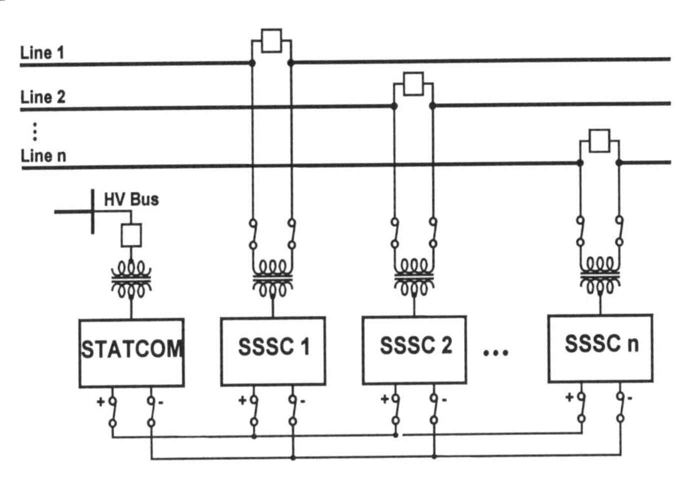

Figure 1.39 Generalized IPFC consisting of n SSSCs and a STATCOM for transmission system power flow management

The constraint for keeping the sum of the real power exchanged with the *n*-lines zero can be circumvented by adding a shunt-connected converter the multi-converter IPFC as illustrated in Figure 1.39. This arrangement is particularly attractive in those cases in which the real power compensation requirement of the "weak" lines exceeds the real power that can be absorbed from the "strong" lines without appreciably impacting their own power transmission or when shunt reactive compensation at the substation is required anyway for voltage support.

1.5 FACTS control considerations

As discussed in this chapter, FACTS controllers generally fall into two discrete families: one comprises mainly the conventional thyristor-controlled SVC, TCSC, and Phase Shifter, the other the converter-based STATCOM, SSSC, UPFC, and IPFC. Although presently a large number of SVC installations exist (due to the SVC's commercial availability over the last 20 years), the converter-based FACTS controllers clearly represent the future trend due to their superior performance and, more importantly, to their much greater functional operating flexibility. (For example, the recently commissioned 160 Mva UPFC installation at the Inez substation of American Electric Power can be reconfigured into a ±320 Mvar STATCOM, or a ±320 Mvar SSSC, or a independently operated ±160 Mvar STATCOM and ±1600 Mvar SSSC. The 200 Mva two converter installation under development for the New York Power Authority will have the comprehensive capability to operate as a STATCOM, an SSSC, a UPFC, or an IPFC.) For these reasons, and also because the converter-based flexible transmission system represents the more general case from the standpoint of system optimization and control, in the following discussion the control activated convertibility of the FACTS controllers, as a potential means of system optimization, will be assumed.

1.5.1 Functional control of a single FACTS controller

The functional control of a single, generalized (converter-based) FACTS controller from the standpoint of the power system can be best explained with the use of the Synchronous Voltage Source (SVS) model. As defined in Section 1.4.2.1, a synchronous voltage source is able to produce a sinusoidal output voltage of a given frequency with controllable amplitude and phase angle. It can exchange both real and reactive power with an ac power system and is able to generate internally the reactive power exchanged. As shown in Section 1.4.2.2, the SVS can be realized by a voltage-sourced, switching converter, in which case the real power exchanged with the ac system must be supplied or absorbed at the dc terminals of the converter. As was also shown, an SVS can be configured and operated as a shunt or series compensator, and can be combined with other SVSs to provide multiple compensating functions.

In order to define the operation of the SVS in the power system, four basic sets of inputs are required: *(1) Operating mode selection, (2) Parameter, gain and limit settings, (3) References,* and *(4) Measured System variables*, as illustrated in Figure 1.40.

The *Operating mode selection* input defines the functional operation of the SVS, i.e., whether it is to function as a STATCOM, SSSC, part of the UPFC, part of the IPFC, or in some other manner. It would also define the particular control mode to be executed within the selected function. For example, the STATCOM could be controlled to regulate bus voltage or reactive power; the SSSC to control line impedance or series compensating voltage or power flow, etc. This input would initiate power circuit configurations appropriate for the selected function, e.g., closing or opening such switches as SW1 and SW2, shown symbolically in Figure 1.40, which facilitate shunt or series connection of the SVS to the transmission line. It would also set the SVS control into the desired operating mode by, for example, activating the appropriate control software, and could select preferred strategies for contingency cases when the desired SVS output is limited by its voltage, current, or Mva rating.

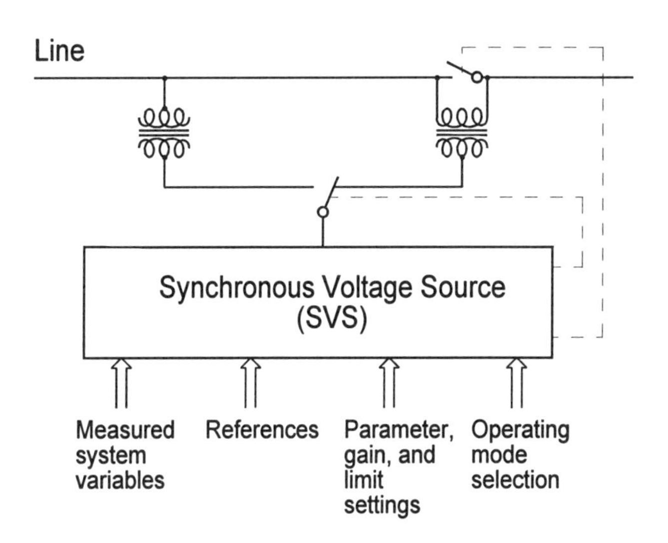

Figure 1.40 Inputs to a generalized converter-based (synchronous voltage source type) FACTS controller

The *Parameter, gain, and limit settings* input identifies the desired operating parameters, such as, set points, regulation slopes, various closed-loop gains, and, most importantly, allowed operational limits, including the maximal values of transmission line voltage, current, and impedance.

The *References* input provides the reference values for the system variables, voltage, current, power, impedance, angle, etc., that the SVS in the selected functional operating mode should maintain. These reference signals may represent steady-state values or real time variables.

The *Measured system variables* are the signals that the SVS controls to match the corresponding references. These signals are either locally measured (present practice) or they may be transmitted from other locations by appropriate communication links.

The control system of the SVS (and that of other types of modern FACTS controllers) is computer-based. It generally consists of two main parts: the *real time control* and the *status monitor, operator, and system interface,* usually with a *CRT display*, as illustrated in Figure 1.41. The status processor is the overall supervisor of the SVS. It collects status data from all SVS subsystems (converter valves, cooling system, coupling transformer, auxiliary components, and support systems, etc.), and receives the measured system variables, reference signals, and all operating instructions. It processes the incoming signals, interfaces with the real time control, and executes the operating instructions, such as start-up, shut-down, and other external switching operations. It monitors and displays system status and operation and provides diagnostics for system error, malfunction, or failure.

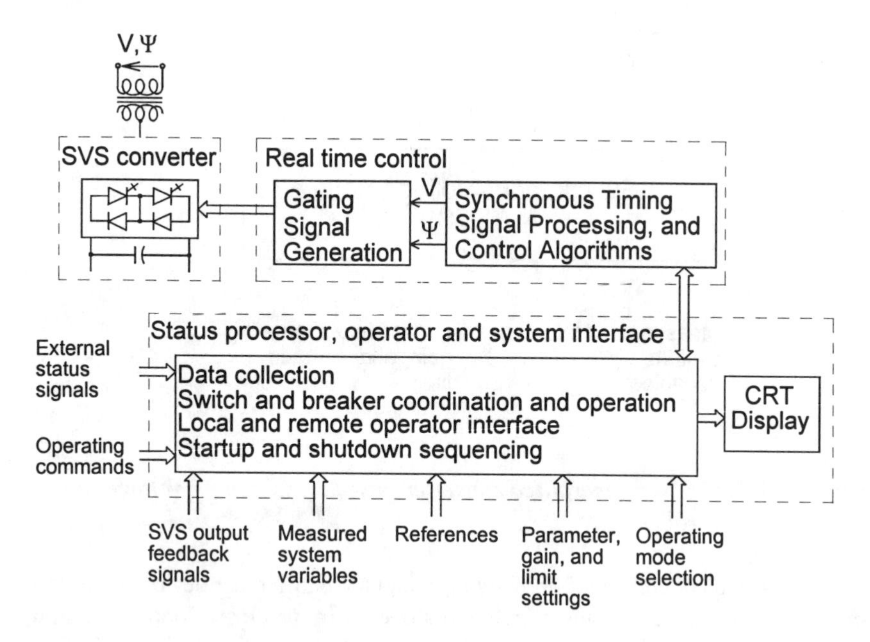

Figure 1.41 General control structure of an SVS type FACTS controller

The real time control operates the SVS converter from scaled and conditioned input signals representing the system variables together with the reference signals and other instructions from the status processor. The digital signal processors (DSPs) in the central control carry out the appropriate algorithms according to the requested operating mode to derive the instantaneous values (magnitude and angle or equivalent) of the desired output voltage to meet the compensation or power flow control demand. The desired instantaneous output values are converted into corresponding gate turn-on and turn-off signals to operate the semiconductor valves in the SVS converter. The real time control (or a coordinated subsystem thereof) also manages the particular operating and protection protocol of the power semiconductors in the converter valves.

1.5.2 FACTS area control: possibilities and issues

The majority of presently installed FACTS controllers are conventional thyristor-controlled Static Var Compensators. In contrast to the generalized converter-based FACTS controller shown in Figures 1.40 and 1.41, these, and the related Thyristor-Controlled Series Capacitors and Phase Shifters, do not have multi-functional capability. They can function only for the single purpose (i.e., shunt compensation, series compensation, or angle control) for which they were installed. They may, however, have the capability to change their operating mode, e.g., the SVC may be set to regulate voltage or reactive power, the TCSC to control line impedance or power flow, etc., similar to, but generally more limited than, their converter-based counterparts. The continuous on-line operation of both types is controlled by the reference signal(s) provided by the system operator.

Until now, the utility practice has been to install a dedicated FACTS controller and operate it in a fixed mode with substantially steady-state reference signals controlling local system variables (voltage, current, impedance, etc.). In other words, the FACTS controller receives no other indication of system problems and contingencies than the change in local variables. Its response is to try to correct immediately regardless of whether or not this correction is appropriate for the new (e.g., post fault) system condition. Similarly, it would have no direct knowledge of the action of other controlled power system devices. In fact, it could preclude the operation of those with slower response, or interact with those of comparable speed. There are possible control measures to provide an acceptable steady-state solution to these problems. However, in order to get predictable and proper dynamic response, as well as maximal steady state benefits, the direct coordination of fast FACTS controllers and other devices, located close enough to interact, seems prudent if not absolutely necessary. Evidently the importance of coordination rapidly increases with the growing number of powerful FACTS controllers deployed.

Beyond the interaction-prompted coordination, the anticipated deployment of FACTS controllers, particularly the converter (SVS) types with multiple functional capability, provides a new basis for reconceptualizing the control of the bulk power transmission system as a dynamic entity. Within this entity, FACTS controllers establish optimal power flow patterns for economic benefits consistent with network security requirements under prevailing system conditions and handle dynamic disturbances by concerted control action.

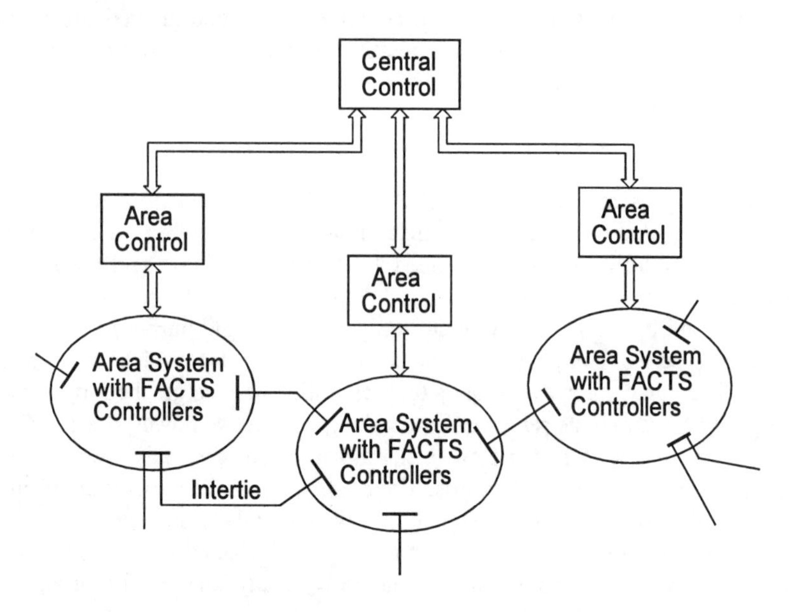

Figure 1.42 Possible hierarchical control scheme for inter-connected power system employing FACTS controllers

From the technical standpoint the most plausible system control structure to manage the transmission grid as a dynamic entity is hierarchical, built up with layers of controls from the level of each individual FACTS controller ultimately to the top level central control that strategically coordinates the operation of the overall network. This hierarchical control, illustrated in Figure 1.42, could be envisioned in several ways. One extreme would be to mandate that the top level central controller direct the operation of each FACTS controller in real time. The opposite extreme would be to operate the FACTS controllers independently from local system data and provide coordination from local system operators in the form of steady-state reference signals (i.e., the method utilities presently use). The first extreme would clearly raise serious questions regarding the security of the total system since errors in the data collection and processing, malfunctions in the central control or in the communication system, could result in major and potentially system wide outages. The second extreme would not completely utilize the inherent flexibility of the FACTS-based power system for full economic or security benefits. A reasonable solution may be similar to that used in organizational managerial structures in which the overall strategic decisions are made at the top level, the tactical decisions at mid-level, and the implementation details are executed at local levels. In this scenario one could visualize that the central control would determine the main transmission paths for the overall system according to demand and economic objectives, taking into account equipment availability and other constraining factors, and would coordinate the operation of neighboring transmission areas. The area controls would optimize the defined power flow through their system by setting, via appropriate communication links, their FACTS controllers into the desired operating mode and by providing reference signals for them for the selected system variables controlled. The area control would supervise the operation of the area transmission system by collecting system status data as well as critical operating and availability information from each FACTS controller. Under area contingency or system flow change demand, the area control would reoptimize its system operation by appropriately changing the functional operating mode of the FACTS controllers and their reference inputs. In the case of an area disturbance, the area control would dynamically change the reference inputs to the appropriate FACTS controllers in order to achieve a coordinated counter action to reestablish rapidly rotational and voltage stability. Within this scheme, the individual FACTS controller would operate self-sufficiently from locally measured system variables and, in the operating mode defined, its output would be controlled by the reference input the area control system provides. In the case of an area control outage or malfunction, the local FACTS controllers would automatically fall back to independent operation determined by the prevailing system steady-state references. Under all conditions, the individual FACTS controllers would accept overriding instruction from authorized operating personnel.

The establishment of a viable hierarchical system control for a large transmission grid will be a gigantic undertaking. It will require fundamental changes in control strategies, development of new system security procedures and wide-area measurement capabilities together with highly reliable communication links and protocols. It will also require the development of analytical (software) tools to process real time information for system security, voltage, and rotational stability assessments.

In addition to the many challenging technical issues, the hierarchical control of FACTS-based power transmission also poses difficult political, ownership, and responsibility problems. If there will be individually owned area sub-systems (presently individual utility systems) forming an overall national grid, who will have the ultimate responsibility to run the central control and determine how the area systems will be utilized (and thus how the areas will economically benefit)? Who will determine which area systems should be upgraded, and at whose expense, to make the overall system more efficient or secure? How will effective competitiveness between the area sub-systems be maintained and, at the same time, the required centrally directed coordination and cooperation be maintained? Would a national transmission grid with single ownership in a large country, like the USA, be feasible and economically practical? There are many questions like these and many different answers, each based on different transmission network and organizational structures. Only time will tell which will succeed. The incentives, however, are clearly established: FACTS with broad application of advanced technologies promises to facilitate deregulation with its potential economic benefits by significantly increasing and expanding power transfers through presently constrained power delivery systems, thus providing power at lower cost to a greater number of customers.

1.6 Summary

An ac power system is a complex network of synchronous generators, transmission lines, and loads. Each transmission line can be represented as a mostly reactive ladder network composed of series inductors and shunt capacitors. The total series inductance, which is proportional to the length of the line, determines at a given voltage, the maximum transmittable power. The shunt capacitance influences the voltage profile along the transmission line.

The transmitted power over a given line is determined by the line impedance, the magnitude of, and phase angle between the end voltages, or in other words, the forcing voltage acting across the transmission line.

The basic operating requirements of an ac power system are that the synchronous generators must remain in synchronism and the voltages must be kept close to their rated values. The capability of a power system to meet these requirements in the face of possible disturbances (line faults, generator and line outages, load switchings, etc.) is characterized by its transient, dynamic, and voltage stability. The stability requirements usually determine the maximum transmittable power at a stipulated system security level.

The power transmission capability of transmission lines can be increased by reactive compensation. The voltage profile along the line can be controlled by reactive shunt compensation, the series line inductance by reactive (capacitive) series compensation. The transmission angle can be varied by phase shifters.

Traditionally, reactive compensation and phase angle control have been applied by fixed or mechanically switched circuit elements (capacitors, reactors, and tap-changing transformers) to improve steady-state power transmission. The recovery from dynamic disturbances was accomplished by generous stability margins at the price of relatively poor system utilization.

Since the 1970s, energy cost, environmental restrictions, right-of-way difficulties, together with other legislative, social and cost problems, have delayed the construction of both generation facilities and, in particular, new transmission lines in the USA (and in other countries). In this time period, there have also been profound changes in the industrial structure and significant geographic shifts of highly populated areas. Recently, in tune with the worldwide movement of deregulation, the Federal Energy Regulatory Commission in the USA ruled to facilitate the development of competitive electric energy markets by mandating open access transmission services.

The economic, social, and legislative developments have demanded the review of traditional power transmission theory and practice, and the creation of new concepts that allow full utilization of existing power generation and transmission facilities without compromising system availability and security.

In the late 1980s, the Electric Power Research Institute formulated the vision of the Flexible AC Transmission System (FACTS) in which various power-electronics based controllers regulate power flow and transmission voltage and, through rapid control action, mitigate dynamic disturbances. The main objectives of FACTS are to increase the useable transmission capacity of lines and control power flow over designated transmission routes.

There are two approaches to the realization of power electronics-based FACTS controllers (compensators): one employs conventional thyristor-switched capacitors and reactors, and quadrature tap-changing transformers, the other self-commutated switching converters as synchronous voltage sources. The first approach has resulted in the Static Var Compensator (SVC), the Thyristor-Controlled Series Capacitor (TCSC), and the Thyristor-Controlled Phase Shifter. The second approach has produced the Static Synchronous Compensator (STATCOM), the Static Synchronous Series Compensator (SSSC), the Unified Power Flow Controller (UPFC), and the Interline Power Flow Controller (IPFC).

The two groups of FACTS controllers have distinctly different operating and performance characteristics.

The thyristor-controlled group employs capacitor and reactor banks with fast solid-state switches in traditional shunt or series circuit arrangements. The thyristor switches control the *on* and *off* periods of the fixed capacitor and reactor banks and thereby, in effect, realize a variable reactive (shunt or series) impedance. Except for losses, they cannot exchange real power with the ac system.

The synchronous voltage source (SVS) type FACTS controller group employs self-commutated dc to ac converters, using gate turn-off thyristors, which can internally generate capacitive and inductive reactive power for transmission line compensation, without the use of ac capacitor or reactor banks. The converter, supported by a dc power supply or energy storage device, can also exchange *real power* with the ac system, in addition to the independently controllable reactive power.

The converter-based SVS can be used uniformly to control transmission line voltage, impedance, and angle by providing reactive shunt compensation, series compensation, and phase shifting, or to control directly the real and reactive power flow in the line by forcing the necessary voltage across the series line impedance.

When used for reactive shunt compensation, the SVS acts like an ideal synchronous compensator being able to maintain the maximum capacitive output current at any system voltage down to zero. This *V–I* characteristic is superior to that obtainable with the conventional thyristor-controlled SVC whose maximum capacitive output current decreases linearly with the system voltage

As a reactive series compensator, the SVS can provide controllable series capacitive compensation without the danger of sub-synchronous resonance. Its capability to maintain the maximum compensating voltage independent of the line current, and to provide capacitive as well as inductive compensation, results in a much wider control range than possible with controlled series capacitor compensation. This makes it highly effective in power flow control, as well as in power oscillation damping.

The arrangement of two synchronous voltage sources, one in shunt-connection and the other in series-connection, results in the novel Unified Power Flow Controller. This arrangement can provide concurrent or individual voltage, impedance, and angle regulation or, alternatively, independent real and reactive power flow control and thus can readily adapt to particular short term contingencies or future system modifications.

The arrangement of two or more synchronous voltage sources, each in series with a different line, results in the unique Interline Power Flow Controller. The IPFC scheme provides, together with the independently controllable reactive compensation in each line, a capability to transfer real power between lines. This capability makes it possible to properly equalize power flow among lines, to transfer power demand from overloaded lines to underloaded lines, as well as to compensate against resistive voltage drops and the corresponding reactive line power. In general, the IPFC provides a highly effective scheme for the control of a multi-line transmission system.

The deployment of increasing number of FACTS controllers will make it necessary to reconceptualize the control of the transmission system as a dynamic entity in order to prevent undesirable interactions and obtain attainable maximum economic and operating benefits. A hierarchical control appears to be a promising candidate scheme. Its ultimate form and operation, however, will depend not only on the successful development of the necessary control and communication technologies and protocols, but also on the eventual structure of the evolving power delivery industry.

1.7 Acknowledgements

I would like to express my gratitude to the Westinghouse Electric Corporation for believing in, and undertaking the practical development of, the converter-based FACTS controllers, and to Siemens Power Transmission and Distribution for the continued support and commercialization of this technology. I am indebted to Mr John P Kessinger, General Manager, FACTS and Power Quality Division for his support of this undertaking. I am also very grateful to the Electric Power Research Institute, the Empire State Electric Energy Research Corporation, the Tennessee Valley Authority, the Western Area Power Administration, the American Electric Power, the New York Power Authority, and other utilities for the support they provided for the development and practical adoption of this technology.

I also wish to acknowledge the significant contributions made by Dr Colin D. Schauder and Mr Eric J. Stacey and other members of the Westinghouse Power Electronics Department and the Energy Management Division (now Siemens FACTS and Power Quality Division) to the development of the converter-based FACTS technology and to the realization of practical hardware. Their work has provided valuable information on the rapid advances of this technology. Special thanks are due to Mr Gary L. Rieger, who read the manuscripts and made valuable suggestions to improve the text, to Mr Miklos Sarkozi, who created many of the illustrations included in my material, to Dr Kalyan K. Sen, who provided computational and simulation assistance, and to Mr Mack R. Lund, who carried out most of the TNA tests and organized the relevant data.

1.8 References

1 "A study of the US electric utility industry," Chapter III, Theodore Barry & Associates, 1980.

2 Miller, T.J., Editor., "Reactive power control in electric systems", Chapter 2, John Wiley & Sons, 1982.

3 Gyugyi, L., "Power electronics in electric utilities: static var compensators", *Proceedings of the IEEE*, 76, (4), April 1988.

4 Hingorani, N.G., "High power electronics and flexible ac transmission systems", *IEEE Power Engineering Review*, pp. 3-4, July, 1988.

5 Gyugyi, L., "Solid-state control of electric power in ac transmission systems", International Symposium on "Electric Energy Conversion in Power Systems", Invited paper, No. T-IP. 4, Capri, Italy, 1989.

6 Gyugyi, L., et al., "Advanced static var compensator using gate turn-off thyristors for utility applications", CIGRE paper 23-203, 1990.

7 Gyugyi, L., "A unified power flow control concept for flexible ac transmission systems", IEE Fifth International Conference on ac and dc Power transmission, London, Conference Publication No. 345, pp 19-26, 1991.

8 Gyugyi, L., "Dynamic compensation of ac transmission lines by solid-state synchronous voltage sources", *IEEE Transactions on Power Delivery*, 9, (2), April 1994.

9 Gyugyi, L., et al., "Static synchronous series compensator: a solid-state approach to the series compensation of transmission lines", *IEEE*

Transactions on Power Delivery, 12, (1), January 1997.

10 Stahlkopf, K. and Wilhem, M.R., "Tighter controls for busier buses", *IEEE Spectrum*, April 1997.

11 Edris, A., et al., "Controlling the flow of real and reactive power", *IEEE Computer Applications in Power*, January 1998.

12 Gyugyi, L., et al., "The interline power flow controller concept: a new approach to power flow management in transmission systems", *IEEE Transactions on Power Delivery*, 14, (3), 1999.

Chapter 2

Power electronics: fundamentals

Geza Joos

2.1 Introduction

Power electronics was first applied in power systems compensation with the advent of power semiconductor diodes and thyristors in the 1970s. Among the first applications were High Voltage DC (HVDC) transmission systems, based on thyristor rectifiers and inverters [1,2]. In parallel, static reactive power compensators were developed and installed: rotating synchronous condensers were replaced by thyristor-based technologies. In typical installations, a thyristor-controlled reactor (TCR) provides variable lagging reactive power (vars), and fixed or thyristor-switched capacitors (TSC) provide the leading vars. The combination of both devices in parallel allows continuous control of vars over a wide range of values, from leading to lagging vars [3,4]. The potential of var compensators based on other static power converter topologies was also recognized and a number of configurations proposed and investigated [4,5].

However, thyristor technology only allows the implementation of lagging var generators, unless complex force-commutation circuits are used. This drawback has been eliminated with the introduction of Gate Turn-Off (GTO) thyristors [6]. This has allowed the development of a number of configurations based on the concept of synchronous voltage sources [7]. Prototype GTO-based shunt var compensators, known as STATCOMs, have been installed and tested by utilities [8]. The STATCOM and other static var compensators have recently been grouped, together with other types of devices used for transmission system control and power flow optimization, under the heading of Flexible AC Transmission System (FACTS) devices [9].

Static power compensators are typically connected in shunt across transmission systems. An alternative connection, the series connection, has recently been the subject of many investigations [10]. Technological solutions have been developed to solve problems associated with insulating the equipment from ground and the full potential of series connections can now be exploited. One of the latest developments in power system compensation technology has been the combination of series and shunt static compensators into one unit, the Unified

Power Flow Controller, or UPFC [11]. In parallel, HVDC systems based on self-commutated power switches have been developed [12].

Static power converters have been successfully applied to a large number of power conversion problems at low and medium power levels. However, adapting these solutions to power transmission and distribution levels raises special issues. Although the capacity of power semiconductor switching devices has gradually increased, large ratings still require combining devices or converters in series and parallel. In addition to the large power handling capacity, static compensators must have very high efficiency, since losses increase both the capital and operating costs of the power system. Switching losses are therefore of primary concern and switching frequencies in general must be kept low. This may result in large harmonic waveform distortion, unless suitable control techniques and power circuit configurations are used.

This chapter presents the principles of power electronic converters as they apply to existing and proposed FACTS devices. It covers the control of passive reactive elements by means of power semiconductor switches and the use of static power converters to synthesize voltage and current sources for power system compensation. Both line-commutated and force-commutated technologies are presented. Topologies suitable for use with self-commutated devices such as GTOs and the more recently available high power IGBTs are discussed. Switch gating issues, including the use of Pulse Width Modulation (PWM) techniques, are addressed. Structures suitable for the design of high power converters for transmission level voltages and currents are presented, particularly multi-level and multi-module topologies.

2.2 Basic functions of power electronics

2.2.1 Basic functions and connections of power converters

The basic functions of static power converters in the area of FACTS devices [9] can be broadly classified as:

- *Controlling the amplitude of the ac voltage applied to a circuit.* Examples of circuits include reactive elements in devices used for power system var compensation. These are connected in either shunt or series, Figure 2.1: an inductor, in Thyristor Controlled Reactors (TCR), a capacitor, in Thyristor Switched Capacitors (TSC), or a capacitor in Thyristor Controlled Series Capacitors (TCSC). The circuit can be another transformer, as in thyristor controlled phase-shifting transformers (TCPST) used for quadrature voltage injection.
- *Producing synthetic voltage or current sources.* These sources can be connected to the transmission system, Figure 2.2, in shunt, as in the Static Compensator (STATCOM), Figure 2.2 (a); in series, as in the Static

Synchronous Series Compensator (SSSC), Figure 2.2 (b); in a combination of shunt and series, Figure 2.2 (c), as in the Unified Power Flow Controller (UPFC).

- *Converting power from one form to another, ac into dc or dc into ac.* This process is used in High Voltage DC (HVDC) transmission and other devices injecting real power into the ac system, such as the UPFC.

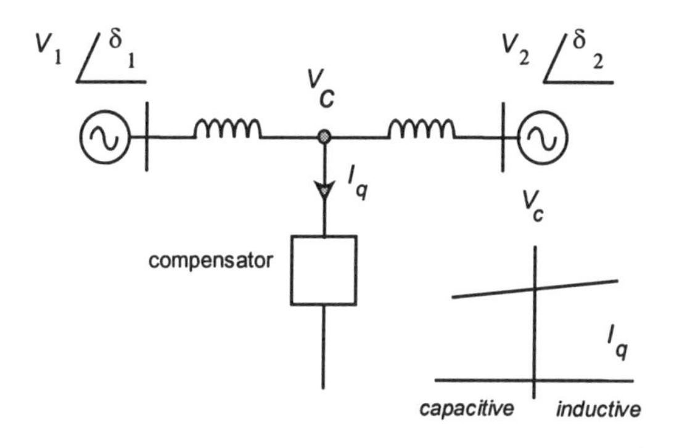

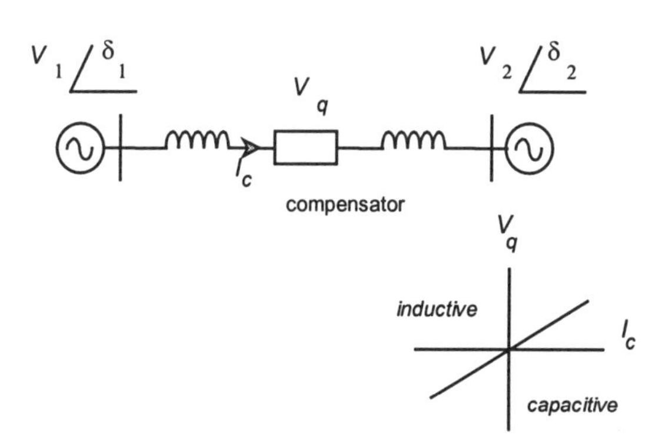

Figure 2.1 Principles of var compensation in transmission systems. (a)Shunt compensation. (b) Series compensation.

2.2.2 Applications of reactive power compensation

The role and potential of static power converters in modifying the characteristics and operation of power systems can be illustrated by means of requirements and applications to power system compensation. Compensation can be viewed as the injection of power, usually reactive power, leading or lagging, into the ac system. In its simplest form, this is achieved by inserting fixed capacitors or inductors in either series or shunt into the ac system. Assuming a compensating reactance X_c is inserted in a transmission system, the generated var Q_c, per phase, is derived as follows:

- *For the shunt connection*, Figure 2.1(a), a reactive current I_q is generated, allowing in particular ac system voltage support at the point of connection, V_c:
  $I_q = V_c/X_c$
  $Q_c = V_c^2/X_c$
- *For the series connection*, Figure 2.1(b) the capacitive impedance X_c partially compensates the line reactance, and a capacitive voltage V_q is inserted in series, the current I_c being the line current:
  $V_q = I_c X_c$
  $Q_c = I_c^2 X_c$

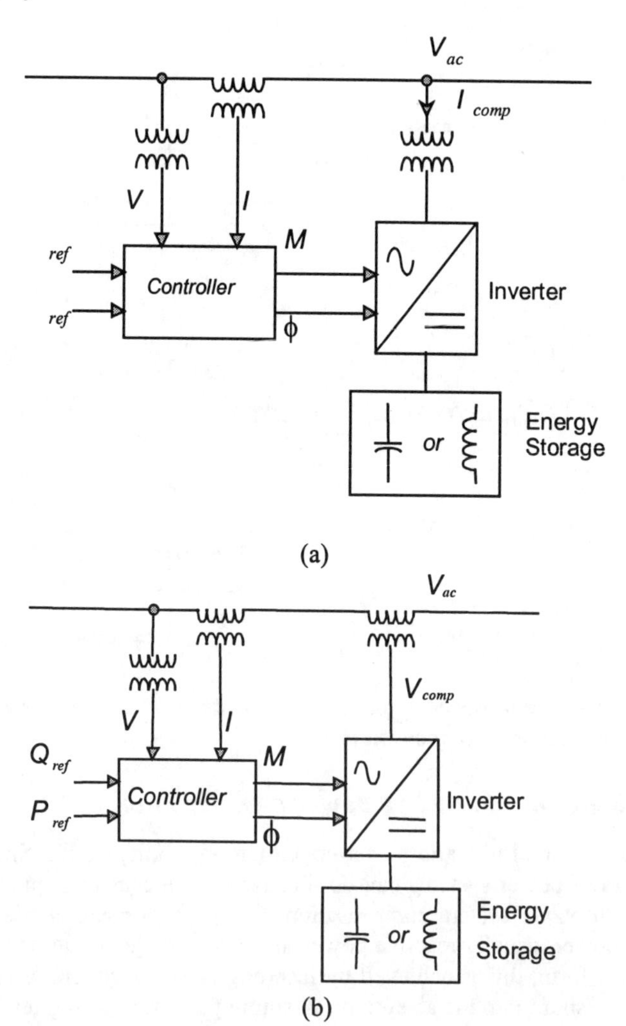

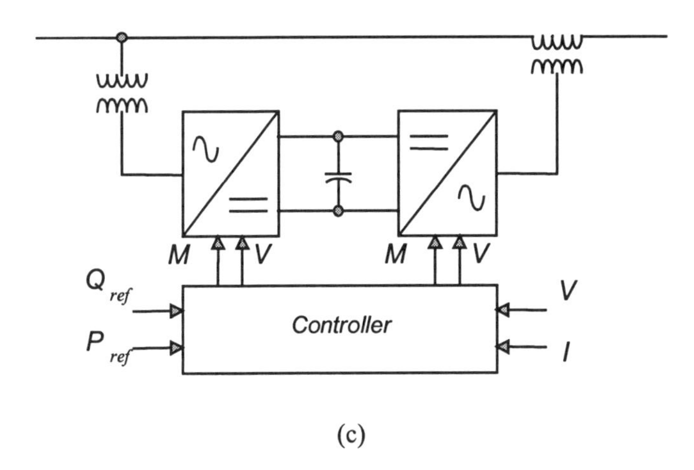

(c)

Figure 2.2 Principles of active compensation based on static power converters. (a) Shunt connection (STATCOM). (b) Series connection (SSSC). (c) Series-shunt connection (UPFC).

In addition to the lack of controllability of the reactive power injection, fixed capacitive compensation may lead to ac system instability, such as in the phenomenon known as Subsynchronous Resonance, or SSR, associated with series compensation [13]. In order to control the amount of reactive power injected, the reactive impedance must be varied. Equivalently, a variable current or variable voltage can be injected into the system, emulating a variable reactance.

The apparent reactive impedance of a fixed element can be varied using ac switches, or ac controllers. Conversely, the current or voltage required to emulate a variable reactance can be injected into the ac system by means of synthetic sources, which can be realized using static power converters, Figure 2.2. In addition to providing reactive power, these active compensators can also be used to supply real power, either transiently, for a given number of periods of the ac supply or on a continuous basis. This real power can be used to dampen power system oscillations or temporarily support the power system voltage under fault conditions. Furthermore, since the compensator is fully controllable, resonant frequencies associated with the use of capacitors are eliminated and the risk of instability eliminated.

Consideration must be given to protecting power electronic converters against power system transients and faults: overvoltages, in the case of shunt connections, or short circuits, in the case of series connections. In addition, series compensators must be isolated from the ac system.

2.3 Power semiconductor devices for high power converters

2.3.1 Classification of devices

Power semiconductor switches applicable to high power static compensators can be classified as follows:

- *Uncontrolled*, unidirectional current capability, unidirectional voltage blocking switch, the diode. The conduction state is dictated by the natural flow of the current.
- *Turn-on controllable switch*, unidirectional current capacity, bidirectional voltage blocking switch, the thyristor. The start of the conduction is initiated by means of a pulse between the gate and the cathode. Reverting to the blocking state requires that the current be extinguished by external means.
- *Fully controllable*, or self-commutated devices. Turn-on and turn-off can be initiated through the gate. These devices are available as either pulse triggered or continuously gated devices.

Table 2.1. Semiconductor switching power devices

Type	Features	Symbol	Implementation
Supply/load commutated	Uncontrolled/ Unidirectional cur.		Diode
	Delay controlled/ Unidirectional cur.		Thyristor
	Delay controlled/ Bidirectional cur.		Thyristors
Self commutated	Bidirectional cur. Unidirectional volt.		IGBT. GTO
	Unirectional current Bidir. volt. blocking		GTO. (IGBT+D)

Figure 2.3 shows the current/voltage characteristics and the commutation features of semiconductor devices. Table 2.1 summarizes the generic types of switches and their implementation using basic devices. It should be noted that conversion at very high power levels demands very high efficiencies, and switch losses should therefore be kept as low as possible. This includes conduction

losses, associated with forward drop, and switching losses, associated with switching frequency.

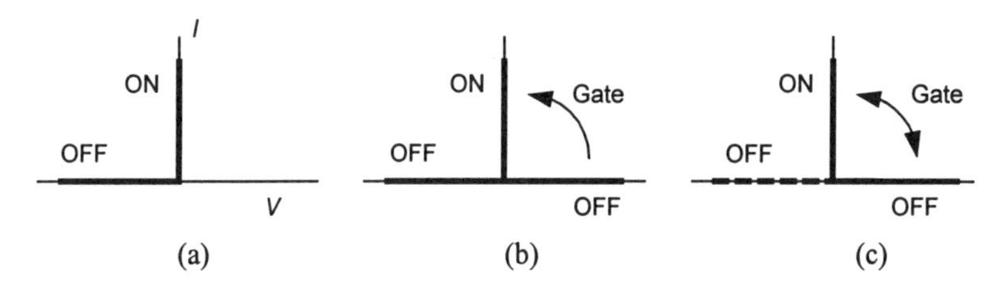

Figure 2.3 Switching characteristics of power semiconductor devices. (a) Diode. (b) Thyristor. (c) Self commutated devices (GTO, IGBT).

2.3.2 Device types and features

The general features of semiconductor switches found in high power static converters are:

- *Thyristors*. They are unidirectional current carrying devices with symmetrical blocking capabilities. They are available in very large power ratings, that is several thousands of volts and amps. They are mostly used today in very high power applications, mainly power systems and very large motor drives. They are very rugged, have very low losses, mainly associated with current conduction, and are triggered by pulses requiring very small amounts of energy. Light-triggered thyristors are also available. However, they have to be turned off using external means: the current flowing through them must be extinguished by means of an external circuit. Current extinction can occur naturally, through voltage reversal, as in ac controllers. It can be initiated by transferring the current flowing through the thyristor to another branch of the power converter. An alternative is the addition of force-commutation circuits, however, these are complex and bulky. Because force-commutation is difficult to implement, thyristors are usually line commutated and switched at line frequency.
- *Pulse triggered self-commutated devices*. These are the Gate Turn Off (GTO) thyristors and are available with either symmetrical voltage blocking capabilities, or more commonly with only forward blocking capabilities. These asymmetrical devices behave as diodes for reverse voltages and are therefore bi-directional current carrying devices. They are pulse triggered, as are thyristors. Blocking is initiated by injecting reverse current into the gate, thereby shunting the main GTO current. Current gain in the blocking mode however is low, of the order of 3, thereby requiring gate drives with large pulse current capabilities. Pulse duration however is short (a few μs). A solution consists in closer integration of the gate drive and the device as in the Integrated Gate Commutated Thyristors (IGCT). In addition, the turn-off time

of a GTO is relatively long, particularly due to the tail current, restricting the switching frequency to a few hundred Hz. In addition, switching losses rise quickly as the switching frequency increases, further restricting operation to low switching.

- *Continuously gated self-commutated devices.* The more common is the Insulated Gate Bipolar Transistor (IGBT). It can be viewed as a hybrid device. It has a behaviour similar to the Bipolar Junction Transistor (BJT) on the power side and to Field Effect Transistor (MOSFET) on the gate side. Gate consumption therefore is small. It must be gated continuously during the conduction period, which may require a separate gate drive power supply. Turn-on and turn-off times are small. The IGBT has forward, but no reverse, voltage blocking capabilities. It is designed to behave as a diode for reverse voltages. It therefore has bi-directional current carrying capabilities. Its availability allows switching frequencies to be increased to the low kHz range and therefore the more effective use of pulse width modulation (PWM) techniques for control of output magnitude and harmonic distortion. It is a relatively new device in high power applications, and has until recently been restricted to voltages and currents in the medium power range. However, larger devices are becoming available. Another hybrid device, still at the preliminary stages of investigation is the MOS Controlled Thyristor (MCT).

2.4 Static power converter structures

2.4.1 General principles

2.4.1.1 Rules of association

In high power applications, static power converters essentially consist of a set or matrix of power semiconductor switches connecting the input circuit to the output circuit, Figure 2.4.

Since the power converter only consists of switches, input and output circuits, that is sources and loads, can only be associated in certain ways. Two voltage sources, typically a source and a capacitor, cannot be connected in parallel through switches, they must be decoupled using an inductor. Loads can also be classified as having voltage or current source features, that is they can be capacitive or inductive. The following basic rules of association must therefore be taken into account in configuring converter systems:

- If the converter is fed from a voltage source, the converter output can be viewed as a voltage source and can only be connected to a circuit having current source features. An inductive load, for example a reactor in a TCR satisfies this requirement. If a capacitor is interfaced with the ac supply through an ac controller (TSC), a series inductor must be added to limit the inrush current and in general continuous control cannot be achieved. Thus, the TSC must be operated with integral cycle control. These constraints are illustrated for ac to dc conversion in Figure 2.4 (a).

- Conversely, if the converter is fed from a current source, the output can only be connected to a circuit having voltage source features, in other words a capacitive load. This is illustrated in Figure 2.4 (b) for a rectifier.

Application of these principles allows the development of the topologies suitable for static power converters.

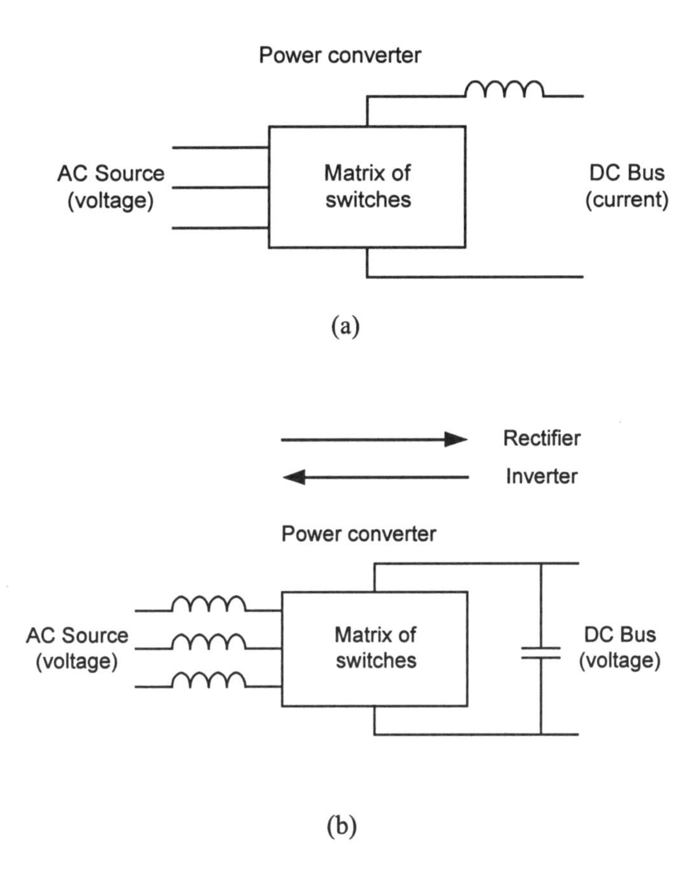

Figure 2.4 Static power converter implementation. AC/DC conversion. (a) Current source converter. (b) Voltage source converter.

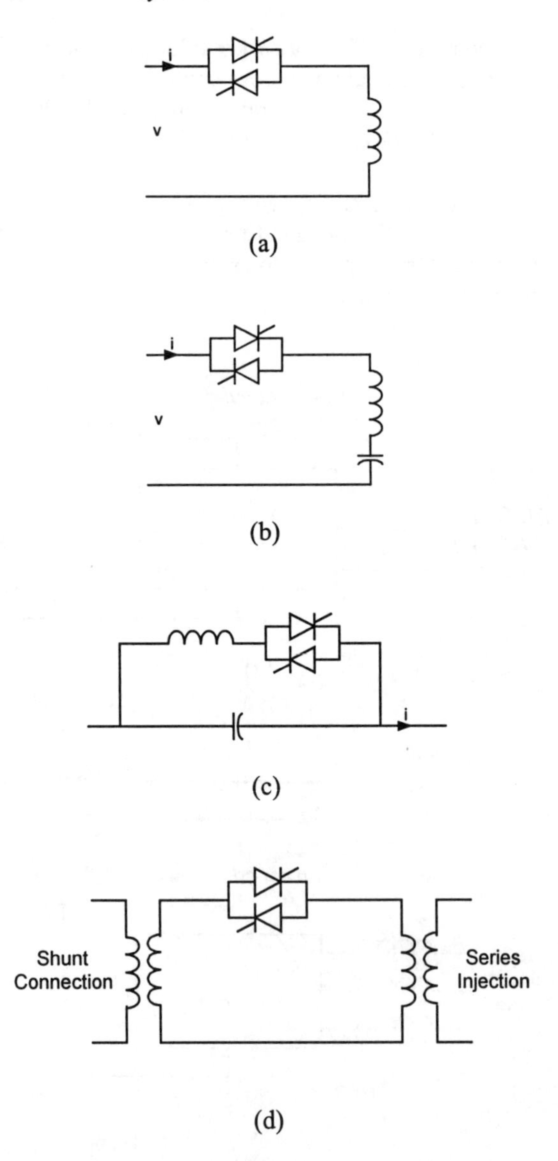

Figure 2.5 Thyristor ac controller circuits. (a) Thyristor controlled reactor (TCR). (b) Thyristor switched capacitor (TSC). (c) Thyristor controlled series capacitor (TCSC). (d) Thyristor controlled phase shifting transformer (TCPST).

2.4.1.2 Sources and converters

In most transmission systems, energy is available in ac form. It can be converted to dc, either for dc transmission, or to create an intermediate dc link for conversion back to ac. Intermediate dc links have a number of advantages. They provide a means to couple two systems operating at different frequencies or with a different phase shift. They also allow the synthesis of voltages and currents of arbitrary amplitude and frequency.

Static power converters can be classified as follows:

- *AC controllers*, which convert fixed amplitude ac to variable amplitude ac, with no change in fundamental frequency, Figure 2.5 (a) and (d).
- *Rectifiers*, which convert ac, of a fixed voltage amplitude and frequency, into dc, Figure 2.6.
- *Inverters*, which convert dc to ac. In power system applications, the ac side is usually connected to the power system, Figure 2.6. Inverters must therefore be synchronized to the ac system.

If the converter is bi-directional, that is, power flow can be reversed, a rectifier can operate as an inverter. This is the case in most converters. The rectifier-inverter nomenclature then refers to the normal power flow mode.

2.4.2 Basic ac/dc converter topologies

Thyristor rectifiers and inverters, because of the thyristor commutation process, require a constant current load on the dc side and an independent voltage source on the ac side. Therefore, an inductor is present on the dc side, Figure 2.6 (a). A convenient and general classification for converters is therefore based on the nature of the dc bus: a current source converter is connected to an inductive dc bus, whereas a capacitive dc bus requires a voltage source converter. The converter can operate as either an inverter or a rectifier, depending upon the nature and the control of the power flow.

It should be noted that in thyristor converters, thyristors are line commutated, that is blocking of one thyristor is produced by the initiation of conduction in the next thyristor triggered. The dc bus current is therefore never interrupted: the line and transformer reactances, Figure 2.6 (a), do not violate the rule of association. They slow the current transition process, a desirable feature for the thyristor. The simultaneous conduction of two or more thyristors is a phenomenon known as commutation overlap.

Applying the rules of association, force-commutated converters can be classified as current source and voltage source converters, Figure 2.6 (b) and (c) respectively. Thyristor converters can be considered current source structures. It should be noted that, since ac supplies are usually inductive due to the line and transformer inductances, a capacitor must be placed directly across the converter input in force-commutated current source converters. This capacitor supplies the instantaneous (ripple) current flowing through the converter switches and produced by switching the dc link current.

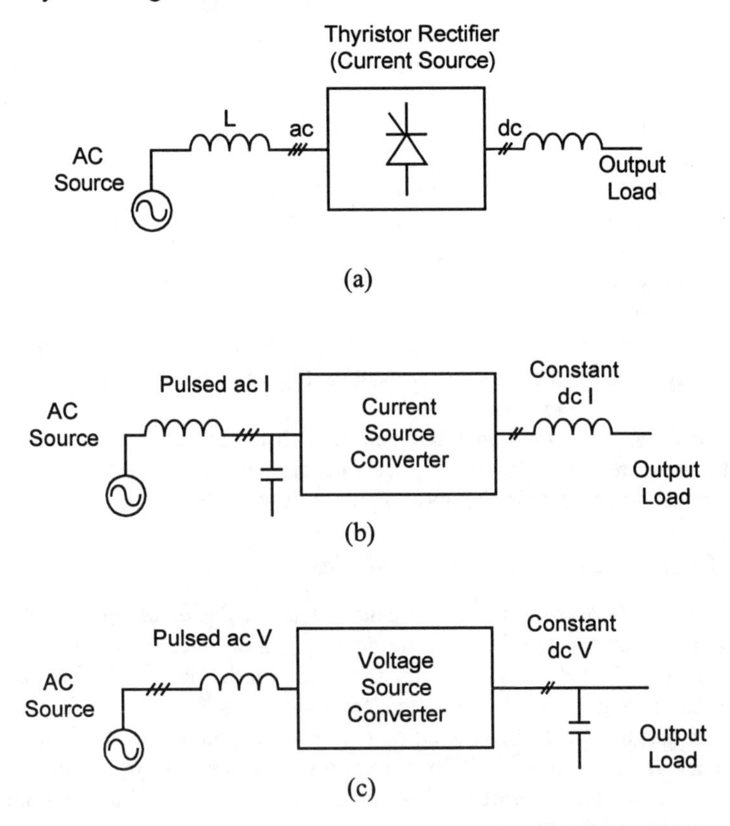

Figure 2.6 Power converter topologies. (a) Thyristor rectifier/inverter. (b) Self-commutated current source converter. (c) Self-commutated voltage source converter.

Current and voltage source topologies can be viewed as dual topologies in terms of current and voltage waveforms. If the dc link current is assumed constant in the current source converter, the ac current is chopped, consisting of one or more pulses per half period. Conversely, if the dc link voltage is assumed constant in the voltage source converter, the ac voltage is chopped, consisting of one or more pulses per half period. Current and voltage waveforms in current and voltage source converters are therefore similar. Furthermore, because they are switching converters, power converters produce quantities that are square waves rather than sinusoidal waves, therefore containing unwanted harmonic components. Techniques are available to manage these harmonics.

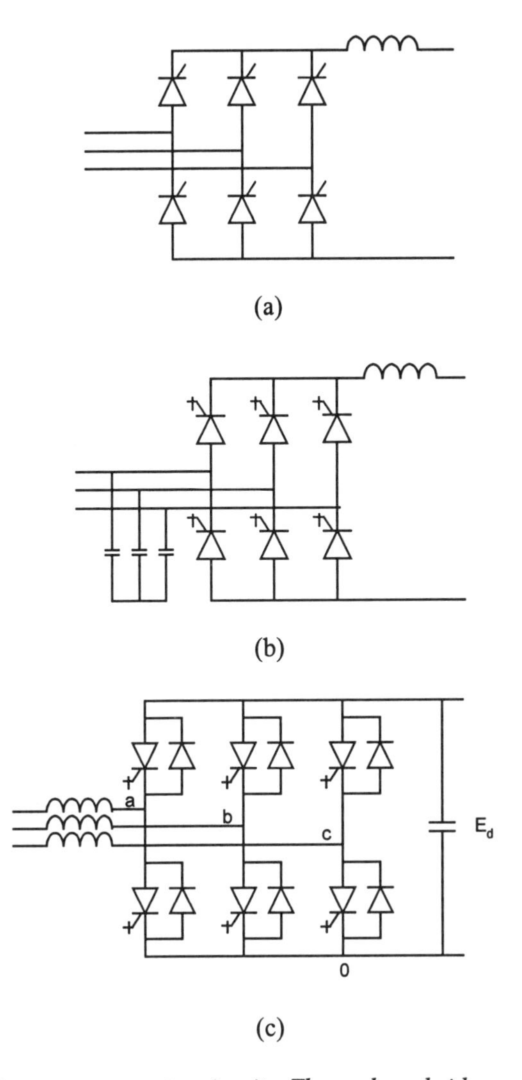

Figure 2.7 Basic power converter circuits. Three phase bridge configurations. (a) Thyristor converter. (b) Self-commutated current source converter. (c) Self-commutated voltage source converter.

Current and voltage source converters require different power switch characteristics:

- In *current source converters*, the dc bus current does not reverse, whereas the dc voltage across the switches can reverse instantaneously, due to the presence of the inductor. The switches are therefore of the unidirectional current, bi-directional voltage blocking capability; thyristors and symmetrical GTOs have this feature, IGBTs require a series diode to satisfy the requirement. Power circuit configurations are given in Figure 2.7 (a) and (b) for thyristor and force-commutated current source converters.

- In *voltage source converters*, the dc bus voltage polarity does not reverse, therefore the dc voltage across the switches remains of the same polarity. The dc current through the switches however can reverse instantaneously: the switches are therefore of the bi-directional current, unidirectional voltage blocking capability; IGBTs and asymmetrical GTOs have this feature. Furthermore, in the general case, voltage source converters require force-commutated switches. The power circuit configuration is given in Figure 2.7 (c) for converters using self-commutated switches.

Circuits and sources connected on the dc side can be of various types:

- A voltage source, typically a capacitor or an energy storage device such as batteries.
- A current source, typically an inductor, which can be of the superconducting type to limit inductor losses.

Alternatively, the dc bus can be supplied by another power converter, as in back to back structures.

2.4.3 Converter power circuit configurations

Combining the available types of power switches, taking into account the nature of the power conversion, and applying the rules of association, allows classification of converters. The following power circuits are common in three-phase power transmission applications:

- *Line commutated thyristor converters*. They can operate either as rectifiers or inverters, depending upon output power requirements and control method, Figure 2.7 (a). Operation assumes that the ac side contains independent voltage sources, usually rotating machines, to define the amplitude and frequency of the ac bus and ensure commutation.
- *Self-commutated converters*. They are typically based on GTOs or more recently IGBTs. Converters can be of the current or voltage source type and can operate as rectifiers or inverters, Figure 2.7 (b) and (c). Since the devices commutate independently of the ac side voltages, the converters can be connected to any ac bus, including buses containing only passive loads. The voltage source converter is particularly interesting as it can be used to produce an ac voltage of controllable magnitude and phase. If synchronized with the ac supply voltage, it can be viewed as a static synchronous generator, with features similar to those of the synchronous machine or of the synchronous condenser. The topology is therefore often referred to as a synchronous voltage source.

Power converters can be configured for a number of applications, the most common being:

- *Static var compensators* (STATCOM). The dc bus is a self-controlled capacitor, Figure 2.2 (a) and Figure 2.7 (c). The voltage of the dc bus is regulated through control of the power flowing into the capacitor. Losses in the converter and capacitor are provided by the ac source. The current source structure using a self-controlled dc link reactor can also be used for var compensation. However, inductor losses, design and control issues make this topology less attractive than the voltage source topology in var applications. The exception is the use of super conducting magnets for storage of energy on the dc bus (SMES system). Therefore, most compensators are based on the voltage source structure.
- *HVDC transmission*. The dc link allows decoupling of two ac systems. The system is based on thyristor converters for very high power applications, or more recently, force-commutated voltage source inverters for smaller installations.
- *Combined shunt and series compensation systems*. These are typically based on voltage source converters (UPFC). The series and shunt-connected converters, Figure 2.2 (c), share the same dc bus. Real power flow is usually controlled through the shunt side, reactive power flow can be controlled independently on the series and shunt sides.

2.4.4 Power flow control

Power flow through static power converter is controlled through the gating of switches and the dc bus load and parameters. There are two basic methods, illustrated with reference to current source rectifier structures, Figure 2.7 (a) and (b). In the discussion, the output power is varied by changing the dc output voltage, the dc current being assumed constant and regulated. The techniques are:

- *Fundamental frequency control*. Gating patterns have a single pulse per ac period, with pattern positioning, or delay angle control. All switches conduct during an equal time interval, 1/3 of a period for a three-phase thyristor rectifier. The current reflected onto the ac side has a fixed waveshape: it has a constant fundamental component and a fixed harmonic content, Figure 2.8 (a). The apparent power is therefore constant. Power can only be controlled by phase-shifting the current by means of the position of the gating pulses. Therefore, as the output power decreases, the phase angle of the fundamental component of the ac current increases, and the power factor decreases. The thyristor rectifier behaves as a reactive load, with a lagging power factor. Current must lag the voltage to permit line commutation. In force-commutated rectifiers, current can be either leading or lagging.
- *Pulse width modulation (PWM) control*. Notches are introduced in the gating patterns, Figure 2.8 (b). This requires force-commutated or self-commutated devices. The fundamental component of the ac current can be reduced by notching, thus decreasing the output power. The phase of the ac current can

be set arbitrarily, and unity power factor operation is possible. AC current amplitude control can be combined with pattern phase shifting.

These general principles also apply to ac controllers, and voltage source converters. Operation in voltage source converters is the dual of that of current source converters. Power control is achieved by phase shifting the voltage pattern or by controlling the amplitude of the converter ac output voltage, or a combination of the two.

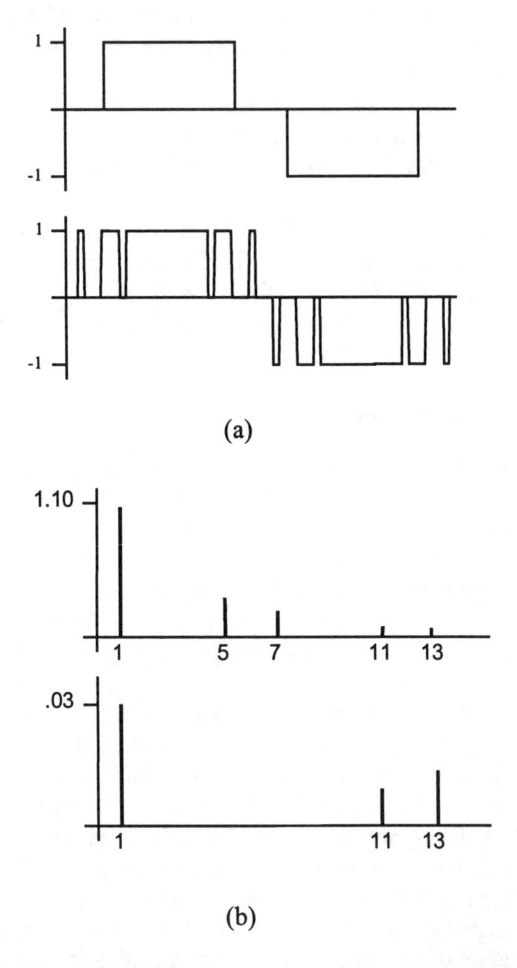

Figure 2.8 Static power converter output waveforms. AC line current for current source converter (ac line voltage for voltage source converter). (a) Waveforms for single pulse patterns and pulse width modulation (SHE) patterns (5th and 7th harmonic elimination). (b) Harmonic spectrum for single pulse patterns and pulse width modulation (SHE) patterns.

2.4.5 Switch gating requirements

The basic requirement of power converters in FACTS devices is the control of either the voltage or the current generated by the converter. Output patterns for current source and voltage source topologies are similar: by virtue of duality, current source inverter (CSI) current patterns are identical to the line to line voltage patterns of the voltage source inverter (VSI). Typical patterns are shown in Figure 2.8.

However, the gating of a VSI is generally simpler than that of a CSI. This is the result of the bi-directional current capability of the switches in the VSI, which ensures that there always is a return path for the dc bus current; therefore, there are no restrictions on the gating pattern. In contrast, switches in the CSI are unidirectional current devices and therefore require complementary gating to satisfy the requirements for continuity of the dc bus current. When the line current is zero, a freewheeling path is provided for the dc bus current in the converter.

2.5 AC controller-based structures

2.5.1 Thyristor-controlled reactor

The basic scheme for the thyristor-controlled reactor (TCR), Figure 2.5 (a), consists in an ac controller which varies the voltage applied to the inductor, therefore its apparent inductance as reflected onto the ac line [3,4]. It provides continuously controllable lagging vars. To cover leading vars, it is biased using fixed, or in most cases thyristor-switched capacitors (TSC), Figure 2.5 (b). The complete system forms a static var compensator (SVC). If a number of TSCs is used, the rating of the TCR can be reduced, the TCR allowing continuous control between two capacitive compensation levels. This limits the amount of lagging vars, however, compensators are mostly operated in the leading var mode. The injected vars can therefore be continuously adjusted from leading to lagging.

Advantages of the system however include ruggedness, high efficiency, good dynamic performance, and a competitive cost. However, the var injection or voltage regulation capability of the TCR is limited by the value of the reactance and is therefore line voltage dependent. The same applies to the TSC. A major disadvantage over synchronous condensers is the injection into the line of large low frequency harmonic currents. The dominant components for a balanced three-phase system are the 5th and 7th components (300 and 420 Hz for a 60 Hz system) for the basic ac controller. Low order harmonics can be removed by paralleling units and using special transformer configurations. Harmonic currents can also be reduced by means of tuned LC filters. These however are costly and can cause voltage oscillations resulting from the added system resonant frequencies.

Performance could be enhanced and the harmonic content reduced by using force-commutated switches and ac chopper-type circuits.

2.5.2 Thyristor-controlled series capacitor

The dual of the TCR is the thyristor-controlled series capacitor (TCSC), Figure 2.5 (c). It uses a thyristor-based ac controller to continuously adjust the apparent reactance inserted in series with the line [14]. The limiting modes of operation are the following: (a) a permanently open thyristor, corresponding to the full capacitive reactance; (b) a continuously gated thyristor, the equivalent reactance being approximately equal to the small inductive reactance. With the gating of the thyristor over only part of the cycle, that is vernier control, the equivalent reactance can be varied continuously from the full capacitive reactance to a small equivalent inductive reactance.

Such units have been successfully tested in transmission systems. However, they have the same limitations as TCRs, including harmonic injection.

2.5.3 Thyristor-controlled phase-shifting transformer

The device, consisting of thyristor controlled phase-shifting transformers (TCPST), allows the control of the load angle between two buses in a transmission line. Voltage phase shift is varied by injection in series of a controlled amount of voltage in quadrature with the line voltage [15]. Mechanical switches in conventional phase shifters can be replaced by thyristors [16]. These can be gated over the complete cycle, that is integral cycle control, emulating mechanical switches. Alternatively, they can be controlled part of the cycle, as in ac controllers used in TCRs, giving continuously controllable phase angle regulation, Figure 2.5 (d).

The advantage of integral cycle control is the absence of harmonics. However, a number of transformer windings or taps are needed to obtain the required steps and incremental control.

2.5.4 Force-commutated ac controller structures

An alternative to the thyristor-based ac controllers is the force-commutated ac controller [17]. The use of self-commutated switching devices allows gating the switches more than once per cycle. Arbitrary gating patterns can be implemented, particularly PWM patterns. Assuming the inductor current is sinusoidal, a pattern with constant duty cycle yields ac line side currents that only contain harmonics around the switching frequency and its multiples. This pattern is simple to implement and allows control of the equivalent inductance; therefore, the amount of vars absorbed can be varied from 0 to a maximum value. In order to reduce the distortion of injected currents, while keeping the switching frequency low, elementary modules are connected in parallel and gated so that harmonics are minimized.

However, compensators based on dc link topologies although more complex to control than ac controller based topologies, offer more flexibility.

2.6 DC link converter topologies

2.6.1 Current source based structures

2.6.1.1 Operation of thyristor converters

Thyristor rectifiers, Figure 2.7 (a), convert an ac voltage into dc. Typical waveforms are shown in Figure 2.9. The output voltage can be adjusted from a maximum positive value, corresponding to operation as a diode rectifier, to a maximum negative value, corresponding to an inverter operation. Inverter operation requires the presence of a dc source feeding power back into the ac supply. This energy can be transiently stored in inductors, particularly super conducting magnets (SMES) or supplied by an independent dc source (HVDC transmission). Since the thyristors can only be gated once per ac cycle, the output voltage E_d is controlled by delaying the turn-on instant by an angle α. The average value is given by:

$$E_d = E_{do} \cos \alpha$$

where E_{do} is the diode rectifier voltage, equal to 1.35 V_L (line to line ac voltage) for a three-phase circuit. This relationship assumes continuous current on the dc side. The output voltage is equal to 0 for $\alpha = 90^\circ$, and power is also 0, both on the dc and ac sides, for any value of the dc bus current, assuming the converter has no losses. Inverter maximum voltage is limited by the thyristor commutation requirements. The maximum phase back angle β is usually of the order of 160°. The converter operates in two quadrants of the E_d–I_d plane, power reversal being obtained by voltage reversal.

The dc side harmonics are of order (6*n*) and increase significantly as the delay angle moves towards 90°.

The ac line current is a square wave with a 120° duty cycle per half cycle. The rms values of the fundamental I_{L1} and of the total current I_L are proportional to the dc bus current I_d and given by:

$$I_{L1} = 0.78\, I_d$$

$$I_L = 0.82\, I_d$$

The line side current harmonics are large, with dominant 5th and 7th harmonics. In general, harmonics are of order ($6n \pm 1$) and of amplitude $1/n$, Figure 2.8 (b).

Real power is controlled directly by the delay angle. For a given dc bus current I_d, the apparent ac power S remains constant and independent of the delay angle α:

$$P = E_{do}\, I_d \cos \alpha$$

$$S = \sqrt{3}\, I_L\, V_L = E_{do}\, I_d / 0.955$$

The thyristor rectifier therefore behaves as an inductive load, with a cos ϕ approximately equal to cosα. The power factor *pf* is lagging for all operating conditions:

$$pf = P/S = 0.955 \cos\alpha = pf_{dist}\, pf_{disp}$$

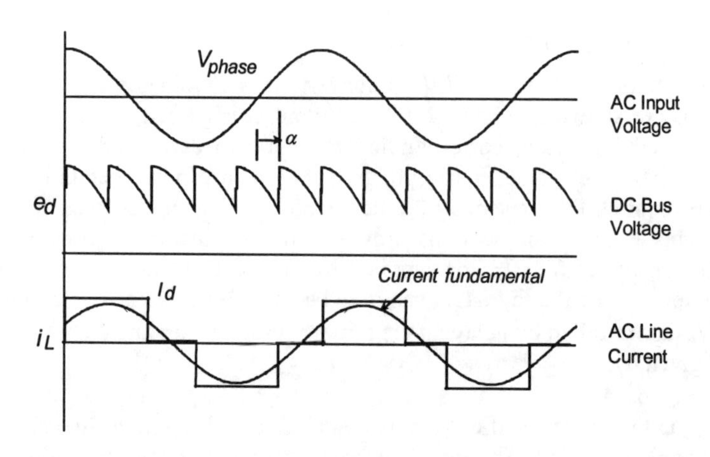

Figure 2.9 Thyristor rectifier waveforms with delay angle (α) control. Input ac voltage, output dc voltage and input line current.

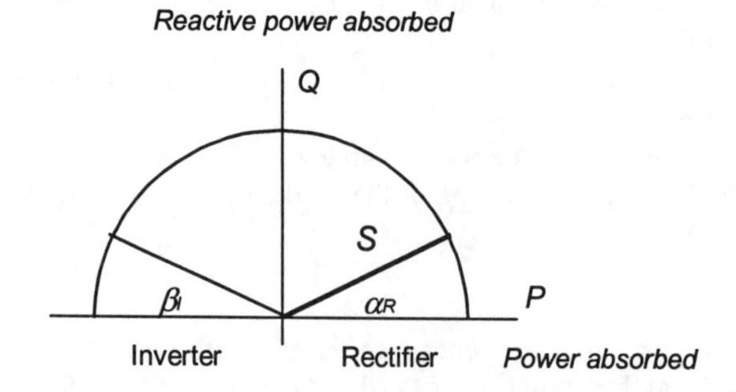

Figure 2.10 Power diagram of a thyristor rectifier. Delay angle (α) control and constant dc bus current.

The delay angle is responsible for the displacement power factor pf_{disp}, line current harmonic distortion accounts for the distortion power factor pf_{dist}. The power *Q–P* diagram, Figure 2.10, illustrates the operation of the rectifier as a function of the delay angle. Operation for most conditions absorbs a significant amount of reactive power, even for small delay angles. For an output voltage of 0.87 E_{do} ($\alpha = 30^{\circ}$), the reactive power requirements are 0.56 p.u. (power factor of 0.83). This reactive power is usually provided by harmonic filters.

2.6.1.2 Operation as reactive compensator

For a delay angle of 90°, the converter operates as a reactive power compensator. If an inductor is the load and is connected directly across the dc bus, only controlled amounts of reactive power are exchanged with the ac line and the converter operates as a var compensator. It only draws from the ac source the real power required to maintain the inductor current constant. This real power covers inductor and inverter losses.

However, thyristor-based units can only provide lagging reactive power, since thyristors are line-commutated. This compensator therefore has limited applications. It has been used in special cases for power system damping, such as in SSR mitigation.

The lagging reactive power limitation is removed when thyristors are either force-commutated, or replaced by self-commutated devices, such as GTOs, Figure 2.7 (b). Leading ac currents can be produced and operation as a capacitive compensator is possible. In addition, the use of such devices also allows the implementation of PWM patterns.

However, the converter requires an input capacitor to produce an input ac voltage source capable of supplying the instantaneous ac current ripple produced by the operation of the current source converter. An additional line reactor may be required for further filtering of the current ripple. This results in a resonant frequency being introduced in the system. Furthermore, in the standard control scheme, the current in the dc inductor is maintained constant at the rated value, resulting in steady state losses, even when the compensator supplies no reactive power. These losses, added to the converter losses, result in a system which is less efficient than alternative structures based on voltage source converters.

Efforts to remove losses in the dc inductor have led to the development of compensators based on superconducting magnet energy storage devices (SMES). In addition to providing leading and lagging reactive power, SMES can also store a sufficient amount of energy that it can operate as a power source under transient operation [8]. Operation is then possible in all four quadrants of the *Q–P* plane, Figure 2.10. This real power can be used to support the power system under fault conditions and damp oscillations on the ac system. This technology however is complex and costly. The preferred compensator configuration is based on voltage source structures.

2.6.2 Synchronous voltage source structures

2.6.2.1 Operation of self-commutated rectifiers

Voltage source converters can be operated as rectifiers and inverters if synchronised with the ac system. The converter generates an ac voltage, Figure 2.11. This voltage is the result of switching the dc bus voltage, assumed constant, so as to create an ac quantity. This process is illustrated in Figure 2.11 for an inverter in which the switches are gated once per ac cycle.

Since the inverter reflects a voltage on the ac line side, an inductor or synchronous link must be used to couple it to the ac system, Figure 2.12. Therefore, such converter systems can be considered as synchronous voltage sources. Operation is similar to that of synchronous machines connected to an ac system [7,20,21].

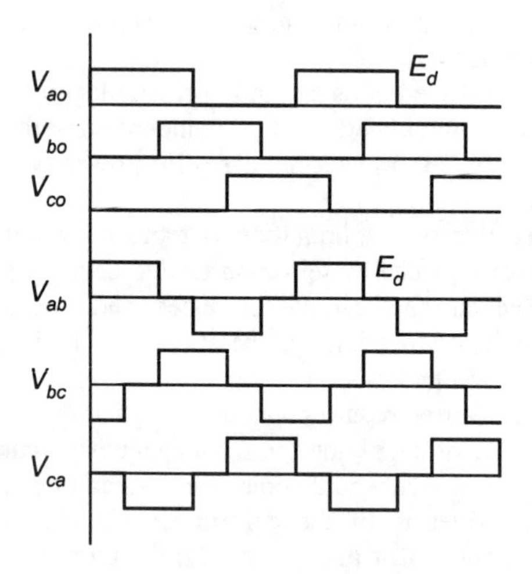

Figure 2.11 AC voltage waveforms of a self commutated inverter, Figure 2.7 (c). Fundamental frequency gating. Constant dc bus voltage.

The voltage reflected on the ac side of the converter depends upon the gating pattern of the switches and the dc bus voltage. Typical waveforms can be derived assuming the dc bus voltage is constant and switches are gated using a single 180° pulse per cycle, alternatively applied to the top and bottom switches. The gating of the switches in one leg must be complementary so that the voltage at the output of a leg is defined. The line to line waveforms at the output of the converter, Figure 2.11, consist of pulses with a 120° width, and are similar to the current waveforms of a current source converter. Dominant harmonic components are the 5th and 7th. In general, harmonics are order ($6n \pm 1$) and amplitude $1/n$, Figure 2.8 (b).

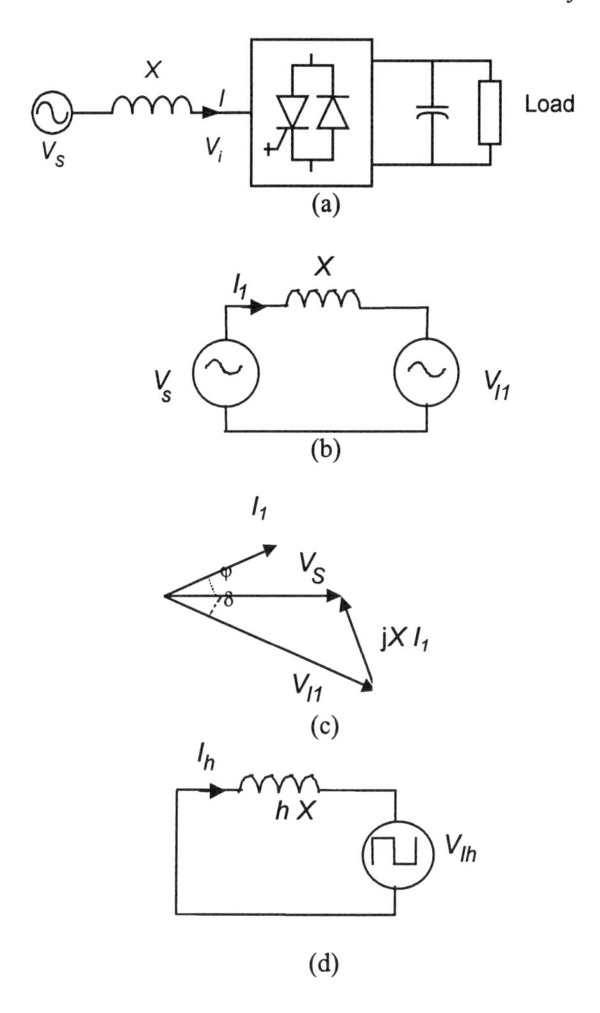

Figure 2.12 Operation of a self-commutated synchronous rectifier. (a) Schematic diagram. (b) Equivalent circuit for fundamental frequency operation. (c) Phasor diagram for fundamental frequency operation. Leading power factor. (d) Equivalent circuit for switching frequency components. Line reactance and synchronous reactance are lumped.

The current flowing into the ac supply and its waveform depend upon the link inductance. This inductance can be provided by the leakage reactance of the coupling transformer. The amplitude of harmonic currents injected into the ac supply can be significant. For example, for a typical 0.1 p.u. reactance at fundamental frequency, corresponding to 0.5 p.u. at the 5 p.u. frequency, the 5th current harmonic component has an amplitude of 0.4 p.u. for a 0.2 p.u. voltage harmonic. This may be unacceptable and methods have to be applied to reduce voltage harmonic components, among others, multi-pulse and PWM techniques.

Since voltage source converters are self-commutated, there is no restriction on the position of the converter current and voltage with respect to the ac line voltage. In the general case and assuming a power source on the dc side, the angle δ between the ac supply V_s and the fundamental component of the converter ac output voltage V_{i1} can be set to any desired value, Figure 2.12 (c). This angle defines the amount of power flowing into or out of the ac supply and it is controlled by the phase shift between the ac supply and the gating pattern and the amplitude of the converter ac voltage. The converter is reversible and the power transfer, neglecting losses, is given by:

$$P = 3\frac{V_{i1}V_s}{X}\sin\delta$$

where voltages are on a per phase quantities. In addition, the converter can be operated at leading or lagging power factor. The reactive power is given by:

$$Q = 3\frac{V_{i1}(V_{i1} - V_s\cos\delta)}{X}$$

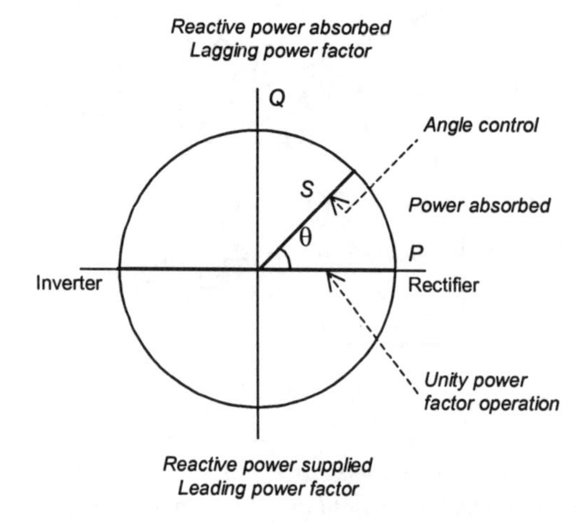

Figure 2.13 Power diagram of a self-commutated ac/dc converter. AC side. Pattern angle and notch control.

Operation is possible in all four quadrants of the *Q–P* plane, Figure 2.13. If the converter load angle is controlled, the power factor is defined by the amount of power flowing through the converter. A leading power factor case is illustrated in Figure 2.12. Operation of the converter is similar to that of a synchronous machine. Equivalent over and under excitation modes are defined by the magnitude of the converter ac voltage.

Voltage and current source converters can be considered dual in terms of waveforms. However, current source rectifiers are buck converters. The dc bus voltage is controllable from 0 to a theoretical value equal to the diode rectifier voltage. Voltage source rectifiers, on the other hand, are in general operated as boost rectifiers, particularly in the PWM mode.

2.6.2.2 Operation as a shunt reactive compensator

For operation as a reactive compensator, no energy storage is required on the dc side. The structure of the Static Compensator (STATCOM), is given in Figure 2.14 (a). A capacitor is used on the dc side to define the dc bus voltage and absorb the dc ripple current. The value of the capacitance is not related to the amount of vars produced as in TSCs. If a battery bank is used on the dc side, energy can be stored and supplied to the ac system on a transient basis [18]. This structure is then the dual of an SMES.

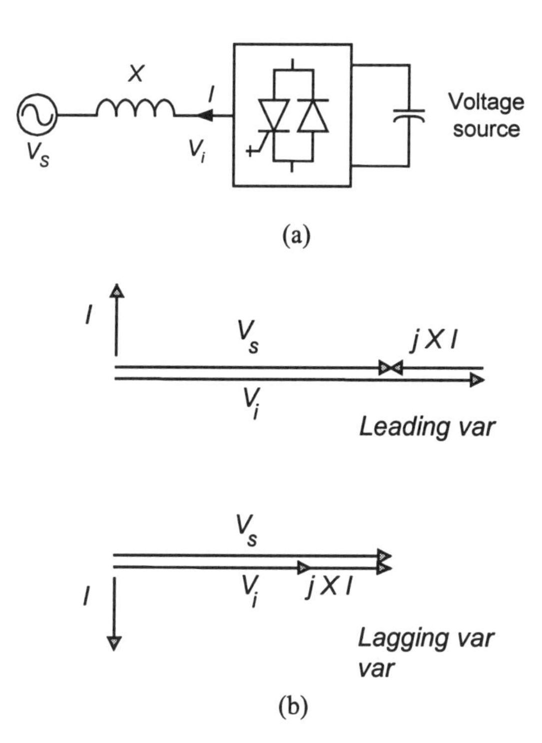

Figure 2.14 Static var compensator based on the principle of synchronous voltage sources. (a) Power circuit configuration based on a voltage source converter. (b) Operation with leading and lagging var injection.

As in a static synchronous condenser, the current injected into the ac supply is controlled by the magnitude of the reflected voltage, Figure 2.14 (b). The angle δ between the supply and converter voltages is 0, assuming a lossless operation. A voltage larger that the ac supply results in injected vars (leading or capacitive); the converse, or lagging reactive power, occurs if the voltage is smaller. The dc bus voltage is controlled by charging or discharging the capacitor through δ angle control. It is independent of the line voltage. The steady state operation of the compensator is therefore independent of the line conditions, contrary to a TCR–TSC compensator.

The converter operates as a reactive power source, and in steady state does not supply or absorb real power. Losses however must be compensated if the capacitor voltage is kept constant. These can be provided form the ac supply for a self-controlled dc bus. This is the preferred solution. In this case an additional dc capacitor voltage control loop must be used. A small amount of real power flows into the converter to cover losses and the angle δ has a small value.

Losses in voltage source converters are lower than for current source structures. However, particular attention must be paid to the risk of dc short circuits, when both switches in a leg conduct. There is also the requirement for dc output short circuit protection. Current source structures are more rugged: the short circuit current is limited, the dc bus being a current source.

Energy storage devices, such as battery banks can be incorporated into the units. This allows the absorption and generation of real power, P_{ref}, which can be used for power system damping and voltage support under fault conditions, Figure 2.2 (c).

The shunt connection has long been the preferred connection: one end of the compensator can be connected to ground and the fault currents do not flow through it.

2.6.3 Other compensator structures

The series connection, the Static Synchronous Series Compensator (SSSC), Figure 2.2 (b), offers a number of advantages over passive series compensation at a cost that is becoming competitive in demanding applications [10,13]. The inverter is similar to that of a STATCOM. It is usually connected to the ac system by means of a transformer.

The combination of series and shunt connected converters, connected to a common dc bus, the UPFC, Figure 2.2 (c) has the advantages of both the series and the shunt configuration and adds another degree of control. Reactive power can be independently set on the series and shunt sides, and the real power flowing through the intermediate dc bus is fully controlled. The dc bus voltage is usually set by the shunt side converter.

In addition to its operation as a var compensator, this scheme can be used as a phase shifter. It also offers the possibility of injecting real power into the system and of modifying the apparent resistance of the line. This can be used to improve the power transmission system characteristics, including the possibility of increasing the *X/R* ratio.

Although the dc bus on the combined series–shunt compensator can be either of the voltage or current source type, most of the structures that are investigated are based on the voltage source configuration.

2.6.4 High voltage dc transmission

Conventional HVDC systems are based on thyristor converters, Figure 2.15 (a). These converters can be separated by a few meters in back to back HVDC links or hundreds of km in HVDC transmission systems. Thyristor-based systems are still prevalent in very high power applications, in the multi-megawatt range. The sending end unit operates as a rectifier, the other as an inverter [2]. Power flow is usually controlled from the rectifier side, the inverter being operated close to the maximum permissible delay angle and defining the voltage. Advantages include very high efficiency and reliability. Disadvantages include large harmonic currents on the ac and dc sides: for a single 6-pulse converter station, ac current harmonics are of order ($6n \pm 1$) and amplitude $1/n$, and $6n$ on the dc side. Low order harmonics, typically 5th and 7th can be eliminated by means of transformers in the typical 12-pulse connection. Alternatively, harmonic filters can be used.

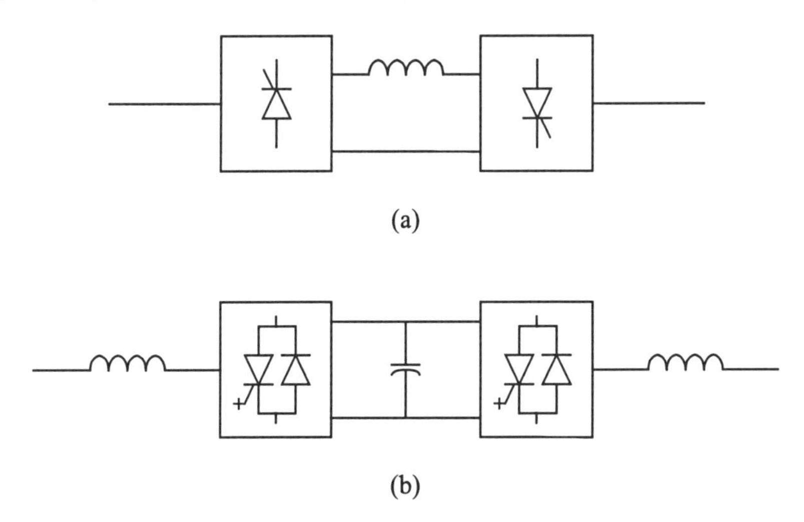

Figure 2.15 HVDC power conversion. (a) Back to back thyristor rectifier/inverter system. (b) Self-commutated voltage source inverter based system.

Experience has shown that operation of thyristor rectifiers and inverters is affected by weak ac systems. Thyristor commutation failures may occur if the ac voltage across the inverter is distorted or there are voltage sags. In particular, implementation of multi-terminal systems is impaired when taps are connected to weak ac systems or faults occur on the ac side, due to problems associated with line commutation. Commutation failure may affect the operation of the complete dc transmission system. Self-commutated switches do not have the limitations of thyristors and can therefore replace thyristors in current source topologies.

With the availability of large self-commutated switching power devices, mainly GTOs and more recently IGBTs, the trend has therefore been to propose and implement HVDC systems based on self-commutated converters [12]. However, voltage source topologies are preferred in HVDC applications, as in the STATCOM and the UPFC, Figure 2.15 (b). Operation of the rectifier is illustrated in Figure 2.12. Power and voltage levels however have not so far reached those of thyristor technologies.

Advantages include better harmonic and power factor control and improved control capabilities. In particular, the possibility of operating the converters at unity power factor removes the need to supply large amounts of reactive power as required in thyristor converters. In addition, since the converters are self-commutated, proper operation is maintained under ac system fault conditions. Furthermore, these HVDC systems can supply loads having no voltage sources, that is no significant rotating machine load, to define the ac bus. Examples include remote loads. Additional functions, such as voltage regulation and power system damping, can be incorporated into the operating features of the self-commutated HVDC systems.

2.7 Converter output and harmonic control

2.7.1 Converter switching

The main gating schemes include fundamental frequency gating and pulse width modulation (PWM) techniques. The following discussion refers to voltage source structures. Conclusions can be extended to current source structures.

If each switch in a converter leg is gated for half the ac supply period, the ac voltage is a square wave voltage, Figure 2.11. This is the conventional means of controlling large static synchronous condensers, Figure 2.14. An important advantage is that this method leads to the lowest switching losses and therefore the highest efficiency. Low switching frequencies are essential if converter efficiency is be kept very high. Thyristor-based converters typically have an efficiency of 98 % or higher, and force commutated converters should attempt to approach this value. If higher switching frequencies are needed, techniques can be devised to switch the devices under zero voltage or zero current conditions, thus reducing switching losses.

In fundamental frequency switching, the pattern is fixed and the fundamental or desired component of the ac voltage depends upon the dc bus voltage. Furthermore, the harmonic content of the output voltage of a single three-phase converter is large, resulting in large injected harmonic currents. Methods to reduce the harmonic content include passive L–C filters, multi-pulse structures and PWM techniques. A combination of the above techniques is also possible.

If the self-commutated devices can be switched at frequencies higher than line frequency, pulse width modulation techniques can be used, Figure 2.8. These techniques allow control of the amplitude of the fundamental and harmonics of the pattern.

2.7.2 Principles of harmonic mitigation

2.7.2.1 Harmonic filters

The drawback of static power converters is the generation of harmonics, particularly harmonic currents, since they flow into the ac system. These currents produce harmonic voltage drops and subsequently voltage distortion in the ac system. Voltage distortion can perturb operation of loads and other equipment connected to the ac system. In addition to low frequency harmonics, converters also produce high frequency interference that may affect control and communication equipment. In this respect, force-commutated converters pose more problems than line commutated thyristor converters.

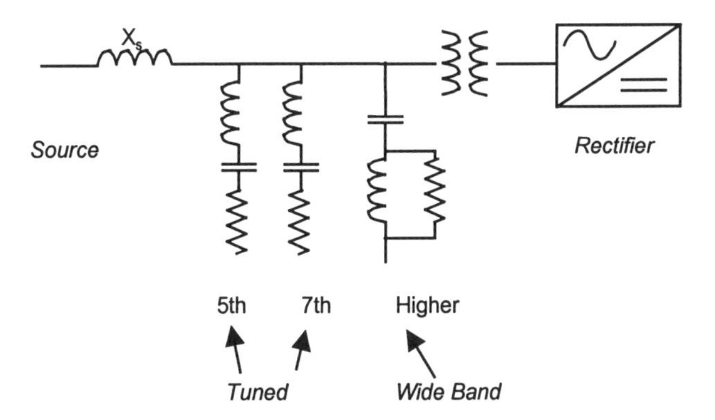

Figure 2.16 Typical passive harmonic filters for rectifiers. Tuned and wide band filters for 6 pulse thyristor rectifiers.

The simplest way to deal with current harmonics generated by converters is to reduce or eliminate them using passive LC filters [1]. Conventional low pass series–parallel LC filters, tuned to low frequencies and designed to eliminate low frequency components, tend to be large. Therefore, notch or tuned series LC filters, connected in shunt, are used for components such as the 5th and 7th, 11th and 13th. A number of components can be combined if a wideband LCR filter is used, a common solution at higher frequencies. A typical installation for a 6-pulse thyristor converter is illustrated in Figure 2.16.

For force-commutated current source rectifiers, an LC input filter is used. For voltage source converters, the link reactance and transformer act as a first order harmonic current filter. The total inductance must be chosen to meet the allowed current harmonic distortion. An additional LC filter may have to be used to further reduce the harmonic components. This can be a tuned shunt LC filter set to the converter switching frequency.

2.7.2.2 Phase shifting transformers

In order to increase the power rating of the compensator to power system levels, typically 100 MVAR or more, available power switching devices must be combined in series and in parallel. However, advantage can be gained by combining converter units rather than individual devices: these units are associated in a way to minimize harmonic injection while maintaining low switching losses.

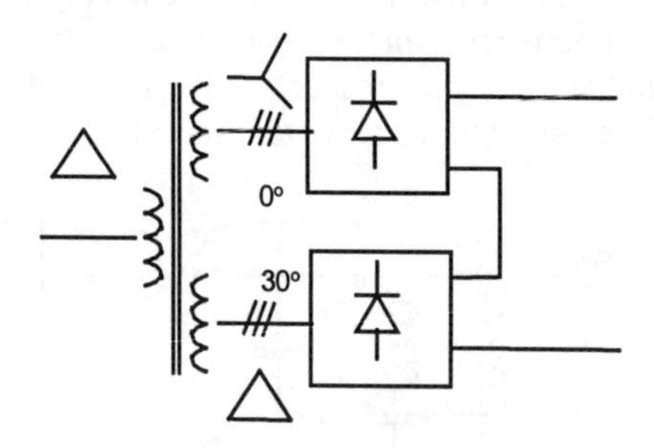

Figure 2.17 Association of rectifiers for harmonic mitigation. 12-pulse operation for a series connection.

The simplest method of cancelling harmonics in a system of converters is by means of coupling transformers. This technique is widely applied in thyristor rectifiers and inverters. It consists in using n transformers with primary or secondary windings phase shifted by $60°/n$. The more widely used is the 12-pulse connection, with one delta connected and one wye connected secondary winding, resulting in a voltage phase shift of 30° between transformer secondaries, Figure 2.17. The primary side consists in either one or 2 wye or delta connected sets of windings. The transformer phase shift results in the elimination of the 5th and 7th current harmonics on the primary side, and the 6th voltage harmonic on the dc side of the rectifier. The elimination can be viewed as the cancellation of harmonic mmfs in the transformer. Although 5th and 7th harmonic current components circulate in the transformer secondary windings, they are not reflected onto the primary side. The process of harmonic elimination leads to multi-stepped current waveforms on the line side of the converter system. Other connections are possible, namely 18 pulse or higher.

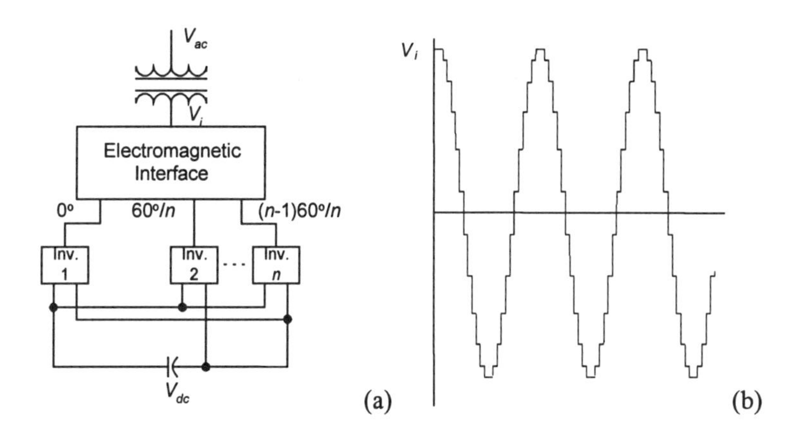

Figure 2.18 Multi-pulse compensator - voltage source topology (STATCOM).
(a) Structure of the multi-pulse system, with phase shifting transformers.
(b) Converter output voltage waveform (24 pulse, n = 4, dominant harmonics at the 23rd and 25th).

The principle of harmonic cancellation by means of phase shifting transformers also applies to synchronous voltage source converters. The output of individual 3-phase converters is added together by means of a magnetic interface circuit. In its simplest form, this interface consists of *n* secondary windings with a phase shift of 60°/*n*, Figure 2.18 (b). The voltage patterns of individual converters are phase shifted by an equivalent angle [21]. The combination of these phase shifts results in the elimination of harmonics and the generation of multi-stepped voltage waveforms that more closely approximate a sinusoidal voltage.

2.7.2.3 Multi-level structures

In addition multi-module structures, multi-stepped voltage waveforms can be obtained also using multi-level converters. By splitting the dc bus capacitor, Figure 2.19 (a), it is possible to generate a stepped voltage similar to that in Figure 2.18 (b). A number of topologies are available among which is the Diode Clamped Capacitor Multilevel Inverter (DCMLI), Figure 2.19 (b) [22,23]. Typically 3 to 7 level converters have been investigated.

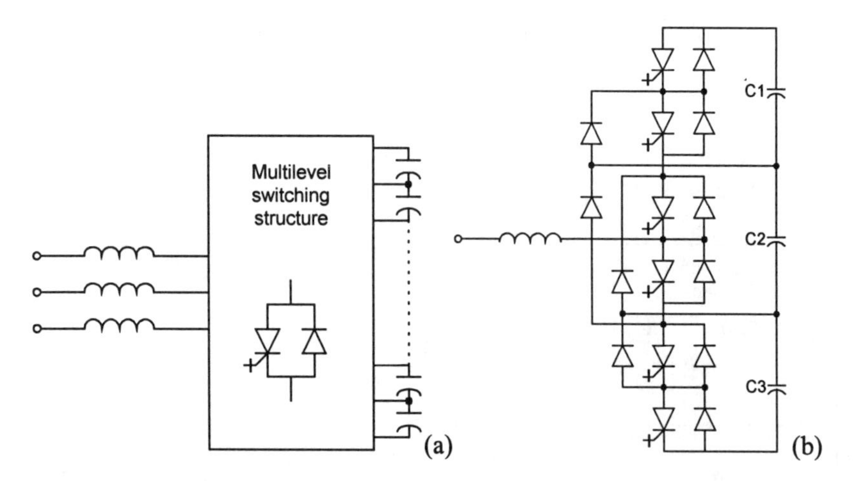

Figure 2.19 Multi-level compensator - voltage source topology (STATCOM). (a) General three-phase structure. (b) Power circuit for one-leg of a 4-level diode clamped converter.

In high power applications, three level converters have been applied. Contrary to the conventional or two level bridge converter, three level converters allow the control of the amplitude of the ac voltage with a fixed dc capacitor voltage.

The number of levels can be chosen to satisfy the waveform distortion and output voltage level requirements [23].

In addition to providing flexibility in the control of the fundamental and harmonic components, these structures allow the increase of the voltage and current rating to levels required for transmission systems. If voltage ratings are sufficient, they also provide the option of removing the coupling transformer used for voltage matching, which may reduce the cost of the installation [22].

2.7.2.4 PWM techniques

An alternative method of eliminating harmonics is by pulse width modulation (PWM). Adding notches in the voltage or current pattern modifies the harmonic content of the waveform and therefore allows control over the amplitude of the harmonic components. The harmonic components of a typical pattern depend upon the position, width, and number of notches. Typically, one notch per quarter cycle allows the control and therefore the removal of one harmonic component. This principle is used in the technique known as selective harmonic elimination (SHE). Implementation of the technique however requires computations: the equations for the angles defining the position of the pulses are transcendental and must be solved by iteration. Solutions are usually obtained off-line and stored, for example in table format.

If the allowable switching frequency can be increased, typically to 1 kHz or higher, various modulation techniques can be used that naturally remove or minimize unwanted harmonic components [24,25]. The techniques apply to both current and voltage source converters.

2.7.3 Output control

2.7.3.1 Principles of PWM

In order to vary the fundamental component of the output voltage for a fixed dc bus voltage, Pulse Width Modulation (PWM) is used. The technique is discussed in reference to voltage source converter structures. However, it is applicable to current source converters.

The principles of pulse width modulation are illustrated in Figure 2.20 (a) for sinusoidal PWM technique based on carrier comparison. The modulation technique consists in converting the reference or desired waveform, the voltage reference in a voltage source converter, into a digital signal corresponding to the gating pattern of one or more switches. Digitizing the reference is carried out in sine PWM by intersecting the reference with a carrier, typically a triangular carrier. The spectrum of the pattern thus created contains the desired or fundamental component, proportional to the reference, and unwanted components or harmonics, Figure 2.20 (b).

The fundamental component of the output voltage is proportional to the reference voltage. The voltage gain of the modulation technique is therefore constant. The maximum voltage in PWM mode is less than the voltage obtained for the single pulse pattern, due to the presence of notches that control the harmonic content. It is equal to 0.67 E_d (rms line to line voltage), compared to 0.78 E_d for the square wave pattern. Operation of the PWM modulator can be extended to a square wave voltage by gradually dropping pulses in the centre of the pattern. This occurs naturally in sine PWM by increasing the amplitude of the reference above the amplitude of the triangular carrier.

If the carrier frequency f_s is sufficiently high compared to the reference frequency f_o, the first harmonic components are around the switching frequency, $f_s \pm pf_o$. Low frequency harmonics are therefore absent. The current harmonics injected into the ac supply are filtered by the link reactance. It is therefore desirable to switch at the highest possible switching frequency in order to reduce the current harmonics.

2.7.3.2 Features of PWM techniques

Many PWM techniques have been developed in addition to sine PWM. Several of them are applicable to high power converters, with some modifications or designed for particular configurations.

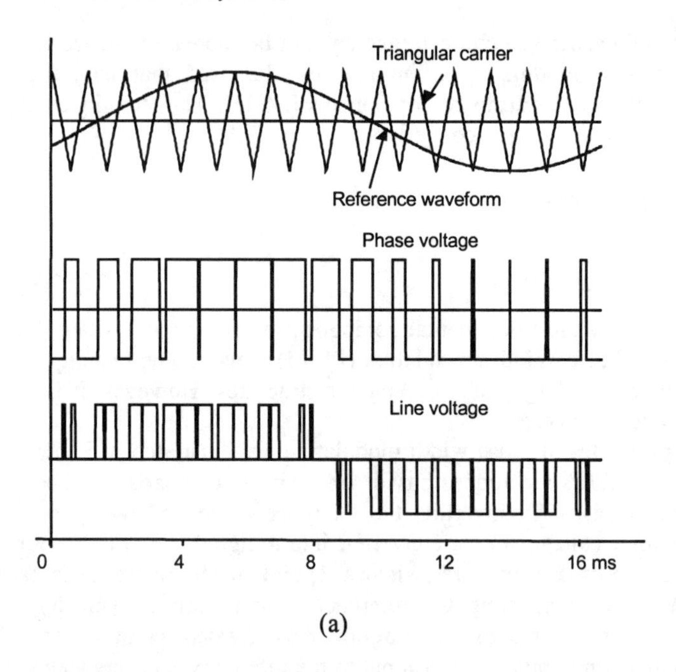

(a)

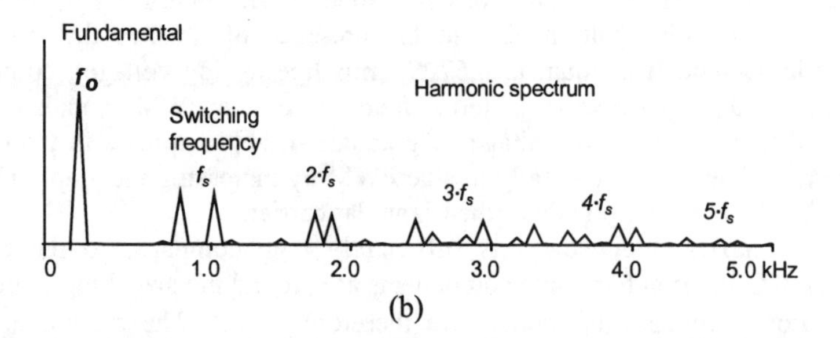

(b)

Figure 2.20 Principles of PWM voltage control. Sinusoidal PWM (SPWM) for Modulation Index M = 1 and switching frequency f_s= 15 p.u. (a) Carrier and modulating signals, switching function and line to capacitor centre-tap voltage, Line to Line voltage. (b) Line to Line voltage frequency spectrum.

Selective harmonic elimination (SHE) can be used for voltage control if a notch is added to define the amplitude of the first harmonic, that is the fundamental component. The technique offers complete control over harmonics since it can be designed to cancel specific components. Therefore, it minimizes switching frequency for a given harmonic content. It has also been used in current source structures, such as SMES systems.

Carrier PWM, the most common being the sine PWM (SPWM), have been shown to be effective in multi-module converter systems, even at low switching frequencies [26]. Carrier techniques are simple to implement both in analog and digital systems. However, there usually is one modulator per converter leg: this leads to redundant switching in a three-phase system. Since the sum of the voltages in a three-phase system must add up to zero, not all three legs can be controlled simultaneously.

Other techniques, such as Space vector modulation (SVM), generate patterns on a three-phase basis, resulting in a reduced switching frequency for a given dominant harmonic frequency. Low frequency SVM techniques can also be adapted to multi-module converter systems [27].

For high power applications, criteria for selecting an appropriate PWM pattern generation scheme can be summarised in terms of specific requirements:

- *Instantaneous on-line control.* SPWM and SVM allow instantaneous control of the inverter output voltage; patterns generated by a SHE algorithm are generally stored in a table or the equivalent. This results in a slower dynamic response.
- *Harmonic minimization.* Direct, selectives, and precise harmonic elimination can only be obtained with the SHE techniques. Other patterns rely on the fact that the dominant harmonics are related to the switching frequency and its multiples. The switching frequency is equal to the carrier frequency for SPWM, as shown in Figure 2.20, or to 2/3 of the cycle time for SVM. The amplitude of harmonic components depends upon the implementation of specific techniques. In SVM in particular, there are a number of techniques for producing the gating patterns, each having a different harmonic content.
- *Maximum inverter output in the controllable range.* Patterns that are generated on a three-phase basis (SVM and SHE) maximize the output voltage for a given dc bus voltage, that is the ac gain. In SPWM, patterns are usually produced on a per leg basis, and the redundant switching results in a reduction in the output voltage. However, this value can be increased by injecting a third harmonic component in the modulating reference voltage.

2.7.3.3 Output voltage and current control

The fundamental component of the output voltage of the inverter, V_{i1} (line to line peak), can be controlled, either by changing the dc bus voltage E_d or the PWM pattern, through modulation index control, M:

$$V_{i1} = K_{ac} M E_d$$

where K_{ac} is the gain of the converter, which depends upon the PWM technique. For a three-phase inverter, the gain K_{ac} varies between 0.86 for SPWM, and 1.03

for SHE, SPWM with third harmonic injection and SVM. The dynamic performance of a static power converter system depends largely on the control technique chosen:

- *DC bus voltage control.* If no PWM or only a fixed SHE pattern designed for harmonic elimination are used, output voltage control is obtained by varying the dc bus capacitor voltage E_d. This is achieved by charging or discharging the dc capacitor, through δ angle control. This may result in a slow response. The dynamic response depends upon the dc capacitor value, the dc voltage and the ac inductor [28].
- *PWM pattern control.* If the PWM pattern is varied to produce a variable voltage from the fixed dc bus voltage, response can be very fast, in the order of the switching period. Instantaneous control of the injected current can then be implemented through a current control loop. The response is much faster than with conventional TCRs.

2.7.4 Multi-stepped converters

2.7.4.1 Multi-module structures

Rather than increasing the frequency of the PWM pattern to reduce the harmonic content of the output voltage, a number of units can be connected in parallel through a magnetic interface circuit, Figure 2.18. Voltage distortion is reduced by harmonic cancellation or minimization. Some multi-pulse structures use PWM. If no PWM is used, output voltage control must be implemented by dc capacitor voltage control.

In general, harmonic cancellation is achieved by phase shifting the harmonic components to be cancelled so that, when the output voltages of individual units are added, these components cancel. This can be obtained by using phase shifting transformers. Alternatively, patterns with phase-shifted harmonics are generated.

One simple technique is to use a phase-shifted carrier, Figure 2.21 (a). For n converters, each carrier is shifted horizontally by T/n, where T is the period of the fundamental reference wave, the line frequency in a STATCOM. The harmonic components produced by each unit are phase shifted and, when voltages are added, cancel if the switching frequency is much higher than the fundamental frequency [26]. The fundamental voltages are in phase if the modulating waveform is common to all converters and the switching frequency high. In the example of Figure 2.21 (b), individual converters have harmonic components around the 9th and multiples. The combined output of 4 units however, Figure 2.21 (e), only contains dominant components around the 36th and multiples. These techniques however only provide harmonic minimization for low switching frequencies. Transformer configurations are similar to that of Figure 2.18, but no phase shift is required and standard transformers can be used. The principles of phase shifted carriers can be applied to force-commutated ac controllers [29].

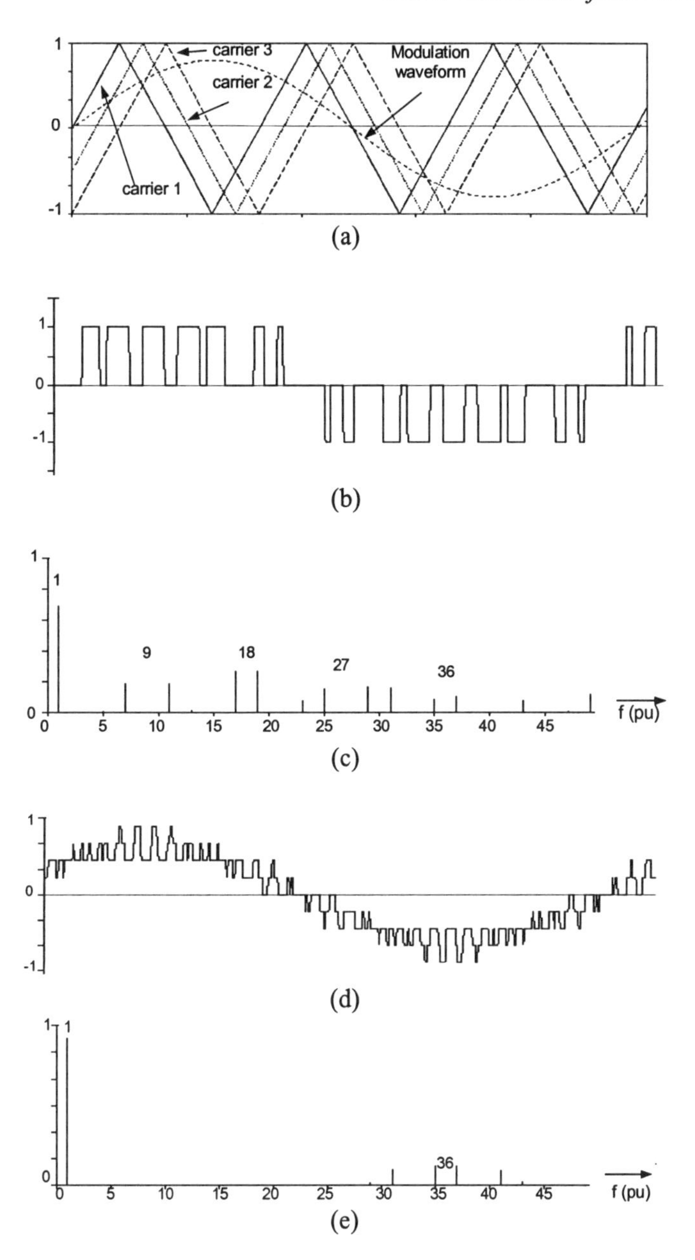

Figure 2.21 Multi-module pattern generation by carrier phase shifting. (a) SPWM with phase shifted carriers (8 carriers, with only 3 shown). (b) Output voltage of individual modules. (c) Harmonic spectrum (carrier to fundamental frequency = 9). (d) Total output voltage. (e) Harmonic spectrum (4 units).

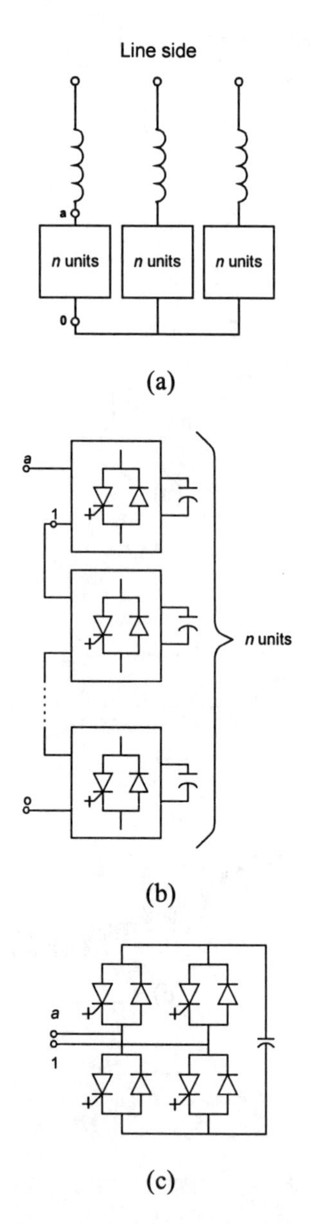

Figure 2.22 Multi-module compensator based on single-phase units. Voltage source topology. (a) Three-phase structure. (b) Multi-module single phase converter, series connection, one leg. (c) Circuit topology of each unit.

2.7.4.2 Multi-level structures

PWM techniques can be applied to multi-level converters. Carrier techniques can be implemented, with one modulator for each converter level. Carriers are then shifted horizontally. Alternatively, space vector modulation can be adapted for low switching frequencies.

2.7.4.3 Single phase structures

The structures presented above are based on three-phase converter units. This is the natural choice for three phase systems. However, implementations based on single-phase converters of the voltage source type have been proposed, Figure 2.22 [30]. Contrary to the three-phase, the output voltage of a single-phase converter operating at fundamental frequency can be controlled by shifting the patterns of the 2 legs. Individually controlled single-phase units are then assembled to form a three-phase converter. Such structures are suitable for series or shunt configurations [31].

Advantages of single-phase structures include: (a) modularity and ease of expansion, obtained by stacking units, particularly to meet voltage ratings; (b) control flexibility, since control is implemented on a per phase basis and unbalances are more easily handled. Output voltage waveforms are similar to that of Figure 2.18 for a single pulse operation. PWM can also be used to control the voltage and the harmonic content.

2.8 Power converter control issues

2.8.1 General control requirements

Control of a static power compensator is carried out at a number of levels, from the local to the system levels:

- Power switch gating. The switch gating characteristics define the gate drive required. Typically, with thyristors and IGBTs, gate power requirements are low. GTOs however require special gate drive circuits to provide the short duration high current pulse required for turn-off. Gating signals must be isolated. Transformers or fiber optic systems are usually employed.
- Gating pulse or pattern generation. Gating patterns are generated by the modulator, which translates a reference into pulses. Single pulse or PWM gating is used.
- Line synchronization. Compensators are connected to the ac system and must therefore be synchronized with the line voltage.
- Voltage or current control. Voltage or current references are generated by the compensator control circuit. Open or closed loop operation is possible.

- Supplementary functions. These include reactive power control and damping of power system oscillations.

It should be noted that static power converters dynamic response is usually limited only by delays associated with the allowable gating instants. For example, in three-phase converters using single pulse gating, delays are of the order 1/6th of an ac period on the dc side. These delays, or transportation lags, are significantly reduced in force-commutated converters. Transportation delays associated with the converter operation are therefore small in comparison with the response required from the converter system. The overall compensator response however depends upon the configuration of the system, namely the control approach, the passive components, such as inductor and capacitor, and the power system.

2.8.2 Line synchronization

Since the voltage and current injected into the ac supply must be in a properly defined angular position, converter gating must be synchronized with the ac supply. A number of options are available, among which:

- Phase locked loops, the conventional means of synchronising converters in HVDC and static var compensators [32]. Advantages of the scheme include: accurate phase information, even with distorted ac system voltage waveforms, presence of the synchronisation signal for short time interruptions of the ac voltages, for example under fault conditions. Disadvantages include slow response.
- Zero voltage crossing detection. The method is fast and accurate, but may be difficult to implement with distorted ac waveforms.

2.8.3 Voltage and current control

The voltage and current injected into the ac supply are dictated by the function of the compensator. If closed loop control is used, the inner loop is a current loop for a shunt compensator, or a voltage loop for a series compensator. Accurate tracking of the current reference requires PWM techniques operating at a reasonable switching frequency. A number of control options are available, among which:

- Control of rms or instantaneous values in the stationary abc frame.
- Control of rms values or instantaneous values in the rotating dq frame. This technique requires a calculator or processor to implement the axis transformation. Advantages include the natural separation of current, and in general power, into real and reactive components.

2.8.4 Supplementary controls

These include:

- Var control.

- Impedance control. In addition to emulating a reactance, there is the possibility of emulating specified values of R and X in series compensators. In particular, with a UPFC, negative resistances can be introduced into the ac transmission lines to cancel line resistance.
- Power system damping. This includes injection of real power, using either STATCOM and SMES units, or UPFC systems to damp power system oscillations.

2.8.5 Operation under non-ideal conditions

Operation of FACTS devices based on static power converters must take into account the following practical constraints present in real systems:

- Unbalances in the gating of the power switches, resulting in additional harmonics being injected into the system.
- AC system unbalanced voltages.
- AC system resonances.
- AC voltage distortion, including the presence of 5th harmonic components.

In addition, dc link topologies do not offer the same flexibility as ac controller structures: gating is usually done on a three-phase basis, whereas devices such as the TCR can be controlled on a per phase basis. It is nonetheless possible to unbalance the gating of three-phase converters to maintain for example balanced compensator currents. This however leads to increased ripple on the dc bus and may require increasing the value and rating of some components, such as the dc capacitor and ac inductor.

2.9 Summary

Advantages of incorporating power electronics in FACTS controllers include unlimited life, fast response, and continuous control and repeatability of the output. In addition, unlike systems based on passive components, static compensators do not introduce potentially damaging resonant frequencies. Power electronic converters are therefore becoming essential to the implementation of most FACTS devices.

The main drawbacks are losses, resulting in additional initial and operating costs, harmonic distortion, and capital costs associated with power semiconductor devices.

Although many devices at this time use structures incorporating thyristors, force-commutated structures offer greater flexibility, in terms of output control, speed of response, harmonic minimisation, and implementation of complex control features. A UPFC is an example of the flexibility provided by back to back voltage sources: in addition to the degrees of freedom available for the control of system performance, it can process real power. Similarly, shunt devices have the capability of injecting real power to support the power system under fault conditions, if equipped with energy storage devices.

Implementation of static power compensators based on force-commutated semiconductor switches should become more attractive as the power handling capability of power devices increases, and methods are developed to enhance reliability, reduce losses, minimize harmonic injection and improve control algorithms. A number of prototype units of the newer static compensator systems have demonstrated their potential.

2.10 References

1 Kimbark, E.W., Direct current transmission, Vol. 1, John Wiley, New York, 1971.

2 Arrillaga, J., High voltage direct current transmission, Peter Peregrinus, London, 1983.

3 Miller, T.J.E., (ed.), Reactive power factor compensators, John Wiley, New York, 1985.

4 Gyugyi, L., "Reactive power generation and control by thyristor circuits", *IEEE Trans. Industry Applications*, **15**, (5), pp. 521-532, 1979.

5 Gyugyi, L., Static frequency changers, John Wiley, New York, 1979.

6 Larsen, E., Miller, N., Nilsson S., and Lindgren, S., "Benefits of GTO-based compensation systems for electric utility applications", *IEEE Trans. Power Delivery*, **7**, (4), pp. 2056-2062, 1992.

7 Gyugyi, L., "Dynamic compensation of AC transmission lines by solid-state synchronous voltage sources", *IEEE Trans. Power Delivery*, **9**, (2), pp. 904-911, 1994.

8 Schauder, C., Gernhardt, M, Stacey, E., *et al.*, "Development of a ± 100 MVAR static condenser for voltage control of transmission systems", *IEEE Trans. Power Delivery*, **10**, (3), pp. 1486-1493, 1995.

9 Edris, A-A., Chair of Task Force, *et al.*, "Proposed terms and definitions for flexible ac transmission system (FACTS)", *IEEE Trans. Power Delivery*, **12**, (4), pp. 1848-1853, 1997.

10 Gyugyi, L., Schauder, C.D. and Sen, K.K., "Static synchronous series compensator: a solid-state approach to series compensation of transmission lines", *IEEE Trans. Power Delivery*, **12**, (1), pp. 406-413, 1997.

11 Gyugyi, L., Schauder, C.D., Williams, S.L., Rietman, T.R., Torgerson, and D.R., Edris, A., "The unified power flow controller: A new approach to power transmission control", *IEEE Trans. Power Delivery*, **10**, (2), pp. 1085-1097, 1995.

12 Suzuki, H., Nakajima, T., Izumi, K., *et al.*, "Development and testing of prototype models for a high performance 300 MW self-commutated ac/dc converter", *IEEE Trans. Power Delivery*, **12**, (4), pp. 1589-1598, 1997.

13 Ooi, B.T. and Dai, S.-Z., "Series-type solid state VAr compensator", *IEEE Trans. Power Electronics*, **8**, (2), pp. 164-169, 1993.
14 Urbanek, J., Piwko, R.J., Larsen, E.V., *et al.*, "Thyristor controlled series compensation – Prototype installation at Slatt 500 kV Substation", *IEEE Trans. Power Delivery*, **8**, (4), pp. 1460-1469, 1993.
15 Iravani, M.R., Dandeno, P.L., Nguyen, K.H., Zhu, D., and Maratukulam, D., "Applications of static phase shifters in power systems", *IEEE Trans. Power Delivery*, **9**, (3), pp. 1600-1608, 1994.
16 Iravani, M.R., and Maratukulam, D., "Review of semiconductor-controlled (static) phase shifters for power system applications", *IEEE Trans. Power Delivery*, **9**, (4), pp. 1833-1839, 1994.
17 Lopes, L.A.C., Joos, G. and Ooi, B.T., "A PWM Quadrature Booster Phase-shifter for FACTS', *IEEE Trans. Power Delivery*, **11**, (4), pp. 1999-2004, 1996.
18 Zhang, Z.-C. and Ooi, B.T., "Multi-modular current source SPWM converters for superconducting magnetic storage system", *IEEE Trans. Power Electronics*, **8**, (3), pp. 250-256, 1993.
19 Walker, L., "10-MW GTO converter for battery peaking service", *IEEE Trans. Industry Applications*, **26**, (1), pp. 63-72, 1990.
20 Sumi, Y, Harumoto, Y., Hasegawa, T., Yano, M., Ikeda, K., and Matsuura, T., "New static var control using force-commutated inverters", *IEEE Trans. Power Apparatus and Systems*, **100**, (9), pp. 4216-4223, 1981.
21 Mori, S., Katsuhiko, M., Hasegawa, T., *et al.*, "Development of a large static var generator using self-commutated inverters for improving power system stability", *IEEE Trans. Power Systems*, **8**, (1), pp. 371-377, 1993.
22 Hochgraf, C. and Lasseter, R.H., "A transformer-less static synchronous compensator employing multi-level inverter", *IEEE Trans. Power Delivery* **11**, (2), pp. 881-887, 1996.
23 Menzies, R.W. and Zhuang, Y., "Advanced static compensation using multilevel GTO thyristor inverter", *IEEE Trans. Power Delivery*, **10**, (2), pp. 732-738, 1995.
24 Moran, L., Ziogas, P., and Joos, G., "Analysis and design of a three-phase synchronous solid-state var compensator", *IEEE Trans. Ind. Appl.*, **25**, (4), pp. 598-608, 1989.
25 Moran, L., Ziogas, P., and Joos, G., "Analysis and design of a 3-phase current source solid-state var compensator", *IEEE Trans. Industry Applications*, **25**, (2), pp. 356-365, 1989.
26 Kuang, J., and Ooi, B.T., "Series connected voltage-source converter modules for force-commutated SVC and dc transmission", *IEEE Trans. Power Delivery*, **9**, (2), pp. 977-983, 1994.
27 Bakhshai, A., Joos, G., and Jin, H., "Space vector pattern generators for multi-module low switching frequency high power var compensators", *IEEE Power Electronic Specialists Conf. Rec.*, St. Louis, USA, **1**, pp. 344-350, 1997.

28 Joos, G., Moran, L., and Ziogas, P.D., "Performance analysis of a PWM inverter var compensator", *IEEE Trans. Power Electronics*, **6**, (3), pp. 380-391, 1991.

29 Lopes, L.A.C., Joos, G. and Ooi, B.T., "A high power PWM quadrature booster phase-shifter based on multi-module ac controller", *IEEE Trans. Power Electronics*, **13**, (2), pp. 357-365, 1998.

30 Lai, J.-S. and Peng, F.Z., "Multilevel converters - A new breed of power converters", *IEEE Trans. Industry Applications*, **32**, (3), pp. 509-517, 1996.

31 Joos, G., Huang, X. and Ooi, B.T., 'Direct-coupled multilevel cascaded series VAr compensators', *IEEE Trans. Industry Applications*, **34**, (5), pp. 1156-1163, 1998.

32 Ooi, B.T. and Wang, X., "Voltage angle lock loop of boost type PWM converter for HVDC application", **5**, (2), *IEEE Trans on Power Electronics*, pp. 229-235, 1990.

33 Hochgraf, C., and Lasseter, R.H., "Statcom controls for operation with unbalanced voltages", *IEEE Trans. Power Delivery*, **13**, (2), pp. 538-544, 1998.

Chapter 3

High voltage dc transmission technology

J. Arrillaga

3.1 Introduction

The invention of the high voltage mercury valve half a century ago paved the way for the development of the HVdc transmission technology. By 1954, the first commercial dc link came successfully into operation and was soon followed by several other schemes orders of magnitude larger. The success of the new technology immediately triggered research and development into an alternative solid state valve, which by the mid-60s had already displaced the use of mercury-arc valves in new schemes. The history and technical development of the HVdc technology are described in detail in references [1–6].

Substantial progress made in the ratings and reliability of thyristor valves has increased the competitiveness of dc schemes by reducing converter costs and break-even transmission distances.

However, the lack of turn-off capability of the thyristor device is an important restriction in terms of reactive power requirement and its control. This limitation has encouraged the development of more controllable power electronic devices, such as GTOs and IGBTs, though at the point of writing, these have not yet challenged the thyristor for HVdc schemes of large power ratings [7]. On the other hand, the ratings of the new devices has permitted the development of the FACTS technology, the subject of this book, to provide solutions to specific problems at a lower cost than HVdc.

The inclusion of dc transmission in this book seems to be a contradiction in terms, as often HVdc and FACTS are perceived as competing technologies. The problem arises from a rather restrictive interpretation of the word 'transmission', which often implies long distance, whereas a large proportion of the existing dc links are 'zero distance' interconnectors.

The present boundaries between HVdc and FACTS relate to the types of solid state devices (which in the HVdc case are at present restricted to the silicon-controlled rectifier) and to the power rating of the schemes. However, as the rating and acceptability of alternative solid state devices improve, the boundaries will gradually become 'blurred'. HVdc will be tempted to use the new devices and FACTS will try and influence power controllability more directly, e.g. by developing the full power asynchronous interconnector, i.e. back-to-back HVdc link. Therefore, the back-to-back link may also be considered as a FACTS device and this is the HVdc application emphasised in the present chapter.

3.2 Ac versus dc interconnection

Below the break-even distances, it is generally preferable to use synchronous (ac) power transmission to interconnect two otherwise independent systems or regions. However, some critical requirements must be met to justify the use of synchronous interconnections[8], among them:

- The link must have enough capacity to maintain the interconnection power flow at schedule levels and to rapidly restore such levels following disturbances.
- The existence or creation of load dispatch centres with reliable and fast communication facilities.
- The capability of each interconnected system to maintain and control a normal frequency and thus maintain adequate short term and long term spinning reserves.

However, the separate regions in many countries face shortages of power especially during the peak hours when the grid frequency remains very low (maintaining spinning reserve is not possible). It is therefore very difficult in such cases to interconnect the regions in the synchronous mode.

For asynchronous interconnection, there are two options – one is through HVdc transmission and the other through an HVdc back-to-back station. The HVdc transmission option is economically viable only when the distance involved is long and the amount of energy to be transferred large. For transfer of surplus power available in one region for a limited period of time, and also to support each other in emergencies, HVdc back-to-back is usually the better choice.

3.3 The HVdc converter[5]

To achieve instantaneous matching of the ac and dc side voltages of the conversion process, Figure 3.1, enough series impedance must be placed either on the ac side or dc side of the converter. The former solution yields predominantly voltage source conversion, with the possibility of altering the dc current by thyristor control. If a large smoothing inductor is placed on the dc side, only pulses of a constant direct current flow through the switching devices into the transformer secondary windings. These current pulses are then transferred to the primary side according to transformer connection and ratio; thus a basic current converter results, with the possibility of adjusting the direct voltage by thyristor control.

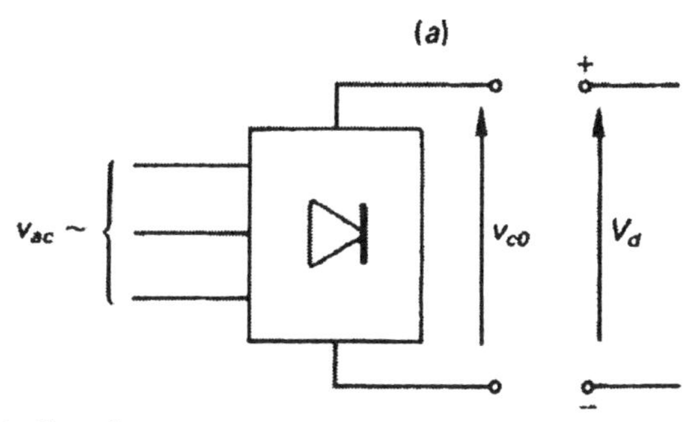

Figure 3.1 Ac/dc voltage conversion.

The use of voltage conversion was rejected in mercury-arc converters due to the impossibility of recovering from arc-back disturbances. With thyristor schemes, rapid changes in the supply voltage can only be accommodated within narrow limits and require the use of large series impedances, which would be uneconomical in terms of reactive power compensation. Therefore, the current conversion principle is still the preferred option for the design of HVdc converters.

In terms of optimal converter utilisation and low peak inverse voltage across the converter valves, the three-phase bridge (shown in Figure 3.2) is used exclusively in HVdc converters

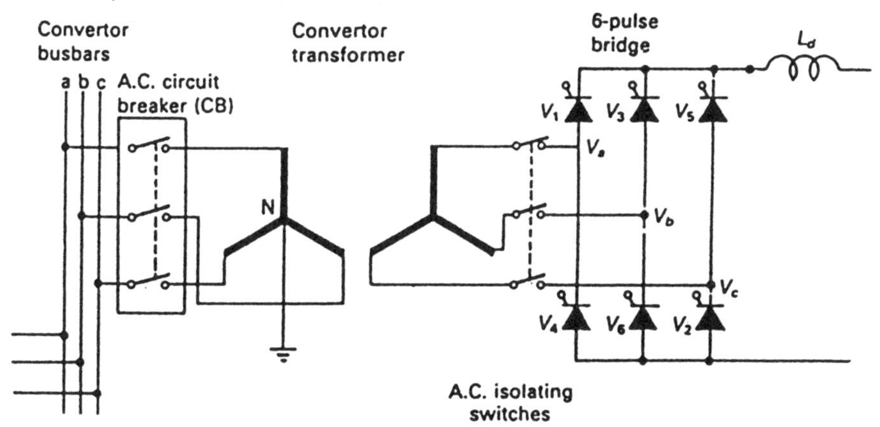

Figure 3.2 The basic three-phase bridge converter.

With HVdc schemes only simple transformer connections are used. This is due to the problem of insulating the transformers so that they withstand the alternating voltages combined with the high direct voltages. A pulse number of 12 is easily obtained with star/star and star/delta transformer connections in parallel (as shown in Figure 3.3) and has become the standard configuration of HVdc converters. With this configuration, the "ideal" phase current is as shown in Figure 3.4; its lowest characteristic harmonic current component is the 11th and the filter costs are significantly reduced.

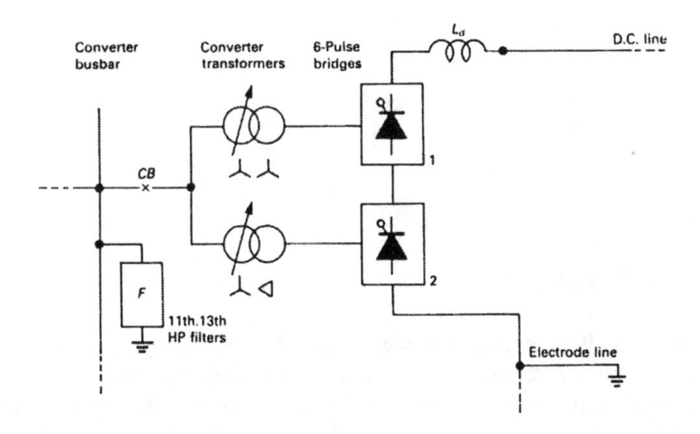

Figure 3.3 12-pulse converter configuration.

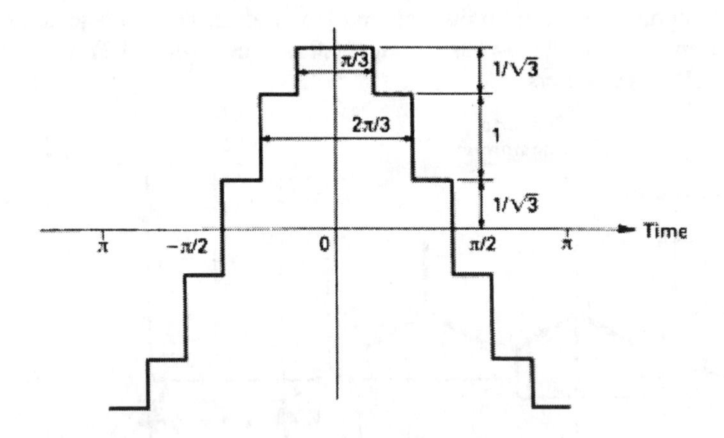

Figure 3.4 Idealised phase current waveform with 12-pulse operation.

3.3.1 Rectifier operation

Typical voltage and current waveforms of a bridge operating as a rectifier with the commutation effect included are shown in Figure 3.5, where P indicates a firing instant (e.g. P_1 is the firing instant of valve 1) S indicates the end of a commutation (e.g. at S5 valve 5 stops conducting), and C is a voltage crossing (e.g. C_1 indicates the positive crossing between phases blue and red).

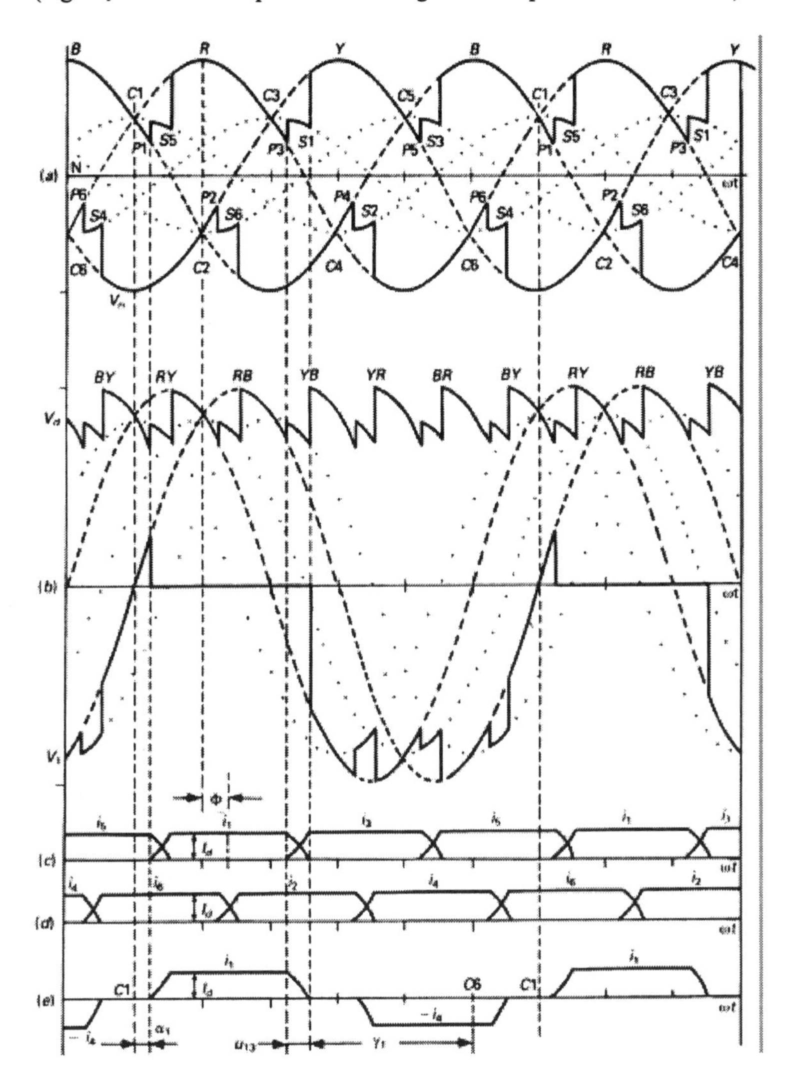

Figure 3.5 Typical six-pulse rectifier operation. (a) Positive and negative direct voltages with respect to the transformer neutral. (b) Direct bridge voltage V_d and voltage across valve 1. (c),(d) Valve currents, i_1 to i_6. (e) AC line current of phase R.

Figure 3.5(a) illustrates the positive (determined by the conduction of valves 1, 3, 5) and negative (determined by the conduction of valves 2, 4, 6) potentials with respect to the transformer neutral. Figure 3.5(b) shows the direct voltage output waveform.

The potential across valve 1, also shown in Figure 3.5(b), depends on the conducting valves. When valve 1 completes the commutation to valve 3 (at S1), the voltage across will follow the red-yellow potential difference until P4. Between P4 and S2 the commutation from valve 2 to valve 4 (see Figure 3.5(a)) reduces the negative potential of phase red and causes the first voltage dent. The firing of valve 5 (at P5) increases the potential of the common cathode to the average of phases yellow and blue; this causes a second commutation dent, at the end of which (at S3) the common cathode follows the potential of phase blue (due to the conduction of valve 5). Finally, the commutation from valve 4 to valve 6 (between P6 and S4) increases the negative potential of valve 1 anode and produces another voltage dent.

Figures 3.5(c) and (d) illustrate the individual valves (1 and 4) and Figure 3.5(e) the phase (red) currents, respectively.

A number of reasonable approximations are made to simplify the derivation of the steady state equations that follow. These are:

The converter valves are treated as ideal switches. When calculating the power loss, the valve resistance can be added to that of the dc transmission line.

The ac systems consist of perfectly balanced and sinusoidal emfs, the commutation reactances are equal in each phase and their resistive components are ignored.

The direct current is constant and ripple-free, i.e. the presence of a very large smoothing reactor is assumed.

Only two or three valves conduct simultaneously, i.e. two simultaneous commutations are not considered. The low ac voltage and/or high dc current required to cause simultaneous commutations are prevented in the steady state; during disturbances, on the other hand, the converter behaviour can only be predicted by dynamic analysis.

Under these assumptions, the following expressions apply to the rectification process:

$$V_d = (1/2)V_{c0}[\cos\alpha + \cos(\alpha + u)] \tag{3.1}$$

$$V_d = V_{c0}\cos\alpha - \frac{3X_c}{\pi}I_d \tag{3.2}$$

$$I_d = \frac{V_c}{\sqrt{2}X_c}[\cos\alpha - \cos(\alpha + u) \tag{3.3}$$

The main effects of non-ideal ac voltage or dc current waveforms are discussed in section 3.6.

3.3.2 Inverter operation

Three conditions must be met to achieve power inversion:

- Availability of an active ac voltage source which provides the commutating voltage waveforms.
- Provision of firing angle control to delay the commutations beyond $\alpha = 90^\circ$.
- A dc power source.

With reference to Figure 3.6(a) and (c), a commutation from valve 1 to valve 3 (at P3) is only possible as long as phase Y is positive with respect to phase R. Furthermore, the commutation must not only be completed before C6 but some extinction angle γ_1 ($> \gamma_0$) must be left for valve 1, which has just stopped conducting, to re-establish its blocking ability. This puts a limit to the maximum angle of firing $\alpha = \pi - (u + \gamma_0)$ for successful inverter operation. If this limit were exceeded, valve 1 would pick up the current again, causing a commutation failure.

Moreover, there is a fundamental difference between rectifier and inverter operations, which prevent an optimal firing condition in the latter case. While the rectifier delay angle α can be chosen accurately to satisfy a particular control constraint, the same is not possible with respect to angle γ because of the uncertainty of the overlap angle u. Events taking place after the instant of firing are beyond predictability and, therefore the minimum extinction angle γ_0 must contain a margin of safety to cope with reasonable uncertainties (values between 15° and 20° are typically used).

The analysis of inverter operation is not different from that of rectification. However, for convenience, the inverter equations are often expressed in terms of the extinction angle γ, i.e.

$$V_{\mathrm{d}} = V_{\mathrm{c0}} \cos\gamma - \frac{3X_{\mathrm{c}}}{\pi} I_{\mathrm{d}} \tag{3.4}$$

3.3.3 Power factor active and reactive power [3]

Owing to the firing delay and commutation angles, the converter current in each phase always lags its voltage (refer to Figures 3.5(e) and 3.6(e)). The rectifier therefore absorbs lagging current (consumes VARs).

In the presence of perfect filters no distorting current flows beyond the filtering point, and the power factor is approximately the displacement factor ($\cos\phi$) where ϕ is the phase angle difference between the fundamental frequency voltage and current components.

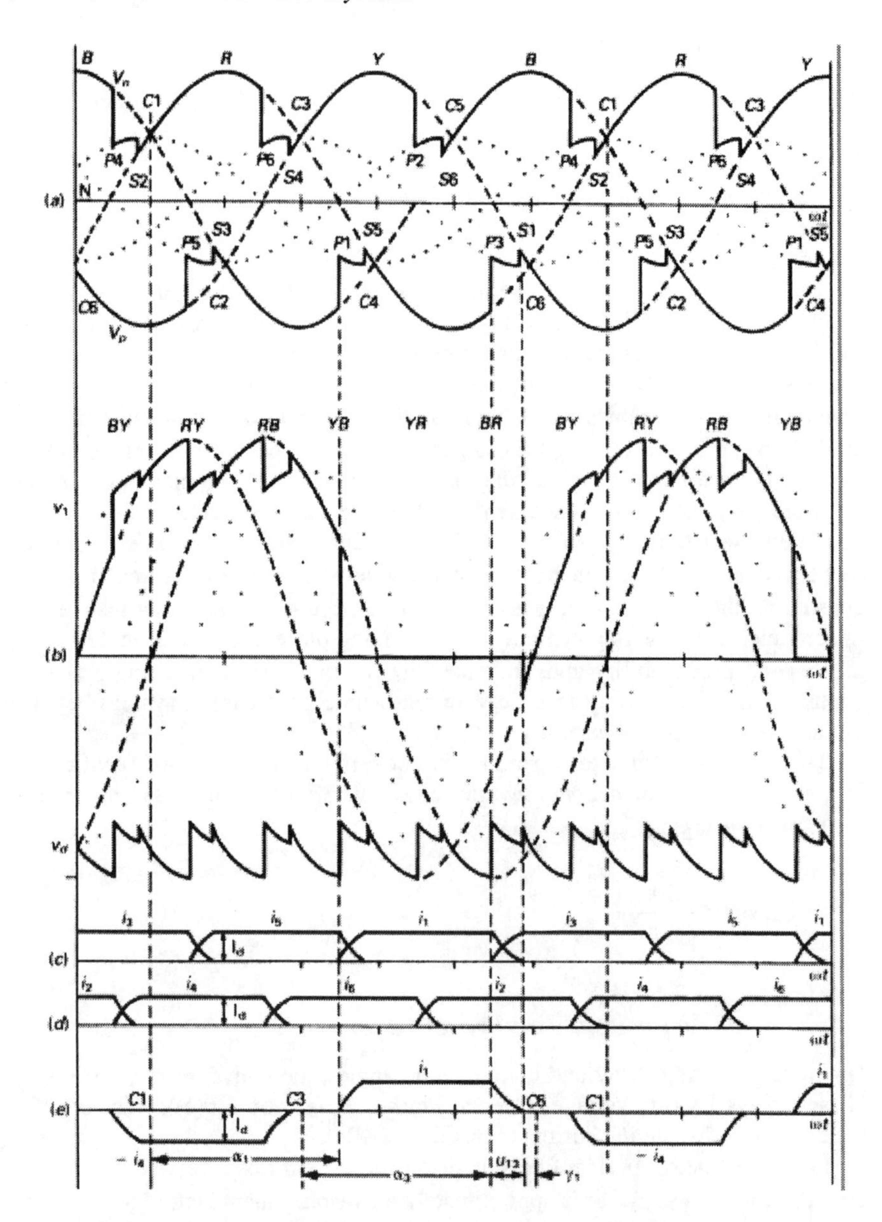

Figure 3.6 Typical six-pulse inverter operation. (a) Positive and negative direct voltages with respect to the transformer neutral. (b) Voltage across valve 1, and direct bridge voltage V_d. (c),(d) Valve currents, i_1 to i_6. (e) AC line current of phase R.

Under these idealised conditions, with losses neglected, the active fundamental ac power (**P**) is the same as the dc power, ie

$$\boldsymbol{P} = \sqrt{3} V_c \boldsymbol{I} \cos \phi = \boldsymbol{V_d I_d} \qquad (3.5)$$

An approximate expression for the power factor is:

$$\cos \phi = {}^1\!/_2[\cos \alpha + \cos (\alpha + u)] \qquad (3.6)$$

The reactive power is often expressed in terms of the active power, i.e.

$$\boldsymbol{Q} = \boldsymbol{P} \tan \phi \ , \qquad (3.7)$$

where $\tan \phi$ is

$$\tan \phi = \frac{\sin (2\alpha + 2u) - \sin 2\alpha - 2u}{\cos 2\alpha - \cos (2\alpha + 2u)} \qquad (3.8)$$

Referring to the ac voltage and valve currents waveforms in Figures 3.6(a) and (e), it is clear that the current supplied by the inverter to the ac system lags the positive half of the corresponding phase voltage waveform by more than 90°, or leads the negative half of the same voltage by less than 90°. It can either be said that the inverter 'absorbs lagging current' or 'provides leading current' both concepts indicating that the inverter, like the rectifier, acts as a sink of reactive power.

Equations (3.5) to (3.8) show that the active and reactive powers of a controlled rectifier vary with the cosine and sine of the control angle, respectively. Thus, when operating on constant current, the reactive power demand at low powers ($\phi \approx 90°$) can be very high.

However, the steady state operation in such conditions is prevented in HVdc converters by the addition of on-load transformer tap-changers, which try and reduce the control angle (or the extinction angle) to the minimum specified. Under such controlled conditions, the reactive power demand is approximately 60% of the power transmitted at full load.

3.4 HVdc system control

3.4.1 Valve firing control

Valve firings are implemented under the principle of equidistant firing control, the basis of which is a voltage-controlled oscillator [9] (shown in Figure 3.7) that delivers a train of pulses at a frequency directly proportional to a dc control voltage V_c. Based on this principle, a variety of control loops are currently used to provide the V_c voltage.

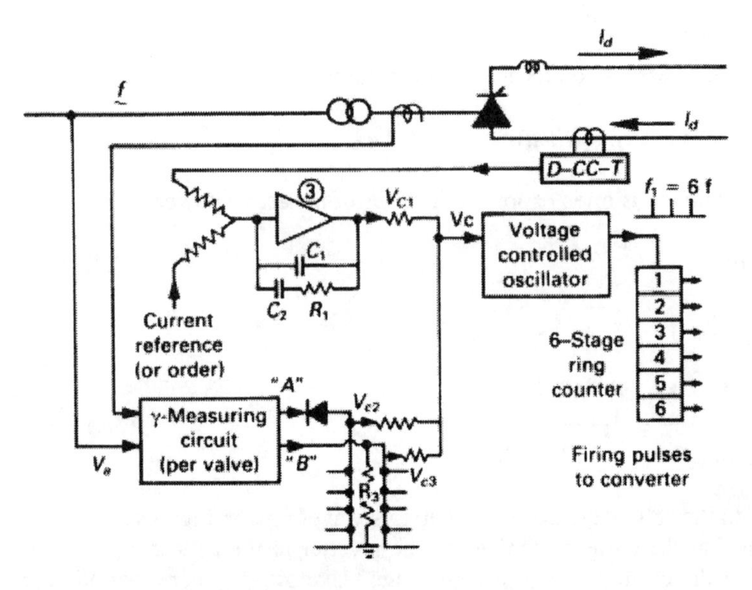

Figure 3.7 Principle of the phase-locked oscillator control system.

The phase of each firing pulse can have an arbitrary value relative to the ac line voltage, i.e. an arbitrary value of converter firing angle α. However, when the three-phase ac line voltages are symmetrical fundamental sine waves, α is the same for each valve.

Some method of phase-locking the oscillator to the ac system is required. This is normally achieved by connecting V_c in a conventional negative feedback loop for constant current or constant extinction angle.

When operating in constant current control, V_c is obtained from the amplified difference (error) between the current reference and the measured dc line current; this forms a simple negative-feedback control loop, tending to hold current constant at a value very close to the reference.

When the current is equal to the reference, the amplified error (V_c) happens to be precisely that value required to give an oscillator frequency of six times the supply frequency. The ring-counter outputs, and thus the valve-gates pulses, will have a certain phase with respect to the ac system voltage. In steady state operation, this phase is identical to firing angle α.

A disturbance such as a drop of back emf in the dc system causes a temporary current increase, which reduces V_c and hence slows down the oscillator, thus retarding its phase and finally increasing the firing angle α. This tends to decrease the current again, and the system settles down to the same current, with the same V_c and oscillator frequency but a different phase, i.e. different α.

The control system will also follow system frequency variations, in which case the oscillator has to change its frequency; this results in different V_c and hence current, but the current error is made small by using high gain amplification.

This constant current scheme is the main control mode during rectification; it is also used during inversion whenever the inverter has to take over the current control, as explained later.

The control system response is fast but, in practice, its effect will be slowed down to the relatively slower response of the dc line, which includes capacitance, inductance, and smoothing reactance.

Inverter extinction angle control is implemented by a negative feedback loop very similar to the current loop. The difference between the measured γ and the γ-setting is amplified and provides V_c as before. However, it differs in that γ is a sampled quantity rather than a continuous quantity. For each valve the extinction angle is defined as the time difference between the instant of current zero and the instant when the anode voltage next crosses zero, going positive.

For each bridge there are six values of γ to be measured, which under symmetrical steady state operation are identical. Under unbalanced conditions, however, the valve with greatest risk of commutation failure is the one having the smallest γ.

3.4.2 Control characteristics and direction of power flow

A hybrid voltage/current philosophy is used in HVdc transmission to suit the needs of the particular operating conditions. This is achieved by adjusting the dc voltage levels on both sides of the link, by means of on-load tap-changer control on the steady state, and by thyristor control following large or small changes of operating conditions at either end of the link. The dc current is only limited by the small resistance of the transmission line and is therefore very sensitive to such variations.

It will be shown in the following sections that the provision of current controllers at both ends, combined with transformer on-load tap changing, offers a perfectly satisfactory solution to this problem; thus the use of current control is universally accepted in HVdc transmission.

In the case of ac transmission, the direction of power flow is determined by the sign of the phase-angle difference of the voltages at the two ends of the line; the power flow direction is in fact independent of the actual voltage magnitudes.

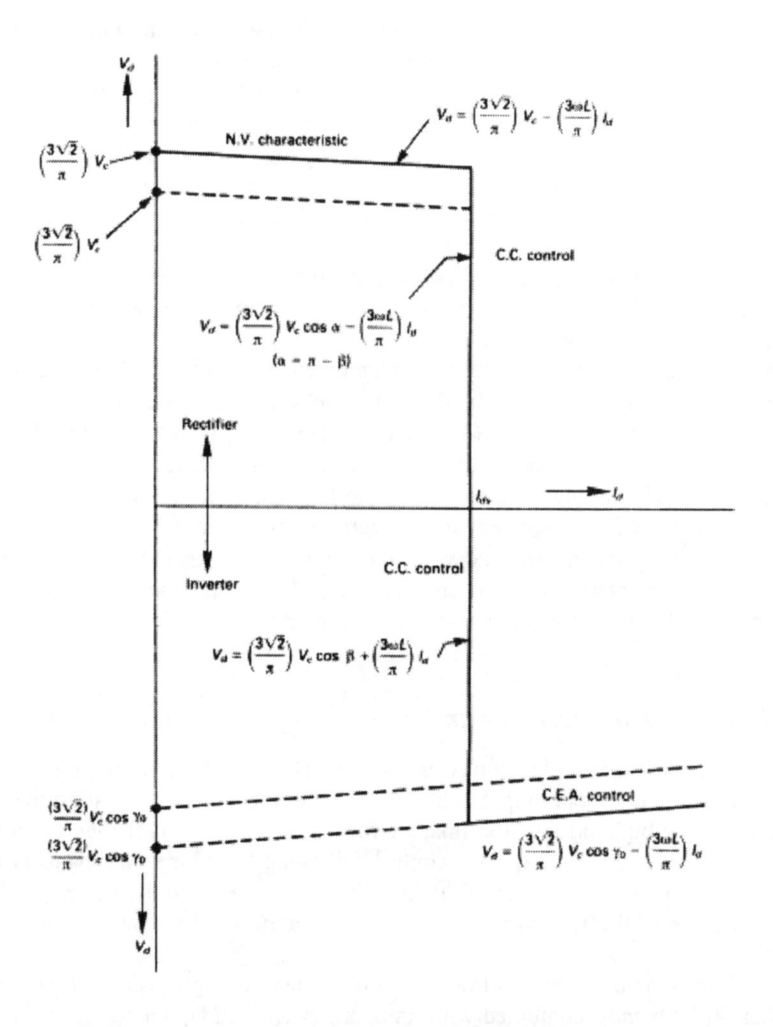

Figure 3.8 Complete control of a converter from inversion to rectification. NV = natural voltage. CC = constant current. CEA = constant extinction angle.

On the other hand, in dc transmission the power flow direction is dictated by the relative voltage magnitudes at the converter terminals and the absolute or relative phase of the ac voltages play no part in the process. However, exercising a type of firing angle control, which can make the power flow direction independent from the terminal ac voltage magnitudes and behave instead like the ac counterpart, can alter this condition.

The basic characteristic of a converter from full rectification to full inversion is illustrated in Figure 3.8. The converter is assumed to be provided with constant current and constant extinction angle controls.

Normally, the total reactive power compensation is least, and the utilisation of the line best, if the rectifier is assigned the current control task while the inverter operates on minimum extinction angle control.

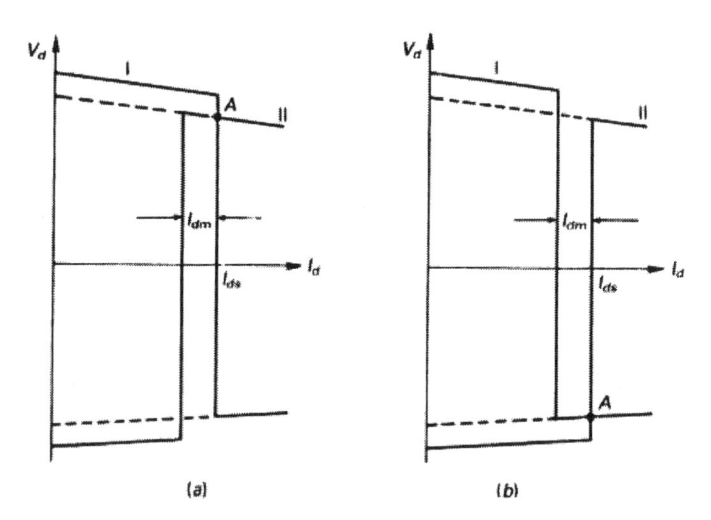

Figure 3.9 (a) Operating point with power flowing from Station I to Station II. (b) Operating point after power reversal.

This combination is achieved in a two-terminal dc link by providing the power sending station with a slightly higher current setting than the power receiving station. The difference between the two settings is termed the current margin (I_{dm}). Its effect can be better understood with reference to Figure 3.9(a) where the operating current is set by the constant current control at the rectifier end. The inverter end current controller then detects an operating current which is greater than its setting and tries to reduce it by raising its own voltage, until it hits the ceiling determined by the minimum extinction angle control at point A. This is the normal steady state operating point, which presumes a higher natural voltage characteristic at the rectifier end, a condition which may require on-load tap change action at the converter transformer.

When a substantial ac voltage reduction occurs at the rectifier end, such that the dc voltage ceiling (the natural voltage) of the rectifier becomes lower than that of the inverter. The inverter current controller will prevent a current reduction below its setting by advancing its firing (i.e. reducing α and hence inverter dc voltage), thus changing from extinction angle to constant current control. A new operating point A′ results at a current reduced by the current margin.

A change in power direction cannot happen naturally as a result of a change in the operating conditions but rather following a control order resulting from the overall requirements of the power system. Figure 3.9(a) shows that the two characteristics do not meet again when Station I increases the delay angle into the inverting region and Station II advances its firing into the rectifying region. In the process, the line current reduces to zero and the complete system is blocked.

Since, as indicated earlier, the rectifying station requires the higher current setting to reverse power, it is necessary to subtract the current margin from the reference value of Station I. This results in the characteristics of Figure 3.9(b), which display one operating point of different voltage polarity and with the roles of the two stations interchanged while the direction of current flow remains unaltered, as required by the semiconductor property.

3.4.3 Modifications to the basic characteristics [10]

During ac system faults at the receiving end there is a big risk of commutation failure and, if the fault is electrically close, the inverter may not be capable of recovering by itself. In such cases, it is important to reduce the stress on the inverter valves, and this is achieved by providing a low-voltage-dependent current-limit to the rectifier control characteristic. The modified characteristics illustrated in Figure 3.10 consists of a branch CD′ at the rectifier and another EF′ at the inverter.

The break points C and E in Figure 3.10 are typically between 70% and 30% of dc voltage, and in special cases even higher depending on ac system requirements. The higher voltage is applied when the receiving ac network is sensitive to disturbances, which could cause large voltage fluctuations.

The terms VDCOL (Voltage Dependent Current Order Limit) and LVCL (Low Voltage Current Limit) are commonly used for the function that reduces the current setting when the voltage decreases.

Also an unwanted voltage reversal of the inverter is generally prevented by limiting α to some minimal value larger than 90° (shown by KK′ of the characteristic). Branches DH and FG represent maximum current limits at low voltage.

Another problem with the basic characteristics occurs when the rectifier voltage ceiling gets very close to that of the inverter, such that the characteristics intersect in the region between I_{ds} and $I_{ds} - I_{dm}$. In this region both current controllers are out of action. In practice, the current will not settle at an intermediate point; instead, the inverter will be periodically entering the current controlling mode. A proven countermeasure against this type of oscillation is a small shape alteration of the inverter characteristic (AB in Figure 3.10 instead of AB′).

As the primary object of HVdc transmission, power control is the main consideration. The power is monitored by multiplying voltage and current (summed from both poles) and fed back directly to the controller. The master controller (at one of the stations) then sends a current order to the pole controls of the two ends of the link. However, limits are built-in to prevent unacceptable current orders (e.g. during start-up)

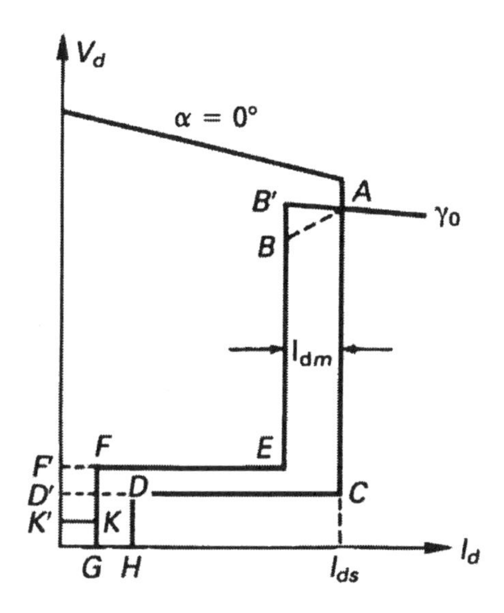

Figure 3.10 Rectifier and inverter modified characteristics.

Normally, power is the control signal applied to the current controller. However, if the frequency goes outside predetermined limits, frequency control can be used to assist the system in difficulty; however, the latter becomes inoperative when the maximum rated transmission power is reached.

3.5 Converter circuits and components

A single-line diagram of the CU HVdc project [11] is given in Figure 3.11.

The main electrical components of each pole of the converter station are shown in greater detail in the circuit diagram of Figure 3.12. All the components enclosed within the thick rectangle are located inside the valve building. The scheme includes two valve groups at each end. Each valve group consists of two series-connected six-pulse bridges supplied from two converter transformers. The transformers are connected in star/star and star/delta, respectively, to provide the necessary 30° phase-shift for 12-pulse operation.

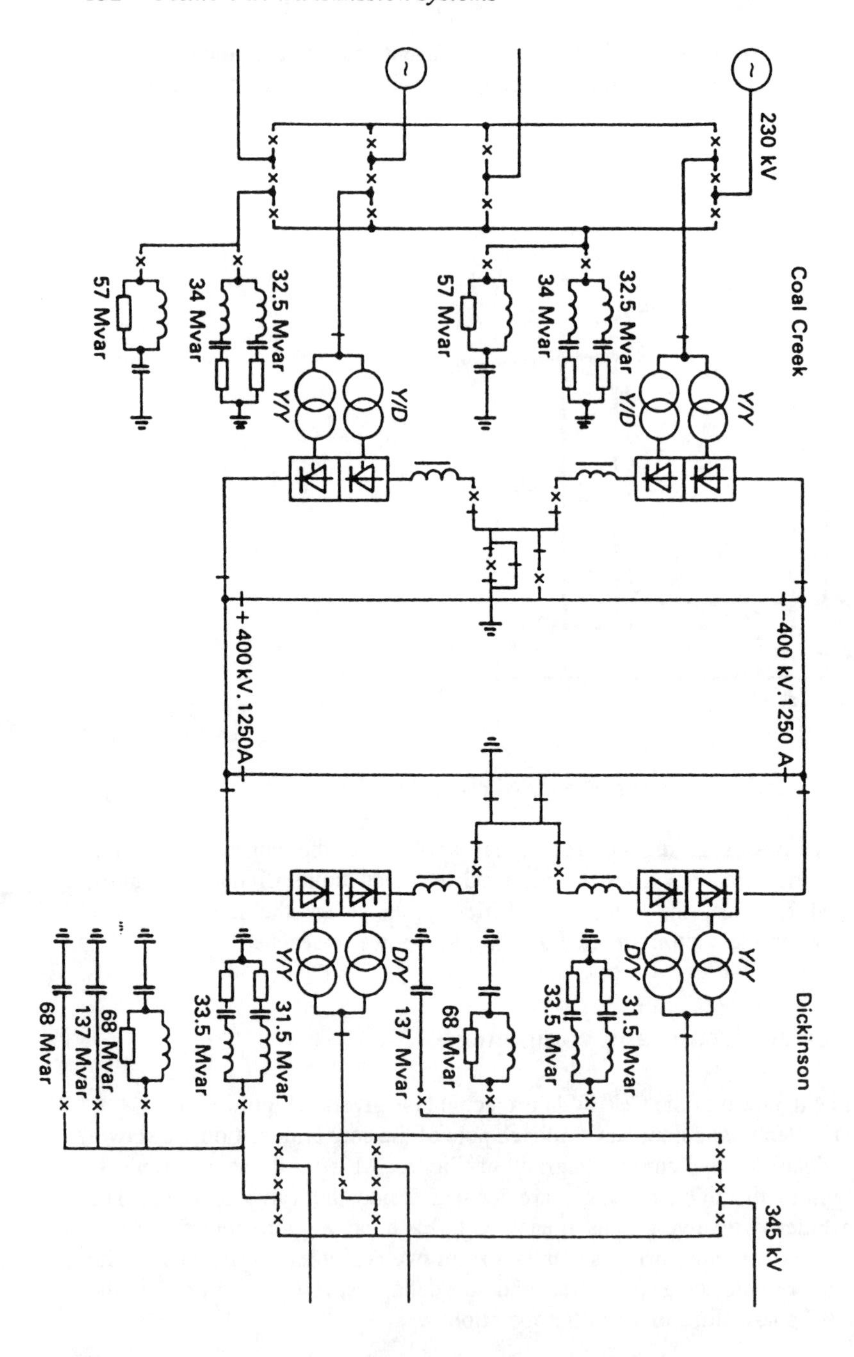

Figure 3.11 Single-line diagram of the main circuit of the CU HVdc scheme.

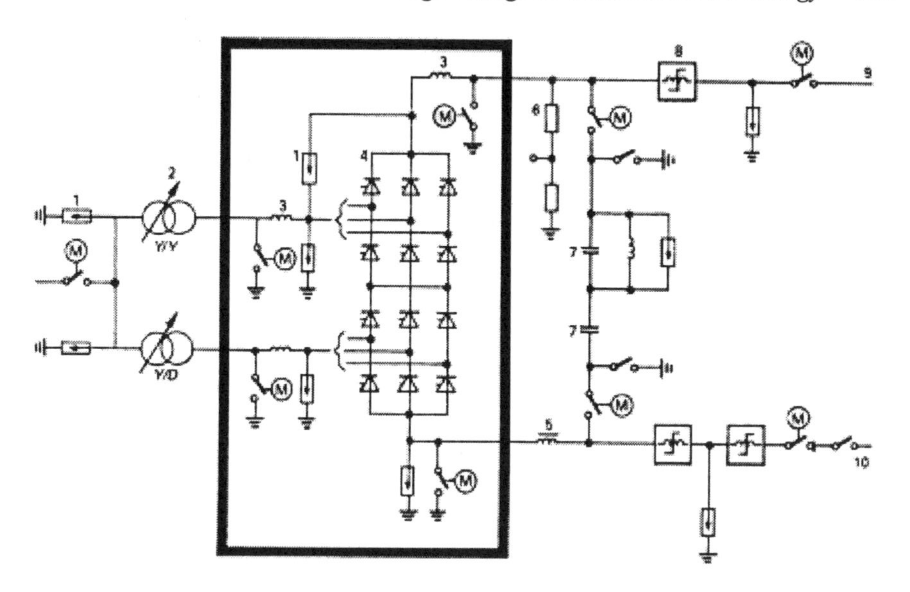

Figure 3.12 Main circuit diagram for one pole of a converter station (ASEA Journal).

1. Surge arrester	*6. Direct-voltage measuring divider*
2. Converter transformer	*7. Dc filter*
3. Air-core reactor	*8. Current measuring transductor*
4. Thyristor valve	*9. Dc line*
5. Smoothing reactor	*10. Electrode line*

At each end of the link there are two sets of harmonic filters, consisting of tuned branches for the 11th and 13th harmonics and a high-pass branch tuned to the 24th harmonic. The harmonic filters are thermally rated for full power operation, including continuous and short-time overload factors. A high-pass dc filter tuned to the 12th harmonic is also placed on the dc side. Extra shunt capacitors are installed at the Dickinson station only, since the generators at Coal Creek can provide the necessary additional reactive power. A dc smoothing reactor is located on the low-voltage side and air-core reactors on the line side of the converters; the latter to limit any steep front surges entering the station from the dc side.

The thyristor valves are protected by phase-to-phase surge arresters. The three top valves, connected to the pole bus, are exposed to higher overvoltages in connection with specific but rare incidents, and they are further protected by arresters across each valve. The indoor arrester connected to the low-voltage side of the valve protects the reactor. Pole and electrode arresters supplement the overvoltage protection.

3.5.1 The high voltage thyristor valve

A large number of thyristors are connected in series to provide the complete valve with the necessary voltage rating. Series connection of thyristors requires components to be added to the valve to distribute the OFF state voltage uniformly between the series-connected thyristors. Thus each thyristor is served by several passive components, not only to ensure that this voltage sharing is achieved but also to protect individual thyristors from overvoltage, excessive rate-or-rise of voltage (*dv/dt*) and rate-of-rise of inrush current (*di/dt*). The thyristor, together with its local voltage grading and thyristor triggering circuits, is known as a *thyristor level*.

The electrical circuit of a thyristor level is shown in Figure 3.13 and forms the basic 'building block' of a valve[12].

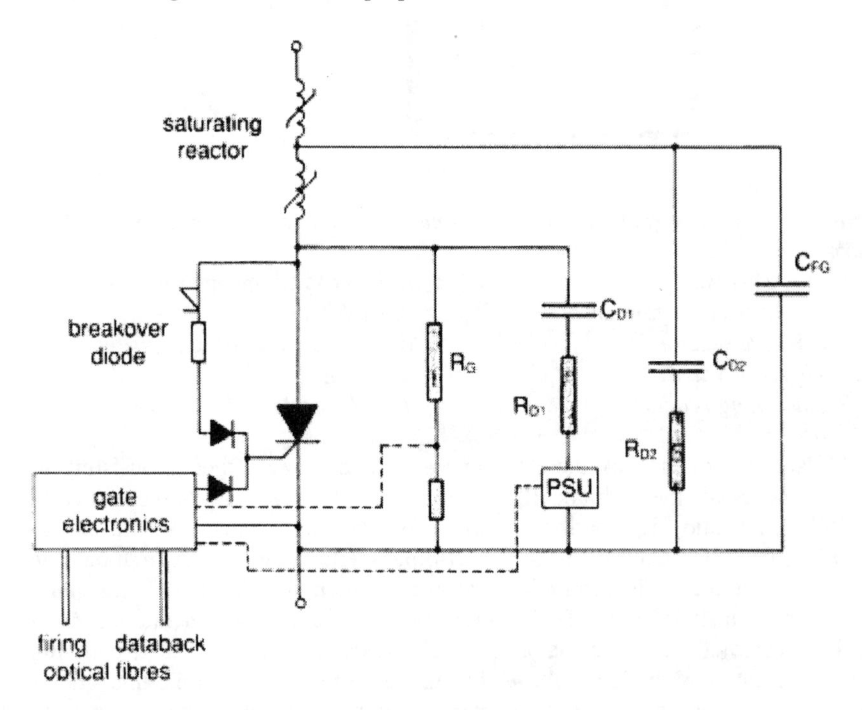

Figure 3.13 Electrical circuit of one thyristor level.

The function of the saturating reactor is to present a large inductance in series with the stray capacitances of the external circuit. This is necessary to protect the thyristors from damage immediately after firing. However, too much inductance is itself undesirable, because it would increase the reactive power absorbed by the converter. The saturating reactor avoids this problem by exhibiting high inductance only at low current. At full load current the effect of the reactor becomes negligible.

Voltage distribution is achieved by several components, acting over different frequency ranges. Direct voltage is distributed by a 'dc grading resistor' (R_G). Voltage distribution in the range from power frequency up to a few kilohertz is controlled by a complementary pair of RC grading circuits (R_D and C_D). This frequency range includes the natural frequency which characterises the voltage oscillation encountered when valves turn off at the end of a conduction interval. The component values are chosen to minimise the magnitude of this voltage overshoot.

Insulation failures within the converter can subject the valve to voltage oscillations of much higher frequency, at which these RC circuits cease to be effective. To ensure that no thyristor level experiences an unduly severe voltage during such events, a capacitive grading circuit is also included. This 'fast-grading capacitor' (C_{FG}) is arranged to discharge via part of the saturating reactor in order to limit its contribution to thyristor inrush current.

Normally the thyristor is triggered into conduction at a particular point-on-wave determined by the control system. The command to fire the valve is sent as an optical signal from a 'valve base electronics' (VBE) cubicle at earth potential to every thyristor in the valve, via individual optical fibres. The optical signals are decoded by a 'gate electronics' unit located adjacent to each thyristor, which then generates a pulse of current to trigger the thyristor. The gate electronics derives the (small) power necessary for its operation from the displacement current in the RC grading circuit during the OFF state interval.

Thyristors could be damaged by excessive forward voltage or forward dv/dt; when the forward voltage threatens to exceed the maximum safe value, a back-up triggering system, based on a breakover diode (BOD) is used to pass a heavy pulse of current to the gate of the main thyristor, triggering it rapidly and safely into conduction.

3.5.2 HVdc configurations

The simplest HVdc scheme, shown in Figure 3.14(a), is a monopolar configuration with ground return. It consists of a single conductor connecting one or more 12-pulse converter units in series or parallel at each end and uses either a sea or earth return. An electrode is required at each end of the line. Because of magnetic interference and corrosion problems, ground return is rarely permitted and metallic return is used instead. Both configurations require a dc smoothing reactor at each end of the HVdc line, usually located on the high voltage side and, if the line is overhead, dc filters are normally required.

A bipolar HVdc system (shown in Figure 3.14(b)) consists of two 12-pulse converter units in series with electrode lines and ground electrodes at each end and two conductors, one with positive and the other with negative polarity to ground for power flow in one direction. For bi-directional flow, the two conductors reverse their polarities. With both poles in operation, the imbalance current flow in the ground path can be held to a very low value.

Multiterminal HVdc operation, although a viable strategy, is rarely used. The two basic configurations are shown in Figures 3.14(c) and 3.14(d), with the converters connected in parallel or series on the dc side.

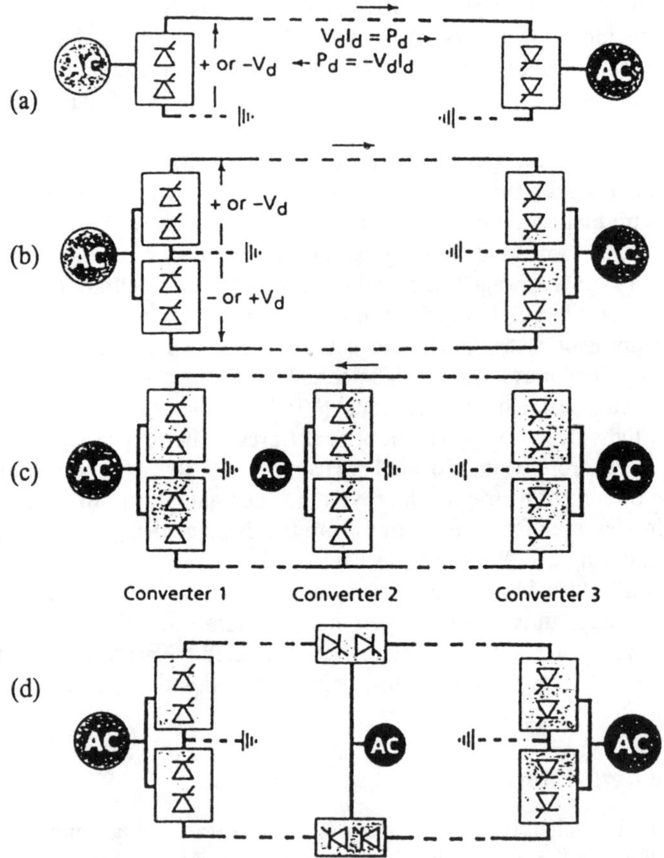

Figure 3.14 Basic HVdc configurations.

3.5.3 Back-to-back configurations

Considering the cost benefit of the back-to-back solution, zero distance interconnections are often preferred when planning HVdc transmission between two asynchronous systems.

North America and India have already made extensive use of the back-to-back solution, as shown in Figures 3.15 and 3.16.

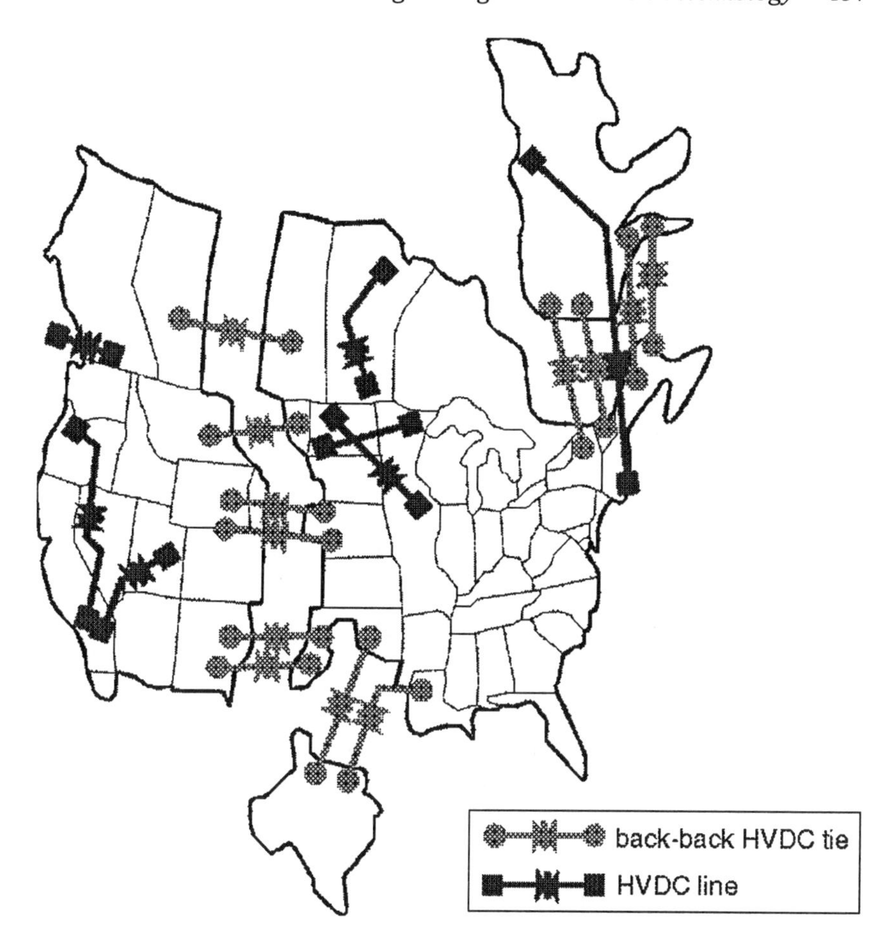

Figure 3.15 Unusual map of North America highlighting four blocks of ac transmission connected together by dc links.

For relatively small ratings (say 50–100 MW) the back-to-back links are of monopolar design and usually include a smoothing reactor. Bipolar configurations with or without smoothing reactors are used for large interconnections (of 500 MW and over).

The valves of both converters of the back-to-back scheme can be located in the same valve hall together with their controls, cooling, and other auxiliary services.

In these systems, the optimisation of the whole station regarding voltage and current ratings for a given power in order to achieve the lowest life cycle cost is a straightforward procedure. Generally, the dc voltage is low and the thyristor valve current high in comparison with HVdc long distance interconnections. The main reason is that, on the one hand, valve costs are much more voltage-dependent as the higher voltage increases the number of thyristors and, on the other, the highest possible current adds very little extra cost to the price of an individual thyristor.

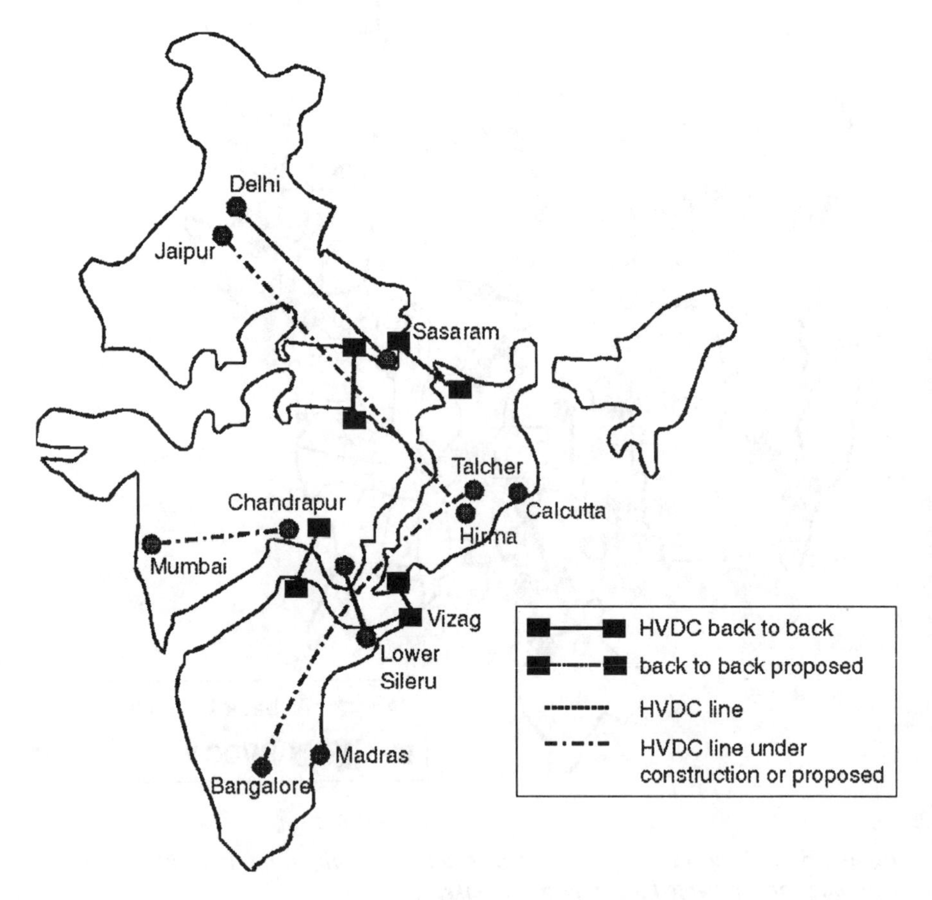

Figure 3.16 Interconnections using HVdc between India's regions.

3.6 Power system analysis involving HVDC converters

In power frequency load flow studies each terminal of the dc link can be represented as a load bus, i.e. by active (P) and reactive (Q) power specifications, subject to capability constraints.

For greater flexibility and improved convergence, some of the dc link variables must be explicitly represented in the power flow solution, replacing the P and Q specifications.

The following set of variables permits simple relationships for all the normal control strategies[13]:

$[x] = [V_d, I_d, a, \cos\alpha, \phi]$

where V_d, I_d are the dc side voltage and current; a is the converter transformer off-nominal tap position; α is the firing angle and ϕ is the fundamental frequency displacement between the ac voltage and current.

This alternative requires either sequential ac and dc iterations or a unified ac/dc Newton-Raphson solution [13].

The so-called characteristic harmonics, i.e. those related to the pulse number, can be easily calculated from the symmetrical steady state converter model. Moreover, these harmonics are normally absorbed by local filters and have practically no effect on the rest of the system.

Three-phase steady state simulation is rarely needed in conventional power systems. However, the dc link behaviour, and specifically the generation of uncharacteristic harmonics, is greatly affected by voltage or components imbalance. Moreover, in the absence of 'perfect' filters, these harmonics can be magnified by parallel resonances between the filters and ac system impedance. Therefore, there is a greater need for three-phase modelling in the presence of ac/dc converters.

The harmonic currents produced by the converters are often specified in advance, or calculated more accurately for a base operating condition derived from a load flow solution of the complete network. These harmonic levels are then kept invariant throughout the solution. That is, the converter is represented as a constant harmonic current injection, and a direct solution is possible. A reliable prediction of the uncharacteristic harmonic content requires more accurate modelling, and often a complex iterative algorithm involving the ac and dc systems[14].

Many methods have been employed to obtain a set of accurate non-linear equations which describe the system steady state. After partitioning the system into linear regions and non-linear devices, the non-linear devices are described by isolated equations, given boundary conditions to the linear system. The system solution is then predominantly a solution for the boundary conditions for each non-linear device. Device modelling has been by means of time domain simulation to the steady state[15], analytic time domain expressions[16][17], waveshape sampling and FFT[18] and, more recently, by harmonic phasor analytic expressions[19]. When the HVdc link is expressed in a form suitable for solution by Newton's method, the separate problems of device modelling and system solution are completely decoupled and the wide variety of improvements to the basic Newton method, developed by the numerical analysis community, can readily be applied.

The use of accurate iterative techniques permits quantifying the response of the HVdc converter to various non-ideal ac or dc system conditions. The main effects are:

- The ac/dc converter acts as a frequency modulator, the relationships between harmonic (or non-harmonic) frequencies on both sides of the converter being as shown in Figure 3.17.

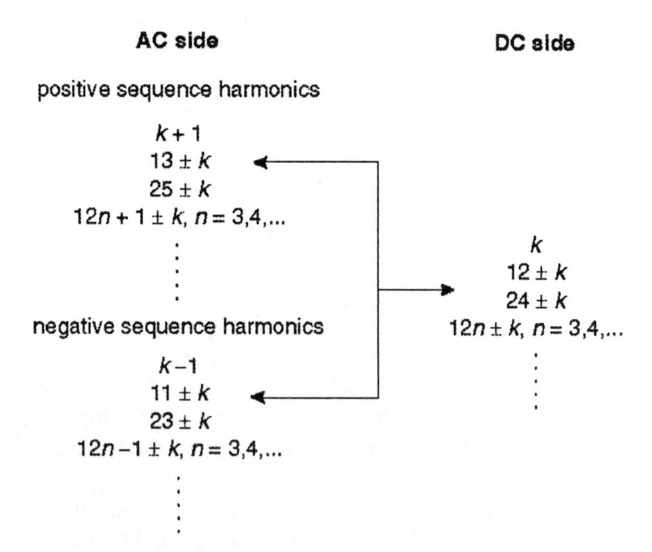

Figure 3.17 AC/DC harmonic interactions.

- The main interactions are between the so-called three port terms. A change in harmonic k on one side of the converter affects harmonics $k + 1$ and $k - 1$ on the other side of the converter.
- The end of commutation is very sensitive to harmonics in the terminal voltage and dc current.
- The individual firing instants are sensitive to harmonics in the dc current.
- The average delay angle, since it relates to the average dc current, is extremely sensitive to changes in the fundamental terminal voltage. There is also sensitivity to harmonics coupled to the fundamental; i.e. the 11th and 13th harmonics on the ac side.

The dc link behaviour during ac, converter, or dc system disturbances cannot be simulated using steady or quasi-steady state algorithms because in general the disturbance will alter the normal operation of the bridge. In the case of the inverter, an ac system fault will generally cause commutation failure. A detailed waveform analysis needs to be carried out using electromagnetic transient simulation. The most popular transient simulation methods are based on the EMTP programs, and these include detailed models of the HVdc converter. In particular, the PSCAD-EMTDC program has been specially designed with HVdc transmission in mind.

The small integration step (typically 50 μs) used by the EMTP programs gives an accurate prediction of the distorted voltage and current waveforms following a disturbance. Realistic electromagnetic transient simulation of HVdc systems always requires that the switching instants are accurately determined as they occur. Invariably, they fall between time steps of the simulation and much work has been directed towards removing any error resulting from this mismatch between switching instant and simulation step. Solutions to this problem first appeared in state variable converter models, where the simulation step itself was varied to coincide with switching instants and to follow any fast transients subsequent to switching. More recently, interpolation techniques have been successfully applied in electromagnetic transients programs. The use of these programs for system simulation, including FACTS and HVdc, is discussed in Chapter 8.

Dynamic and transient stability studies use quasi steady state component models at each step of the electromechanical solution. Similarly to fault simulation, the presence of dc links prevents the use of these steady state models for disturbances close to the converter plant. Unlike fault simulation, however, the stability studies require periodic adjustments of the generator's rotor angle and internal emfs, information that the electromagnetic transient programs cannot provide efficiently. Thus, in general, the ac/dc stability assessment involves the use of the three basic programs discussed in previous sections, i.e. power flows, electromagnetic transient and multi-machine electromechanical analysis. The interfacing of the component programs is considered in Chapters 10 and 11.

3.7 Applications and modern trends

The main historical reason for the introduction of HVdc was the economy of power transmission at long electrical distances which, in the case of transmission by cable, occur at relatively small geographical distances (of 50 km or less). However, it is the absence of synchronous restrictions that is finding increasing applications for dc interconnections. There are currently 20 back-to-back asynchronous interconnections spread all over the world, with the larger number being located in North America.

Attention is now turning towards the prospective application of gate turn-off devices to create a four quadrant voltage-sourced back-to-back asynchronous interconnector. In this respect, GTOs are attractive for dc power conversion into ac systems, which have little or no voltage support. In such cases, the synchronous condensers or static VAR systems used in some present schemes will not be needed. Before GTOs can be regularly used for dc transmission, confidence must be increased in their series connection and the switching losses reduced. If we accept the inevitability of GTO converters for future dc transmission, the system designer is then faced with determining a control strategy which must control dc power, ac voltage and ac system frequency. At present, however, the GTOs ratings are much lower than those of thyristors, and their cost and losses are almost twice those of thyristors; these three aspects (GTOs ratings, cost and losses) have a large influence on the cost of the other equipment in a complete converter station.

In Japan, several power companies and the Central Research Institute of the Electric Power Industry, are developing an 800 MW back-to-back self-commutated asynchronous interconnector[20]; the configuration used is shown in Figure 3.18.

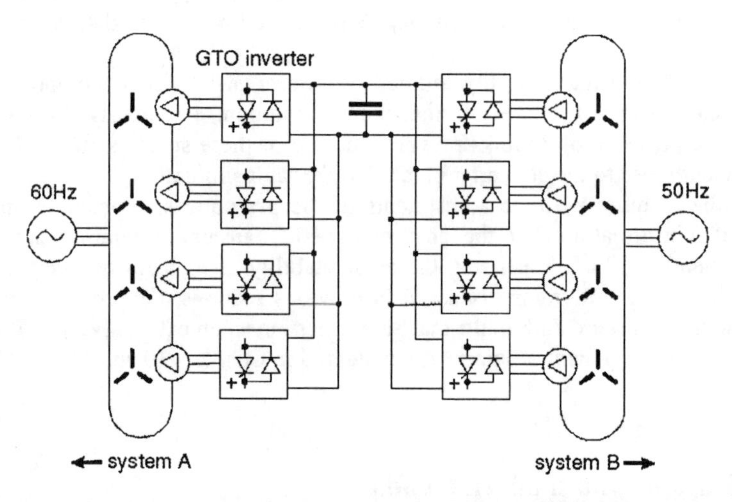

Figure 3.18 System configuration of a 800 MW back-to-back interconnector.

The voltage source converter technology can also be used to extend the economical power range of HVdc transmission down to a few megawatts and even feed totally passive industrial or distribution systems with no other sources of supply. The dc feeder then controls the frequency and voltage levels. A pilot 3 MW, ±10 kV, 10 km long scheme has recently been completed by ABB (from Hellsjön to Grängesberg) in Sweden. This project uses series-connected Insulated Gate Bipolar Transistor (IGBT) converters without transformers at either end. Possible applications of the HVdc light concept include the supply of power to islands, infeed to cities, remote small-scale generation, and off-shore generation.

In the face of competition from the new semiconductor devices, the thyristor-based converter technology is not remaining standstill. Under the banner 'HVdc 2000'[21], ABB has recently proposed a new generation of HVdc converter stations, incorporating the latest developments; these affect a number of technical areas and are aimed at improved performance, design simplicity, and reduced construction time.

The key features of HVdc 2000 include:

series capacitor commutated converter (CCC),

actively tuned ac filters;

air-insulated outdoor thyristor valves;

active dc filters.

Through dc transmission, with or without distance, new generating plant can be added without increasing the system fault level and without the risk of synchronous instabilities.

A new concept termed 'unit connection'[22] has been investigated by a Working Group of CIGRE SC11 and SC14 study committees. This configuration, illustrated in Figure 3.19, eliminates the generator transformer and the ac busbar. The generators are directly connected to the converter transformers and the harmonic currents produced by the 12-pulse unit-connected scheme are absorbed by the generator so that the need for ac filters is eliminated. Moreover, voltage control can be exercised entirely by the generator excitation, and transformers on-load tap changers are no longer needed.

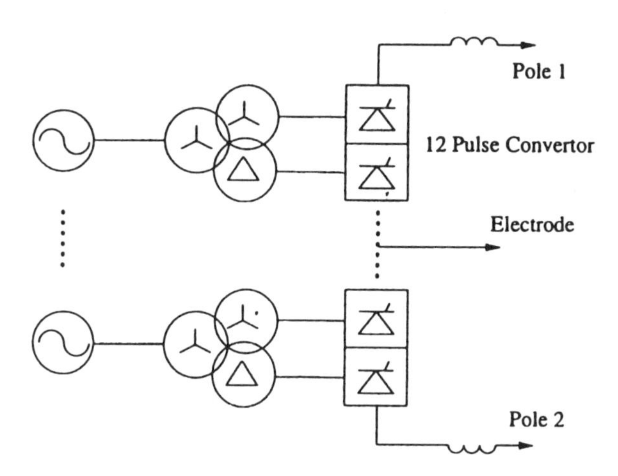

Figure 3.19 Unit connected HVdc power station, in which generators and converters are integrated into single units. Series parallel and/or parallel combination of units at the dc side is also allowed.

The direct connected scheme is considered an attractive proposition for electrical generation from remote sources of power, such as hydro and low grade coal fields, when a new development supplies little or no local load. Its potential for variable speed operation can be used to optimise the efficiency of hydro sets under different load conditions and under varying water heads. This property can also be useful in pump storage and wind power applications.

3.8 Summary

After almost half a century of reliable operation, HVdc transmission is now well established in long distance and asynchronous power systems interconnection, even though some countries still look at it with suspicion.

Considered as a FACTS device, the HVdc interconnector involves all the ingredients of present-day high technology and provides a higher order of controllability.

This chapter has highlighted the basic components of the HVdc transmission technology as well as the main simulation techniques presently used for their integration in power system analysis. Some of the new concepts under consideration for future HVdc links have also been introduced.

3.9 References

1 "The history of high voltage direct current power transmission", Direct Current: Part I, December 1961, p.260; Part II, March 1962, p.60; Part III, September 1962, p.228; Part IV, January 1963, p.2; Part V, April 1963, p.89.
2 Adamson C. and Hingorani N.G., *High voltage direct current power transmission*, Garraway Ltd, London, 1960.
3 Kimbark E.W., *Direct current transmission*, Wiley Interscience, New York, USA, 1971.
4 Uhlman E., "Power transmission by direct current", Springer-Verlag, Berlin/Heidelberg, Germany, 1975
5 Arrillaga J., "High voltage direct current transmission", *IEE Power Engineering Series,* 1983.
6 Padiyar K.R., "HVDC power transmission systems – technology and system interactions", New Delhi-Eastern Ltd, 1990
7 Lips P., "Semiconductor power devices for use in HVDC and FACTS controllers", *CIGRE WG 14.17 Report,* 1996.
8 Dwivedi P.K., Jha J.S., and Gupta S., "Planning of interconnections for disparate regional grids – A challenge", *AC and DC Power Transmission, IEE Publ. no.423*, pp.19-25, 1966.
9 Ainsworth J.D., "The phase-locked oscillator – a new control system for controlled static convertors", *Trans. IEEE,* **PAS-87**, no.3, pp.859-65, 1968.
10 Jotten R., Bowles J.P., Liss G., Martin C.J.B., and Rumpf E., "Control of HVDC systems – The state of the art", *CIGRE Paper 14-10,* Paris, 1978.

11 Flisberg G., and Funke B., "The CU HVDC transmission", *ASEA Journal,* **54**, no.3, pp.59-67, 1981.

12 Brough C.A., Davidson C.C., and Wheeler J.D., "Power electronics in HVDC power transmission", *IEE Power Engineering Journal*, **8**, no.5, pp.233-40, October 1994.

13 Arrillaga J., Arnold C.P., and Harker B.J., *Computer modelling of electrical power systems*, John Wiley & Sons Ltd, UK, 1983.

14 Arrillaga J., Smith B.C., Watson N.R., and Wood A.R., *Power system harmonic analysis*, John Wiley & Sons Ltd, UK, 1997.

15 Arrillaga J., Watson N.R., Eggleston J.F., and Callaghan C.D., "Comparison of steady state and dynamic models for the calculation of ac/dc system harmonics", *Proc. IEE,* **134C**(1), pp.31-37, 1987.

16 Yacamini R., Oliveira J.C., "Harmonics in multiple converter systems: a generalised approach", *IEE Proc. B,* **127**(2), pp.96-106, 1980.

17 Carpinelli G., *et al.*, "Generalised converter models for iterative harmonic analysis in power systems", *Proc. IEE Generation, Transn. Distrib.,* **141**(5), pp.445-451, 1994.

18 Callaghan C.D., Arrillaga J., "A double iterative algorithm for the analysis of power and harmonic flows at ac-dc converter terminals", *Proc. IEE,* **136**(6), pp.319-324, 1989.

19 Smith B. *et al.*, "A Newton solution for the harmonic phasor analysis of ac-dc converters", *IEEE PES Summer Meeting "95,* **SM 379-8**, 1995.

20 Horiuchi S. *et al.*, "Control system for high performance self-commutated power converter", *Paper 14-304, CIGRE,* 1996.

21 Carlsson L. *et al.*, "New concepts in HVdc converter station design", *Paper 14-102, CIGRE,* 1996.

22 Campos Barros J.G. *et al.*, "Guide for preliminary design and specification of hydro stations with HVdc unit connected generators', *CIGRE WG 11/14-09,* 1997.

Chapter 4

Shunt compensation: SVC and STATCOM

H. L. Thanawala, D. J. Young and M. H. Baker

4.1 Introduction: principles and prior experience of shunt static var compensation

The principle of shunt reactive power (var) compensation is illustrated in Figure 4.1. The static var compensator (SVC) generates or absorbs shunt reactive power at its point of connection, shown here as in the middle of a high voltage (HV) ac transmission line. Before the advent of static var compensators in the 1960s, such a compensation function was performed by rotating synchronous generators operating without prime movers (turbines) as synchronous compensators and capable of generating or absorbing reactive power only, with their active power losses being supplied from the ac network itself.

If the equipment is arranged to perform in such a way that the voltage V_M is controlled at all levels of power flow the line is effectively split into two half sections (of reactive impedance $X/2$), the synchronous stability of each of which is then governed by the equation given on the diagram. Figure 4.1(b) also compares the power–angle (P–δ) curves with and without the SVC [1–3].

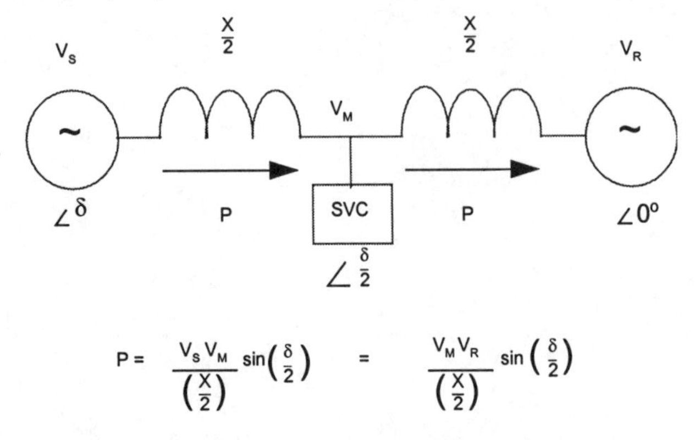

Figure 4.1(a) Shunt static var compensation

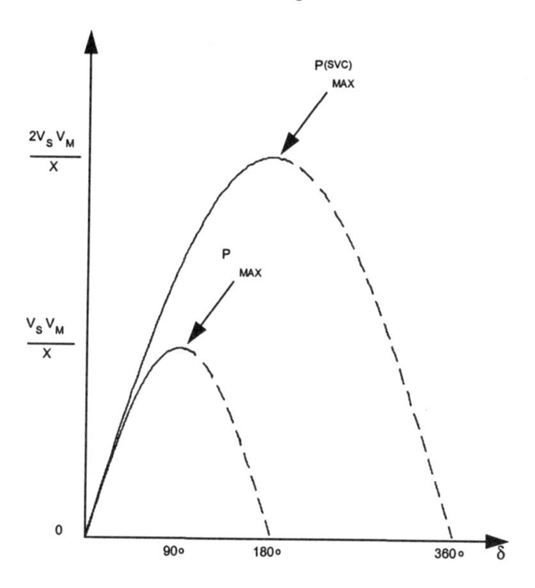

Figure 4.1(b) Effect of shunt compensation on power–angle diagram

The P–δ curves show that the theoretical maximum stable power limit P_{MAX}(svc) is doubled with the SVC present. The line total angle is higher at this power level, but because the voltage magnitude is held constant dynamically, the synchronous stability is maintained ($dP/d\delta > 0$) even at angles greater than 90°. The line stability limit as well as voltage stability and voltage collapse limit are thus improved. The reactive power rating of the SVC will, in practice, not be infinite and the practical size will depend on cost versus the requirement for increase in stable power transfer capability, taking due account of the actual dynamic performance of the SVC and its controls.

Before the wider availability of static var compensators (SVC) the control of voltage was only possible either at generator terminals, i.e. V_S and V_R (Figure 4.1). The alternatives of mechanical switching of shunt reactors (MSR) and capacitors (MSC) are not sufficiently smooth or rapid to influence the power–angle relationships dynamically, although they can limit any continuous overvoltages and undervoltages.

The earliest form of continuously variable static shunt reactive compensator was based on a saturated iron-cored reactor, pioneered in the UK during the early 1950s and 1960s [3–5]. The simplest form of saturated reactor (Figure 4.2a) is an iron-cored inductor with one winding per phase. The inherent rms voltage-current (V–I) characteristics shown in Figure 4.2(b) follows the flux-current (B–H) relationship of a saturating iron-core and thus the reactor effectively has two values of reactance. The basic component of voltage is of fixed amplitude corresponding to the 'saturation voltage' and the upper sloping part determines the change in current for a further increase in voltage and corresponds to the air-core saturated inductance of the winding, i.e. the 'slope reactance'.

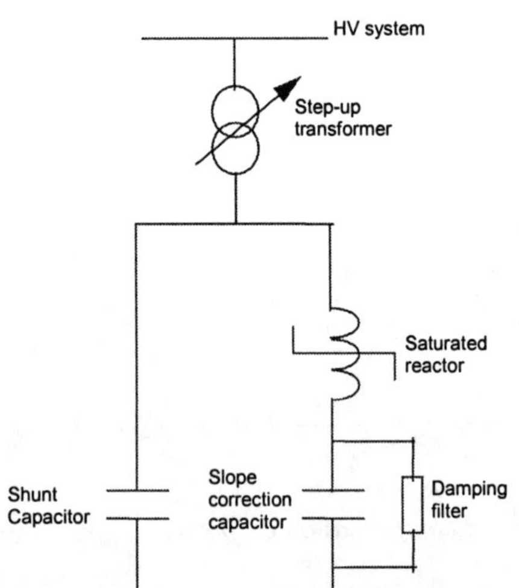

Figure 4.2(a) The saturated reactor compensator (SR-SVC)

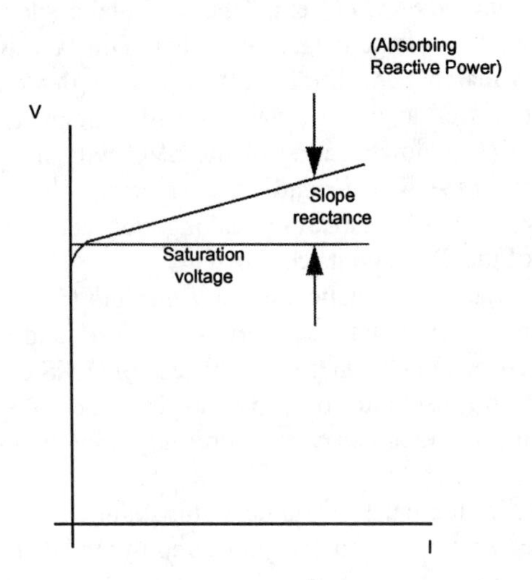

Figure 4.2(b) SR-SVC voltage-current characteristic

The basic single-phase saturated reactor (SR) is inherently able to respond rapidly to maintain an approximately constant voltage (to within 10% say) at its terminals without any feedback control system, but its current has substantial harmonic components which are usually undesirable. The use of a three-phase interconnected winding scheme reduces the harmonic content substantially to that illustrated in Figure 4.3(b). The use of a slope correction capacitor (Figure 4.2(a)) in series with the three-phase reactor enables better constancy of voltage (e.g. 2 to 3%).

Figure 4.3(a) Core and windings of a multi-winding saturated reactor

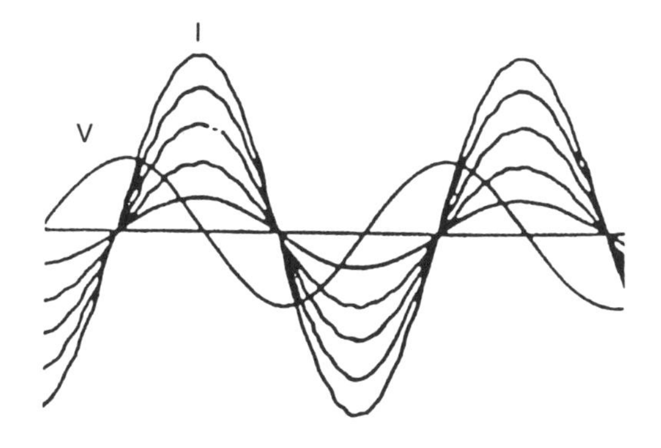

Figure 4.3(b) Waveforms of voltage and currents at 0, 0.25, 0.5, 0.75, 1.0, and 1.25 pu current levels

Such a multi-winding reactor (Figure 4.3(a)) has been widely used in the UK and abroad [5,6] in utility transmission and distribution networks (and in industrial networks for fluctuating load compensation) until the economic availability of semi-conductor devices in the 1970s. The saturated reactor can be economically designed to operate at voltages up to 66 kV. It is usual to employ a step-up transformer for connection to higher voltages (Figure 4.2(a)) which has an added advantage that by use of its on-load tap changer, the constant 'saturation reference voltage' level can be adjusted, if desired, for transmission system operation. The use of a shunt capacitor or mechanically switched shunt capacitor (MSC) (incorporating a residual harmonic filter if required) enables the complete Static Var System (SVS) to cover capacitive as well as inductive operation.

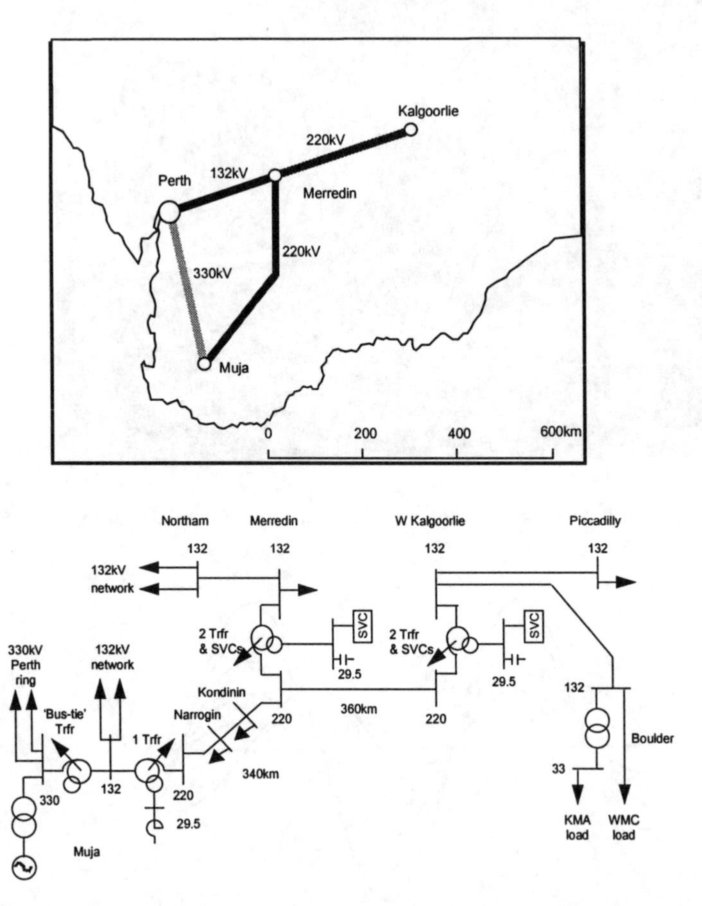

Figure 4.4 Eastern goldfields transmission system in Australia employing four SVCs

Figure 4.4 illustrates an application in the early 1980s of four SVCs on a 654 km long 220 kV multiple-compensated line, each SVC having a rating of about +45/-30 Mvar at its 29.5 kV tertiary connection. The SVCs enabled power transfer of more than the surge impedance power level, which is about 125 MW, to be achieved over such a long single-circuit line into an economically important gold-mining area in Western Australia [6]. Even larger (±150 Mvar, 56 kV) saturated reactor SVCs designed in the late 1970s are in service for the control of voltage, particularly load-rejection overvoltages, two on the 400 kV system at the converter station at the UK end of the Cross-Channel England–France 2000 MW HVDC Scheme [5] and one at a nearby substation.

The development of thyristor technology has, since the 1980s, superseded the use of saturated reactors. Thyristor valves are cheaper, less lossy and give direct control over the instant of the start of current conduction in each half-cycle [3,7,8,9]. The essential functional features illustrated in Figure 4.2(b), such as reference voltage, V–I slope, shunt capacitors (and filters) and step-up transformers are, however, similar to the saturated reactor. With the advent of gate turn-off (GTO) thyristors (and similar semi-conductors) additional features become available in the STATCOM described in Section 4.3.

4.2 Principles of operation, configuration and control of SVC

4.2.1 Thyristor Controlled Reactor (TCR)

The thyristor controlled reactor comprises a linear reactor, connected in series with a thyristor valve made up of inverse-parallel, or 'back-to-back' connected pairs of high power, high voltage thyristors which are themselves connected in series to obtain the necessary total voltage and current rating for the valve [7–9].

The simplest design of TCR uses three, single-phase, valves connected in delta giving a six-pulse unit (Figure 4.5). The variation of current is obtained by control of the gate firing instant of the thyristors and thus of the conduction duration in each half cycle, from a 90° firing angle delay as measured from the applied voltage zero for continuous conduction, to 180° delay for minimum conduction (Figure 4.6(a)). An illustration of the voltage control scheme is shown in Figure 4.6(b) and it employs the reset-integrator type of phase-locked feedback control loop [10].

The design of the high voltage thyristor valves has generally similar features to those for a HVDC valve. The six-pulse scheme (Figure 4.5) produces no significant third harmonics but substantial 5th and 7th harmonic currents are present (up to around 5% of the rated TCR current) and in most cases these require the use of harmonic filters [9] to minimise distortion in the power transmission and distribution voltage. For large ratings it can be economic to connect the TCR directly to the power system busbar at a voltage up to about 33 kV. For most applications, however, a step-up transformer is required to match the optimised valve design voltage to the higher voltage of the power system. The transformer typically has one delta winding and one earthed star one (usually the higher voltage).

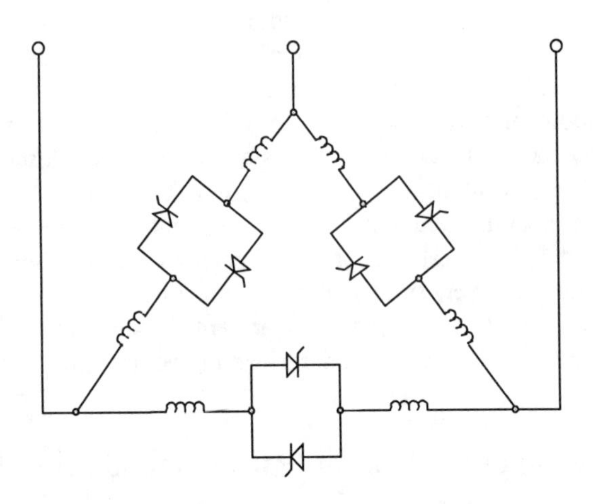

Figure 4.5 Thyristor-controlled reactor six-pulse configuration

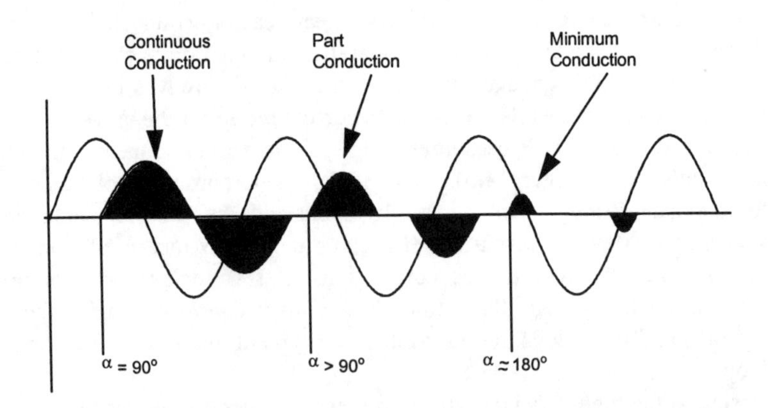

Figure 4.6(a) Waveforms illustrating the effect of TCR firing angle control

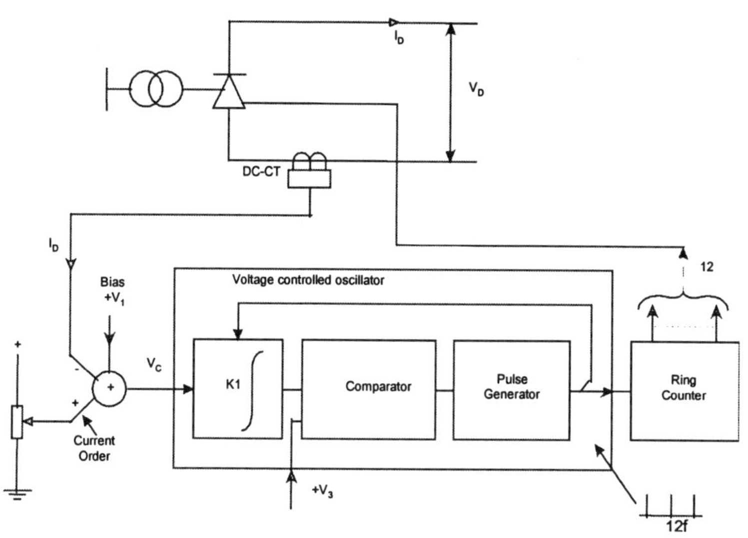

Figure 4.6(b) SVC control principles

For large installations it may be economic to use a transformer with two phase-displaced secondary windings connected in star and delta, each with delta-connected thyristor valves and reactors. This 12-pulse design gives better harmonic compensation, the principal harmonics being 11th and 13th which are of relatively low current magnitudes (around 1%) [11] and harmonic filters may be avoided.

The control system of a TCR can produce any desired voltage/current characteristic within an overall envelope AEFBJHA, derived from various input signals (Figure 4.7). When the valve reaches maximum conduction the characteristic follows the slope (FD) set by the total impedance value of the linear reactors plus the coupling transformer (typically the total being 0.7 to 1.0 pu based on the full load TCR rating). This impedance line applies up to the point of the thyristor current thermal limit(D). Points on the envelope may be chosen to suit the system requirements (including overload capability for a specified short duration). For digital computer modelling of a controlled TCR in system planning and design studies, simplified models of the characteristic are usually adequate as discussed in [12–16], although for SVC equipment design more detailed computer models, or physical and digital real-time simulator methods, are often employed.

For transmission and utility applications closed-loop voltage control is usually suitable. With this control, for a sudden change of system voltage the corresponding var change along the controlled slope line can have a speed of response of one to three cycles, depending on the system short-circuit level. This control scheme helps to maintain voltage and synchronous stability or to avoid voltage collapse after a major disturbance. By suitable auxiliary control features it may also enhance damping of subsequent system swings and oscillations.

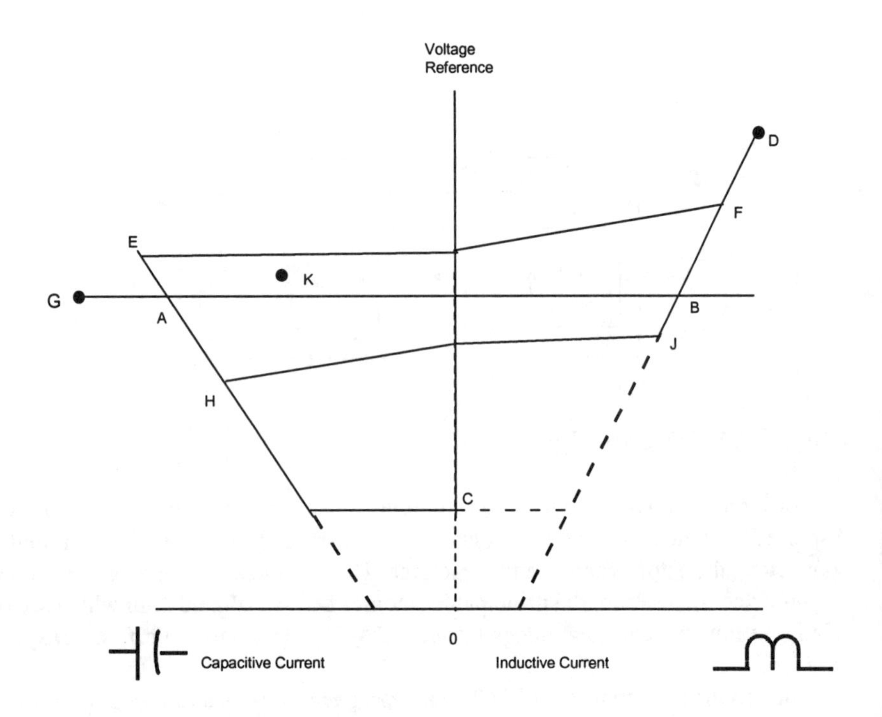

Figure 4.7 TCR-SVC voltage–current characteristics

A faster response can be obtained by using an open-loop current control; this is used for certain applications such as for industrial fluctuating loads where the highest speed is sometimes required and some sacrifice of voltage control accuracy is acceptable. The control may be on a separate per phase basis to permit network or load imbalance to be compensated, or it may be on an equal three-phase basis. The control equipment uses analogue electronics or digital software.

A thyristor switched reactor (TSR) has similar equipment to a TCR, but is used only at fixed angles of 90° and 180°, i.e. full conduction or no conduction. It therefore generates no harmonics but, being only on/off controllable, it has a narrower application.

4.2.2 Thyristor Switched Capacitor (TSC)

This is the only type of SVC (apart from the STATCOM) which can give directly the effect of a variable capacitance although, for reasons given below, the variation is not smooth but in steps. Thyristor valves consisting of inverse-parallel connected thyristors, generally similar to those used for the TCR, are used to give rapid switching of three-phase delta-connected blocks of capacitors (Figure 4.8(a)). Combined with a TCR, the range from 150 Mvar capacitive to 75 Mvar inductive, for example, is given by the arrangement of Figure 4.8(b) [14].

The TSC has one important difference from the TCR. After the capacitor current through the thyristor ceases at current zero, unless re-gating occurs, the capacitor remains charged at peak voltage while the supply voltage peaks in the opposite polarity after a half cycle [9]. This can impose a doubled voltage stress on the non-conducting thyristors and an increase can be necessary in the number of thyristors in series in a TSC compared with a TCR valve of equivalent voltage. The very high inrush currents that can occur on energising the capacitors make it necessary to choose the point-on-wave for gating the thyristors with accuracy. Ideally gating should take place when the voltage across the thyristor is zero and the supply voltage is at its peak (i.e. when $dV/dt = 0$) to eliminate transients. Such ideal conditions, however, cannot always be achieved in practice due to mis-matches between the voltage of the capacitor (partially discharged from the previous cycle) and the system voltage (which may have changed over the same period). One of the functions of the control system therefore is to choose the precise gating instant of each valve for minimum transients as illustrated in Figure 4.9. The rate of discharge has been substantially increased in the latest TSC installations in the UK [17] by means of a saturable discharge transformer connected across the capacitor, e.g. about 10ms compared with 100s in earlier designs, thus permitting a significant reduction in the switching delay and in the voltage stress on the thyristor valve and hence in its size.

There will be occasions when severe transients occur, for instance during initial energisation of an uncharged capacitor. Series reactors are included to reduce the magnitude of the inrush current under the worst case to a value within the thyristor capability. With the exceptions of these transients, however, the TSC does not produce harmonics. If there are other sources of harmonic currents (e.g. in the TCR or in the system itself), care in circuit design is necessary to ensure that they are not unduly magnified by resonance effects in the TSC circuit.

Once gated, the thyristor conducts for a half-cycle and, unless again gated, the current then ceases. The TSC therefore can only give a variation of capacitance in discrete steps for a discrete period of an integral number of half-cycles. The number of steps and hence the fineness of control is dictated by economics. The effective number of steps can be increased by using different sizes, e.g. 1, 2, and 4 in binary switching sequence (see Section 4.4.2).

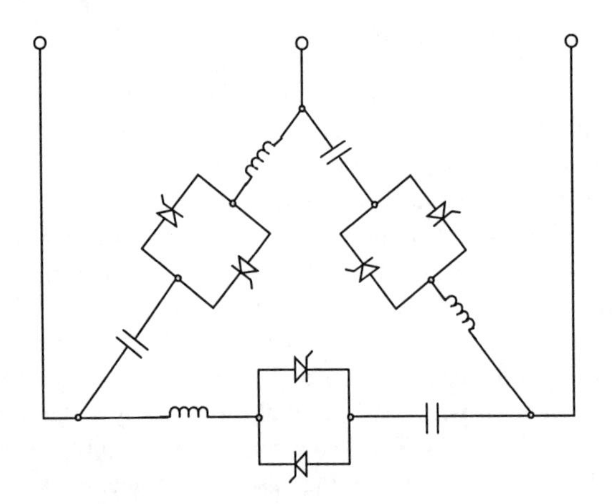

Figure 4.8(a) Thyristor Switched Capacitor Circuit

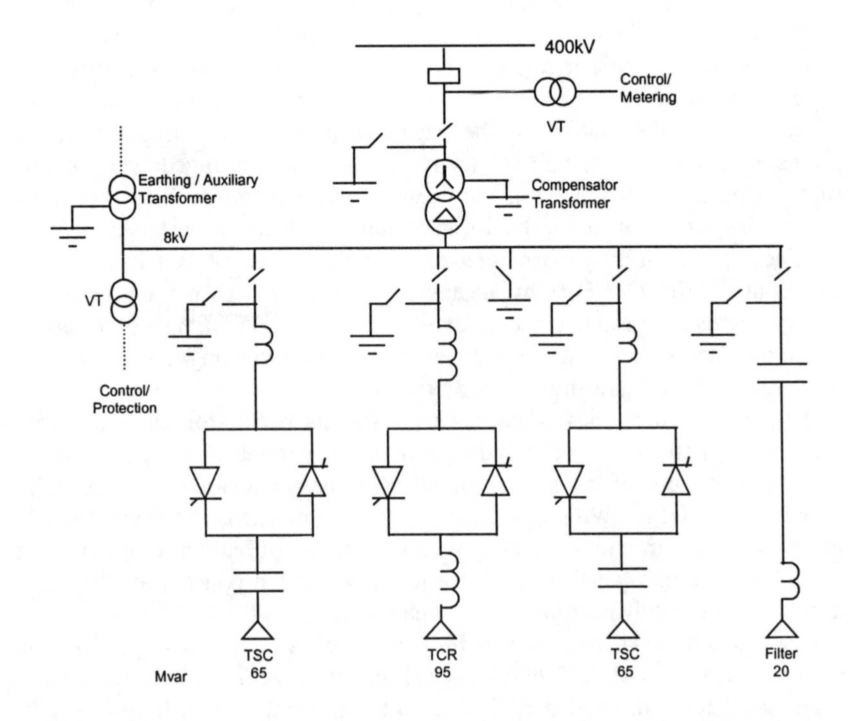

Figure 4.8(b) Configuration of a Combined TCR/TSC to give +150/-75 Mvar Continuous Control

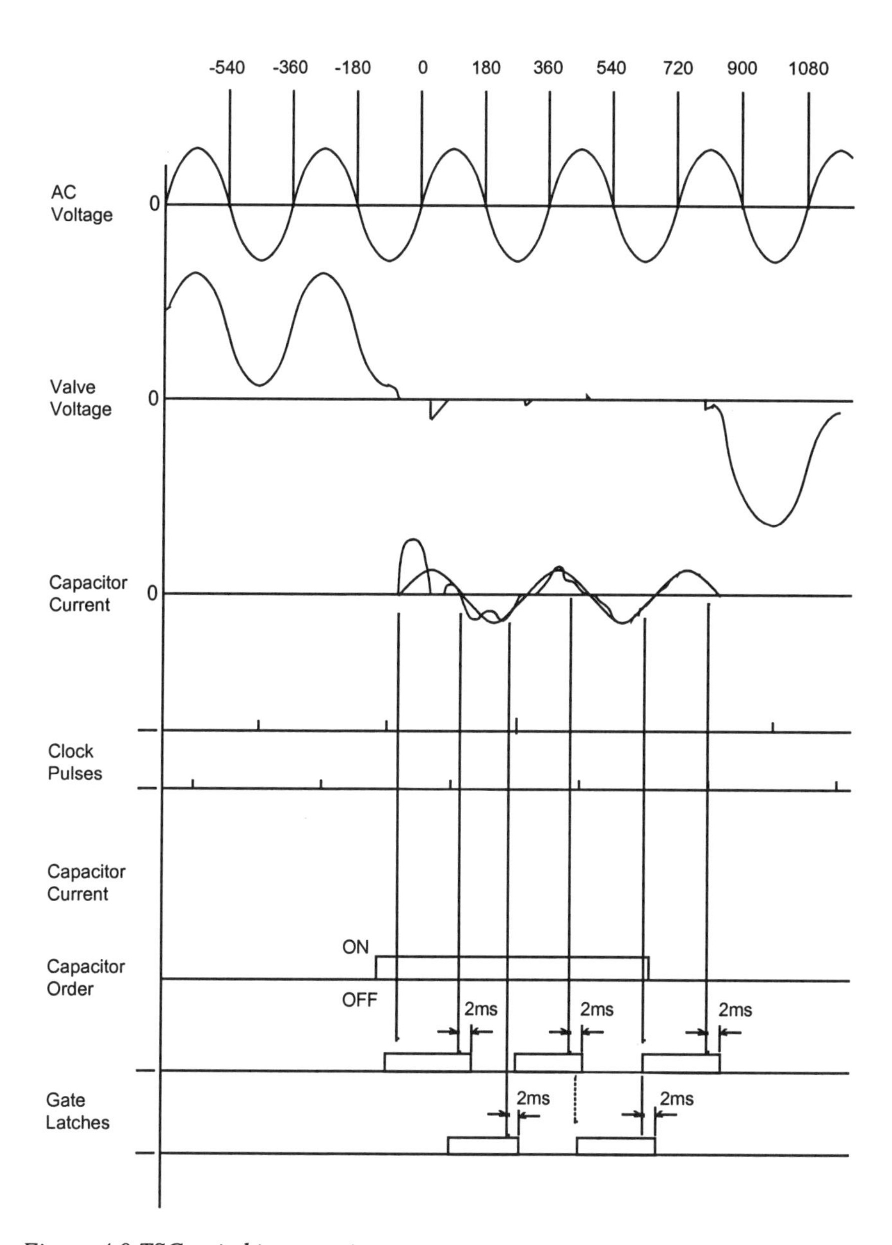

Figure 4.9 TSC switching transients

Thyristors can be used to switch on a large bank of capacitors faster than is possible using a mechanical switch and therefore the TSC can be used for fast boosting of the transmission voltage to maintain stability following a disturbance, such as sudden line or generator or load outage, or major rapid swing in system power flow. Frequent switching by such electronic means causes no mechanical wear.

4.2.3 Combined TCR/TSC

TCRs are often used in conjunction with TSCs as shown in Figure 4.8(b) in order to give reduced power losses at zero var output (the float condition) compared with the losses of a scheme using a fixed capacitor and a larger TCR (FC-TCR). The former can also provide an increased operating range in the var generation region where speed of capacitor energisation is of importance, or where frequent changes in output are required. In these applications the TSC provides the coarse steps, seldom exceeding four, whereas the TCR gives a fine continuous control in between.

In some system applications, although the SVC must be capable of operating under particular system conditions at the extremes of its var range, it need not be capable of swinging in a continuous manner from one extreme to the other. Its 'dynamic' or 'swing' range, which is the range over which it can give an instantaneous or near instantaneous response, may be only a proportion of its total range. In such a design the size of the rapidly variable element, which is more expensive per unit of rating, can be reduced to deal only with the rapid swing range whilst mechanically (breaker) switched shunt reactors and capacitors (MSR and MSC) can be controlled more slowly but still automatically to give the total var range. The overall cost of such static var systems is lower than a TCR/TSC [7].

Figure 4.10 shows the losses versus the var output for the TSC-TCR arrangement of Figure 4.8(b), but excluding the losses in the transformer, compared with a single TSC and fixed capacitor. The TCR/FC scheme requires the least capital equipment, particularly thyristor valves, and incurs lowest loss at maximum capacitive var generation (leading Mvar) but highest loss at maximum inductive var absorption (lagging Mvar) when the TCR carries a large current to cancel out the fixed capacitor output. When TCRs are included, the TSC steps can be arranged to switch off at appropriate output levels (Figure 4.10), so that in the lagging region only a smaller FC and TCR carry current and the losses are correspondingly much lower. Depending on the operational duty cycle of the SVC and the evaluated capitalisation cost rate for losses, the overall cost (i.e. life-time cost) can be substantially lower for the TCR/TSC scheme.

It is possible with an SVC incorporating a TCR to modulate the reference voltage signal in order to improve power system damping. The modulating signal can be the line power flow, the voltage phase angle(s) or their rates of change. A TSC having several steps can also be used on its own to provide some swing damping in a transmission system. SVCs can also be used to provide a degree of phase balancing if the controls are arranged to be on a per-phase basis, but for a TCR scheme this can introduce third harmonic currents which must be allowed for in the harmonic filter design.

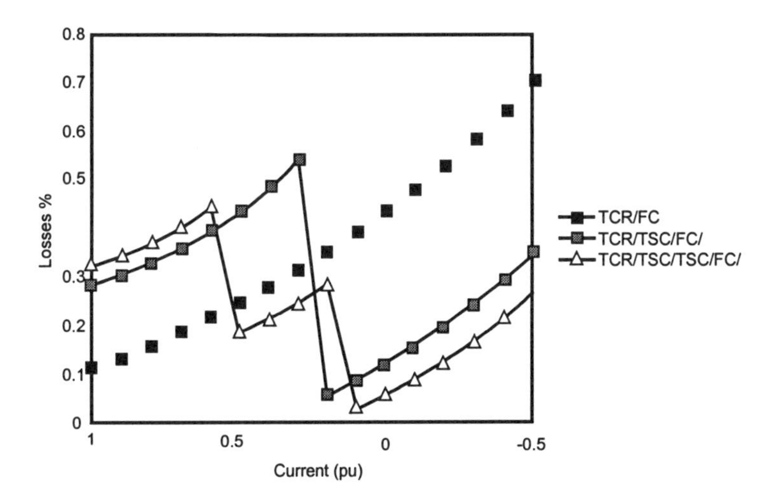

Figure 4.10 Typical SVC power loss curves for various arrangements

4.3 STATCOM configuration and control

4.3.1 Basic concepts

The description of a STATCOM may helpfully start with the rotating machine, the synchronous compensator, which was referred to in Section 4.1.

The dynamic behaviour of a synchronous compensator depends on the voltage which is developed in its ac winding by the dc excitation of the field winding. The reactive current flowing into or out of a synchronous compensator depends on the difference between the voltage of the supply system and the excitation voltage of the machine. This is illustrated in Figure 4.11. The synchronous compensator, M, connected to the system busbar B, with a voltage V_b, has an excitation voltage V_e and an equivalent inductive reactance X_e. When V_e is smaller than V_b, the machine is "under-excited" and the current flowing into it lags behind the system voltage; the machine then acts as a inductive impedance, absorbing Mvar from the system as shown in Figure 4.11(b). When V_e exceeds V_b, the machine is "over-excited" and acts as a shunt capacitor, generating Mvar (Figure 4.11(c)).

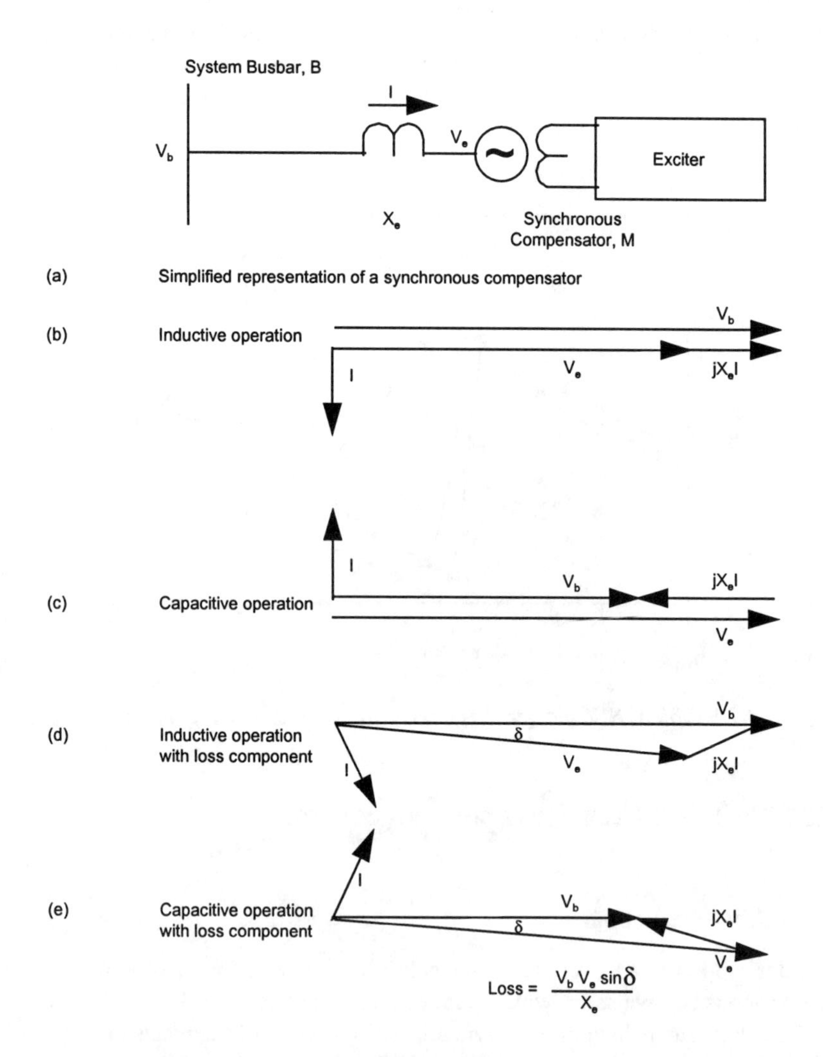

Figure 4.11 Simplified representation of a synchronous compensator and its vectorial relationship between voltage and current

A synchronous compensator is not driven by a prime mover; therefore all its losses must be supplied from the system. Consequently the vector of V_e lags very slightly behind V_b, by an angle δ, so that the necessary energy can be supplied to the machine from the system, as indicated in Figures 4.11(d) and (e). The form of the equation describing the losses will be seen to be the same as that describing power flow between voltage sources in Figure 4.1(a).

As described earlier, the behaviour of conventional SVCs is often represented by a model of an equivalent, but inertialess, machine incorporating an equivalent excitation voltage (reference voltage) and an equivalent reactance (V–I slope). However, there is a class of SVCs which is based on dc to ac converters in which a real alternating voltage (or current) can be produced from a direct voltage (or current) by the process of inversion in a solid-state DC to AC converter; the converter can be controlled to behave as if it were an idealised rotating machine (Figure 4.12). The class of SVCs which uses this principle has been assigned the name STATCOM (a shortened form of STATic COMpensator).

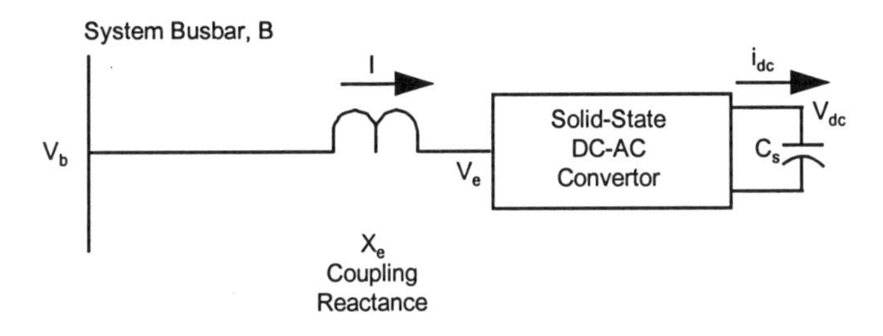

Figure 4.12 Simplified representation of a STATCOM

The basic behaviour of a STATCOM is very similar to that of a synchronous compensator. If the voltage generated by the STATCOM is less than the voltage of the system busbar to which it is connected, the STATCOM will act as an inductive load, drawing Mvar from the supply system. Conversely, a STATCOM will act as a shunt capacitor, generating Mvar into the supply system, when its generated voltage is higher than the system voltage. The losses of a STATCOM are normally supplied from the system, in the same way as for a machine, and not from the source of direct voltage or current.

4.3.2 Voltage-sourced converters

Various types of inverter circuit and source have been suggested and examined. The dc voltage-sourced converter (VSC) is the type which has received most attention in the practical realisation of the STATCOM principle. A very simple inverter (Figure 4.13) produces a square voltage waveform as it switches the direct voltage source on and off. Clearly this waveform has a large content of low order harmonic components.

The direct voltage source can be a battery, whose output voltage is effectively constant, or it may be a capacitor, whose terminal voltage can be raised or lowered by controlling the inverter in such a way that its stored energy is either increased or decreased.

The inverter must use either conventional thyristors with forced commutation, or it must use devices which can be turned off as well as turned on, such as gate turn-off (GTO) thyristors, which have been used for many years in drives for traction and industrial applications. A new generation of devices which require less energy for the switching process includes Integrated Gate Bipolar Transistor (IGBT), Integrated Gate Commutated Thyristors (IGCT) and MOS-Controlled Thyristors (MCT). These devices are becoming available with ratings that can be used for STATCOMs; devices with a higher voltage rating, using silicon carbide (SiC), are expected to become commercially available within the next few years.

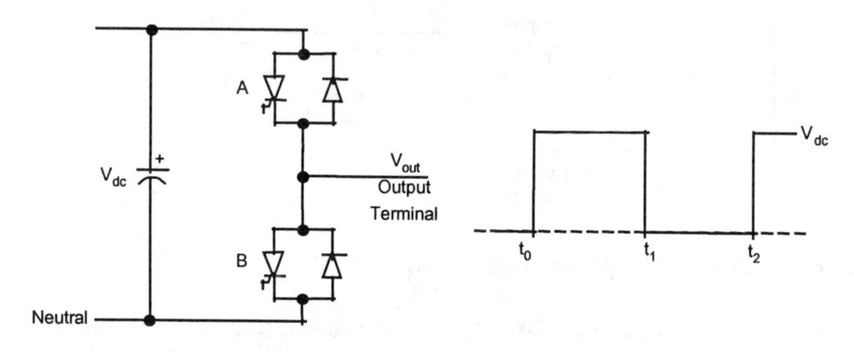

Figure 4.13 Two-level voltage-sourced converter

The pattern of switching of the semi-conductor devices in poles A and B of Figure 4.13 depends on whether the current leads or lags the voltage. Thus, if during the interval t_0 to t_1, V_{out} is in phase with and greater than the system voltage, the converter will act as a capacitor (as seen from the system) and the current flowing INTO the converter at the junction of poles A and B (i.e. into the output terminal) will lead the voltage by 90°. Just before t_0, the GTO of pole B is conducting the line current, the GTO of Pole A is blocked and the output voltage, V_{out}, is zero. There is no current through pole A or through the diode of pole B. At time t_0, GTO B is forced to turn off, by control action, and the current through it (now at its peak value) ceases to flow. However, the line current itself is not interrupted because it now finds a low impedance path through the diode of pole A and through the dc voltage source back into the system. The voltage at the output terminal, V_{out}, becomes equal to the dc source voltage. Because V_{out} exceeds the system voltage, the line current will steadily be forced to zero. The flow of current will cease unless GTO A is now turned on, because there will be no path for a reverse current to flow through the blocked converter; at zero current the voltage of the output terminal must follow the value of the system voltage.

Therefore in order to ensure continuation of the current, GTO A is deblocked shortly after t_0 so that the reverse current can flow out, from the voltage source, into the ac system; V_{out} remains equal to the dc source voltage until time t_1, when V_{out} must be collapsed to zero. This is achieved by blocking GTO A so that current continues to flow out of the converter, at zero output voltage, through diode B. The cycle continues by firing GTO B at, or before, the instant of zero current, so that the positive half cycle of current can flow again.

In contrast, when V_{out} is in phase with but less than the ac system voltage, the converter will act as an inductor (as seen from the system) and the current INTO the converter at the output terminal will lag the voltage by 90°. In this case, just before time t_0, current will be conducted through the diode of pole B and GTO A and GTO B will both be in a blocked state. At time t_0, GTO A will be deblocked, the output voltage will rise to the dc source voltage and current flow will transfer (at its peak value) from diode B to GTO A. Because the system voltage exceeds V_{out}, current flow will be forced to zero and will then reverse, flowing through diode A. GTO A can now easily be blocked because it is not carrying current. The sequence will continue by turning on GTO B at peak current to take over the current and collapse the voltage at time t_1.

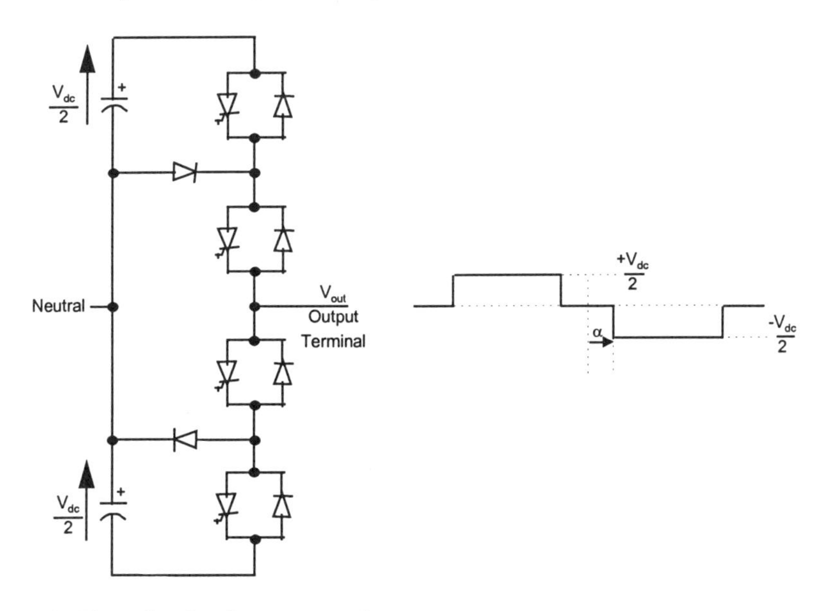

Figure 4.14 Three-level voltage-sourced converter

Thus, when acting in capacitive mode, the GTO devices must be controlled to block the alternating current at its peak value, but are turned on without an immediate flow of current. The opposite happens in inductive mode; the devices turn off by natural commutation and are then blocked. When they are de-blocked, the maximum current flows immediately. Clearly GTOs must never be turned on simultaneously so that they do not short-circuit the dc source. The control of the GTO switching must therefore be accurate and extremely reliable.

With the neutral terminal as shown in Figure 4.13, the output voltage of the two-level converter is asymmetrical. One way in which the basic circuit can be developed into a 3-level converter (of twice the rating) is shown in Figure 4.14.

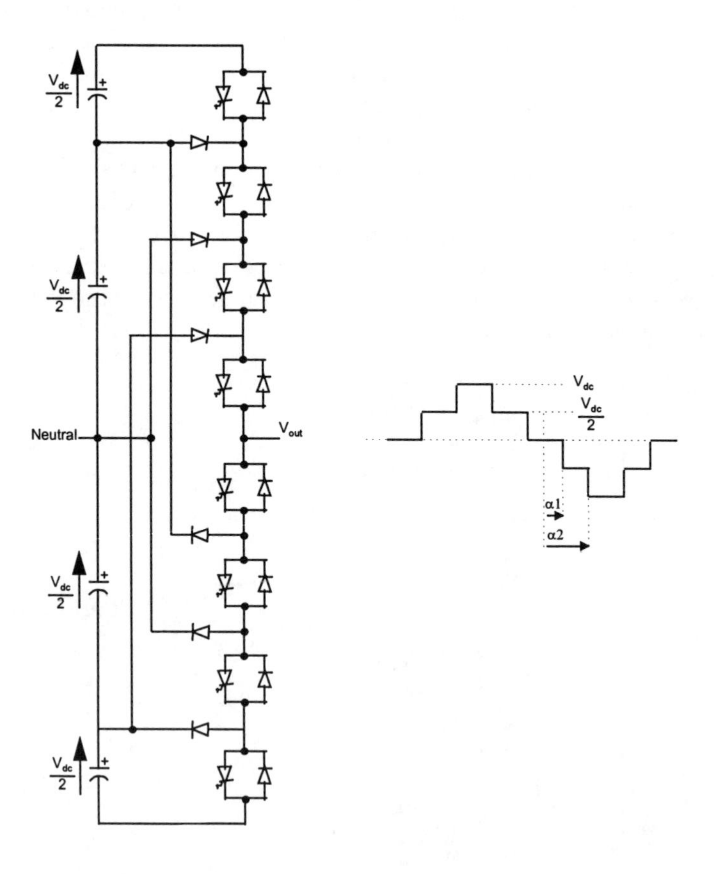

Figure 4.15 Five-level voltage-sourced converter

This converter is more complex and requires the dc voltage source to be split or centre-tapped in order to provide a zero voltage reference. It can generate symmetrical positive and negative half cycles. The switching angles (such as α in Figure 4.14) for turn-on and turn-off are normally chosen to be symmetrical with respect to the instant at which the system voltage passes through zero and the output voltage waveform is a sequence of rectangular blocks. By varying the switching angles, α, both the fundamental magnitude and the harmonic spectrum can be varied. The fundamental voltage component can also be changed by keeping α constant and changing V_{dc}, when a capacitor is used as the voltage source; in this case the harmonic spectrum remains independent of the magnitude of the fundamental component.

By adding further sets of GTOs and diodes, for example in Figure 4.15, additional steps can be added to the output voltage waveform, increasing the controllability of the fundamental and harmonic content of the waveform by co-ordinated control of the switching instants $\pm\alpha_1$ and $\pm\alpha_2$.

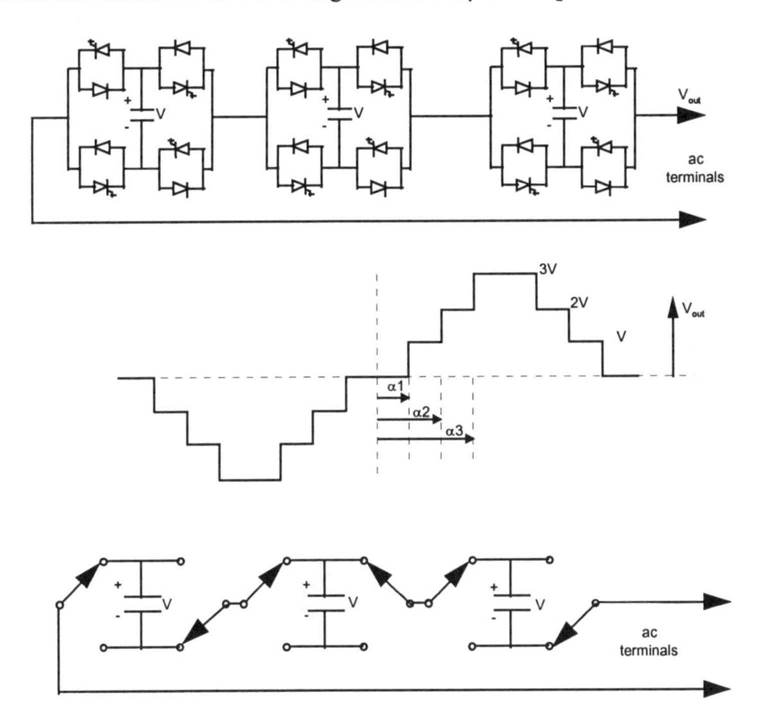

Figure 4.16 The chain circuit with mechanical switch analogue

Alternative forms of multi-level converters can be built up using other arrangements of GTOs, diodes, and source voltage components. One important form of converter is the chain circuit in which several converter bridges, each with its own source capacitor, are connected in series as illustrated in Figure 4.16. By appropriate switching of the GTO and diode pairs, the dc sources are connected into the circuit with either positive or negative polarity or are bypassed to give zero voltage contribution. In principle there is no limit to the number of converter bridges (or "links") that can be connected in series in a chain to generate, directly, a high value of alternating voltage.

4.3.3 Three-phase converter

Three single-phase converters, for example of the type illustrated in Figure 4.13, can be connected, one per phase, to form a three-phase converter. The three single-phase converters can be controlled in a co-ordinated way to generate a balanced three-phase set of voltages. There is also freedom, under system fault conditions — or for compensating unbalanced loads — to control each phase independently, in order to assist in balancing the system while avoiding converter overload. When using capacitors for the dc voltage sources, the voltage in each phase can easily be changed independently of the other two phases if necessary.

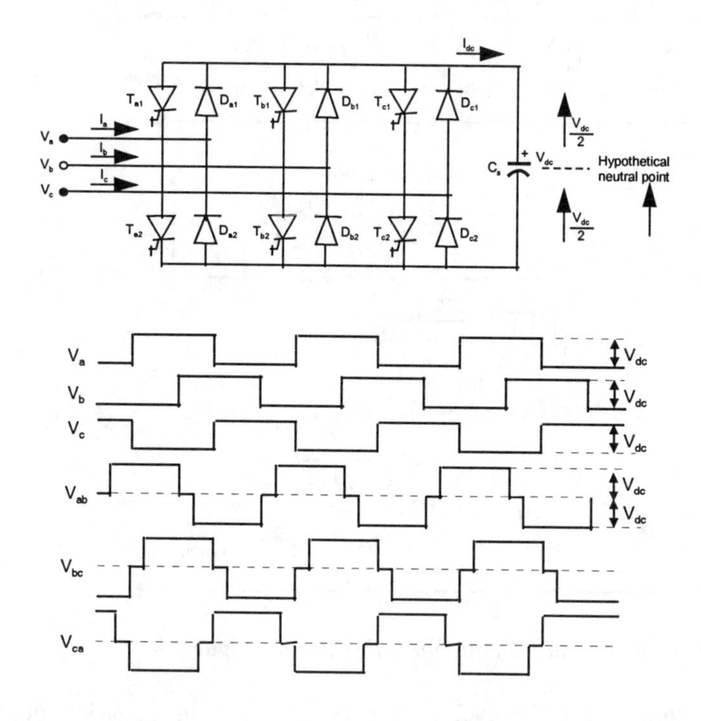

Figure 4.17 Three-phase Graetz-bridge converter and its output voltages

When the primary objective of the STATCOM is to respond to balanced load conditions, it is convenient to use the same voltage source for all three phases; the three converters can take the form of a Graetz bridge as shown in Figure 4.17(a).

The three phase voltages are shown in Figure 4.17(b), with respect to the apparent neutral of the three-phase system. It will also be seen that the line-to-line voltages reach twice the peak magnitude of the phase voltages but are of shorter duration (120° conduction angle rather than 180°, due to the commutating action of the individual phases). As described later, this inherently eliminates third harmonic components from the line-to-line voltages (under balanced system conditions). During fault and unbalanced conditions, particular care is needed in the control and protection of the converters because of the use of a common dc source voltage.

4.3.4 Reduction of harmonic distortion

4.3.4.1 Harmonics in a square, or block, waveform

The simple, single-phase, two-level inverter (Figure 4.13) produces a nominal square voltage waveform with a peak-to-peak magnitude equal to the dc source voltage V_{dc}. The fundamental component, V_1, has a peak-to-peak value of $\frac{4}{\pi} V_{dc}$. Due to symmetry, there are no even harmonics and the odd harmonics, V_n, have a peak-to-peak magnitude of $\frac{4}{\pi n} V_{dc}$ for the nth harmonic, i.e. $V_3 = \frac{V_1}{3}$, $V_5 = \frac{V_1}{5}$, etc.

In any three-phase circuit with isolated neutral (or delta-connected converters), the triplen order harmonics, 3rd, 9th, 15th etc, will be only of zero sequence and will not flow in the line currents, unless the supply voltage or the converter become unbalanced, in which case the triplen harmonics will include positive and negative sequence components, which will then flow into the system.

Figure 4.18 shows the basic harmonic spectrum for a square wave with single-phase triplet harmonics shown dashed.

For the three level converter shown in Figure 4.14, by introducing a switching angle of α before and after the nominal zero voltage crossing point, the waveform consists of repeated symmetrical rectangular blocks. The fundamental voltage V_1 is reduced to $\frac{4}{\pi}\cos\alpha . V_{dc}$ and the harmonic magnitudes become $\frac{4}{\pi} . \frac{\cos n\alpha}{n} . V_{dc}$ for the nth harmonic. Consequently, individual harmonics can be forced to have zero magnitude by appropriate choice of switching angle α.

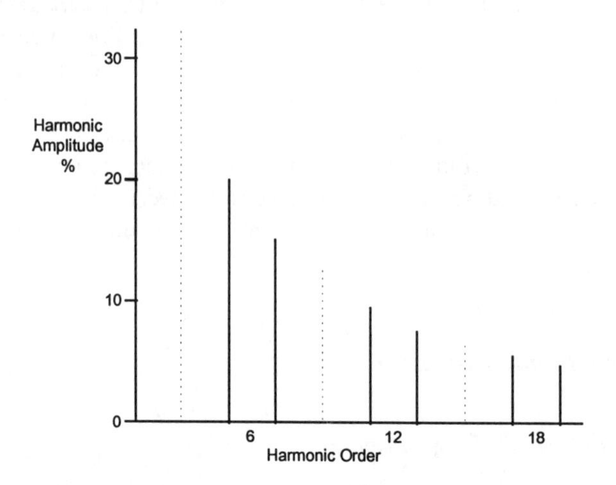

Figure 4.18 Harmonic spectrum of a square wave

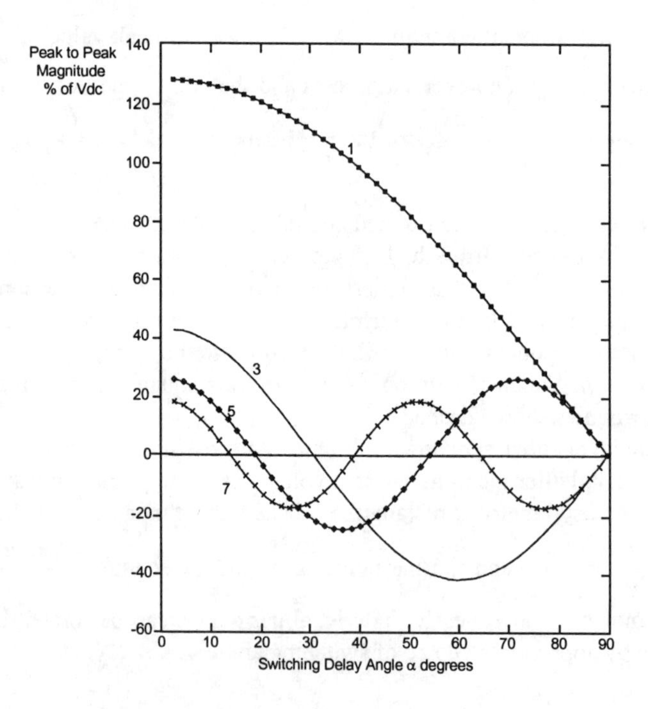

Figure 4.19 Effect of variation of switching delay angle on fundamental and harmonics in a rectangular block wave

Thus the 3rd harmonic becomes zero when $\alpha = 30°$, because $\cos 90° = 0$; similarly the 5th harmonic becomes zero when $\alpha = 18°$ (and 54°) and 7th harmonic becomes zero when $\alpha = 12.8°$ (and also 38.6° and 64.3°). Figure 4.19 indicates the relative magnitudes of fundamental and low order odd harmonics for a rectangular block waveform as the switching angle is changed.

4.3.4.2 Pulse width modulation

Pulse width modulation (PWM) is an extension of this simple concept of harmonic control. The GTOs are repetitively turned on and blocked, turned on and blocked, turned on and blocked again, during each half cycle. The sequential switching instants are selected in a co-ordinated manner, to satisfy simultaneous requirements, i.e. to develop the desired fundamental voltage and to eliminate selected low order harmonics. Figure 4.20 illustrates how two poles of a converter can be controlled, each with 5 on/off actions per cycle, to eliminate both 5th and 7th harmonics together. Additional, correctly timed, switchings can eliminate additional higher harmonic components, but there are two disadvantageous aspects. The usable fundamental rating and the converter efficiency are reduced because GTOs require significant energy (resulting in increased thyristor losses and heating effects) for each switching operation. In contrast, devices such as IGBTs and IGCTs require much lower switching energy and are better suited to the use of PWM techniques.

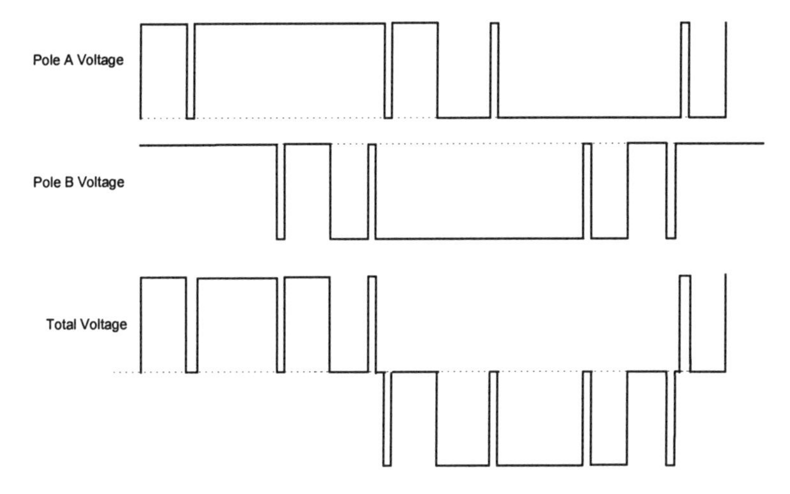

Figure 4.20 Pulse width modulation waveforms to eliminate 5th and 7th harmonic components

Whereas PWM is essentially a technique that depends on control action and lends itself particularly to converters of relatively low rating and having few power components, the chain circuit of Figure 4.16 requires larger numbers of power electronic components and lends itself to good harmonic compensation and higher ratings. The natural switching frequency for the active devices is once per two cycles of power frequency and therefore GTO devices, with their high power handling capability, are well suited to this type of converter.

4.3.4.3 The chain circuit – a multi-level arrangement

Figure 4.16 shows that a single-phase chain with only three links can produce seven output levels. Each link in the chain produces its own rectangular block of voltage, with an amplitude equal to its own source voltage (normally all these source voltages are controlled to have the same value). Each voltage block therefore makes a contribution to the total fundamental voltage and to the total spectrum of harmonics. The more links there are in the chain, the more closely the overall waveshape approximates to a sinusoid, as seen even for 5 links (11 levels) in Figure 4.21. This stepped waveform appears broadly similar to the current waveform of a conventional multi-pulse ac to dc rectifier. However, the rectifier waveform consists of a series of pulses of varying height but equi-distant in time, whereas for the chain circuit, the pulses are of equal height and the interval between switching instants varies.

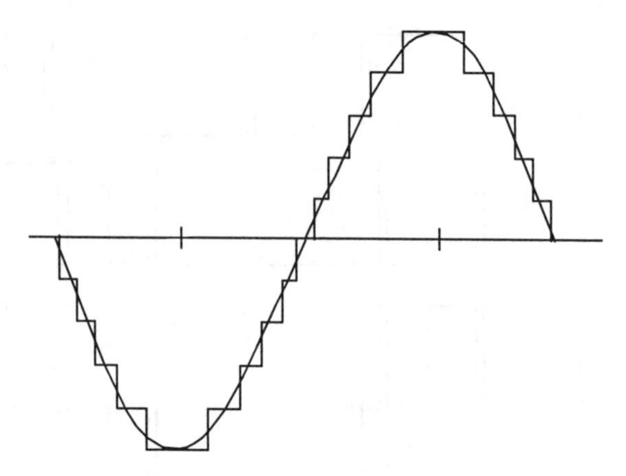

Figure 4.21 Output voltage of chain circuit with five "links"

The successive switching angles of the chain circuit may be controlled in various ways. One technique is to co-ordinate the switching instants so as to minimise the total harmonic content e.g. by a least squares method. This will usually result in a residual but small content of the low-order harmonics i.e. 5, 7, 11 and 13. Alternatively the switching may be used to null several specific harmonics; N links can be controlled to eliminate completely N harmonics per phase. In a three-phase arrangement this approximates to "6N-pulse" operation.

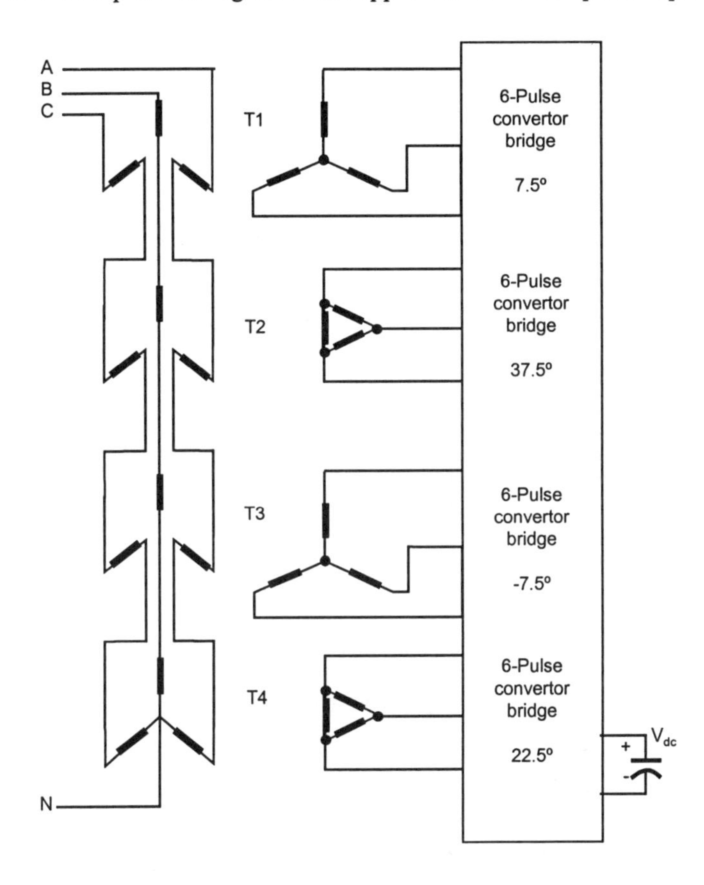

Figure 4.22 Converter arrangement providing quasi 24-pulse operation

4.3.4.4 Phase displacement of pulses

Multi-pulse operation can also be achieved, as for conventional rectifiers, by connecting identical three-phase bridges (such as in Figure 4.17) to transformers which have outputs that are phase-displaced with respect to one another. Star- and delta-connected windings have a relative 30° phase shift and a 6-pulse converter bridge connected to each transformer will give an overall 12-pulse operation, eliminating 5th and 7th harmonics from the line currents for balanced operating conditions. The transformers may be operated in parallel, in which case 5th and 7th harmonic currents will circulate between the converters, limited by the transformer reactance. Alternatively the transformers can be connected in series so that their 5th and 7th harmonic voltages will cancel out and so prevent any flow of the corresponding currents.

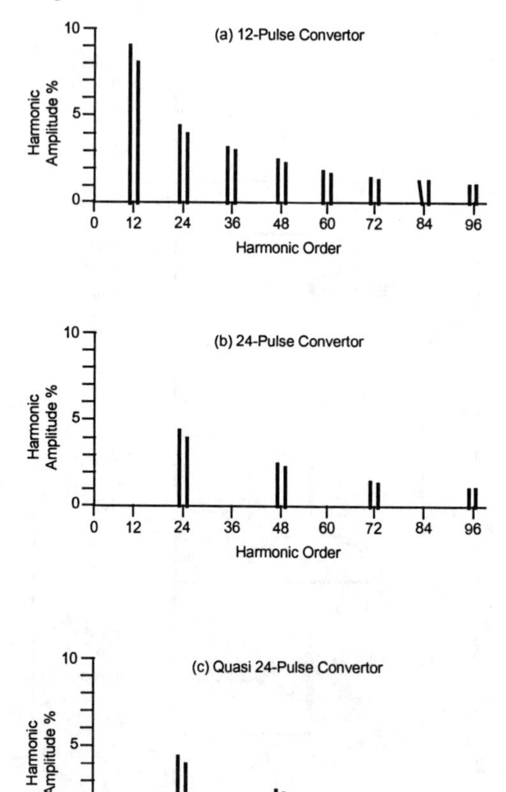

Figure 4.23 Harmonic spectra for various multi-pulse converter arrangements

This principle can be extended: with four transformers having relative phase displacements of 15°, 24-pulse operation can be obtained; with eight transformers having 7.5° phase shift, 48-pulse operation results. A practical disadvantage of this principle is that relatively complex phase-shifting transformers are needed. It is usual to employ a simple transformer to provide the step-down duty from the high voltage system to an appropriate intermediate voltage, but several different designs of phase-shifting transformers are then required to supply the individual converter bridges.

A compromise is illustrated in Figure 4.22. Two star-star and two star-delta transformers are connected with their primary windings in series and each secondary winding is connected to a 6-pulse bridge converter. All four of the converters utilise the same dc voltage source but they are operated effectively with relative phase-displacements of 15° to produce a quasi 24-pulse configuration.

The 12-pulse harmonics which are characteristic for each pair of converters are not perfectly cancelled, but the residual magnitude is acceptably small, as illustrated in Figure 4.23. For practical applications of this kind of STATCOM, the Graetz bridges must use several GTOs in series in each leg of the bridge to obtain an adequate rating. The GTOs in each leg must be arranged to turn on and off at precisely the same instant (within microseconds) to ensure voltage sharing between the individual GTOs.

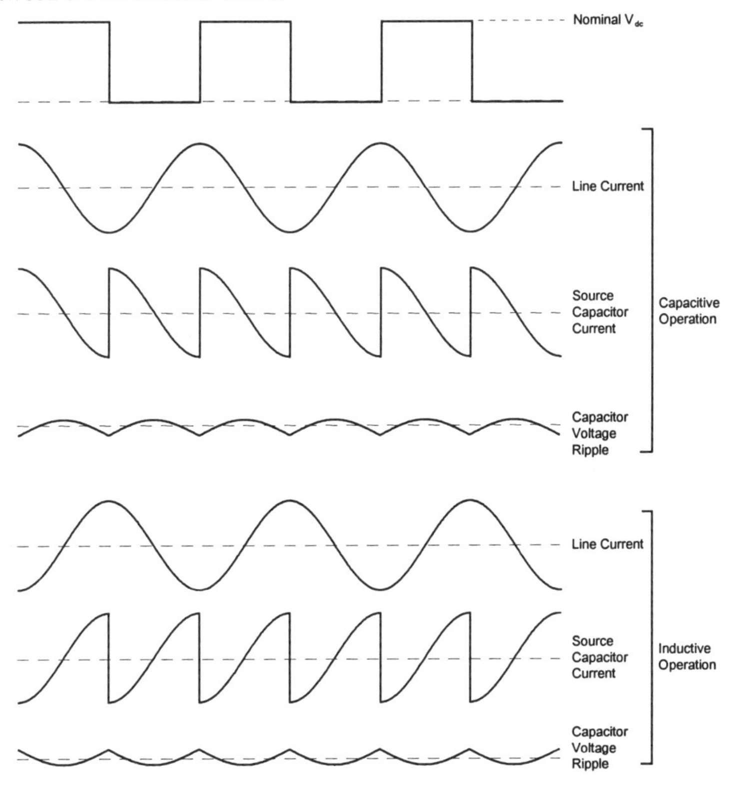

Figure 4.24 Converter output voltage and current waveforms and dc capacitor current and voltage during var generation and absorption

4.3.5 Source voltage ripple

Ideally, the dc source should be so strong that its voltage remains effectively constant at the chosen level, under steady state conditions. In practice, especially for capacitor voltage sources, this would require extremely large, bulky and expensive dc capacitors. A compromise is necessary to allow the capacitor to charge and discharge to some extent between each switching operation, i.e. a constant average voltage can be maintained but with a super-imposed ripple voltage of a few percent, Figure 4.24. This ripple must be taken into account in selecting switching instants and in evaluating the overall harmonic behaviour of the converter system.

4.3.6 Snubber circuits

GTO devices are available in a wide range of voltage and current ratings including, in particular, the current turn-off capability. A widely used GTO has a peak voltage rating of 4.5kV and a peak turn-off current of 4kA. As with conventional thyristors, it is important to protect individual GTO devices against both forward and reverse overvoltage and against excessive rates of change of inrush current and of voltage at turn-off. Figure 4.25 illustrates a typical snubber circuit arrangement.

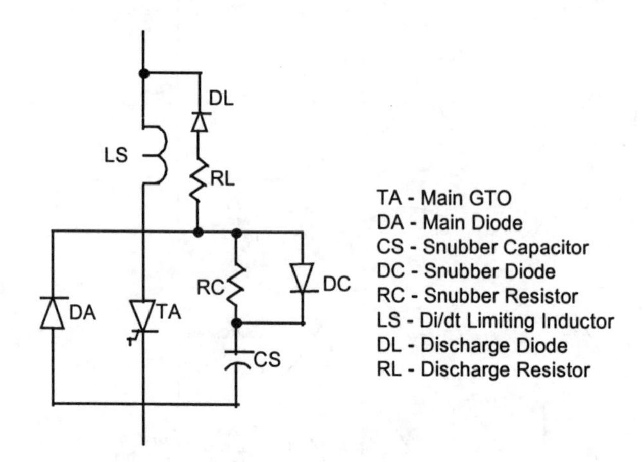

Figure 4.25 Typical snubber circuit arrangement for GTOs in a STATCOM

In order for the GTO to turn off safely at 4kA, the snubber capacitor, CS must have a high value, about 6mF. The energy stored in this capacitor must be dissipated after every switching. If a smaller capacitor is used, say 3mF, the switching losses are substantially reduced but the safe turn-off current is reduced to about 3kA.

Energy stored in the di/dt limiting inductor of the snubber circuit at turn-off is dissipated via the discharge resistor and diode. Some of the dv/dt and di/dt circuit energy can be recovered by additional circuits. The added complexity and cost of these energy recovery techniques must be weighed against the saving in losses and the possibility that they might enable simple PWM techniques to be applied to the GTO converters.

4.3.7 Some practical implications

It would clearly be inappropriate to connect a voltage-sourced converter directly to the supply system, which will generally have a very much higher short-circuit power than the rating of the converter. As shown in Figure 4.12, there is always a coupling reactance between the system busbar and the converter terminals. When the converter is connected to the system via a transformer, this may have sufficient inherent reactance for satisfactory operation of the STATCOM.

In some applications where there is a stringent harmonic performance requirement, it may be necessary to include shunt capacitors or harmonic filters on the converter busbar, in which case buffer reactors (separate from the stepdown transformer) may be needed to limit the flow of harmonic currents from the converter into the capacitors.

As well as limiting harmonic currents, the coupling reactance limits fault currents into the STATCOM; it also "softens" the transient current response from the STATCOM when there is a fault on the supply system which collapses the voltage, or when there is a large magnitude dynamic overvoltage. Inevitably, the coupling reactance requires the voltage generated by the converter to be higher than that of the system when the STATCOM is acting as a capacitor and to be lower than the system voltage when the STATCOM is acting as an inductor. The vars absorbed by the coupling reactance (and the supply transformers) must be taken into account in the design and rating of the converters.

4.3.8 STATCOM operating characteristics

The "natural" voltage–current characteristic at the terminals of a STATCOM, Figure 4.26(a), is entirely dependent on the converter source voltage V_e and on the coupling reactance, X_e. In general the coupling reactance has a typical value between 10% and 20%, i.e. the voltage drop (or rise) across it is about 10 to 20% of the nominal system voltage at the rated current of the STATCOM. The continuous current rating of a GTO, IGBT, or IGCT is almost independent of whether the current lags or leads the voltage. These devices usually also possess a short-time, or transient, overcurrent rating, which may exceed the safe turn-off (or turn-on) current for the STATCOM valve components. If, by accident or design, the safe turn-off current is exceeded, then turn-off signals must be prevented from being issued until the current returns to below the safe switching level.

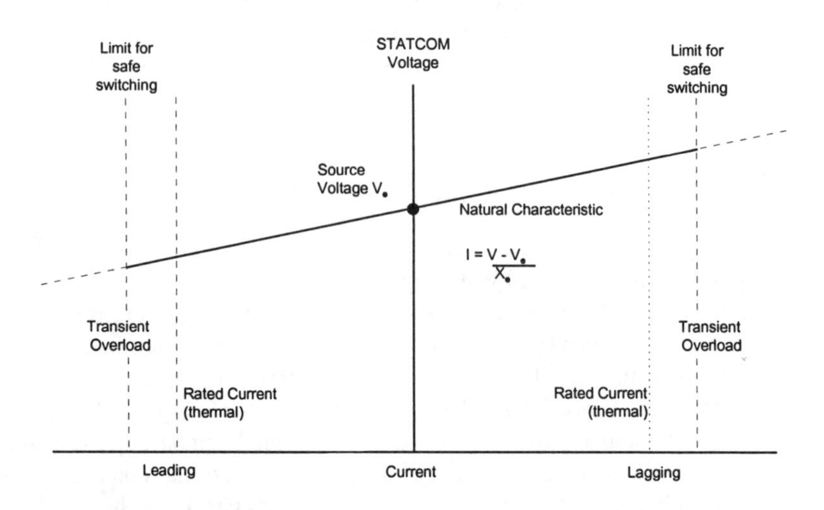

Figure 4.26(a) Natural V–I characteristic of a STATCOM

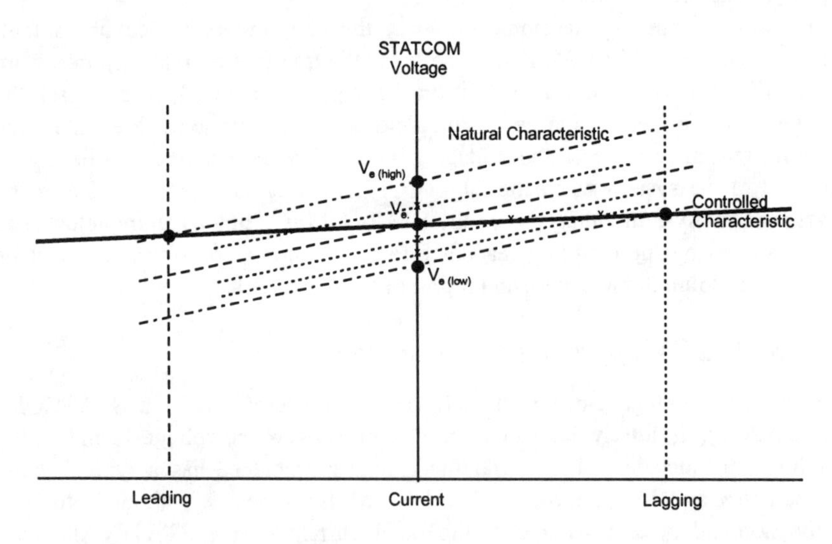

Figure 4.26(b) Controlled V–I characteristic of a STATCOM

The nominal maximum steady state voltage for a supply system is 1.1p.u. of its nominal value and this presents no difficulties for the design and rating of a STATCOM. However, a STATCOM must also withstand dynamic overvoltages and transient overvoltages up to the protective level provided by the STATCOM's surge arresters. During transient conditions in which the instantaneous applied voltage exceeds the dc source voltage, the diodes of the STATCOM will allow current to flow, and this current will charge the dc source capacitors to a higher voltage.

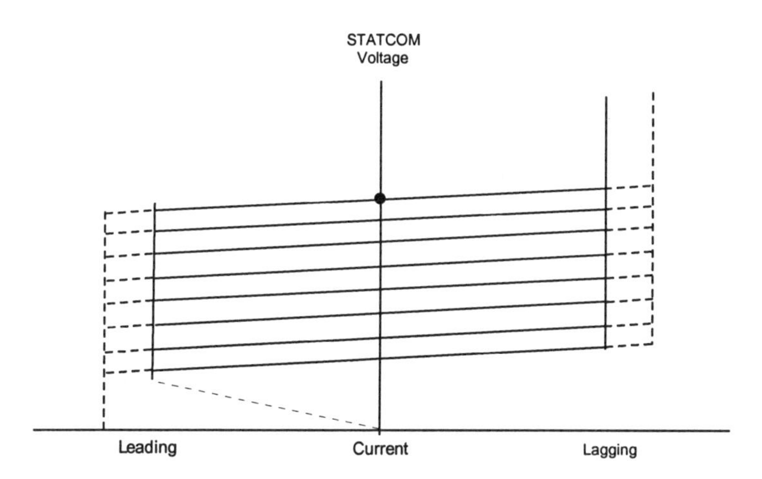

Figure 4.26(c) Family of V–I characteristics for a STATCOM

Most practical applications require any type of SVC or STATCOM to operate with a slope reactance typically of between 2% and 5%, which is much lower than the value of the coupling reactance. Because the coupling reactance is fixed, the converter source voltage must therefore be changed as shown in Figure 4.26(b), raising it to V_e(high) to obtain the desired leading (capacitive) current conditions or reducing it to V_e(low) for lagging (inductive) conditions. This can be done very quickly, initially by changing the switching pattern and then followed, if appropriate, by changing the magnitude of the dc source voltage. The desired control characteristic is thus achieved by changing the source voltage in precisely the same way as for a synchronous compensator, but very much faster.

Both target voltage and slope reactance can be set in the STATCOM control. A set of voltage–current characteristics, for a range of target voltage settings with constant slope, is shown in Figure 4.26(c). For comparison, the V–I characteristics of a conventional SVC are shown in Figure 4.27. At reduced voltage, the STATCOM can continue to be operated at rated leading (or lagging) current, with a constant transient overload current margin. These capabilities are available down to very low voltages. In contrast, the current limits for conventional SVCs are proportional to voltage. A STATCOM is better able to provide reactive/current support for a supply system whose voltage is severely depressed, whereas a conventional SVC can generally do more than a STATCOM to limit dynamic overvoltage.

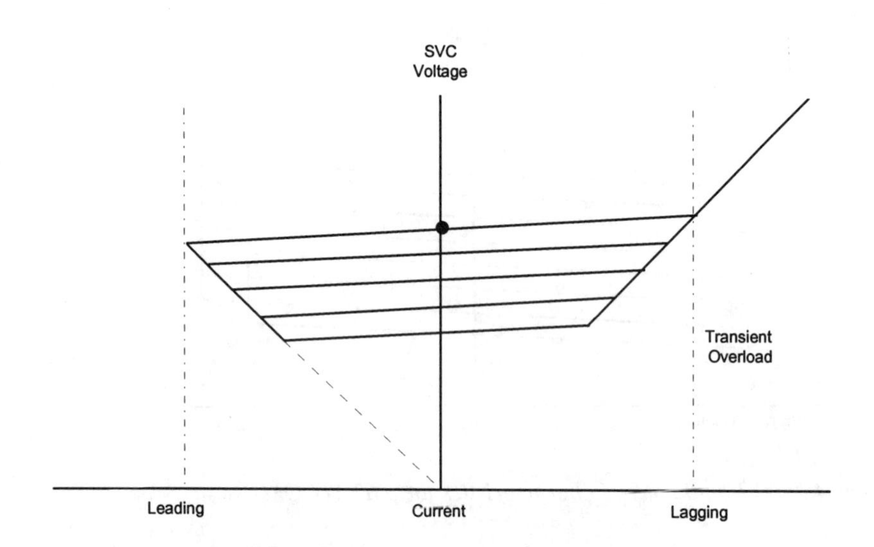

Figure 4.27 Family of V–I characteristics for conventional SVC

4.3.9 Transient response

Because the operation of a STATCOM is based on the generation of a sinusoidal voltage, its response to transient disturbances is inherently good and extremely rapid. The steady state operating condition of a STATCOM is dependent on the system voltage (and impedance) and the STATCOM source voltage and its coupling impedance. Thus in Figure 4.28(a), with an open circuit system voltage V_{s1} slightly lower than the target voltage of the STATCOM steady state characteristic, the STATCOM draws a small capacitive current I_1. In order to generate this current the STATCOM source voltage V_{e1} must be slightly higher than the target voltage.

If now the system voltage is depressed, due to a fault, to a value V_{s2}, the point of intersection of the system characteristic and the STATCOM controlled characteristic demands a current I_2. Initially, before there has been any change of STATCOM source voltage, the STATCOM current increases substantially from I_1 to I_2' (given by the intersection of the system characteristic and the natural STATCOM characteristic); this is increased by control action to the required value I_2 by an increase of source voltage to V_{e2}, normally within one half cycle.

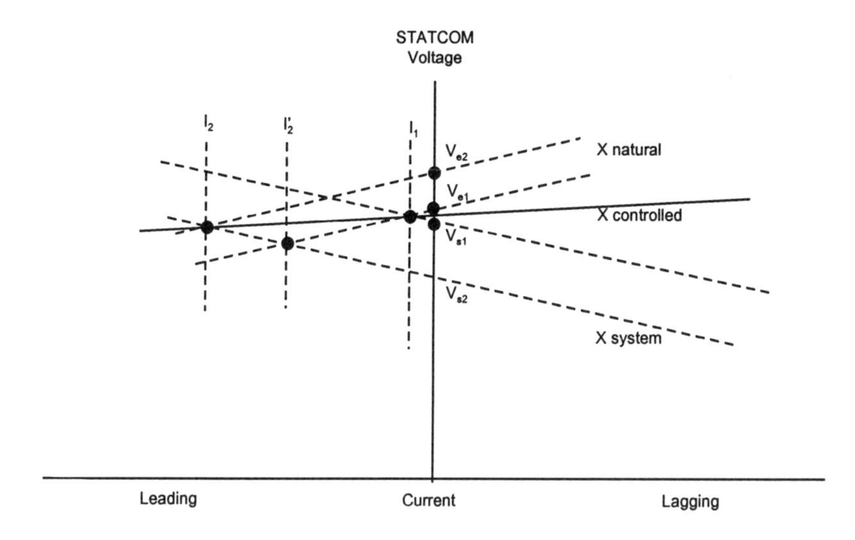

Figure 4.28(a) Response of a STATCOM to a system voltage change

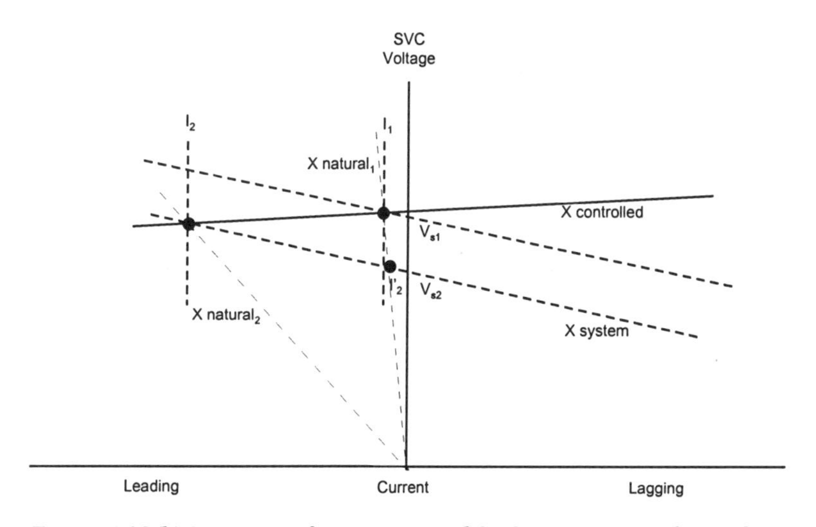

Figure 4.28(b) Response of a conventional SVC to a system voltage change

In contrast, the controlled characteristic of a conventional SVC is obtained by adjusting the susceptance of the SVC elements. The target current I_1 in Figure 4.28(b) is therefore obtained by controlling the SVC to a small capacitive value, $X_{\text{natural 1}}$. In this case, when the system voltage falls, the susceptance is unchanged and the SVC output current initially falls – exacerbating the voltage depression on the system! Control action generally requires a few cycles of power frequency before the required steady state susceptance $X_{natural2}$ is obtained to provide the target current I_2.

Figure 4.29 illustrates how a STATCOM responds to a voltage disturbance. Prior to the voltage dip, the STATCOM is operating at about its rated lagging current. A dip of system voltage suddenly occurs, to about 50% of its steady state value. The STATCOM inherently responds to this disturbance by generating a capacitive current to support the system voltage but, even on the natural characteristic, there would be a capacitive overload current. To prevent this, the STATCOM control system detects the sudden change and reduces the target voltage to limit the STATCOM current to its rated capacitive value.

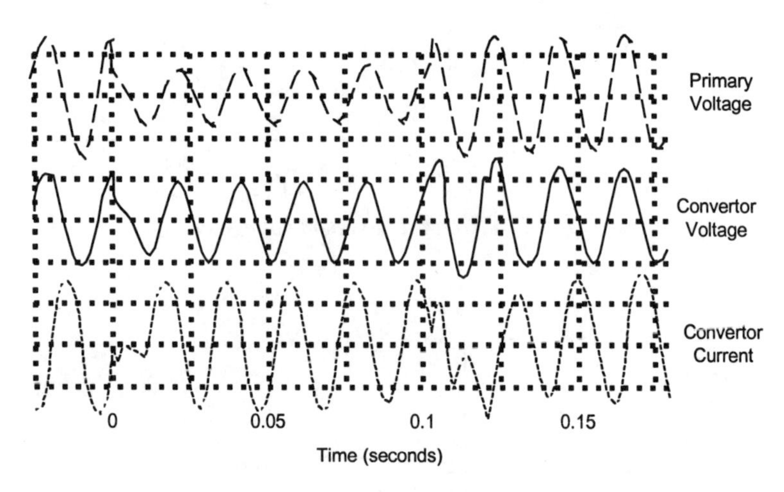

Figure 4.29 Response of a STATCOM to a depression in system voltage

When the fault is cleared and the system voltage recovers to its pre-fault value, this will tend to cause an inductive overload current in the STATCOM. Again, the STATCOM control system is able to detect the change and adjust the target voltage appropriately to reach rated lagging current. Although there is unavoidable transient distortion of the STATCOM current at each step change, it can be seen from Figure 4.29 that the changes from inductive to capacitive and capacitive to inductive current take place within a half cycle.

4.3.10 STATCOM losses

The forward voltage drop of GTO thyristors is greater than that of conventional thyristors because of the more complex system of semi-conducting junctions and the energy required for the turn-off duty. Figure 4.30 shows the approximate variation of STATCOM losses (% of rated current) through the operating range from rated leading to rated lagging current.

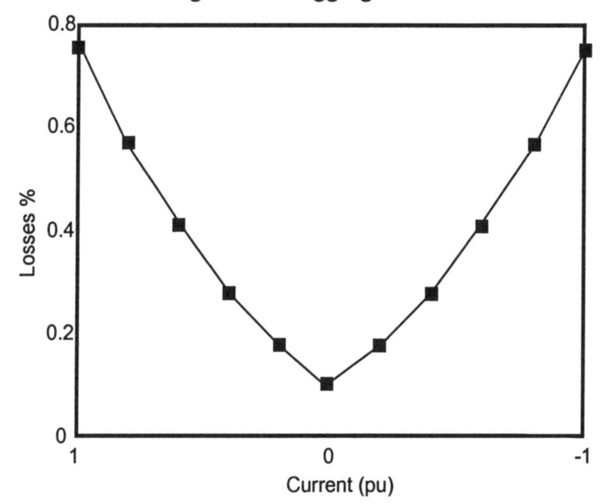

Figure 4.30 Typical loss curve for a STATCOM

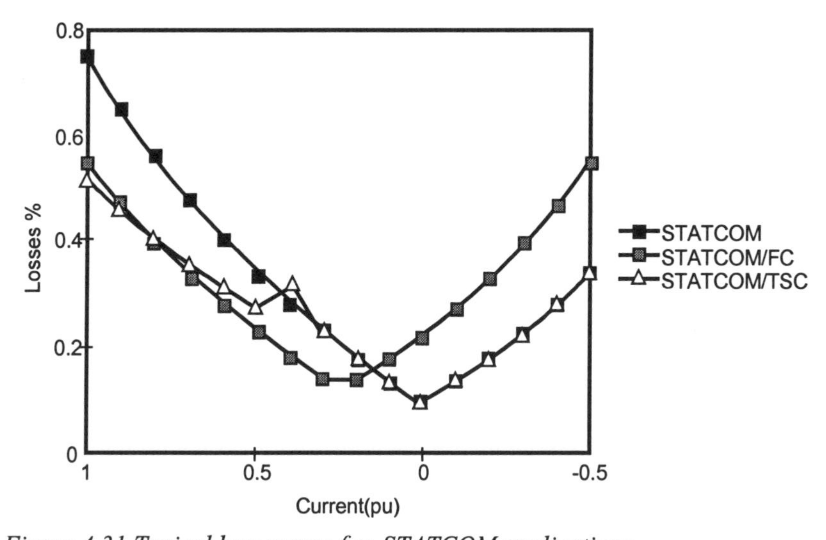

Figure 4.31 Typical loss curves for STATCOM applications

In many cases, the STATCOM output will need to be biased, generally towards the capacitive side for SVC applications. Figure 4.31 shows the loss patterns for the same output range (+1.0 to –0.5pu current) as was illustrated for conventional SVCs in Figure 4.10.

If the STATCOM is rated for an output of ± 1.0pu current, for this range, the upper half of the inductive range is not used. The losses in the float condition (0 Mvar) and within the lagging range are quite low, but become quite high in the upper part of the capacitive range. These capacitive losses can be reduced by halving the rating of the STATCOM and combining it with a TSC of about 0.6pu to reduce the losses (and probably the overall costs).

An intermediate option is also illustrated with a STATCOM of ± 0.75 rating, to cover the total dynamic range, biased by a fixed capacitor (or filter bank) of 0.25pu output. This may give a better overall optimisation of cost and losses especially if the predominant range of operation of the compensator is from about 0.1 to 0.6pu capacitive current.

4.3.11 Other types of STATCOM source

In addition to voltage sources using batteries and capacitors, STATCOMs can be operated with an inductor, which provides a source of direct current rather than voltage, Figure 4.32. A three-phase, current-source converter then generates a set of three-phase output currents which, by appropriate switching action, lag or lead the system voltages. The basic output current is a square or block wave and harmonic reduction requires PWM, or multi-level or multi-pulse techniques, and/or harmonic filters. The energy in the current source can be sustained by drawing energy from the supply system or by using an external energy source. However the losses of a current-sourced converter tend to be higher than those of a voltage-sourced converter.

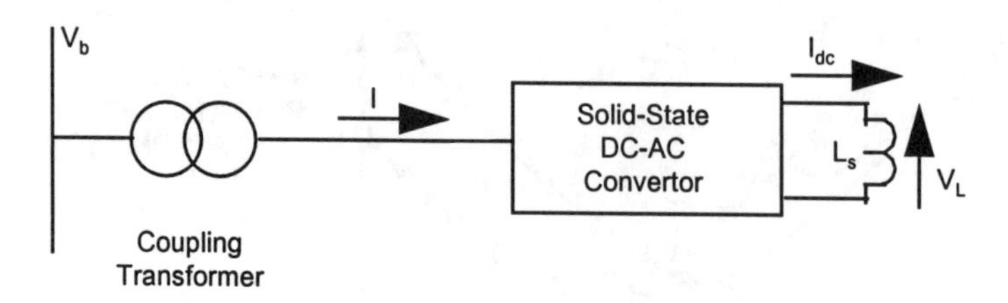

Figure 4.32 Current-sourced converter

A further interesting concept is to design the converter as an ac to ac frequency changer. The "source" can be a three-phase high frequency generator, a resonant circuit (a parallel capacitor and inductor in each phase), or even a transformer which is itself connected to the supply system (Figure 4.33).

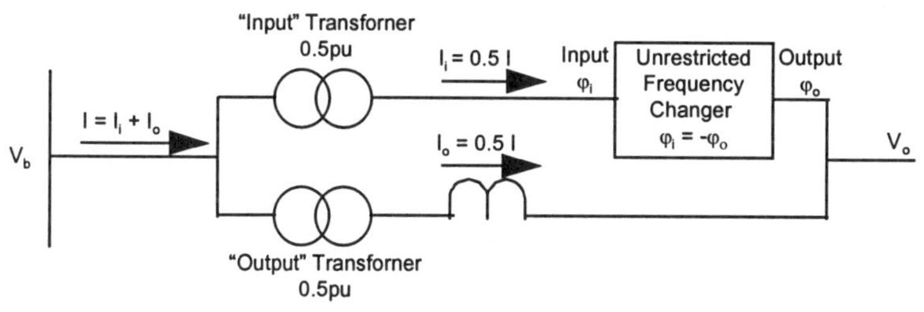

Figure 4.33 Power doubling converter arrangement

When the output voltage of the converter is higher than the voltage of the system, vars are exported to the system from the output terminals i.e. the output current lags behind the output voltage by 90°. The converters are controlled so that the phase displacement of the input current with respect to input voltage, (φ_1), has the same magnitude as the phase displacement between output current and voltage, (φ_0), but the opposite sign. Thus in this example, the input current will lead the input voltage and hence the converter input will also act as a capacitor, exporting vars to the system. As the total output at the converter terminals is approximately double that of the converter itself, this has been described as a "power-doubling" scheme, though the effective output to the system is reduced by the vars absorbed in the coupling transformers.

In this power-doubling arrangement, there are no energy storage components as such; the output connection to the output transformer can be considered to behave as a current source for the converter. However, the input terminals then need to behave as a voltage source and therefore an input filter needs to be connected to the input terminals of the converter. This type of converter requires special devices which have both bi-directional current carrying and forward and reverse voltage blocking capabilities. Suitable devices to implement the power-doubling arrangement are not yet commercially available, so this scheme is only of theoretical interest at present.

4.4 Applications

4.4.1 Some practical SVC applications

Figure 4.34 illustrates valves of two phases of a TCR, using 56 mm diameter thyristors [6,17]. Figure 4.35 shows part of a view of a larger, three-phase valve, constructed in a trefoil manner using 100mm diameter thyristors for a higher rating and suitable for TCR or TSC application[14]. Both thyristor stack designs permit easy replacement of any failed thyristor without disturbing the cooling water circuit or power connections.

Figure 4.34 Thyristor valve arrangement (showing two phases)

Figure 4.35 A 16 kV thyristor valve

Figure 4.36 shows a perspective view of the entire SVC for a +150/–75 Mvar as in Figure 4.8(b), where the valves are indoor-mounted and the main capacitors, reactors, and transformers are outdoor[14]. A typical thyristor valve de-ionised-water cooling plant, shown in Figure 4.37, has only the air-blast coolers outdoors.

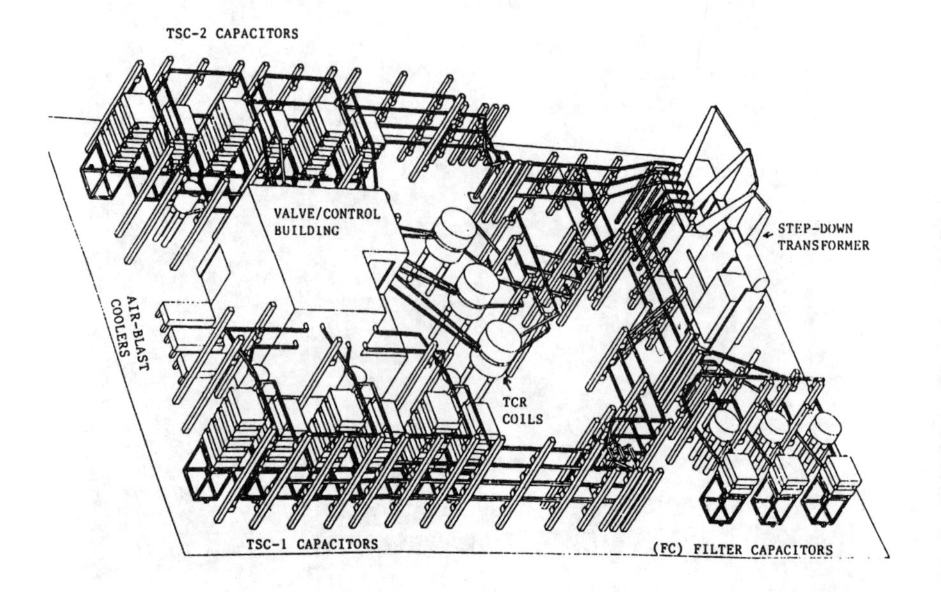

Figure 4.36 Perspective view of SVC

Many hundreds of SVCs are currently in service throughout the world for voltage control and in some cases for stability enhancement [7,9,13]. About 30 of these installations (mostly thyristor-controlled) are in the 400/275 kV network in England as shown in the diagrammatic map of Figure 4.38, in ratings ranging up to +150/–150Mvar [14,16].

Some applications have employed SVC valves and controls in a transportable cabin, as shown in Figure 4.34. This type of construction helps not only initial transport and installation but also permits relocation of the SVC (as described in Section 4.4.2) when the changing needs of a power transmission system require reactive support at a different network location.

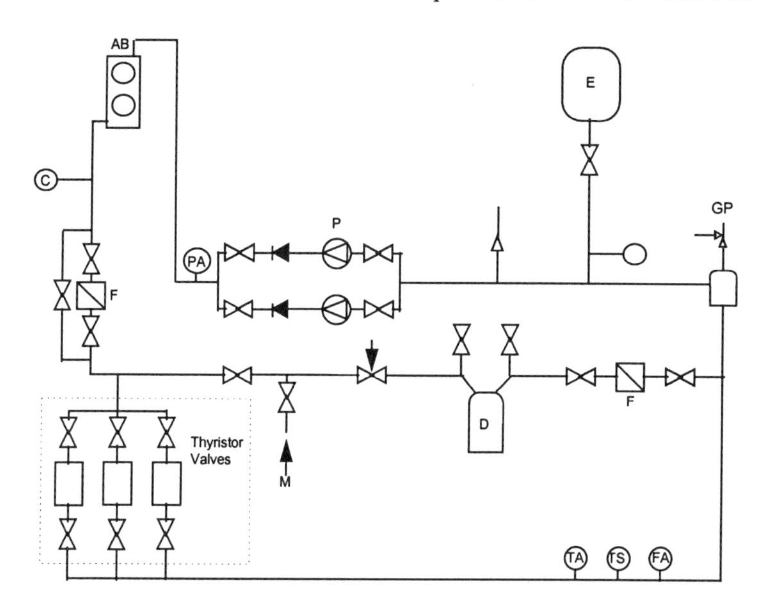

Figure 4.37 Thyristor valve cooling plant

AB	*Air-blast cooler, outdoors*
C	*Conductivity meter*
D	*Deioniser*
E	*Expansion tank*
FA	*Flow signal*
F	*Filter*
GP	*Gas purge*
M	*Make-up tank*
P	*Pumps*
PA	*Temperature signal*
TS	*Temperature switch*

4.4.2 Recent relocatable SVC applications in UK practice

The privatisation of electricity supply systems is now commonly implemented by the unbundling of generation and transmission functions into separate entities. When these two parts of the system are in different hands, the previously existing link between the planning of generating stations and transmission lines is broken. Not only is the active power (energy) generation less easily co-ordinated but so also the reactive power generation and absorption by power stations and thus var supply becomes an ancillary service to be co-ordinated economically. This presents transmission companies with much less predictable future conditions, and they must now be ready to cope with major changes in load flow patterns at timescales which are, for hardware investments, very short. A case in point is the situation in England and Wales, which has led to a need for Relocatable Static Var Compensators (RSVCs) (Figure 4.39).

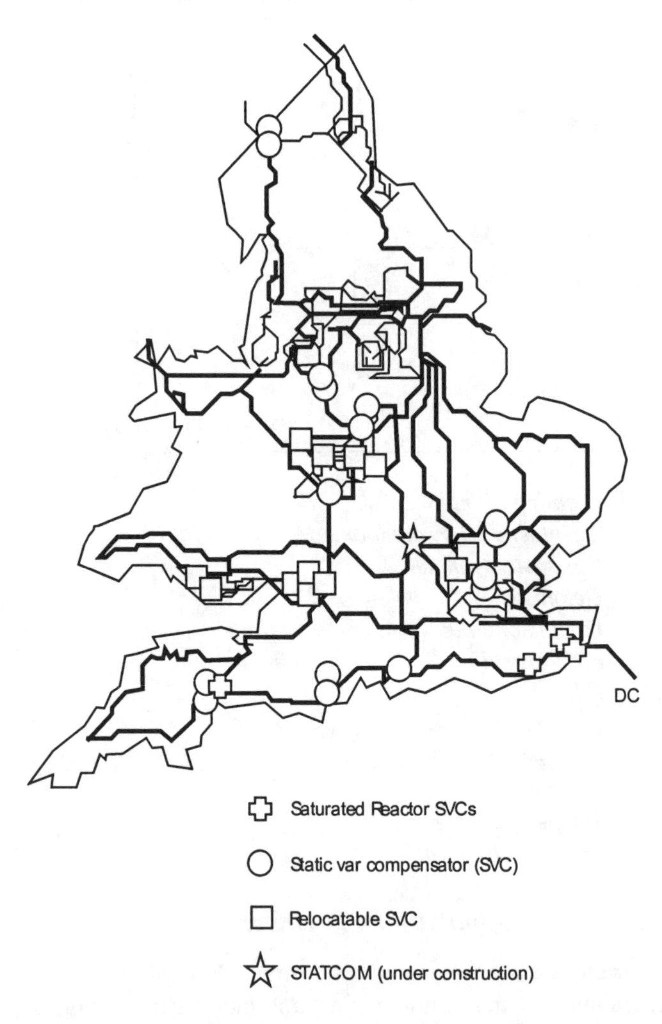

Figure 4.38 SVC applications in the grid network of England and Wales

Since the privatisation and unbundling of the UK electricity supply industry in 1990, The National Grid Company (NGC) has had little control over the siting of new generating stations and no control over the closure of existing generating stations in England and Wales. On closure of a generator not only is reactive power generation lost but power flows change in the transmission network leading to the need for yet further voltage support. The requirements to maintain security and quality standards, coupled with the need to sustain system flexibility, have led to a strategic need for var compensation devices that will be able to provide, within short lead times, voltage support at those points where it is needed.

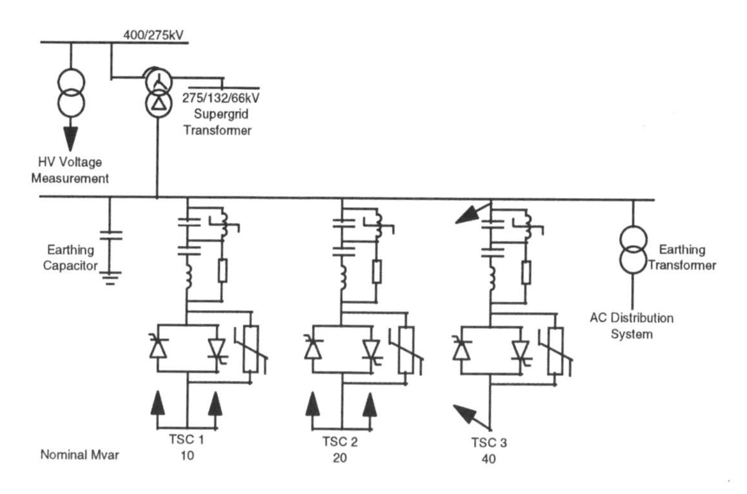

Figure 4.39 A relocatable static var compensator

Figure 4.40 Three TSC arms, incorporating C-filters, form one 60 Mvar relocatable SVC for connection to the tertiary winding of a SGT

Most of NGC's Supergrid Transformers (SGTs) are equipped with delta-connected 13 kV tertiary windings of 60 MVA nominal rating which can be brought out to bushings and are therefore available for the connection of compensation equipment. No dedicated step-up transformer is needed and there are benefits in terms of improved flexibility, ready relocatability and reduced capital cost. The nominal voltage on the high voltage (HV) side of a SGT is either 400 kV or 275 kV; the nominal voltage on the low voltage (LV) side is 275 kV, 132 kV or 66 kV. The rating of the transformers lies in the range 180 MVA to 1000 MVA, with short-circuit levels (three-phase) of up to 1125 MVA (50 kA) being possible at the tertiary terminals. The RSVC must be suitable for connection onto the tertiary winding of any transformer within this range [17].

Because the SGTs are not dedicated to the RSVCs there is no direct relationship between the tertiary voltage and the voltage on the HV system. The tertiary voltage can be strongly influenced by the power and vars exchanged between the HV and LV systems. The range of operating voltage must therefore be wider for tertiary-connected SVCs than for conventional SVCs.

In order to utilise the available tertiary capacity, the RSVC is required to generate 60 Mvar of capacitive reactive power at 0.9 pu of the 13 kV nominal tertiary voltage, but is not required to provide inductive capability. Due to voltage variations on the LV and HV systems, the voltage on the tertiary winding can vary between 0.8 pu and 1.3 pu, although the RSVC is not required to generate above 1.2 pu.

The HV and LV networks have their own independent harmonic characteristics and are coupled together, and to the RSVCs, via the SGTs. Due to the requirements for relocatability, the conditions at very many substations were examined to establish the range of harmonic self-impedances of the grid system which could be seen from the tertiary terminals of any SGT. The self-impedances are used in conjunction with the specified pre-existing harmonic distortion on the tertiary busbar to evaluate the harmonic currents that may be absorbed by the RSVC in determining equipment rating.

In addition to the above requirements for equipment rating, the RSVC must not contribute significant levels of harmonic distortion on either the HV or LV side of the transformer to which it is connected. The specified levels have been determined assuming that a maximum of three RSVCs will be connected at the same site on different transformers. The calculation of harmonic distortion on the HV and LV systems is very complex due to the wide range of system transformer impedances, but the utility has resolved this difficulty by deriving, from extensive studies, sets of harmonic transfer impedances (from tertiary to HV and tertiary to LV systems) for different values of self-impedance.

TSCs, which do not generate harmonics under balanced conditions, were preferred to TCRs and a multiple thyristor-switched capacitor arrangement was selected. This consists of three TSCs shown in Figures 4.39 and 4.40 having ratings in the binary sequence of 1:2:4 to give eight equal steps from 0 to 60 Mvar at 0.9 pu tertiary voltage.

TSCs are conventionally delta-connected. For the largest TSC, a delta-connection yielded a maximum fundamental valve current of 1300 A suitable for a compact thyristor valve design. The surge arrester (employing non-linear metal-oxide resistors) across each valve helps limit transient voltage levels thus avoiding excessively large valve voltage ratings. For the two small TSCs star-connection is used instead of delta-connection, which gives higher currents: 564 A and 1128 A respectively but permits 2 phase switching to be used. These smaller TSCs operate in an unearthed star arrangement, because a low impedance earth connection is not available to provide phase voltages of 13√3kV from a delta-connected tertiary winding. The corresponding asymmetry of recovery voltage when the valves are blocked requires an increased voltage rating. Each TSC circuit includes a series damping reactor to limit the magnitude of peak inrush currents and rates of change of current which flow in the TSC due to switching operations and faults.

The supply system seen from the tertiary terminals of a SGT can appear as either a capacitive or an inductive impedance with extremely low damping at all the major harmonic frequencies. Whenever the system impedance and TSC impedance form a series resonant condition, the background voltage distortion can cause harmonic currents of extremely high values to flow within the RSVC. Such resonant conditions need to be taken into account for each harmonic frequency up to the 31st, either by sufficient component rating or by adding damping to limit the severity of the resonant conditions. The damping is preferably frequency selective using the C-type filter. This circuit arrangement is indicated in the single line diagram of Figure 4.39.

4.4.3 STATCOM applications

4.4.3.1 Transmission applications

In order to demonstrate the features of STATCOMs and to gain practical operating experience, several prototype STATCOM installations have been put into service in transmission networks. In addition, several smaller STATCOM installations have been used, particularly in Japan, to reduce flicker caused by electric arc melting furnaces. However, most of the conventional applications of SVCs do not need to take advantage of many of the particular features of STATCOMs, such as equality of lagging and leading outputs, faster response time, possible active harmonic filtering capability, and smaller footprint.

One prototype STATCOM installation of ±80 Mvar was installed for service in Japan in 1991. The main converter circuit configuration is given in Figure (4.41). This STATCOM uses eight voltage-sourced converters, each of 10 MVA rating, connected to a main STATCOM transformer via eight converter transformers producing 7.5° phase angle displacement from each other, resulting in 48-pulse operation [21].

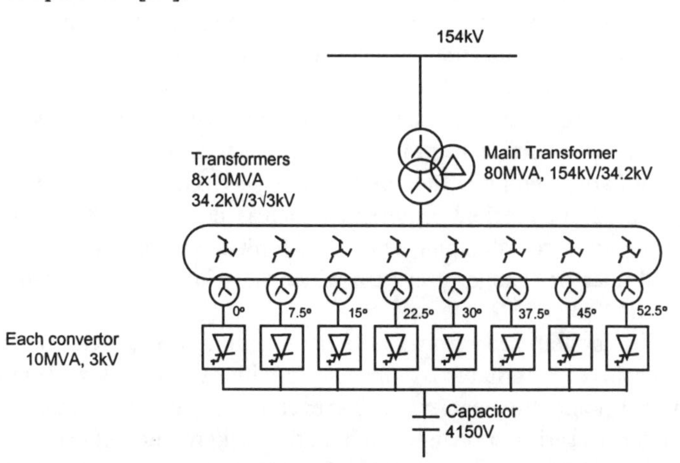

Figure 4.41 ± 80Mvar STATCOM in Japan

The control system incorporates power system voltage control, power oscillation damping and constant reactive power output control. The control system varies the width of the rectangular output voltage of each converter to achieve voltage magnitude control and to ensure low losses.

A gapped-core design is used for the eight phase-displacement transformers to reduce the effects of dc magnetisation, decrease magnetic impedance, and improve the uniformity of voltage sharing between windings. This is especially important when the STATCOM is energised from the power system during start-up sequences or following system faults. Initially the converter start-up system used a relatively large, separate, "start-up converter" to supply dc voltage to the STATCOM main converter. This method was found to be slow and a new system now allows the STATCOM converter to start immediately after the energisation of the STATCOM transformer and the eight converter transformers from the power system.

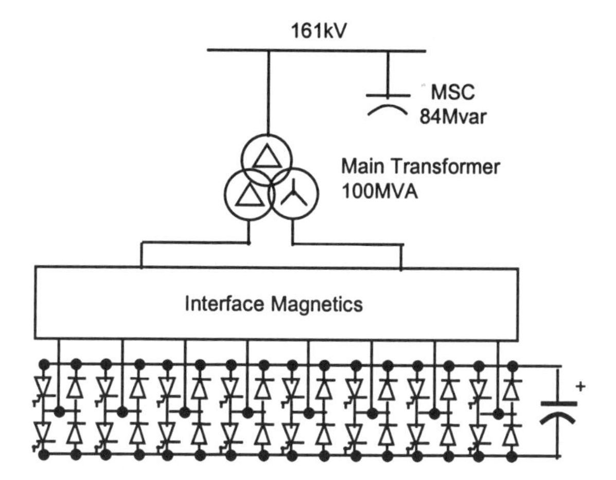

Figure 4.42 ±100Mvar STATCOM in USA

A ±100 Mvar prototype STATCOM has also been commissioned in the USA [22]. The main converter circuit configuration is given in Figure (4.42). Eight 12.5 MVA voltage-sourced converters are connected to the substation 161 kV busbar through a main STATCOM transformer and through "interface magnetics", which include phase shifting transformers to give quasi 48-pulse operation. The full reactive power output range of the STATCOM and an associated MSC of 84 Mvar are used to regulate the 161 kV busbar voltage and minimise transformer tap changer operation on the 500 kV/161 kV transformer feeding the substation. Apart from the normal voltage control on the 161 kV busbar the STATCOM also helps to prevent voltage collapse under system contingency conditions such as the loss of a 500 kV circuit.

Confidence in the STATCOM principle has now grown sufficiently for some utilities to consider them for normal commercial service.

In 1996, the National Grid Company plc of England and Wales sought relocatable dynamic reactive compensation equipment for its 400 kV transmission network, capable of generating 0 to 225 Mvar at 0.95 p.u. system voltage, with a particular reference to the inclusion of a STATCOM of 150 Mvar range. The design adopted includes a ±75 Mvar STATCOM in conjunction with a 127 Mvar TSC and 23 Mvar harmonic filter to provide a full controlled range of output +225 to –52 Mvar, Figure 4.43 [18].

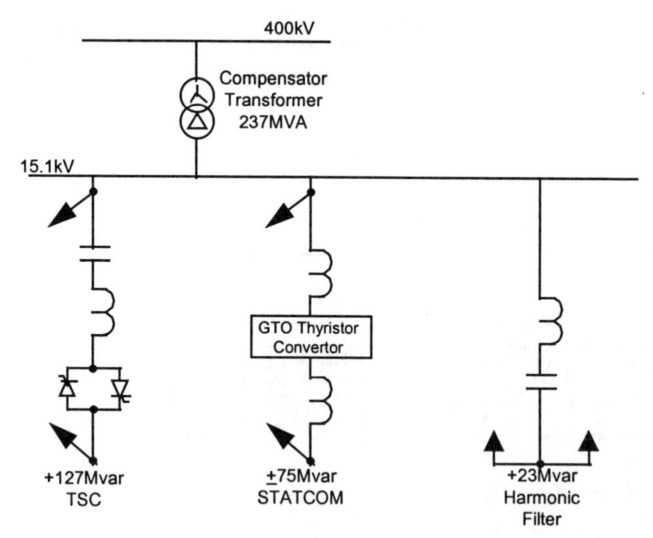

Figure 4.43 +225/–52 Mvar SVC including ±75 Mvar STATCOM in England

This STATCOM design is required to meet stringent harmonic emission levels and immunity to existing and future prospective harmonic levels. It uses multi-level converters in a chain circuit configuration. The control system incorporates voltage control, reactive setpoint regulation, and a co-ordinating control for the STATCOM and the associated TSC. Provision is also made to include power oscillation damping control in the future.

All the controls and power electronic equipment are housed in weatherproof, transportable GRP (glass reinforced plastic) cabins and the outdoor components are grouped together on frameworks to satisfy the requirement for easy relocation to another substation when this is required.

4.4.3.2 Specialised STATCOM applications

Other applications of smaller STATCOMs, in service or under consideration, are for the reduction of lamp flicker [24] due to arc furnaces, for voltage control for windfarms [26] and for balancing of single-phase traction loads. These smaller units generally use PWM to obtain a satisfactory harmonic performance.

4.4.3.3 Energy storage applications

Some manufacturing processes require absolute continuity of supply to maintain product quality and/or safety, for example, float glass, paper, semi-conductor devices, and some chemical and nuclear processes. The cost of disruption may be so great that auxiliary or emergency power sources are economically justified.

Uninterruptible Power Supplies (UPS) of moderate ratings are now very widely used to enable some processes to ride through brief voltage dips and interruptions. Where larger power ratings are required, STATCOMs, with enhanced energy storage or auxiliary power sources, offer a potential solution [25].

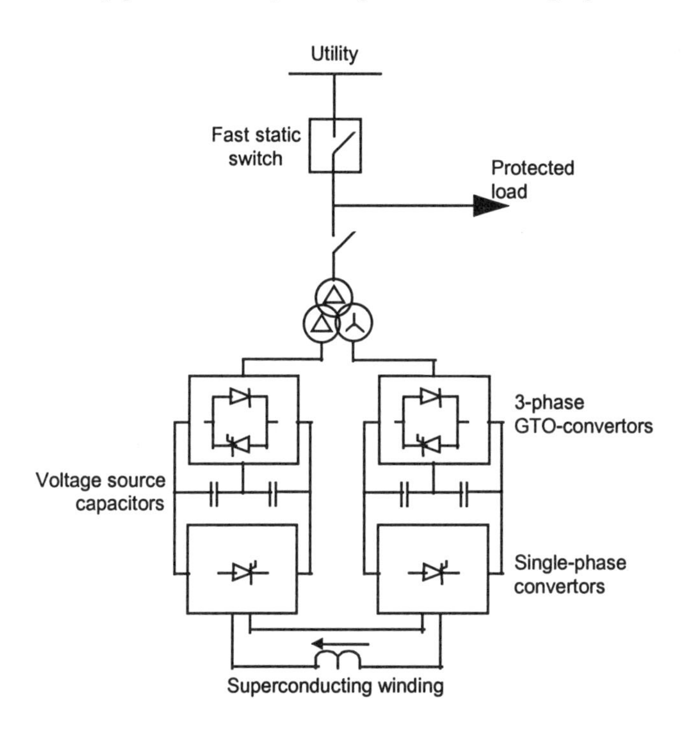

Figure 4.44 STATCOM with superconducting magnetic energy storage

Batteries have already been mentioned as voltage sources, but for stored energies of many mega-watt seconds, present designs of battery are very large and expensive. Very large capacitor banks might be used but are not efficient for bulk energy storage. Super-conducting storage sources (SMES), together with current-sourced converters have also been used in prototype format (Figure 4.44) [23] and, in the future, might be very suitable for energy storage [27]. Yet another possible future energy source is the fuel cell.

4.5 Summary

Both SVCs and STATCOM employ static equipment for inherently or automatically varying shunt reactive power in transmission, distribution, and industrial power networks. The principal equipment may be based on special (saturated) reactors, thyristors, or gate turn-off thyristor (GTO) in a variety of configurations and control modes. The equipment responds rapidly to changes in the network and/or loads to supply or absorb reactive power so as to achieve enhanced system performance more economically than otherwise possible. When required the equipment has also been designed for relocatability at different points in networks which require more flexibility as system generation and load patterns vary over the years.

The various uses of SVCs are :

- To compensate for flicker; the world's first industrial SVC was in Ethiopia, commissioned in 1964.
- To extend stability limits by providing an intermediate voltage boost within a transmission system: the world's first such SVC was commissioned in Zambia in 1968.
- To provide dynamic voltage support at weak points in a meshed transmission network.
- To avoid a new transmission line if voltage boost is needed on the loss of a line: for the James Bay transmission in Quebec, the use of SVCs reduced the number of 735 kV transmission lines from eight to five.
- To extend the transmission distance while maintaining acceptable operating voltages at any load, which is of benefit in remote developing or rural regions, as well as in developed regions.
- To contain the risk of temporary overvoltage on load rejection.
- To permit greater flexibility in planning the supply of reactive power in power systems by the ability to relocate static compensators.
- To balance single-phase loads; adding reactive loads in two phases can offset an active load in the other phase.
- To contribute to the reduction of harmonic distortion in a system.

4.6 Acknowledgment

The authors acknowledge the National Grid Company plc for the use of certain information contained herein.

4.7 References

1 Thanawala, H. L. and Young D. J., "Saturated reactors – some recent applications in power systems", *Energy International*, 7, (11), Nov 1970.
2 Thanawala, H. L., Williams, W. P., and Young, D. J., "Static reactive compensation for ac power transmission – ten years experience". *GEC Journal of Science and Technology*, 45, (3), 1979.
3 CIGRE Working Group 38-01 Task Force No 2, "Static var compensators" (Ed. I A Erinmez).
4 Thanawala, H. L., Ainsworth, J. D., and Williams, W. P., "The operating characteristics of static compensators using saturated and thyristor-controlled reactors". *GEC Journal of Science and Technology*, 47, (3), pp. 142-148, 1981.
5 Dabbs, M. J., Horwill, C., Thanawala, H. L., and Young, D. J. , "The Static compensators for the 2000 MW HVDC Cross Channel link". *IEE Conference on ac and dc Power Transmission*, Pub No 255, pp. 189–194, 1985.
6 Lowe, S. K., "Static var compensators and their applications in australia", *IEE Power Engineering Journal*, p. 247, Sept 1989.
7 Miller, T. J. E., "Reactive power control in electric systems", John Wiley, New York, 1982.
8 Mathur, R. M. (Ed.), "Static compensators for reactive power control", *Canadian Electrical Association*, 1984.
9 Thanawala, H. L., "Reactive power control" Chapter 16 of Electrical Engineers Reference Book (Ed. Laughton, M. A., Say, M. G.), 14th Edition, Butterworth-Heinemann, 1990.
10 Ainsworth, J. D., "Developments in the phase-locked oscillator control system for HVDC and other large converters". IEE Conf Pub No 255, pp. 98, 1986.
11 Crawshaw, A. M., Thanawala, H. L., and Mukhopadhay, S. B., "Design and application of TCR static var compensator for Paraguay". IEE Pub No 255, pp. 195-200, 1985.
12 CIGRE Working Group 31-01, "Modelling of static shunt var systems (SVS) for system analysis", *ELECTRA*, 51, March 1977.
13 CIGRE Task Force 38.05.04 , "Analysis and optimisation of SVC use in transmission systems", CIGRE Booklet No 77, 1993.
14 Baker, M. H., Thanawala, H. L., Young, D. J., and Erinmez, I. A. , "Static var compensators enhance a meshed transmission system". CIGRE Paper 14/37/38-03, 1992.
15 Muttik, P. K., Taylor, P. L., Thanawala, H. L., and Sadullah, S. , "Planning and design studies for the Far North Queensland Georgetown and Normanton

ac transmission scheme". *IEE Conference on ac and dc Power Transmission*, Pub No 345, pp. 44–49, 1991, London, UK.

16 Gardner D, Haddock J. L., Thanawala H. L., and Young D. J., "Digital computer studies and some test results on a static var compensator in the UK transmission network". IEE Conference Pub No 345, pp. 319-324, 1991.

17 Knight, R. C., Young, D. J., and Horwill, C., "Relocatable static var compensator help control unbundled power flows". *Modern Power Systems*, p. 49, Dec 1996.

18 Knight, R. C., Young, D. J., and Trainer, D. R., "Relocatable GTO-based static var compensator for NGC substations". CIGRE Paper 14.06, 1998.

19 CIGRE Working Group 14.19 "Static synchronous compensator (STATCOM)", (Ed. Erinmez and Foss), 1998.

20 Gyugyi. L. et al., "Advanced static var compensator using gate turn-off thyristors for utility applications", CIGRE Paper 23-203, 1990.

21 Mori, S. et al., "Development of a large static var generator using self-commutated inverters for improving power system stability", *IEEE Trans on Power Systems*, 8, (1), Feb 1993.

22 Schauder. C. et al., "Development of a ±100Mvar static condenser for voltage control on transmission systems", *IEEE Trans on Power Delivery,* 10, (3), July 1995.

23 Hassenzahl, W., "Superconducting magnetic energy storage", *IEEE Trans on MAGN*, 25,(2), p. 750, March 1989.

24 Takeda, M. et al., "Arc furnace flicker compensator with static var generator, *Mitsubishi Electric Engineering Review,* 65, (6), 1991.

25 Walker, L. H., "10 MW GTO converter for battery peaking service", *IEEE Trans Vol PAS-26,* (1), P. 63, Jan 1990.

26 Bergmann, K. et al., "Application of GTO based SVCs for improved use of the Rejsby Hede wind farm", *9th National Power Systems Conference, Kaupur, India, NPSC96*, Vol 2, p.391,1996.

27 Price, A. et al., "A novel approach to utility scale energy storage", *IEE Power Engineering Journal*, p.122, June 1999.

Chapter 5

Series compensation

M. Noroozian, L. Ängquist, and G. Ingeström

5.1 Introduction

Series capacitors have been successfully used for many years in order to enhance the stability and loadability of high-voltage transmission networks. The principle is to compensate the inductive voltage drop in the line by an inserted capacitive voltage or in other words to reduce the effective reactance of the transmission line.

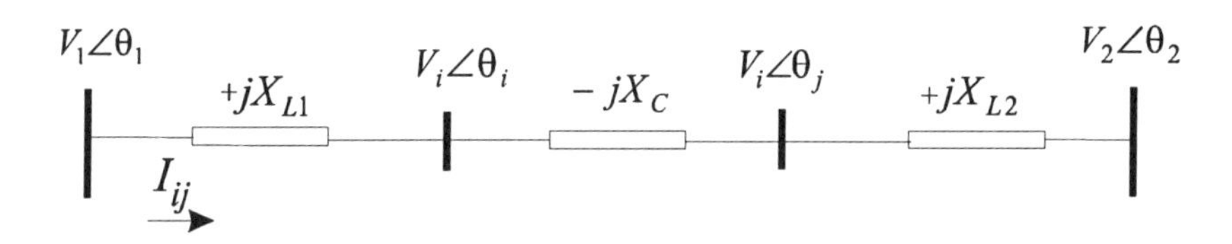

Figure 5.1 A series compensated transmission system

The inserted voltage provided by a series capacitor is in proportion to and in quadrature with the line current. Thus, the generated reactive power provided by the capacitor is proportional to the square of the current. This means that a series capacitor has a self-regulating impact. When the loading of the system increases, the generated reactive power from the series capacitor also increases. The impact of series compensation on a power system can be summarized as below:

5.1.1 Steady state voltage regulation and prevention of voltage collapse

A series capacitor is capable of compensating for the voltage drop of the series inductance in a transmission line. During low loading the system voltage drop is lower and at the same time the series compensation voltage is lower. When loading increases and the voltage drop becomes higher, the contribution of the series compensation increases and therefore the system voltage will be regulated as desired. Series compensation also expands the region of voltage stability by reducing the reactance of the line and therefore might be valuable for prevention of voltage collapse. Figure 5.2 shows that the voltage stability limit has increased from P_1 to a higher level P_2.

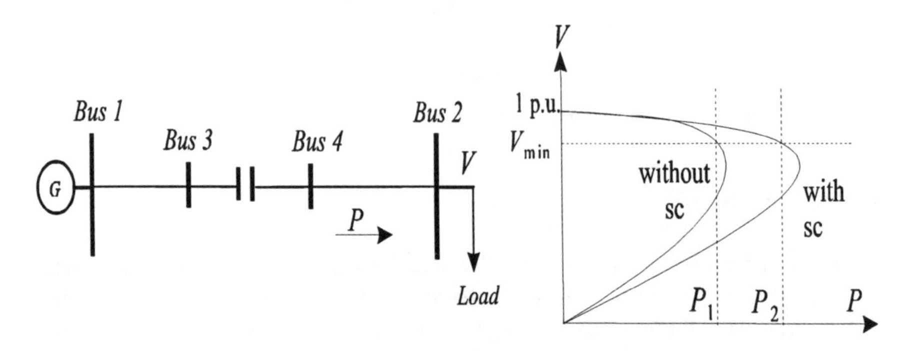

Figure 5.2 Voltage profile for a simple power system

5.1.2 Improving transient rotor angle stability

Consider the basic one-machine infinite-bus system shown in Figure 5.3. The equal-area criteria is used to show the effectiveness of a series capacitor for improvement of transient stability. In the steady state $P_e=P_m$ and the generator angle is δ_0. If a 3-phase fault occurs at a point near the machine, the electrical output of the generator reduces to zero. At the time of clearing the fault the angle has increased to δ_c and the trajectory will move to point D and will follow the $P–\delta$ curve. The stability of the system will remain if $A_{dec}>A_{acc}$. Figure 5.3 shows that a substantial increase in the stability margin is obtained by shifting upwards the $P–\delta$ curve by installing a series capacitor.

5.1.4 Power flow control

Series compensation can be used in power system for power flow control in the steady state. Compensation of transmission lines with sufficient thermal capacity can relieve the possible overloading of other parallel lines.

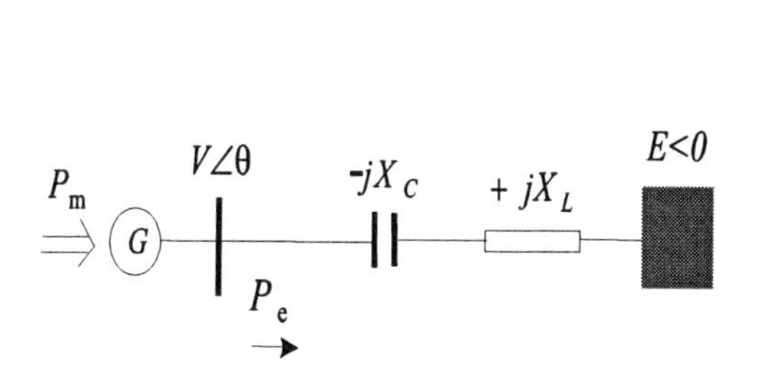

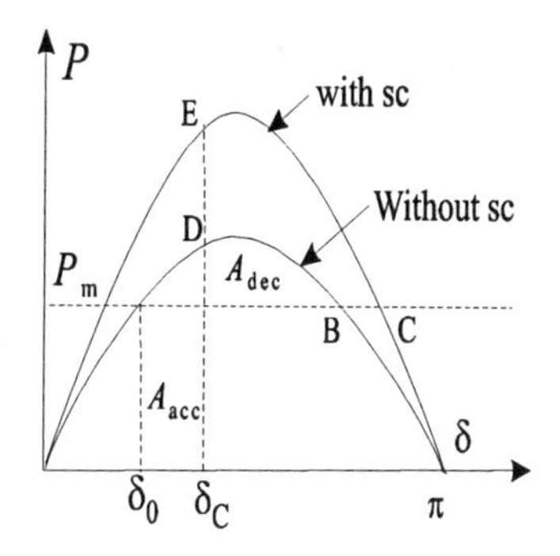

Figure 5.3 Enhancing the transient stability margin by use of a series capacitor

5.1.5 Series compensation schemes

Transmission lines can be compensated by fixed series capacitors or more effectively by controllable series capacitors using thyristor switches. A controllable series capacitor can be realized in two ways:

- Thyristor Controlled Series Compensation (TCSC)
- Thyristor Switched Series Compensation (TSSC)

TCSC configurations use thyristor-controlled reactors (TCRs) in parallel with segments of a capacitor bank. The TCSC combination allows the fundamental frequency capacitive reactance to be smoothly controlled over a wide range and switched to a condition where the bi-directional thyristor pairs conduct continually and insert an inductive reactance in the line. TSSC applications uses thyristor switches in parallel with a segment of the series capacitor bank to rapidly insert or remove portions of the bank in discrete steps. Figure 5.4 shows typical outlines of a series compensation schemes.

This chapter focuses entirely on TCSC. The outline of this chapter is as follows:

- Section 1 describes the basic theory and the principle of operation.
- Section 2 describes the application of TCSC for damping of electromechanical oscillations.
- Section 3 explains the application of TCSC for mitigation of sub-synchronous resonance.
- Section 4 describes the layout design and protection of TCSC

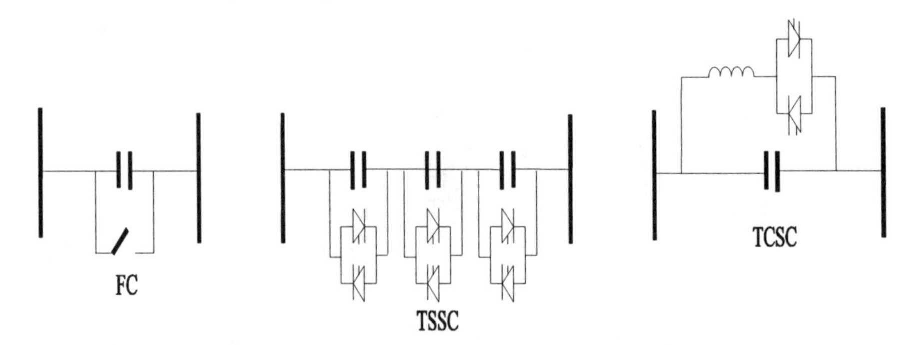

Figure 5.4 Different configuration of series compensation

5.2 Principle of operation

The main circuit of the TCSC consists of a capacitor bank and a thyristor controlled inductive branch connected in parallel as shown in Figure 5.5. The capacitor bank may have a value of e.g. 10–30 Ω/phase and a rated continuous current of 1500–3000 A rms. The capacitor bank for each phase is mounted on a platform providing full insulation towards ground. The valve contains a string of series-connected high-power thyristors with a maximum total blocking voltage in the range of hundreds kV. The inductor is an air-core inductor with a few mH inductance. A metal-oxide varistor (MOV) is provided across the capacitor in order to prevent overvoltages across the capacitor bank.

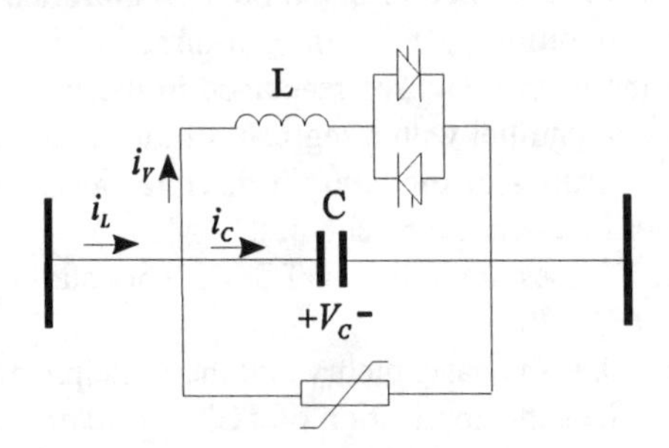

Figure 5.5 TCSC main circuit components.

The characteristic of the TCSC main circuit depends on the relative reactances of the capacitor bank $X_C = -\dfrac{1}{\omega_n C}$ and the thyristor branch $X_V = \omega_n L$. The resonant frequency of the LC circuit formed by the inductance in the thyristor branch and the capacitance in the series capacitor bank is given by

$$\omega_{\mathrm{R}} = \frac{1}{\sqrt{LC}} = \omega_{\mathrm{n}} \sqrt{\frac{-X_{\mathrm{C}}}{X_{\mathrm{V}}}} \tag{5.1}$$

A parameter λ is defined as the quotient between the resonant frequency and the network frequency

$$\lambda = \frac{\omega_{\mathrm{R}}}{\omega_{\mathrm{n}}} = \sqrt{\frac{-X_{\mathrm{C}}}{X_{\mathrm{V}}}} \tag{5.2}$$

Reasonable values of λ fall in the range of 2 to 4. Thus the reactance of the inductor is much smaller than that of the capacitor bank at rated the frequency. The TCSC can operate in a number of modes which exhibit different values of apparent reactance. In this context the apparent reactance simply is defined as the imaginary part of the quotient between phasors representing the fundamental of the capacitor voltage and the fundamental of the line current at rated frequency

$$X_{\mathrm{app}} = \mathrm{Im}\left\{\frac{\boldsymbol{U}_{\mathrm{C1}}}{\boldsymbol{I}_{\mathrm{L1}}}\right\} \tag{5.3}$$

Further it is practical to define a boost factor, K_{B}, as the quotient between the apparent and the physical reactance of the TCSC:

$$K_{\mathrm{B}} = \frac{X_{\mathrm{app}}}{X_{\mathrm{C}}} \tag{5.4}$$

5.2.1 Blocking mode

When the thyristor valve is not triggered and the thyristors are kept in non-conducting state the TCSC is operating in blocking mode. The line current passes only through the capacitor bank. The capacitor voltage phasor is given by the line current phasor according to the formula:

$$\boldsymbol{U}_{\mathrm{C}} = jX_{\mathrm{C}}\boldsymbol{I}_{\mathrm{L}} \qquad X_{\mathrm{C}} < 0 \tag{5.5}$$

In this mode the TCSC performs like a fixed series capacitor with boost factor equal to one. Reference directions for voltage and current are defined in Figure 5.5.

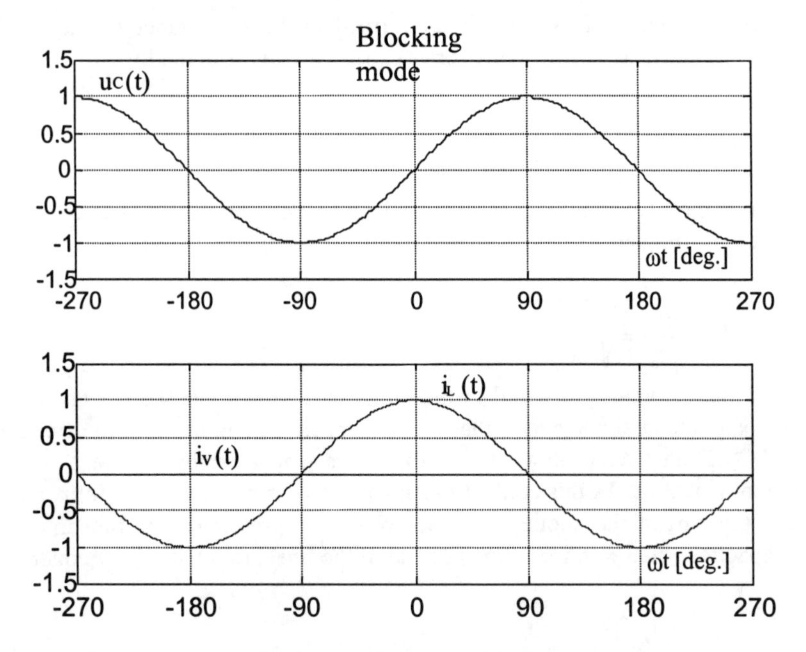

Figure 5.6 Waveforms of voltage and currents in block mode

5.2.2 Bypass mode

If the thyristor valve is triggered continuously the valve stays conducting all the time and the TCSC behaves like a parallel connection of the series capacitor bank with the inductor in the thyristor valve. The resulting voltage in the steady state across the TCSC is given by:

$$\boldsymbol{U}_{\mathrm{C}} = \frac{-jX_{\mathrm{C}}}{\lambda^2 - 1}\boldsymbol{I}_{\mathrm{L}} \qquad X_{\mathrm{C}} < 0 \tag{5.6}$$

This voltage is inductive and the boost factor is negative.

$$K_{\mathrm{B}} = -\frac{1}{\lambda^2 - 1} \tag{5.7}$$

The waveforms in Figure 5.7 show that the valve current is somewhat bigger than the line current due to the current generation in the capacitor bank. When λ is considerably bigger than unity the capacitor voltage at a given line current is much lower in bypass than in blocking mode. Therefore the bypass mode is utilized as a means to reduce the capacitor stress during faults.

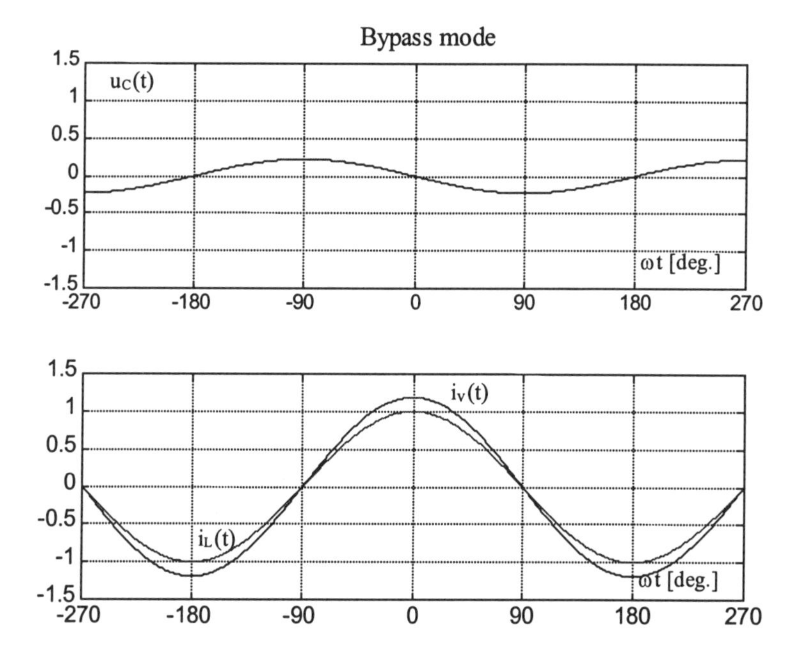

Figure 5.7 Waveforms of capacitor voltage and currents in bypass mode (λ=2.5)

5.2.3 Capacitive boost mode

If a trigger pulse is supplied to the thyristor having forward voltage just before the capacitor voltage crosses the zero line a capacitor discharge current pulse will circulate through the parallel inductive branch. Voltage and current waveforms are shown in Figure 5.8.

The discharge current pulse adds to the line current through the capacitor bank. It causes a capacitor voltage that adds to the voltage caused by the line current. The capacitor peak voltage thus will be increased in proportion to the charge that passes through the thyristor branch. The fundamental voltage also increases almost proportionally to that charge. The charge depends on the conduction angle β, which is defined in Figure 5.8, and the main circuit parameter λ according to equation 5.2. A simple mathematical formula can be obtained when the losses are neglected and assuming that the line current is stiff and sinusoidal:

$$K_{\mathrm{B}} = 1 + \frac{2}{\pi} \frac{\lambda^2}{\lambda^2 - 1} \left[\frac{2\cos^2 \beta}{\lambda^2 - 1} (\lambda \tan \lambda\beta - \tan \beta) - \beta - \frac{\sin 2\beta}{\beta} \right] \qquad (5.8)$$

Due to the factor $\tan(\lambda\beta)$ this formula has an asymptote at:

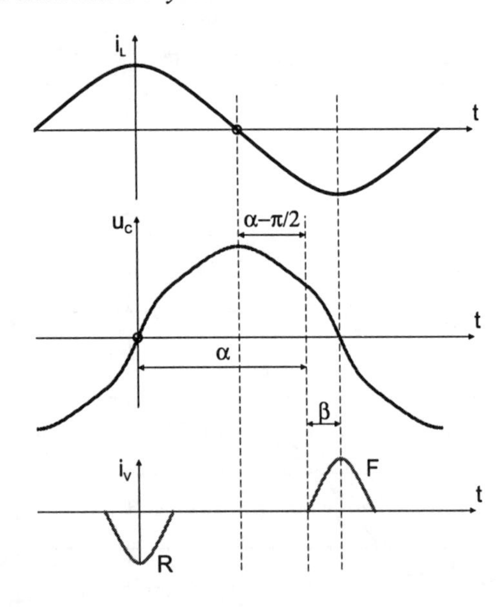

Figure 5.8 Voltage and current waveforms in the capaitive boost mode

$$\beta_{\infty} = \frac{\pi}{2\lambda} \tag{5.9}$$

The TCSC operates in the capacitive boost mode when:

$$0 < \beta < \beta_{\infty} \tag{5.10}$$

Often the characteristics are presented using the trigger angle α as a variable instead of the conduction angle β that appears in the formula. The trigger angle is defined as the delay angle after the first occurrence of forward voltage across the thyristor. We have:

$$\beta = \pi - \alpha \tag{5.11}$$

The left-hand branch of the curve in Figure 5.9 illustrates the characteristics according to equation 5.8.

The TCSC is provided with a controller with the means to influence the conduction angle β and means for synchronization of the triggering of the thyristors with the line current. Figure 5.10 shows examples of waveforms for different boost levels.

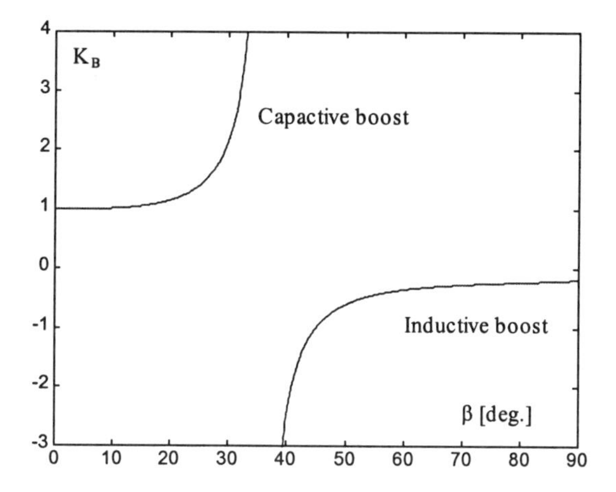

Figure 5.9 Boost factor versus conduction angle for TCSC with λ=2.5

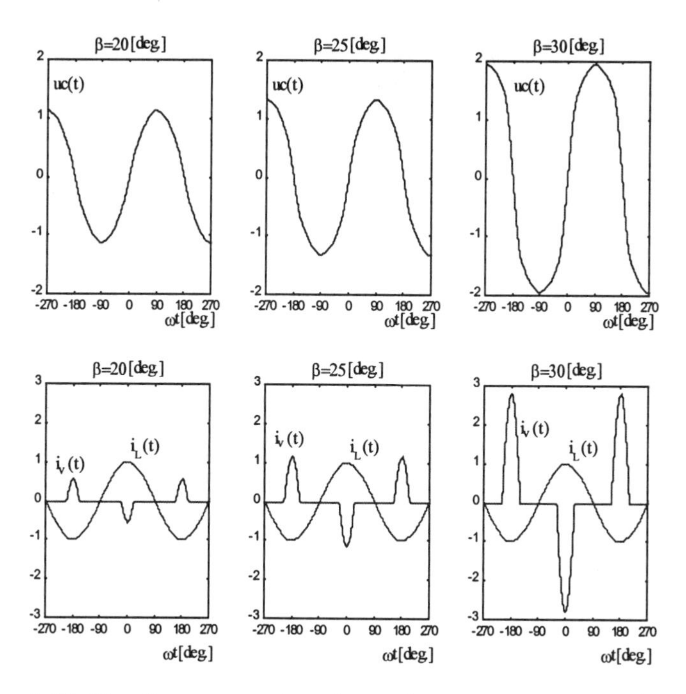

Figure 5.10 Waveforms at various boost factors in capacitive boost mode for λ=2.5

The main circuit of the TCSC is designed for a certain maximum capacitor voltage, which defines the maximum boost factor available at a certain line current. When the line current exceeds a certain value the boost factor thus must be reduced along a hyperbola giving constant capacitor voltage as is shown in Figure 5.11. Most often also a maximum boost factor K_{Bmax} is defined for lower line currents. Different curves are applied for continuous duty and short-time overloads. Beside the maximum capacitor voltage two other factors limit the operating area for the TCSC in the capacitive boost mode. First it should be noted that the range with $K_{\mathrm{B}} < 1$ is not available within the controlled area. Secondly a limitation normally is provided at very low line current. This limitation has several causes, which are related to both the main circuit (auxiliary power) and the control system (noise and resolution in sensors).

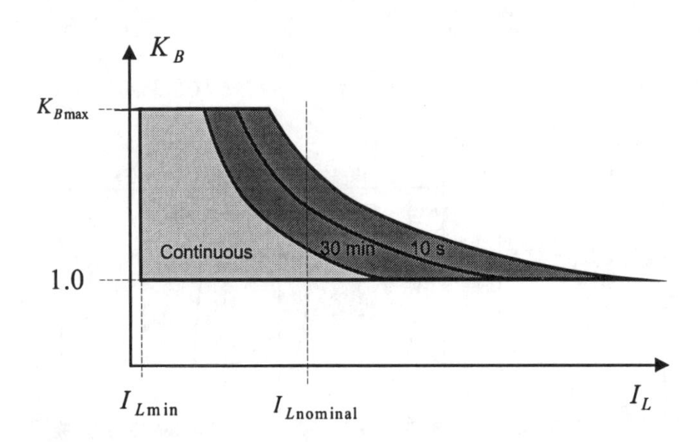

Figure 5.11 Operating area for capacitive boost mode

5.2.4 Inductive boost mode

In a different mode of operation the circulating current in the TCSC thyristor branch is bigger than the line current. This occurs if:

$$0 < \beta < \beta_{\infty} \tag{5.12}$$

The formula for the boost factor given in (5.8) still applies. The boost characteristics as a function of the conduction angle β appear in the right hand side branch of the curve in Figure 5.8. The voltage and current waveforms are shown for various boost factors in Figure 5.12.

It can be seen from the curves that large thyristor currents result in inductive boost mode. Further the capacitor voltage waveform is very much distorted from its desired sinuoidal shape. The peak voltage appears close to the turnon. The poor waveform and the high valve stress makes the inductive boost mode less attractive for steady state operation.

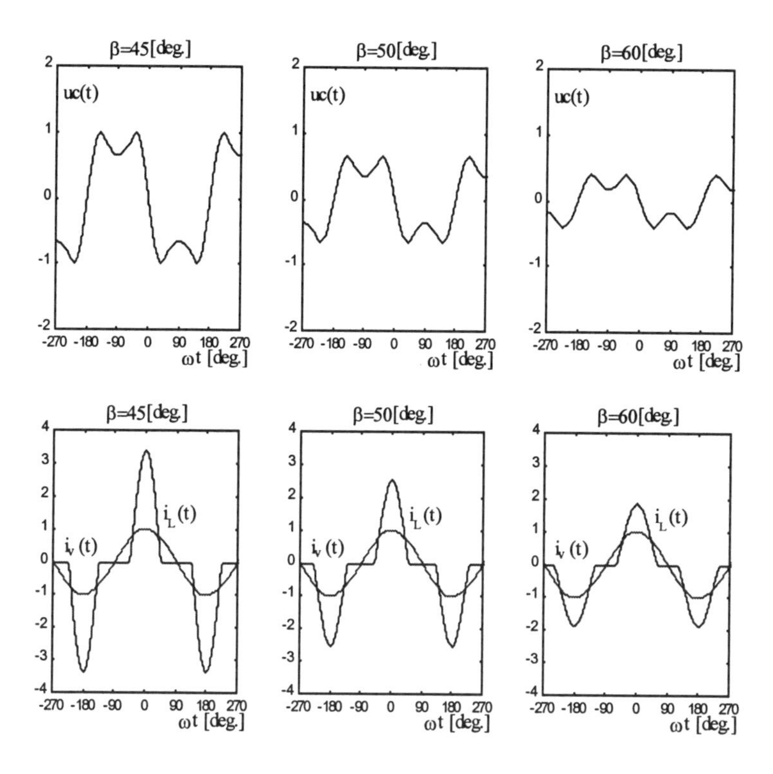

Figure 5.12 Waveforms in inductive boost mode for TCSC with λ=2.5

5.2.5 Harmonics

The harmonics caused by a TCSC emerge from the harmonics in the thyristor branch current. Figure 5.13 shows the model that can be used for study of harmonic distribution.

In Figure 5.13:

Z_A and Z_B : are the corresponding harmonic impedances seen from Bus A and Bus B respectively.

Z_{lineA} and Z_{lineB} : are the corresponding harmonic impedances of two section of the compensated line.

In the figure, the thyristor branch is modelled as a current source which injects harmonic current into the series capacitor bank. The series capacitor bank provides a low impedance path for the harmonics and very little current will leak into the transmission line. Generally the interest is focused on the harmonic content of the voltages at Bus A and Bus B and currents flowing towards Bus A and Bus B.

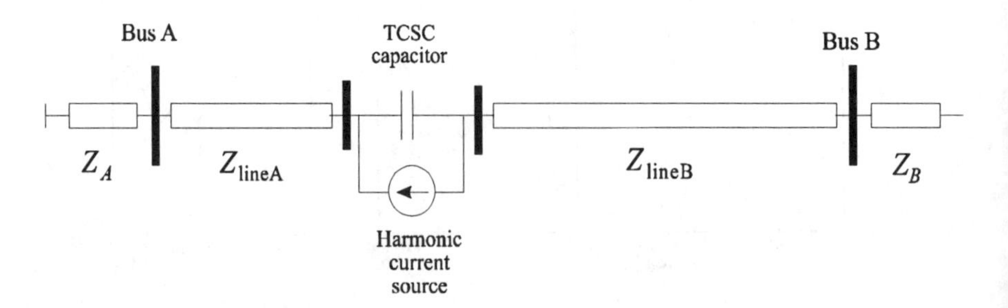

Figure 5.13 Model for evaluation of harmonic distortion.

The lower harmonics of the inserted voltage are proportional to $K_{\text{B}}-1$ and to the line current amplitude. Normally the TCSC operates in the capacitive boost mode with a boost factor close to one. Under these circumstances only the lowest order harmonics, like the third and fifth harmonics, have any practical importance.

5.2.6 Boost control systems

The boost control system provides trigger control signals to the thyristors so that the desired boost level is obtained. The control system may be arranged in different ways and some main principles will be described in the following.

5.2.6.1 Open loop boost control

The most straightforward approach simply uses the steady-state characteristics as described above in (5.8) and in Figure 5.9. It translates a boost reference value K_{Bref} into a corresponding conduction angle β and further into a trigger angle α as defined in Figure 5.14. The synchronization normally will be derived from the line current rather than from the capacitor voltage as this will provide much better stability. Figure 5.14 illustrates this boost control approach.

The step response time to boost reference steps is of the order of several hundred ms when the open loop boost control is used. It is important that the control system maintains sufficient margin from the asymptote in order to avoid overvoltages. Specifically for main circuits having high λ the characteristic is very steep at high boost levels.

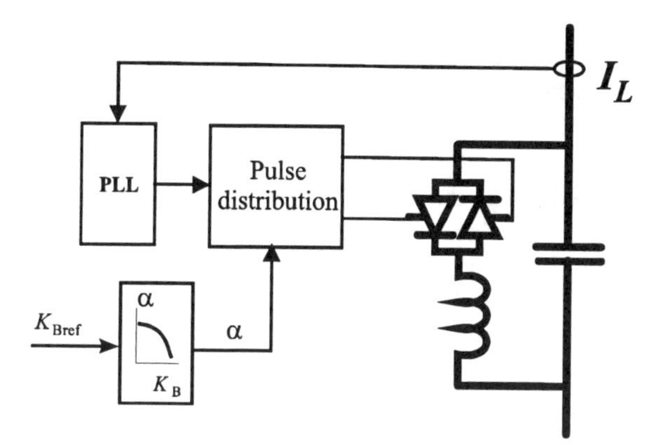

Figure 5.14 Open-loop boost control system

5.2.6.2 Feedback boost control

An external feedback loop may be provided in order to speed up the control system described above. The capacitor voltage and/or line current is measured and a feedback signal is sent to a regulator, which provides the trigger angle to the thyristors. Several feedback signals may be used e.g. boost level (equivalent to apparent reactance) or line current amplitude or line power transfer.

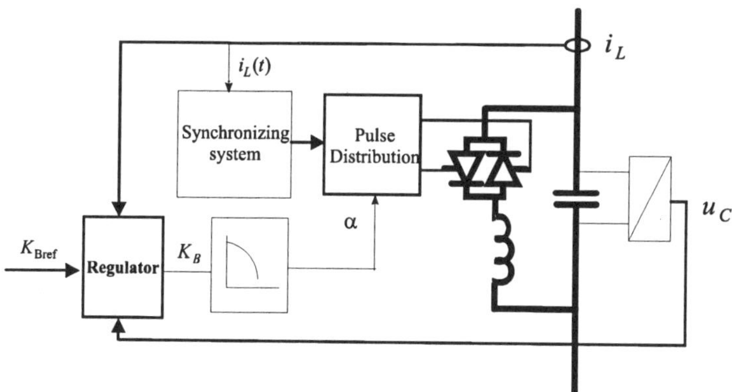

Figure 5.15 Feedback loop

The response time for boost control may be reduced by a factor of 2–3 as compared with the open loop. Further the risk of causing overvoltage when operating close to the asymptote is eliminated when the capacitor voltage is continuously measured and fed back to the regulator.

5.2.6.3 Boost control based on instantaneous capacitor voltage and line current

A different type of boost control is based on a layered control structure with an inner control loop which utilizes instantaneously measured capacitor voltage and line current. Two different approaches use different inner control targets:

The first approach aims at controlling the charge passing through the thyristor branch at a referenced value (the charge is proportional to the boost voltage).

The second approach aims at controlling the timing of the capacitor voltage zero crossing. In both cases a synchronization system provides phase information which is derived from the line current. An algorithm that uses the measured capacitor voltage and line current as input, determines the thyristor-triggering instant The first system is shown in Figure 5.16.

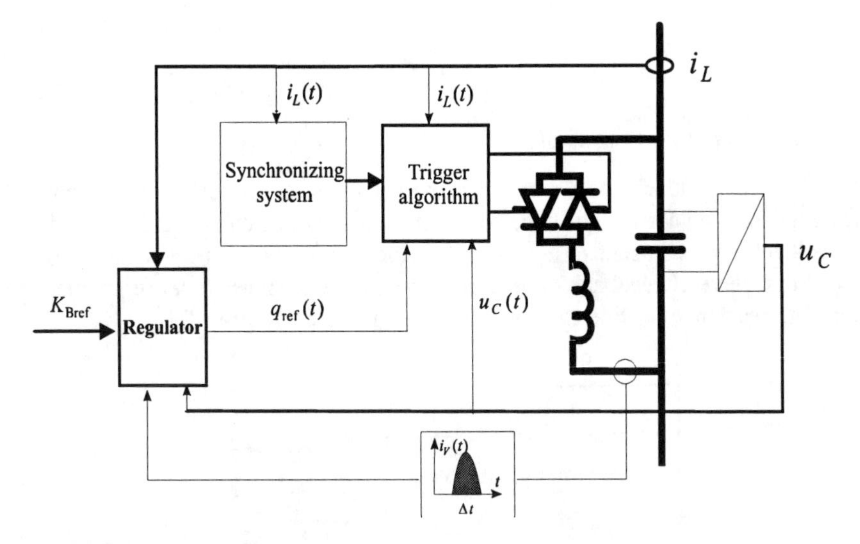

Figure 5.16 Boost control system using subordinated thyristor charge control

The second system works with timing control of the capacitor voltage zero-crossing (equivalent to timing of thyristor current peak). The block diagram is shown in Figure 5.17.

The idea of using the timing is illustrated in Figure 5.18, which shows that the same boost will be obtained independent of the main circuit parameters if the capacitor voltage zero-crossing remains at a certain instant. In a main circuit having negligible inductance an instantaneous capacitor voltage reversal would occur. Figure 5.18 shows that the boost caused by thyristor action can be represented by an instantaneous capacitor voltage reversal at the time when the capacitor voltage zero-crosses.

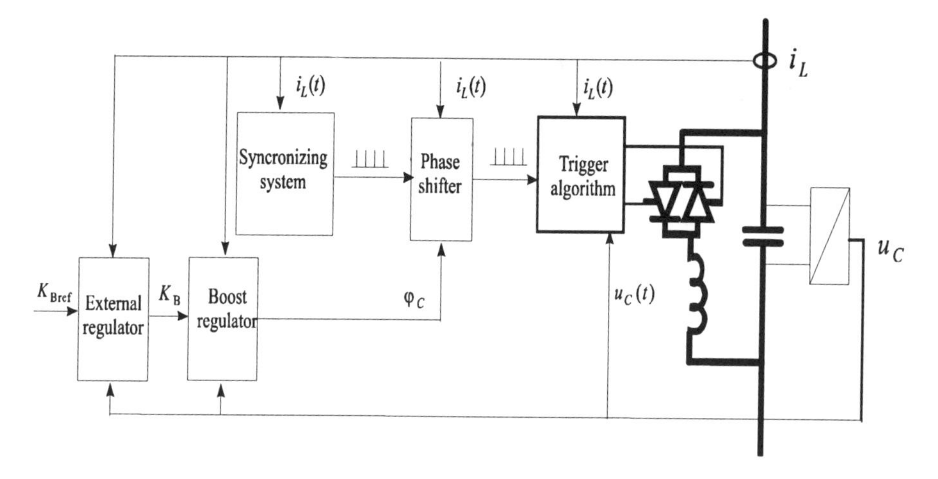

Figure 5.17 Boost control system using subordinate timing control

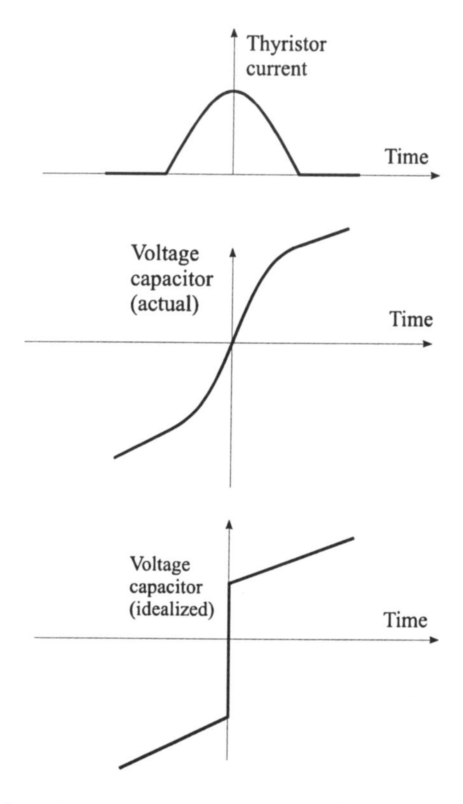

Figure 5.18 Equivalent, instantaneous capacitor voltage reversal

If the reversal instants can be directly controlled, the TCSC boost can easily be controlled as is shown in Figure 5.19. In the steady state only one equilibrium reversal instant exists. This unique instant is defined by the condition that the charge received from the line current must vanish between two consecutive reversals. This condition is illustrated in Figure 5.19a. When the timing of the reversal is phase advanced relative the equilibrium position an increase of the boost will occur in each reversal, Figure 5.19 b. Similarly the boost will decrease when the timing of the reversals is delayed beyond the equilibrium position, Figure 5.19(c).

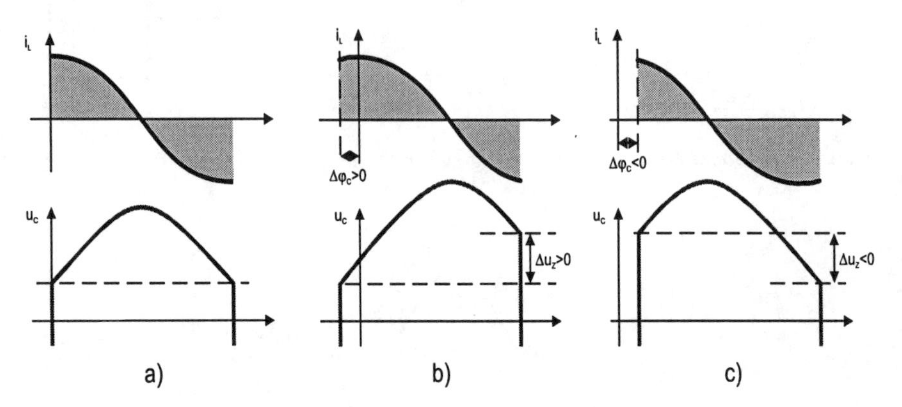

Figure 5.19 Boost control using phase control of reversal instants.

A simple linear regulator can be used to provide phase reference for the pulse trains that controls the timing of the reversals in the underlying system so that the desired boost factor is achieved. The boost control systems that utilize instantaneously measured capacitor voltage and line current do have faster responses than the systems based on steady-state characteristics. The response time may be in the range of some tens of ms.

5.3 Application of TCSC for damping of electromechanical oscillations

Damping of electromechanical oscillations has been recognized as an important issue in electric power system operation. Application of power system stabilizers (PSSs) has been one of the first measures to enhance the damping of power swings. With increasing transmission line loading over long distances, the use of conventional power system stabilizers might not, in some cases, provide sufficient damping for inter-area power swings. In these cases, other effective solutions need to be studied.

The basic power flow equation through a transmission line shows that modulating the voltage and reactance influences the flow of active power. In principle, a thyristor-controlled series capacitor (TCSC) could provide fast control of active power through a transmission line. The possibility of controlling the transmittable power suggests the potential application of these devices for damping of power system electromechanical oscillations.

A question of great importance is the selection of the input signals for the TCSC in order to damp power oscillations in an effective and robust manner. From control design and practical consideration, a desirable input signal should have the following characteristics:

- The swing modes should be observable in the input signal.
- A desirable level of damping should be achieved.
- The damping effect should be robust with respect to changing operating conditions.
- The input signal should preferably be local.

5.3.1 Model

For power swing damping studies, a TCSC can be modeled as a variable reactance. Figure 5.20 shows the general block diagram of the TCSC model used for power swing studies:

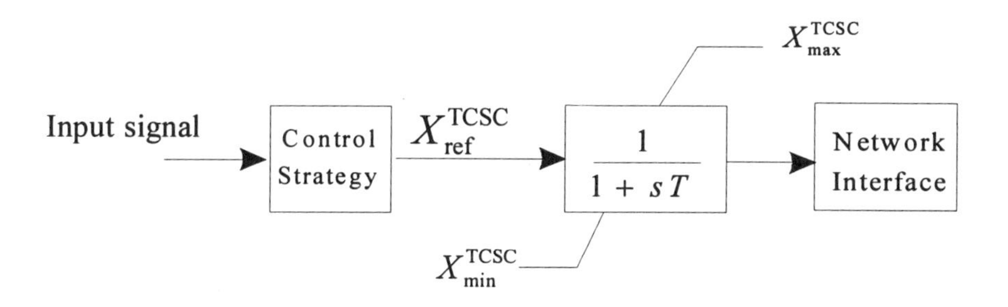

Figure 5.20 TCSC model as damping controller

Based on a control strategy a reference reactance X_{ref}^{TCSC} is determined. This signal is passed through a delay block. The time constant T approximates the delay due to the main circuit characteristics and control systems. The output of the model is restricted by two limits:

1. Static reactance limit

2. Dynamic reactance limit

Static reactance limits are suggested by the planner for a desirable level of TCSC contribution to damping of power swings. *Dynamic reactance limit* is a constraint, which is decided by the voltage across the TCSC. It means that whenever the voltage across the TCSC exceeds the maximum permissible value $V_{\max}^{\mathrm{TCSC}}$, the reactance should be lowered. If I_{Line} is the line current, then:

$$X^{\mathrm{TCSC}} \leq \frac{V_{\max}^{\mathrm{TCSC}}}{I_{\mathrm{Line}}} \tag{5.13}$$

Mathematically, the TCSC reactance is modeled as:

$$X^{\mathrm{TCSC}} = X_{\mathrm{C}}(1 + K_{\mathrm{B}}) \qquad X_{\min}^{\mathrm{TCSC}} < X < X_{\max}^{\mathrm{TCSC}} \tag{5.14}$$

Where X_{C} is the reactance of the fixed capacitor, K_{B} shows the degree of the voltage boost across the capacitor and X^{TCSC} is the apparent reactance of TCSC.

5.3.2 TCSC damping characteristics

In this section the fundamental control signal required for damping of electromechanical oscillations is discussed. To facilitate the analysis, a one-machine infinite-bus system is considered and the classical model for a synchronous machine is used.

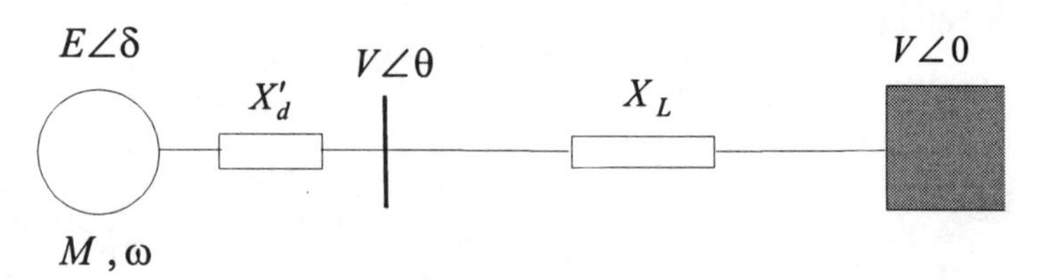

Figure 5.21 System for study of damping control

Assuming a constant mechanical power, the linearized equation of the system in the state space form is:

$$\begin{bmatrix} \Delta\dot{\omega} \\ \Delta\dot{\delta} \end{bmatrix} = \begin{bmatrix} 0 & \frac{K_{\mathrm{S}}}{M} \\ 1 & 0 \end{bmatrix} \begin{bmatrix} \Delta\omega \\ \Delta\delta \end{bmatrix} \tag{5.15}$$

Where K_{S} is the synchronizing coefficient:

$$K_{\rm S} = \frac{VE}{X'_{\rm d} + X_{\rm L}} \cos\delta \tag{5.16}$$

Equation 5.16 shows that with this model the system response is purely oscillatory. To damp the power oscillations, supplementary power is needed to modulate the generated power. If the modulated power is selected as:

$$\Delta P = K_{\omega} \Delta\omega + K_{\delta} \Delta\delta \tag{5.17}$$

The controlled matrix is:

$$\begin{bmatrix} \Delta\dot{\omega} \\ \Delta\dot{\delta} \end{bmatrix} = \begin{bmatrix} \frac{K_{\omega}}{M} & \frac{K_{\rm S} + K_{\delta}}{M} \\ 1 & 0 \end{bmatrix} \begin{bmatrix} \Delta\omega \\ \Delta\delta \end{bmatrix} \tag{5.18}$$

The system has at most two distinct eigenvalues and in the case of an oscillatory response:

$$\lambda_{1,2} = \frac{K_{\omega}}{M} \pm j \left(\frac{4(K_{\rm S} + K_{\delta})}{M} - \frac{K_{\omega}^2}{M^2} \right)^{\frac{1}{2}} \tag{5.19}$$

The equation above shows that only the component of $\Delta\omega$ in (5.17) contributes to the damping and $\Delta\delta$ affects only the frequency of oscillation (synchronizing torque).

5.3.3 Damping of power swings by TCSC

It is assumed that a TCSC is located on the intertie to enhance the damping of power swings:

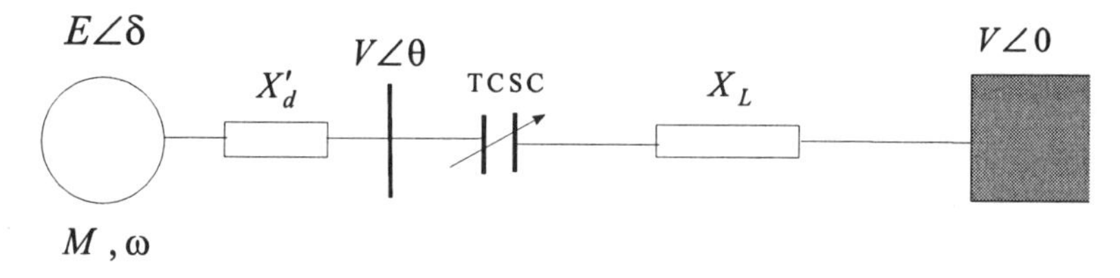

Figure 5.22 Study of TCSC for damping control

No damping torque is assumed in the system, which means that the transmission system will oscillate by itself without damping and only TCSC can contribute to the damping. The control signal is selected as the difference between the speed of the machine and the infinite bus. For the reason discussed in the previous section, this signal is appropriate for damping of power swings. Thus the control law is:

$$\Delta X_{C} = K_{C}\Delta\omega \tag{5.20}$$

where K_C is the gain. The linearized machine equations are:

$$\begin{aligned} M\Delta\dot{\omega} &= -\Delta P \\ \Delta\dot{\delta} &= \omega \end{aligned} \tag{5.21}$$

where $P = bEV\sin\delta$ with $b = \dfrac{1}{X_L - X_C + X'_d}$. The linearized controlled system matrix is:

$$\begin{bmatrix} \Delta\dot{\omega} \\ \Delta\dot{\delta} \end{bmatrix} = \begin{bmatrix} \dfrac{-K_C b^2 E\sin\delta}{M} & \dfrac{-bE\cos\delta}{M} \\ 1 & 0 \end{bmatrix} \begin{bmatrix} \Delta\omega \\ \Delta\delta \end{bmatrix} \tag{5.22}$$

It is seen the damping term depends both on K_C and $\sin\delta$. This reveals the following conclusions:

- A TCSC can enhance the damping of electromechanical oscillations.
- The damping effect of a TCSC increases with transmission line loading. This is a very important feature of a TCSC, since the damping of a system normally is lower with heavily loaded lines.

5.3.4 POD controller model

Figure 5.23 shows a suggested structure for the POD regulator.

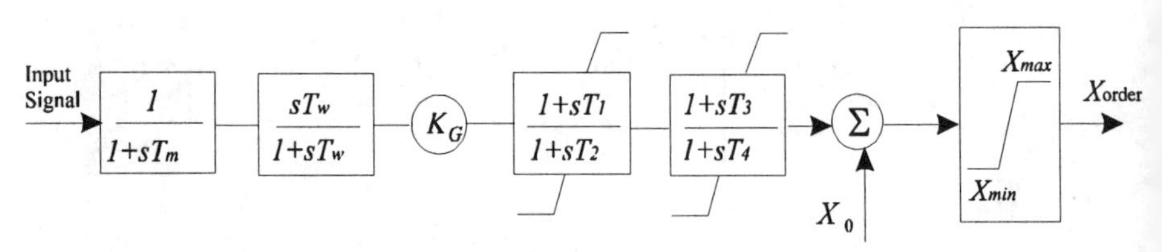

Figure 5.23 Structure of POD regulator

Where

T_m is the measuring device time constant

T_w is the washout time constant

K_G is the regulator gain

T_1, T_2, T_3, T_4 are the parameters of the phase compensation block

5.3.5 Choice of POD regulator parameters

This section describes a method based on the sensitivity of eigenvalues of the linearized power system for determination of the POD regulator parameters. Figure 5.24 shows the model of a power system with a regulator:

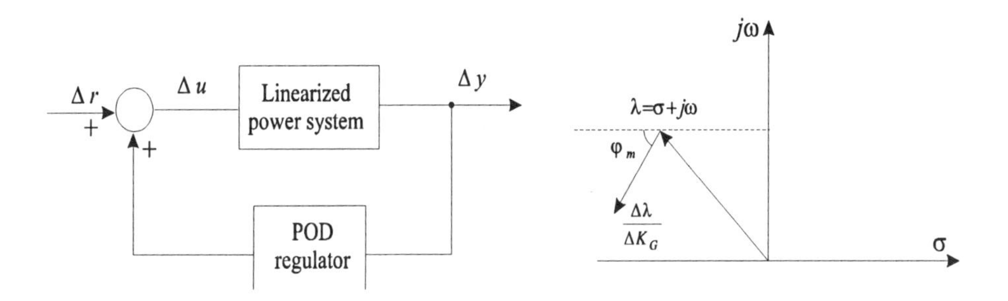

Figure 5.24 Linearized power system and demonstration of eigenvalue displacement

Suppose the uncontrolled power system has a critical eigenvalue $\lambda = \sigma + j\omega$. The sensitivity of this eigenvalue with respect to the gain of the regulator shows how the eigenvalue will change after the introduction of the regulator. Figure 5.24 shows the displacement of the eigenvalue after the POD control action. The phase angle φ_{m} shows the compensation angle which is needed to drive the eigenvalue to the −180 line. These angles can be obtained by a lead–lag function.

$$H(s) = \left(\frac{1 + s\tau / \alpha}{1 + s\tau} \right)^2 \tag{5.23}$$

Where

$$\alpha = \frac{1 - \sin\varphi_{\mathrm{m}}}{1 + \sin\varphi_{\mathrm{m}}} \qquad \tau = \frac{1}{\omega\sqrt{\alpha}} \tag{5.24}$$

5.3.6 Numerical examples

In this section, the application of a TCSC is demonstrated through a model power system.

5.3.6.1 Example 1

Figure 5.24 shows a two area system connected via an intertie.

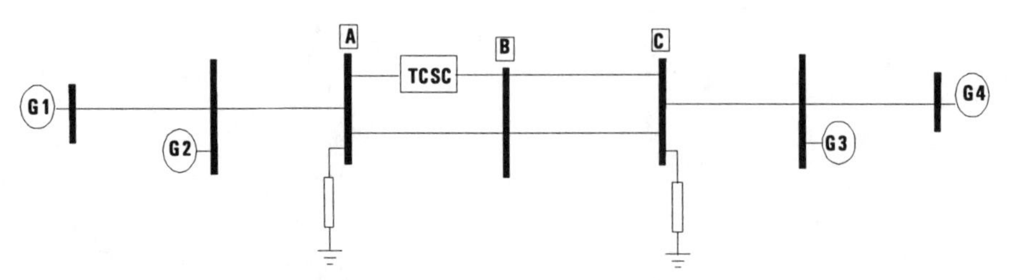

Figure 5.25 Four-machine test system

The two-area system exhibits three electromechanical oscillation modes:

- An inter-area mode with a frequency of 0.65 Hz in which the generating units in one area oscillate against those in the other area.
- A local mode in area 1, with a frequency of 1.13 Hz, in which generator G1 and G2 oscillate against each other.
- A local mode in area 2, with a frequency of 1.16 Hz, in which generator G3 and G4 oscillate against each other.

The method of analysis is based on eigenvalue analysis of the linearized power system. The eigenvalues related to the electromechanical modes are shown in Table 5.1.

Table 5.1 The eigen-values of the uncontrolled system

	Eigenvalue		Damping
Mode	1/s	Hz	ratio (%)
Inter-area	0.00	0.65	0.00
Local G1, G2	-0.81	1.13	11.3
Local G3, G4	-0.83	1.16	11.3

It illustrates that while the damping of local modes is rather good, the inter-area mode has no damping. A controllable series capacitor is assumed to be located on the inter-tie to enhance the damping of the inter-area mode. The local variables based on the principles discussed in the previous section, are used for input signals. The following input signals are selected:

- P_E : Active power flow through the compensated line
- V_{AB} : Voltage across the compensated line

The regulator parameters are designed for the two inputs and are given below:

$T_w = 1 \quad T_1 = 0.1274 \quad T_2 = 0.5907 \quad T_3 = 0.1274 \quad T_4 = 0.5907$

The impact of TCSC on damping of electromechanical modes are shown in Table 5.2 for the two input signals and for $K_G = 1$:

Table 5.2 Damping of electromechanical modes with different inputs

	Eigenvalue		Damping
Mode	1/s	Hz	ratio (%)
Inter-area	-0.14	0.59	3.8
Local G1, G2	-0.81	1.13	11.3
Local G3, G4	-0.82	1.16	11.2

Line power as input

	Eigenvalue		Damping
Mode	1/s	Hz	ratio (%)
Inter-area	-0.13	0.65	3.1
Local G1, G2	-0.81	1.13	11.3
Local G3, G4	-0.82	1.16	11.3

Line voltage as input

The simulation results show that a TCSC can contribute to the damping of the power swings in complex systems where many modes are present. In this example, the regulator has been designed to damp the inter-area mode. It is noted that the damping of the local modes is not degraded.

5.3.6.2 Example 2

In this example two TCSCs and one SVC are assumed in the system. The TCSCs are intended for power oscillation damping and the SVC is intended for voltage regulation.

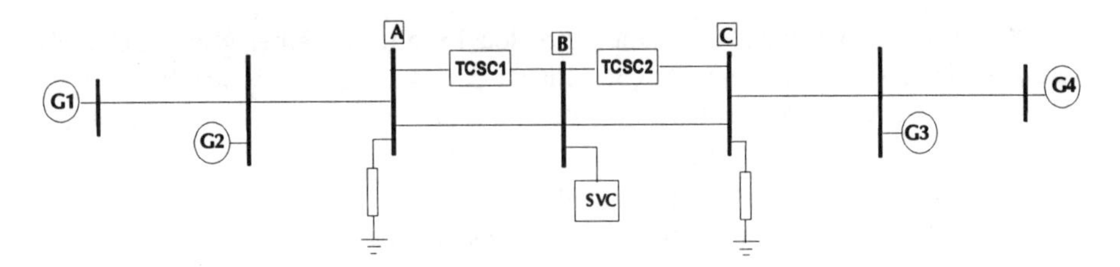

Figure 5.26 Four-machine system with several controllers

The size of SVC and TCSC is 200 Mvar and 261 MVAR respectively. The maximum boost of TCSC apparent reactance is ±15% of the intertie reactance. Table 5.3 shows different studied cases. In case I both TCSCs act as fixed capacitors, i.e. in blocking mode. In the second case TCSC1 is controlled but TCSC2 is fixed. In the third case TCSC1 is fixed and TCSC2 is controlled. In Case IV both TCSCs are controlled.

Table 5.3: Simulation scenarios for the four-machine system

Case	TCSC1	TCSC2
I	Fixed	Fixed
II	Controlled	Controlled
III	Fixed	Controlled
IV	Controlled	Controlled

The following disturbance is considered: a three-phase fault occurs at node C and is cleared after 100 ms with opening of the uncompensated line between Bus B and Bus C. The speed of the generator 1 is taken as reference and the speed between generator 3 and generator 1 (representative of the inter-area mode) and the difference between the speed of generator 2 and 1 (representative of the local mode) for different cases are shown in Figures 5.27 to 5.30.

The simulation results show the effectiveness of the TCSCs for power swing damping. Figure 5.28 (Case II) and Figure 5.29 (Case III) show that each TCSC has improved the power swing damping. Figure 5.30 (Case IV) shows that the contribution of both TCSCs is higher than each individually. In other words the damping effects of the two TCSCs are additive.

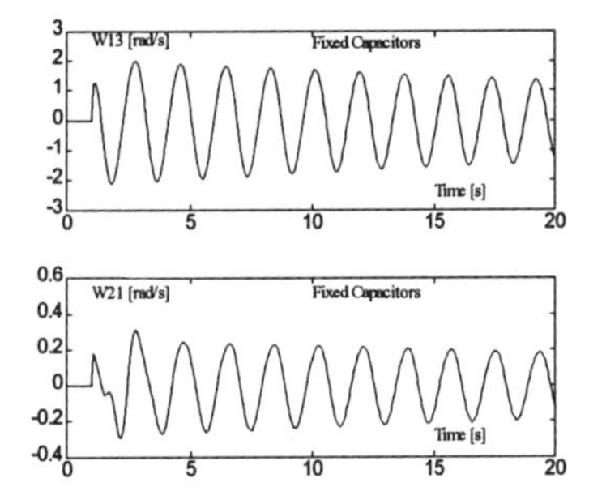

Figure 5.27 System response with fixed capacitors

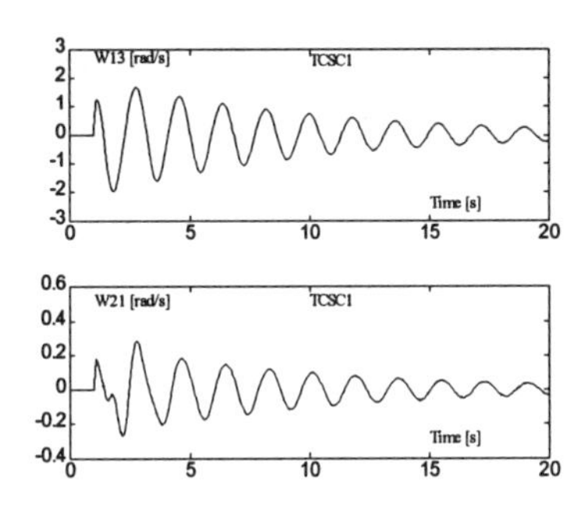

Figure 5.28 System response with TCSC1 controlled

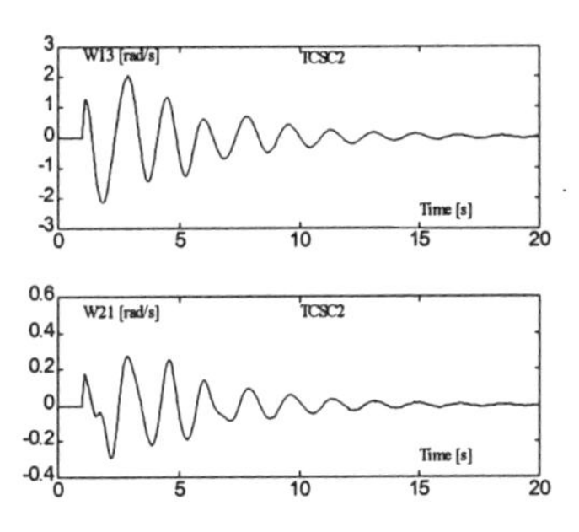

Figure 5.29 System response with TCSC2 controlled

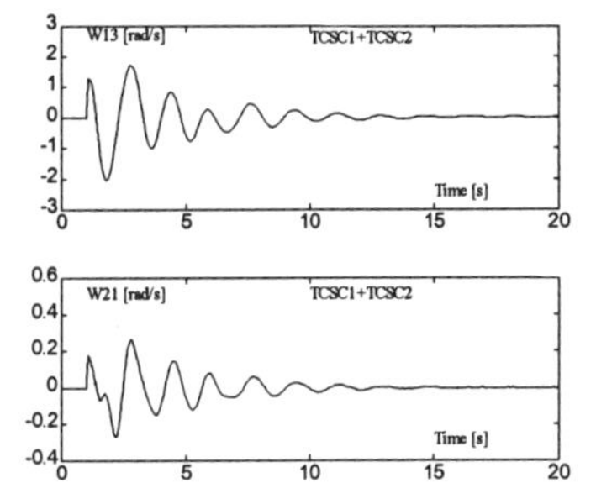

Figure 5.30 System response with both TCSCs controlled

5.4 Application of TCSC for mitigation of subsynchronous resonance

The introduction of series compensation improves the transmission system with respect to voltage stability and angular stability. However, at the same time an electrical resonance is introduced in the system. Experience has shown that such an electrical resonance may under certain circumstances interact with mechanical torsional resonances in turbine-generator shaft systems in thermal generating plants. This phenomenon is known as (one form of) Subsynchronous Resonance (SSR). Today the SSR problem is well understood and taken into account when series compensation equipment is designed. Sometimes SSR conditions may limit the degree of compensation needed for better power system performance. The use of TCSC may alleviate such restrictions as will be explained in the following sections.

5.4.1 The subsynchronous resonance (SSR) phenomena related to series compensation

The simplest electrical representation of a transmission line is a RL circuit exhibiting the impedance $Z = R + jX$. Inserting a series capacitor with the reactance $X_C = kX$ (k degree of compensation) causes a series resonance at frequency:

$$\omega_R = \omega_n \sqrt{k} \tag{5.25}$$

When the compensation level is below unity the electrical resonance frequency will be located below the network rated frequency. The consequence of the existence of the electrical resonance is that substantial currents at the resonance frequency may be caused by voltage sources having low amplitude. Figure 5.31 shows the impedance variation of a typical 250 km 400 kV line for uncompensated and 50% compensated line.

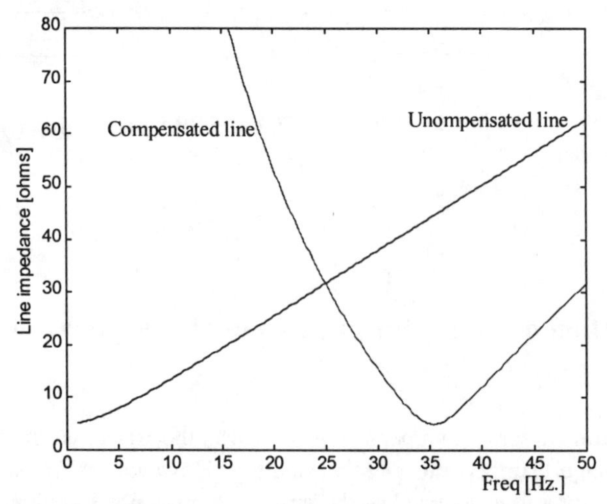

Figure 5.31 Line impedance with and without series compensation

Three different types of interactions in the sub-synchronous frequency range, which may arise between generators connected to the network and the transmission system, are normally classified as Sub-Synchronous Resonance (SSR) phenomena:

5.4.1.1 Induction generator effect SSR

Figure 5.32 shows a simplified equivalent circuit of a synchronous machine and the transmission line at sub-synchronous frequency. The machine has been represented by a circuit similar to that used for induction motors.

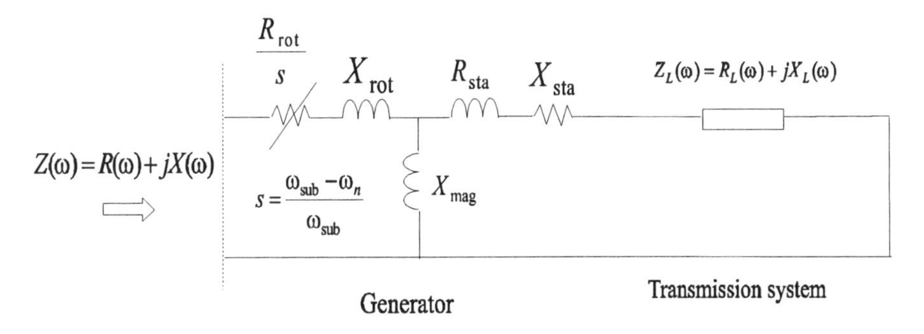

Figure 5.32 Illustration of the induction generator effect

The slip is negative at any sub-synchronous frequency since $\omega_{sub} < \omega_n$. This means that the machine contributes with a negative resistance to the total effective resistance of the machine and transmission system. Self-excitation of a sub-synchronous current might occur if an electrical resonance frequency exists (i.e. total reactance vanishes at that frequency). In this case the total resistance, i.e. the resistance in the generator plus the transmission line, is negative. In such a case the machine might pick up energy from the mechanical system and build up increasing subsynchronous currents. Such subsynchronous currents will create a pulsating torque together with the main flux in the generator. If the condition persists sufficiently long the pulsation may reach destructive levels which might result in damage to the shaft. This type of SSR is called induction generator effect. It is a purely electrical phenomenon and it may occur for all kinds of generators. However, the crucial point is that the contribution of negative resistance from the equivalent circuit rotor side must exceed the positive resistance in the lines.

As the negative resistance is inversly proportional to slip and in view of (5.25) the electrical resonance comes close to the synchronous speed only for a very high degree of compensation. Therefore it appears that sufficient conditions for this type of SSR are not very likely to occur in the normal power systems with realistic degrees of compensation.

5.4.1.2 Torsional interaction SSR

A thermal power plant consists of a turbine-generator set containing a number of components having substantial inertia (turbines and generator) linked together by a mechanical shaft system. This system exhibits mechanical torsional resonances with low damping at certain frequencies below the rated speed of the generator. The torsional mechanical oscillations will be excited by any torque transient that occurs. The different oscillation modes fade away with time constants up to of ten seconds. The subsynchronous line current together with the main flux in the machine creates an electrical torque pulsation. Calculations show that the torque generated may further increase the amplitude of the torsional oscillation in the turbine-generator shaft system, so that an exponentially increasing amplitude of the oscillation results. Such a condition might lead to shaft destruction. The phenomenon is known as the Torsional Interaction SSR.

Figure 5.33 shows a simple power system which consists of one generator and a transmission line. The machine is represented by a fixed rotor flux which rotates in the stator with rated frequency ω_n. The machine is driven through an elastic shaft with the spring constant K by a prime mover. The stator is connected to a network that is represented by the impedance Z_L.

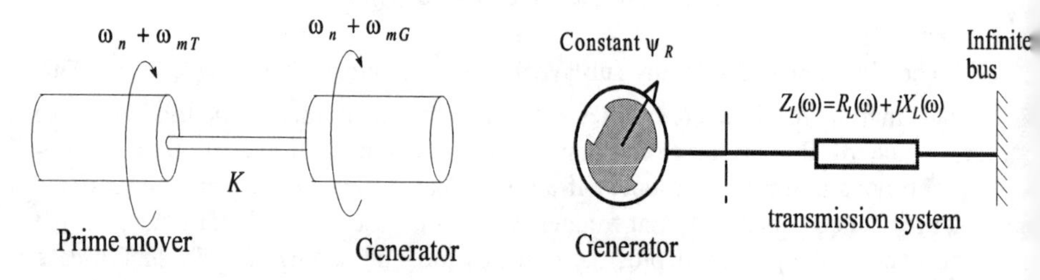

Figure 5.33 Model to determine torsional interaction SSR conditions

Assume that the rotor rotates at the nominal angular velocity $\omega_n = 2\pi f_n$, where f_n is the nominal frequency. If disturbance with small amplitude is superimposed on the nominal oscillation, it results in a mechanical torsional resonance between generator and the prime mover. From Figure 5.33 we have:

$$\omega_m \overset{\Delta}{=} (\omega_n + \omega_{mG}) - (\omega_n + \omega_{mT}) \tag{5.26}$$

This mechanical oscillations has the following impacts:

- A sub-synchronous current I_{sub} with angular frequency $\omega_{sub} = \omega_n - \omega_m$ and a super-synchronous current I^{sup} with angular frequency $\omega^{sup} = \omega_n + \omega_m$ will be produced. The phase and amplitude of these currents are determined by the impedance of the grid and the generator at that particular angular frequency.

- Both these currents will interact with the rotor flux and give rise to instantaneous contributions that can either amplify or damp the mechanical interactions.
- The sub-synchronous current I_{sub} will give a damping contribution:

$$D_{\text{sub}} = -\frac{\psi_{\text{R}}^2}{2\omega_{\text{m}}}\left[\frac{\omega_{\text{sub}}}{\omega_{\text{n}}}\frac{R_{\text{sub}}}{R_{\text{sub}}^2 + X_{\text{sub}}^2}\right] \tag{5.27}$$

Note that as long as R_{sub} is positive, D_{sub} gives a negative contribution. The magnitude of D_{sub} has a maximum value when $X_{\text{sub}} = 0$ or $X_{\text{sub}} = R_{\text{sub}}$.

- The super-synchronous current I^{sup} will give a damping contribution:

$$D^{\text{sup}} = +\frac{\psi_R^2}{2\omega_m}\left[\frac{\omega^{\text{sup}}}{\omega_{\text{n}}}\frac{R^{\text{sup}}}{(R^{\text{sup}})^2 + (X^{\text{sup}})^2}\right] \tag{5.28}$$

Note that as long as R^{sub} is positive, D^{sup} gives a positive contribution.

- If the natural damping of the mechanical system is D_{n}, then the determining criterion to avod the risk of SSr is:

$$D_{\text{n}} + D_{\text{sub}} + D^{\text{sup}} > 0 \tag{5.29}$$

5.4.1.4 Transient torques

Transient torques are those that result from system disturbances. For networks that contain series capacitors, the transient currents will contain one or more oscillatory frequencies that depend on the network capacitance as well as the inductance and resistance. If any of these subsynchronous network frequencies coincide with one of the natural modes of a turbine-generator shaft, there can be peak torques that are quite large since these torques are directly proportional to the magnitude of the oscillating current.

5.4.2 Apparent impedance of TCSC

From the discussion above it has been clarified that the conditions for SSR depend on the network impedance that the machine sees at the sub and supersynchronous frequencies corresponding to its torsional resonance frequency ω_{m}.

The reactance of a fixed series capacitor varies inversely with frequency and once its reactance at rated frequency has been selected, its reactance at all frequencies are determined. This is not the case for the TCSC as the boost of the TCSC depends on control actions that may change the triggering of the thyristors for each half-cycle of the line current. The behaviour of the TCSC at various frequencies depends on the response of the control system to different frequency components of the line current.

At a given steady-state operation point the apparent impedance of the TCSC at frequency ω_{m} is defined as follows. Assume a steady-state line current with rated-frequency passing through the TCSC. If a small disturbance current with frequency ω_{m} is added to the nominal current, we have:

$$i_{\mathrm{L}}(t) = \mathrm{Re}\left\{\boldsymbol{I}_{\mathrm{L}} e^{j\omega_{\mathrm{N}} t} + \Delta\boldsymbol{I}_{\mathrm{L}} e^{j\omega_{\mathrm{m}} t}\right\} \tag{5.30}$$

The capacitor voltage will be influenced by the disturbance. Through Fourier analysis of the voltage one will also find a component that has the same frequency as the exciting current:

$$u_{\mathrm{C}}(t) = \mathrm{Re}\left\{jX_{\mathrm{C}}\boldsymbol{I}_{\mathrm{L}} e^{j\omega_{\mathrm{N}} t} + \Delta\boldsymbol{U}_{\mathrm{C}} e^{j\omega_{\mathrm{m}} t} + \ldots\right\} \tag{5.31}$$

The apparent impedance of the TCSC then can be defined as the complex quotient:

$$Z_{\mathrm{app}}(\omega_{\mathrm{m}}) = R_{\mathrm{app}}(\omega_{\mathrm{m}}) + jX_{\mathrm{app}}(\omega_{\mathrm{m}}) = \frac{\Delta\boldsymbol{U}_{\mathrm{C}}}{\Delta\boldsymbol{I}_{\mathrm{L}}} 7 \tag{5.32}$$

It should be noted that the apparent impedance is a property of the TCSC main circuit *together with its control system* and not of the main circuit only. In general the apparent impedance for a specific TCSC in a specific network must be determined by simulation or measurement. For several control schemes it has been reported that the apparent impedance is of resistive–inductive type. A general discussion will be performed here for two cases

- Constant thyristor charge in each reversal
- Synchronous voltage reversals

5.4.2.1 Constant thyristor charge control

In Figure 5.16 a subordinated control loop of the TCSC is arranged to keep the charge that passes through the thyristor branch constant in each reversal. To perform this, measured values of the instantaneous capacitor voltage and line current are used as inputs to an algorithm that determines the exact trigger instant. If the algorithm functions perfectly no variations in the charge passing through the thyristor branch will occur even if the line current contains a small current with subsynchronous frequency in addition to the steady state line current component with rated network frequency. This means that the subsynchronous current must pass entirely through the capacitor bank and the apparent impedance of the TCSC will be the physical reactance of the capacitor at subsynchronous frequency. The differences with respect to a fixed capacitor bank from a subsynchronous standpoint are:

- The physical reactance of the capacitor bank, which is the one that should be used for apparent impedance calculations at subsynchronous frequencies, will be reduced in proportion to the boost factor that the TCSC uses in the steady state operation. Thus less capacitive reactance will occur at subsynchronous frequencies.
- The charge reference may be modulated by the external regulator in order to increase the damping of the electrical resonance, which is beneficial in order to avoid torque amplification when disturbances occur in the system

5.4.2.2 Synchronous voltage reversal (reversal timing control)

Figure 5.17 shows a system where the timing of the thyristor reversals are controlled directly by a subordinate control loop using instantaneously measured capacitor voltage and line current values as inputs. If the pulse train that gives reference for the reversals is only influenced by the steady-state line current, the reversal instants will remain unchanged when a small subsynchronous line current is added to the steady state line current. The TCSC then operates in "synchronous voltage reversal" mode. A simplified calculation, assuming instantaneous, equidistant capacitor voltage reversals at twice the rated frequency and neglecting losses, reveals the apparent impedance of TCSC to be

$$X_{\mathrm{app}}(\omega_{\mathrm{m}}) = -X_{\mathrm{C}} \frac{\omega_{\mathrm{n}}}{\omega_{\mathrm{m}}} \frac{1-\cos\left(\dfrac{\omega_{\mathrm{m}}}{\omega_{\mathrm{n}}}\dfrac{\pi}{2}\right)}{\cos\left(\dfrac{\omega_{\mathrm{m}}}{\omega_{\mathrm{n}}}\dfrac{\pi}{2}\right)} \tag{5.33}$$

The function is positive in the whole subsynchronous frequency range, showing that the apparent reactance is inductive as illustrated in Figure 5.34. It is remarkable that this general result can be achieved from very few and simple assumptions. E.g. the boost level in steady state does not appear in the formula. However, in reality the subsynchronous added line current will always have some influence on the timing through the action of the synchronization, the boost regulator and the external regulator. At frequencies close to the rated frequency therefore the apparent impedance will leave the undisturbed characteristics and get capacitive.

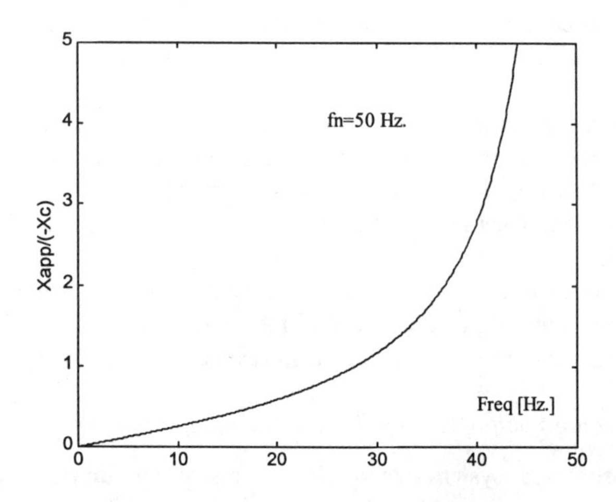

Figure 5.34 Ideal apparent reactance of TCSC operating in synchronous voltage reversal mode

5.4.3 Application example

The Forsmark nuclear power plant situated in mid Sweden is interconnected with the north of the country by means of a number of 400 kV lines of varying length, all series compensated. However, one of the generator units at Forsmark, rated 1300 MW, is subject to SSR under certain conditions in the power grid. In normal operation the SSR modes do not have negative damping, but at certain network conditions involving specific line outages, a risk of SSR exists. The frequency of the critical torsional mode is 21.1 Hz, corresponding to an electrical frequency of 28.9 Hz. The SSR in this case is related to parallel resonance in the network. Another mechanical mode exists at 16.1 Hz, corresponding to the electrical frequency 33.9 Hz. This mode however is less critical in the system.

Part of the existing series capacitor at Stöde (30% of its totally installed reactance) was consequently rebuilt into a TCSC. The TCSC uses the synchronous voltage reversal control approach. The nominal voltage boost is $K_B = 1.2$. The apparent impedance of the TCSC is shown in Figure 5.35. It can be noted that the reactance is inductive for all frequencies below 40 Hz.

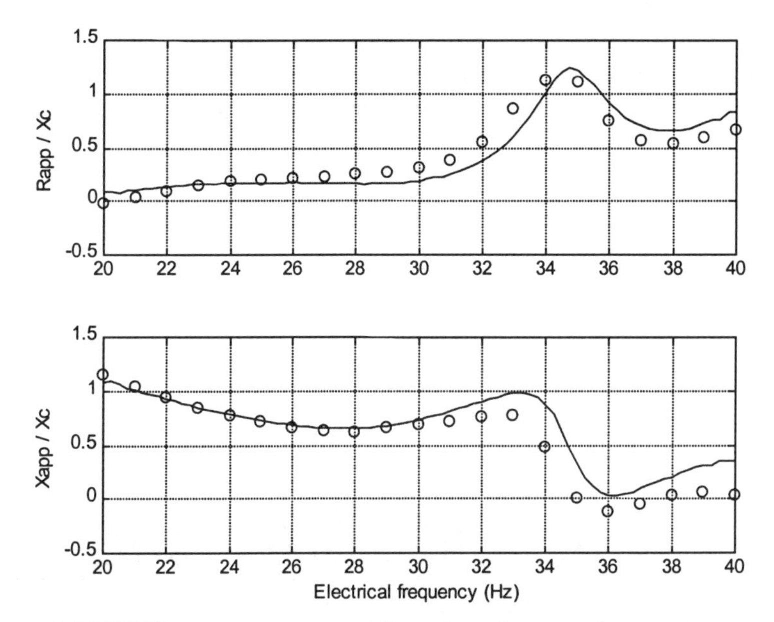

Figure 5.35 TCSC apparent reactance for subsynchronous frequencies (O=measured values)

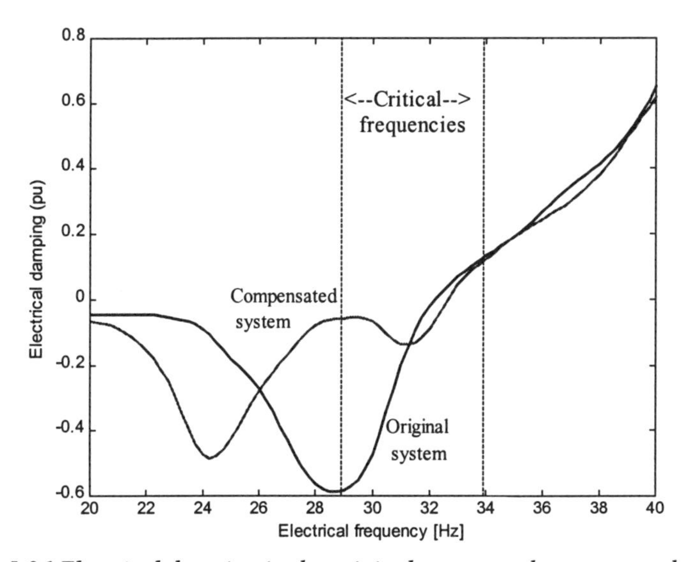

Figure 5.36 Electrical damping in the original system and compensated system by TCSC

Figure 5.36 shows the electrical damping versus electrical frequency in the original system and after the redesign of the series capacitor. It can be seen that as a consequence of the change of apparent impedance at subsynchronous frequency the damping at the critical frequency has been moved towards a safe area.

The SSR conditions were investigated in a power simulator during the development phase of the project. Figure 5.37 shows an experiment where critical network conditions with line outages were setup. Initially the TCSC was blocked and an SSR oscillation spontaneously built up. Deblocking the TCSC control system changed the SSR conditions and the oscillation faded away. The figure shows the oscillation between different parts of the turbine.

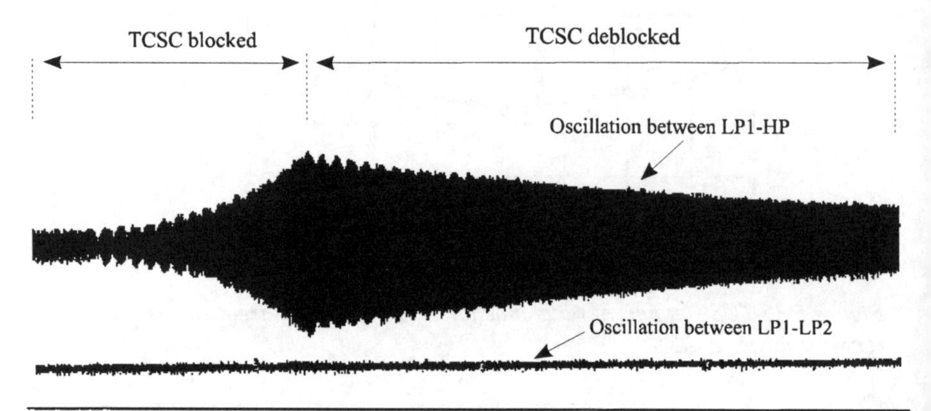

Figure 5.37 Impact of TCSC control in power simulator test.

5.5 TCSC layout and protection

The basic layout of a TCSC consists of an inductor placed in parallel with the series capacitor. The current through the inductor is controlled with a thyristor valve. A possible TCSC configuration is shown in Figure 5.38. The capacitor, ZnO-varistor, and the thyristor valve are placed on a platform that is insulated from the ground. The insulation between the platform and ground corresponds to the system voltage. One side of the capacitor bank is grounded to the platform. This means that a second insulation system is built up where the platform acts as a reference for the insulation levels of the equipment mounted on the platform.

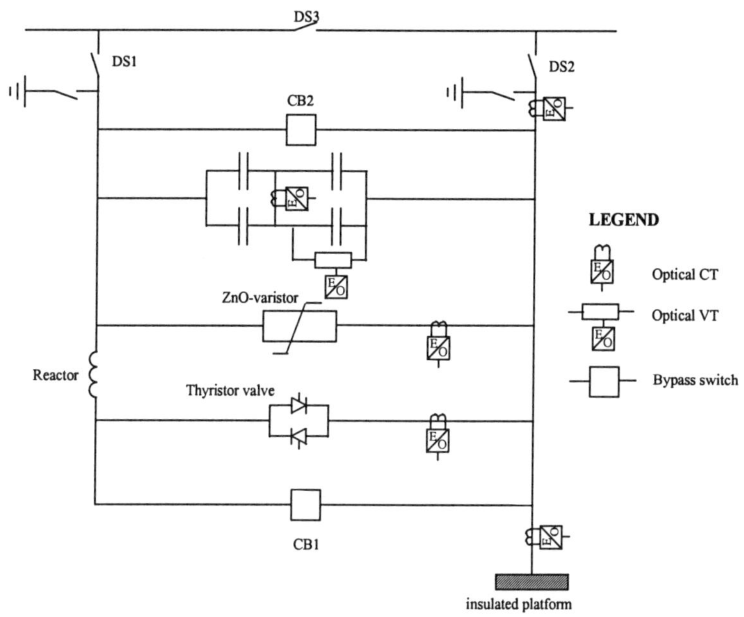

Figure 5.38 A TCSC layout

5.5.1 TCSC reactor

A TCSC reactor is relatively large compared to the reactor of a conventional series capacitor. The purpose of the reactor in a conventional series capacitor is mainly to limit the maximum current due to capacitor discharge, i.e. when the series capacitor is bypassed during severe line faults.

The situation in a TCSC is somewhat different. Here the reactor takes an active part in the capacitor voltage boost process, thus the inductance must be chosen with respect to the thyristor valve stresses together with the required speed of the voltage reversals. A lower inductance would create faster voltage reversals with correspondingly higher capacitor discharge current amplitudes while a large reactor would lower the current amplitude at the cost of the voltage reversal time. The thyristor voltage stress and the snubber losses are connected to the current derivative just before the thyristor commutation, i.e. the choice of inductance is a trade off between voltage reversal speed and thyristor valve stress/losses.

5.5.2 Bypass breakers

Two bypass breakers have been included in Figure 5.38. This is a consequence of the relatively large reactor. The purpose of the breaker CB2, is to eliminate the voltage drop across the TCSC during the energization.

5.5.3 Capacitor overvoltage protection

The capacitor is protected from overvoltages by means of a ZnO-varistor that is connected in parallel with the capacitor. During steady state conditions, all line current flows through the series capacitor. Following fault initiation in the surrounding network, high current flowing through the series capacitor will result in a correspondingly high voltage across the capacitor and the ZnO-varistor. The high voltage causes the varistor to conduct excess current and thereby limit the voltage across itself and across the capacitor to the protective level of the varistor. During this conduction period, the energy is absorbed by the ZnO-varistor. If the energy duty of the varistor is exceeded, due to the occurrence of extreme system faults, the series capacitor must be bypassed in order to prevent the energy capability of the varistor from being exceeded. For a conventional series capacitor a sparkgap offers a fast way to bypass the capacitor during extreme system faults, but for a TCSC this function it can be handled by the thyristor valve.

The thyristor valve is a fast device and in contrast to a sparkgap, no deionization time is required before the TCSC can be reinserted upon fault clearing. This offers a way to reduce the energy requirement of the varistor.

5.5.4 Thyristor valve

There are certain design requirements that must be noted for a TCSC valve. One of the most important issues, is the ability to fire the thyristors during all possible operating conditions including:

- Normal voltage reversals for a wide range of line currents
- Bypass of the varistor if the energy duty is exceeded
- Avoid blocking of the valve against high voltage
- Avoid blocking of the valve within the recovery time of the thyristors
- Continuos bypass of the series capacitor for a wide range of line currents.

The thyristor valve must be able to maintain the capacitor voltage boost for a wide range of line currents. At low line current and low voltage boost, this means that it must be possible to fire the thyristors from a low voltage without any substantial time delay.

The thyristor valve must be designed to handle the worst combination of short circuit current on the line and discharge current, that arise when the capacitor is discharged from the highest possible voltage, i.e. the protective level of the varistor. The valve must be able to handle this current for a time corresponding to the operating time of the parallel bypass switch, normally 3-4 periods of the system frequency. In cases where the TCSC is located at the end of a transmission line, the short circuit currents could be very high. Combined with the relatively high voltages across a TCSC and the objective of minimizing the number of thyristors in series, this calls for thyristors with both high voltage ratings and high surge current capability.

Figure 5.39 A thyristor valve

5.5.5 Measuring system

The measuring system of a TCSC should detect signals at high potential, i.e. the actual system voltage and deliver these signals down to ground potential. One way of doing this is to use optical fibers between the remote device mounted in the primary circuit and the local device mounted at ground potential. The system indicated in Figure 5.38, consists of optical current transformers (OCT) and an optical voltage transducer (OPVT). The optical fibers from the platform down to ground potential are hosted inside insulators and the signal column for the thyristor valve control fibers and cooling pipes.

5.5.6 Capacitor voltage boost

The most important feature of a TCSC is the ability to boost the capacitor voltage. An important question is how much boost capability a TCSC should have and how this will affect the cost of the equipment. Suppose that a TCSC has an apparent reactance X_{app} and a physical capacitor reactance X_C and let I_L be the line current. We have:

$$X_{app} = X_C K_B \tag{5.34}$$

The utilized TCSC reactive power from a system point of view is:

$$Q_{Utilized} = 3X_{app} I_L^2 = 3X_C K_B I_L^2 \tag{5.35}$$

Installed capacitor power:

$$Q_{\text{Installed}} = 3\frac{U_{\text{C}}^2}{X_{\text{C}}} = 3\frac{(X_{\text{C}}K_{\text{Boost}}I_{\text{L}})^2}{X_{\text{C}}} = 3X_{\text{C}}K_{\text{Boost}}^2 I_{\text{L}}^2 \tag{5.36}$$

Hence, for a certain apparent reactance the relation between installed and utilized capacitor power can be expressed as:

$$\frac{Q_{\text{Installed}}}{Q_{\text{Utilized}}} = K_{\text{B}} \tag{5.37}$$

High requirements regarding capacitor voltage boost and/or overload capability also affect the lowest possible varistor protection level. The varistor protection level must be chosen so that the operating voltage of the TCSC during normal operation and overloads is below the knee-point of the varistor characteristics. Otherwise the varistor will start to conduct and this conduction will result in heating of the varistor.

5.5.7 Fault handling

The strategies for thyristor control action during the duration of faults and the control sequences executed thereafter will greatly impact how fast the TCSC can establish normal operation after fault clearing. When the fault current is very high and exceeds the maximum line current due to external faults, mechanical breaker bypass will always be commanded and also the line breakers will operate (as an internal fault must be present). Thus the obvious strategy is to command the thyristor valve to bypass as soon as this condition is recognized. If, on the other hand, a fault causes low line current (below the maximum overload current level), the TCSC will continue to operate in its capacitive boost mode and also in this case the strategy is obvious. Therefore the discussion should focus on faults that create line fault current that is beyond the maximum overload current level but below the maximum external fault current level.

Basically two approaches may be considered:

The TCSC is commanded to **bypass** the fault current and await fault clearing. As soon as the line currents have become normal the TCSC will return into CAP boost mode.

The TCSC is commanded to **block** during the duration of the line fault and await the fault clearing. When the line current returns to normal operating level the TCSC will restart in CAP boost mode.

The strategy using valve bypass during the fault duration appears the most appropriate one for the following reasons:

Consistent actions for *all* over-current situations; the TCSC must be bypassed during internal faults with heavy surge currents.

No accumulated energy in the MOVs will occur during faults.

No charges will be trapped in the capacitors when the line breaker opens; no capacitor discharge required.

Capacitor voltage will have less offset voltage after fault clearance thereby facilitating the reentering of the capacitive boost mode of the TCSC.

Figure 5.40 The TCSC in Stöde, Sweden

5.6 Static synchronous series compensator (SSSC)

The use of a shunt-connected voltage source inverter has been described in Chapter 3. The device which is called a static synchronous compensator (SSC) or STATCOM is primarily used for voltage control in power systems. A voltage-source inverter could be used in series in transmission. This device is called static synchronous series compensator (SSSC).

5.6.1 Principle of operation

Figure 5.41 shows a voltage source inverter connected in series via a transformer to a transmission line. A source of energy is needed to provide the DC voltage across the capacitor and supply the losses of VSI.

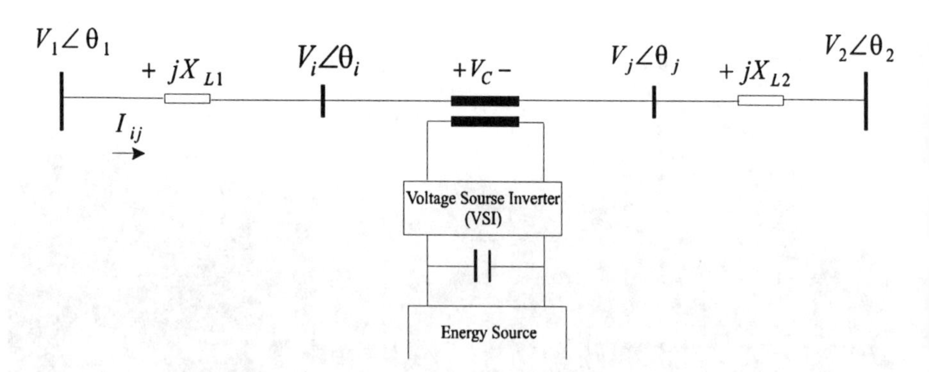

Figure 5.41 The basic configuration of an SSSC

In principle, an SSSC is capable of interchange of active and reactive energy with the power system. However if only reactive power compensation is intended, the size of the energy source could be quite small. The injected voltage could be controlled in magnitude and phase if sufficient energy source is provided. For the reactive power compensator function, only the magnitude of the voltage is controllable since the vector of the inserted voltage is perpendicular to the line current. In this case the series injected voltage can either lead or lag the line current by 90 degrees. This means that the SSSC can be smoothly controlled at any value leading or lagging within the operating range of VSI. Thus the behaviour of an SSSC can be similar to a controllable series capacitor and a controllable series reactor. The basic difference is that the voltage injected by SSSC is not related to the line current and can be independently controlled. The importance of this characteristic is that an SSSC is effective for both low and high loading.

5.6.2 SSSC model for load flow and stability analysis

We consider a general case when SSSC can interchange both active and reactive power. Obviously, the special case of a reactive compensation scheme could be derived from the general equations by considering the right angle between the inserted series voltage and the line current.

Suppose a series-connected voltage source is located between nodes i and j in a power system. The series voltage source can be modeled with an ideal series voltage V_s in series with a reactance X_S. In Figure 5.42, V_s models an ideal voltage source and V_i' represents a fictitious voltage behind the series reactance. We have:

$$V_i' = V_s + V_i \tag{5.38}$$

The series voltage source V_s is controllable in magnitude and phase, i.e:

$$V_s = rV_i e^{j\gamma} \tag{5.39}$$

where $0 < r < r_{max}$ and $0 < \gamma < 2\pi$.

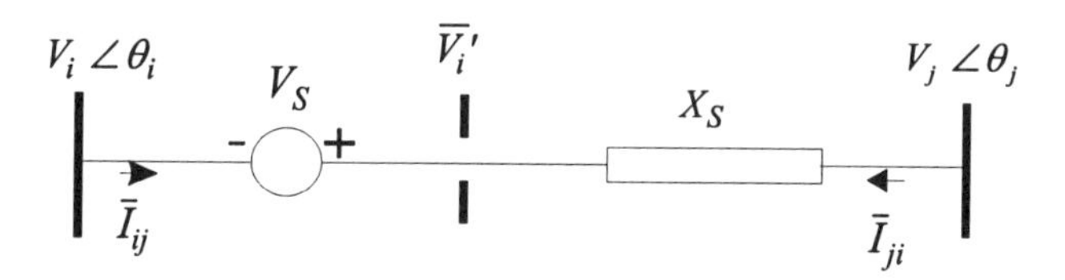

Figure 5.42 Representation of a series connected voltage source

The equivalent circuit vector diagram is shown in Figure 5.43:

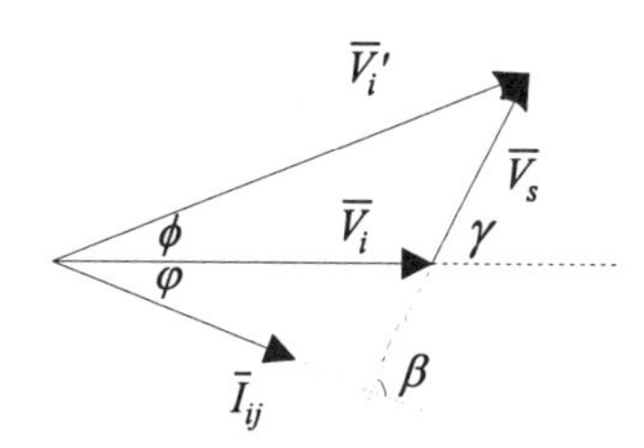

Figure 5.43 Vector diagram of the equivalent circuit of VSC

The injection model is obtained by replacing the voltage source V_s by the current source $I_s = -jb_s V_s$ in parallel with the line where $b_s = \dfrac{1}{X_s}$:

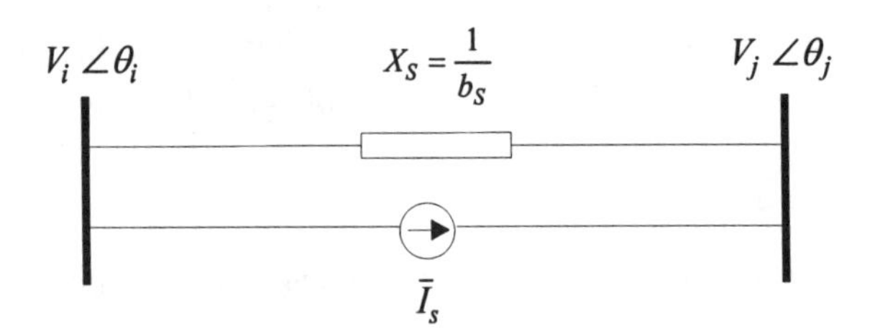

Figure 5.44 Replacement of a series voltage source by a current source

The current sources $\boldsymbol{I}_s$ corresponds to the injection powers $\boldsymbol{S}_{is}$ and $\boldsymbol{S}_{js}$, where:

$$\boldsymbol{S}_{is} = \boldsymbol{V}_i(-\boldsymbol{I}_s)^* \tag{5.40}$$

$$\boldsymbol{S}_{js} = \boldsymbol{V}_j(\boldsymbol{I}_s)^* \tag{5.41}$$

The injection power $\boldsymbol{S}_{is}$ and $\boldsymbol{S}_{js}$ are simplified to:

$$\boldsymbol{S}_{is} = \boldsymbol{V}_i\left[jb_s r\boldsymbol{V}_i e^{j\gamma}\right]^* = -b_s rV_i^2 \sin\gamma - jb_s rV_i^2 \cos\gamma \tag{5.42}$$

If we define: $\theta_{ij} = \theta_i - \theta_j$, we have:

$$\boldsymbol{S}_{js} = \boldsymbol{V}_j\left[-jb_s r\boldsymbol{V}_i e^{j\gamma}\right]^*$$
$$= b_s rV_i V_j sin(\theta_{ij} + \gamma) + jb_s rV_i V_j cos(\theta_{ij} + \gamma) \tag{5.43}$$

Based on the explanation above, the injection model of a series connected voltage source can be seen as two dependent loads as shown in Figure 5.45.

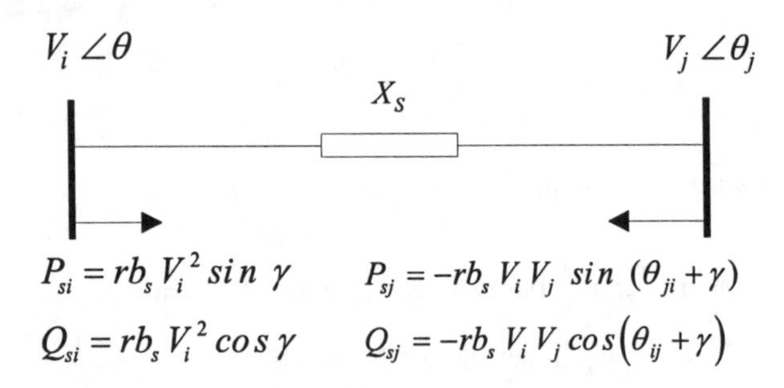

Figure 5.45 Injection model for a series connected VSC

The injection model could easily be incorporated in loadflow and stability programs.

5.6.3 Power interchange

The apparent power supplied by he series voltage source is calculated from:

$$
\begin{aligned}
S_{\text{VSI}} &= \boldsymbol{V}_{\text{s}} \boldsymbol{I}^{*}_{\text{ij}} \\
&= r e^{j\gamma} \boldsymbol{V}_{\text{i}} \left(\frac{\boldsymbol{V}_{\text{i}}' - \boldsymbol{V}_{\text{j}}}{j x_{\text{s}}} \right)
\end{aligned} \tag{5.44}
$$

Active and reactive power supplied by series VSI are distinguished as:

$$P_{\text{VSI}} = r b_{\text{s}} V_{\text{i}} V_{\text{j}} \sin(\theta_{\text{i}} - \theta_{\text{j}} + \gamma) - r b_{\text{s}} V_{\text{i}}^{2} \sin\gamma \tag{5.45}$$

$$Q_{\text{VSI}} = -r b_{\text{s}} V_{\text{i}} V_{\text{j}} \cos(\theta_{\text{i}} - \theta_{\text{j}} + \gamma) + r b_{\text{s}} V_{\text{i}}^{2} \cos\gamma + r^{2} b_{\text{s}} V_{\text{i}}^{2} \tag{5.46}$$

5.6.4 Applications

The general application of a controllable series capacitor is valid for SSSC. Namely, power flow control, voltage, and angle stability enhancement. The fact that an SSSC can induce both capacitive and inductive series voltage on a line, widens the operating region of the device. For power flow control, an SSSC can be used both for increasing and decreasing the flow. For stability it gives a better possibility for damping of electromechanical oscillations. However the inclusion of a high-voltage interphasing transformer in the scheme, causes a big cost disadvantage as compared to controllable series capacitors. The transformer also decreases the performance of the SSSC due to introducing extra reactance. This deficiency might be overcome in future by introducing transformerless SSSC. The scheme also calls for a protecting device that bypasses the SSSC at high fault currents through the line.

5.7 References

1. HAUER, F., "Reactive power control as a means for enhanced inter-area damping in the western US power system", *IEEE Tutorial Course 87THO187-5-PWR*, 1987.
2. LARSSON, E.V., et al., "Characteristics and rating considerations of thyristor controlled series compensation". *IEEE Transactions on Power Delivery*, 9, (2), pp. 992–1001, April 1994.
3. NOROOZIAN, M., HALVARSSON, P., and OTHMAN, H., "Application of controllable series capacitors for damping of power swing". *Proceedings of V Symposium of Specialists in Electric*

Operational and Expansion Planning, vol. 1, pp. 221–225, Recife, Brazil, May, 1996.

4 NOROOZIAN, M., and ANDERSSON, G., "Damping of power system oscillations by controllable components", *IEEE Transactions on Power Delivery*, 9, (4), pp. 2046–2054, Oct. 1994.

5 NOROOZIAN, M., and ANDERSSON, G., "Damping of inter-area and local modes by controllable components", *IEEE Transactions on Power Delivery*, 9, (4), pp. 2046–2054, Oct. 1994.

6 KIMBARK, E.W., "Improvement of system stability by switched series capacitors," *IEEE Transactions on Power Apparatus and Systems*, PAS-85, pp. 180–188, Feb. 1966.

7 ÖLWEGÅRD, Å., et al., "Improvement of transmission capacity by thyristor control reactive power", *IEEE Transactions on Power Apparatus and Systems*, PAS-100(8), pp. 3933–3939, Aug. 1981.

8 ÄNGQUIST, L., LUNDIN, B., and SAMUELSSON, J., "Power oscillation damping using controlled reactive power compensation", *IEEE Transactions on Power Systems*, pp. 687–700, May 1993.

9 ROUCO, L., and PAGOLA, F.L., "An eigenvalue sensitivity approach to location and controller design of controllable series capacitors for damping power system oscillations", *IEEE Transactions on Power Systems*, 12, (4), pp. 1660-1666, Nov. 1997.

10 KUNDUR, P., "Power system stability and control", McGraw-Hill, New York, 1993.

11 ANDERSSON, P.M., AGRAWAL, B.L., and VAN NESS, J.E., "Subsynchronous resonance in power systems", IEEE PRESS, New York, 1990.

12 HOLMBERG, D., DANIELSSON, M., HALVARSSON, P., and ÄNGQUIST, L., "The Stöde thyristor controlled series capacitor", Cigre Session 1998, paper 14–105.

13 GYUGYI, L., "Dynamic compensation of AC transmission lines by solid-state synchronous voltage source", *IEEE Transactions on Power Delivery*, 9, (2), pp. 904-911, April 1994.

Chapter 6

Phase shifter

M.R. Iravani

6.1 Introduction

Power transformers are widely used in electric power systems for in-phase and/or phase-angle voltage regulation [1]. Voltage regulation is carried out either under-load (on-load) or off-load conditions. In this document, the term 'phase shifter' refers to an assembly of one or more power transformers that provides regulation on magnitude and/or phase-angle of voltage under both on-load and off-load conditions, for a three-phase electric power system.

Conventional applications of phase shifters are for steady-state (1) power flow regulation and (2) voltage regulation. Availability of power semiconductor switches and converter topologies, for implementation in phase shifters' assemblies, widens the scope of applications of phase shifters beyond their original roles. In addition to steady-state power flow and voltage regulation, semiconductor-controlled phase shifters can also be used for (1) power quality enhancement, (2) dynamic voltage control, (3) mitigation of small-signal dynamics, and (4) transient stability enhancement.

Main objectives of this chapter are to describe (1) principles of operation, (2) operational characteristics, (3) technical merits and limitations, (4) various circuit configurations, and (5) applications of phase shifters that use semiconductor technology in their power circuit assemblies.

The rest of this chapter is organized as follows. Section 6.2 introduces a generic configuration for a phase shifter and describes functions of its fundamental components. The configuration is used to describe principles of operation of a phase shifter. Section 6.3 describes a steady-state representation and mathematical model for a phase shifter. Based on the model, operational characteristics of a phase shifter under steady-state conditions are deduced in Section 6.4. Various power circuit configurations for phase shifters are illustrated and described in Section 6.5. Applications of phase shifters in electric power transmission and distribution voltage levels are briefly described in Section 6.6. A summary of the chapter is given in Section 6.7. Section 6.8 provides a list of references that provide further details of the materials presented in this chapter.

6.2 Principles of operation of a phase shifter

Figure 6.1 shows a schematic representation of a phase shifter. The phase shifter is installed in a transmission line between buses E and B. The power system external to the phase shifter is represented by voltage phasors V_S and V_R and the corresponding impedances Z_S and Z_R respectively. Power circuitry of the phase shifter is comprised of:

- Exciting Transformer (ET),
- Boosting Transformer (BT),
- and converter circuit.

The exciting transformer provides input voltage to the phase shifter. The boosting transformer injects a controlled voltage in series in the system. Magnitude and/or phase-angle of the injected voltage is controlled by the converter. Figure 6.1 illustrates that, depending on the magnitude and phase-angle of the injected voltage V_B, magnitude and/or phase-angle of the system voltage V_P is varied.

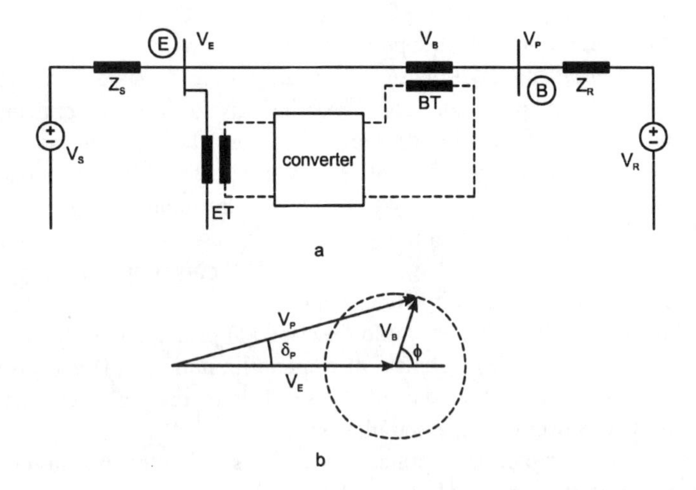

Figure 6.1 (a) Schematic diagram of a phase shifter; (b) Voltage phasor diagram

Phasor relationship among V_E, V_P and V_B is also illustrated in Figure 6.1. The circle identifies a region where the tips of V_P and V_B can be located. Magnitude and/or relative phase-angle of the injected voltage, i.e. $|V_B|$ and ϕ, are used to control (1) voltage at bus B and (2) real power transfer (P) of the line.

$$P = (|V_S||V_R|/X_{eq})\sin(\delta_S - \delta_R \pm \delta_P) \tag{6.1}$$

where X_{eq} is the net equivalent reactance of line, and δ_S and δ_R are phase-angles of phasors V_S and V_R respectively. Based on equation (6.1) the angle δ_p is the dominant variable for power flow control. The range of angle δ_P that a phase shifter can provide, primarily depends on the characteristics of the converter circuitry.

The converter section of a conventional phase shifter comprises mechanical switches which are usually embedded within the exciting transformer and may not be readily identifiable as a separate unit [1]. A conventional phase shifter can vary angle δ_p approximately within ±30 degrees in discrete steps of about 1 to 2 degrees. Figure 6.2 illustrates a simplified diagram of a conventional phase shifter [2]. Phase-angle and magnitude of the injected voltage V_B in Figure 6.2 are adjusted by mechanical switches SW_1 and SW_2.

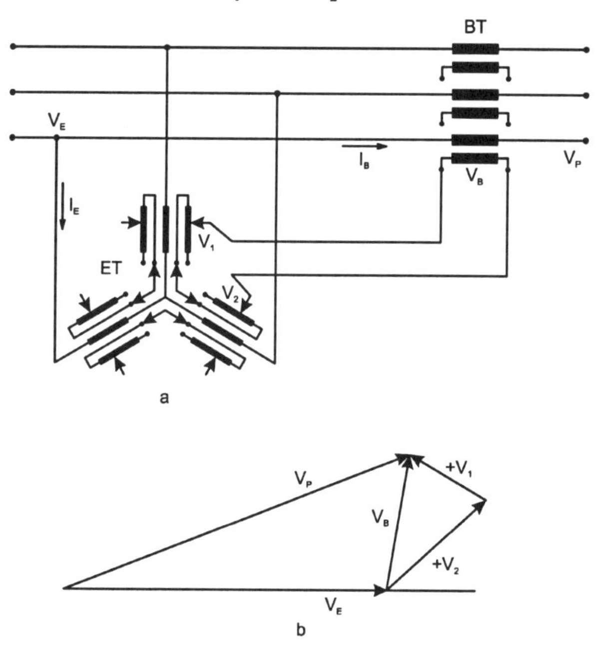

Figure 6.2 (a) Schematic diagram of a conventional phase shifter; (b) Voltage phasor diagram

Various circuit configurations for the realization of conventional phase shifters are described in References 1, 2 and 3. Major technical drawbacks associated with a conventional phase shifter are:

- Slow response due to the inherent inertia of mechanical switches
- Limited life-time and frequent maintenance requirement due to mechanical wear-out and oil deterioration

The first technical drawback limits the applications of conventional phase shifters only for steady-state power-flow and voltage regulation. The latter drawback is partially overcome by means of semiconductor-assisted phase shifter configurations [4]. Conventional and semiconductor-assisted phase shifters are not the subject of this chapter and are not discussed any further.

Technical drawbacks of a conventional phase shifter are conceptually overcome if its mechanical converter (mechanical switches) is substituted by a semiconductor (static) converter. Hereinafter the term Static Phase Shifter (SPS) is used to distinguish semiconductor-controlled phase shifters from conventional (mechanical) phase shifters. The IEEE FACTS Working Group [5] defines a Thyristor Controlled Phase Shifting Transformer (TCPST) as "a phase-shifting transformer, adjusted by thyristor switches to provide rapidly variable phase angle". It should be noted that TCPST defines a sub-group of apparatus under the more general category introduced by the term SPS.

Direct substitution of mechanical switches of a phase shifter by semiconductor switches does not lead to the optimum technical/economical utilization of the overall phase shifter circuit. This has resulted in an ongoing investigation of various power electronic circuits to be used as SPS converters.

6.3 Steady-state model of a Static Phase Shifter (SPS)

Figure 6.3 is a three-phase schematic representation of a SPS. Figure 6.4 is a single-phase representation of the system under a steady-state operating conditions. System S (R) is approximated by a voltage source V_S (V_R) behind impedance Z_S (Z_R). Z_E and Z_B represent impedances of ET and BT respectively. V_P is the injected voltage by BT in the system.

For the sake of simplicity (without the loss of generality) assume that in the system of Figure 6.4:

- ET and BT are ideal, i.e. Z_B and Z_E are negligible.
- The converter is lossless and also does not exchange reactive power with the system.

Thus the converter is an ideal device that provides the required magnitude and/or phase-angle conversion between the corresponding windings of ET and BT.

Based on the above assumptions one can deduce

$$V_B = kV_E \exp(j\phi) \tag{6.2}$$

where $k = |V_B|/|V_E|$ and

ϕ is the phase angle between V_B and V_E.

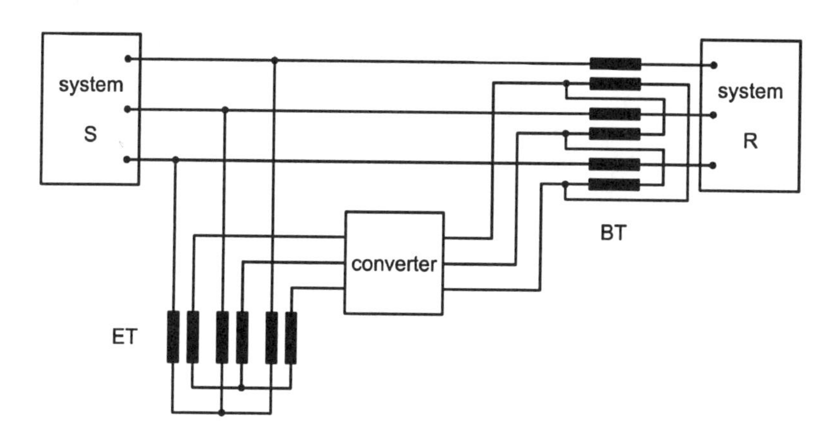

Figure 6.3 Three-phase schematic diagram of an SPS

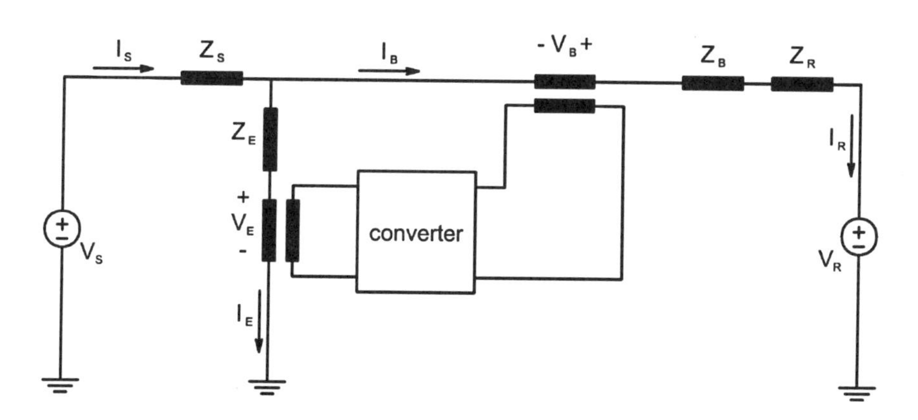

Figure 6.4 Single-phase schematic diagram of Figure 6.3

The above assumptions also imply that the SPS configuration neither absorbs nor injects power with respect to the power system.

$$S_E + S_B = 0 \tag{6.3}$$

or

$$V_B I_B^* = V_E I_B^* \tag{6.4}$$

$$Q_B = Q_E \tag{6.5}$$

$$P_B = P_E \tag{6.6}$$

Substituting for V_B in Equation (6.4) from Equation (6.3) yields

$$I_E = kI_B \exp(-j\phi) \tag{6.7}$$

It should be noted that if the converter is capable of reactive power exchange with the system through ET and/or BT, e.g. converters corresponding to SPSs of Sections 6.5.4 and 6.5.5, then equation (6.5) is not necessarily valid unless the SPS control system imposes the necessary condition(s).

Equations (6.3) to (6.6) imply that a SPS can be represented by two mutually-dependent voltage sources as illustrated in Figure 6.5

Equivalent models illustrated in Figure 6.5 can be further generalized to include converter losses and impedances Z_B and Z_E [6].

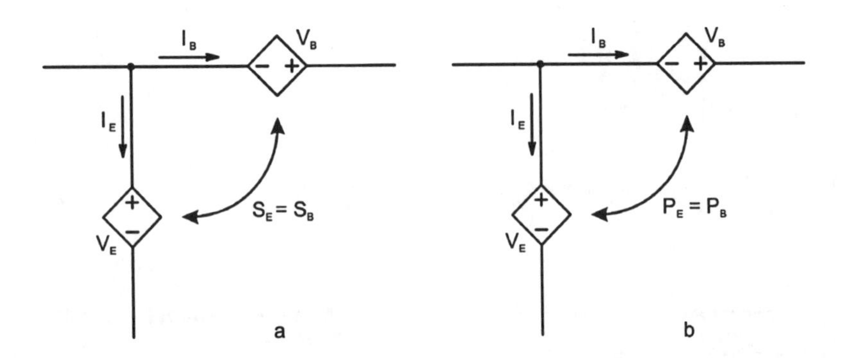

Figure 6.5 (a) Equivalent Circuit of an SPS (SPSs described in Sections 6.5.1 to 6.5.3); (b) Equivalent circuit of an SPS (SPSs described in Sections 6.5.4 and 6.5.5)

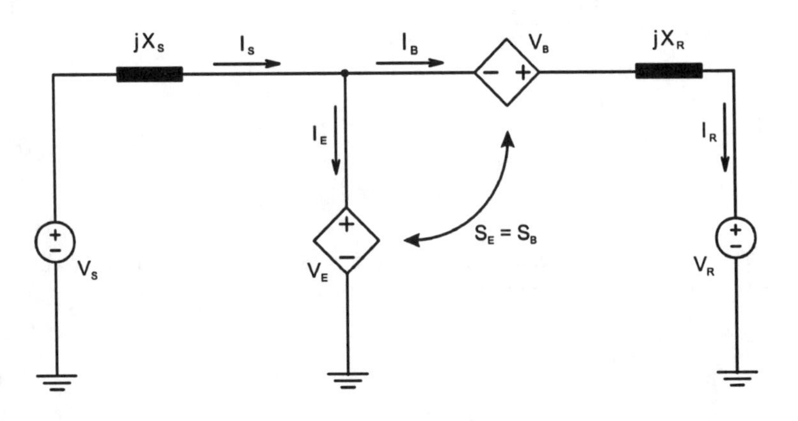

Figure 6.6 Equivalent circuit of a radial line and an SPS

6.4 Steady-state operational characteristics of SPS

The main objective of this section is to highlight the impact of an SPS on electrical quantities, e.g. reactive power flow and the line voltage profile. Figure 6.6 is a single-line diagram of a transmission line which is equipped with an SPS. The SPS is represented by the model illustrated in Figure 6.5(a), and it neither absorbs nor injects reactive power.

The SPS is located at the middle of the line (X_S=X_R=0.40 per unit). The sending and the receiving end voltages are

$$V_S = 1\angle 40° \text{per unit}$$

$$V_R = 1\angle 0° \text{per unit}$$

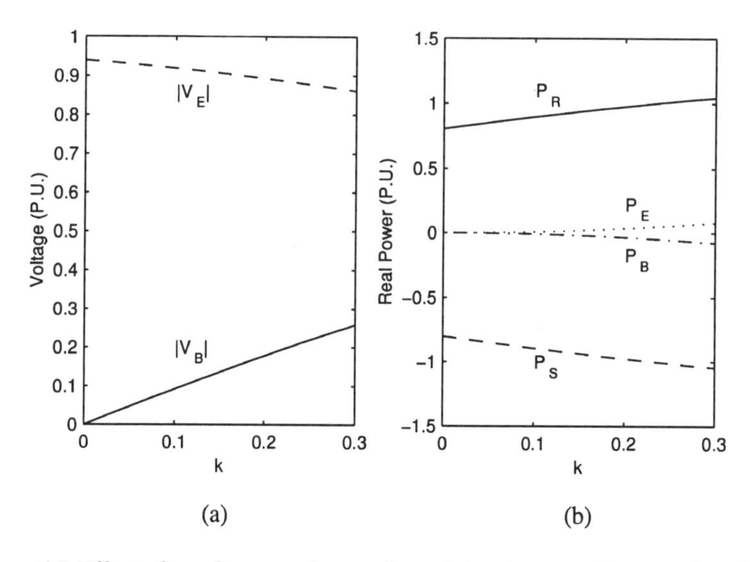

Figure 6.7 Effect of quadrature-phase voltage injection on voltage and real power ($X_S = X_R = 0.4$ p.u., $\phi = 90°, V_S = 1\angle 40°$ p.u., $V_R = 1\angle 0°$ p.u.)

The equations that govern the system electrical variables are

$$V_B = kV_E \exp(j\phi) \tag{6.8}$$

$$I_E = kI_B \exp(-j\phi) \tag{6.9}$$

$$I_S = I_B + I_E \tag{6.10}$$

$$I_R = I_B \tag{6.11}$$

$$V_S - jX_S I_S - V_E = 0 \tag{6.12}$$

$$V_E + V_B - jX_R I_R - V_R = 0 \tag{6.13}$$

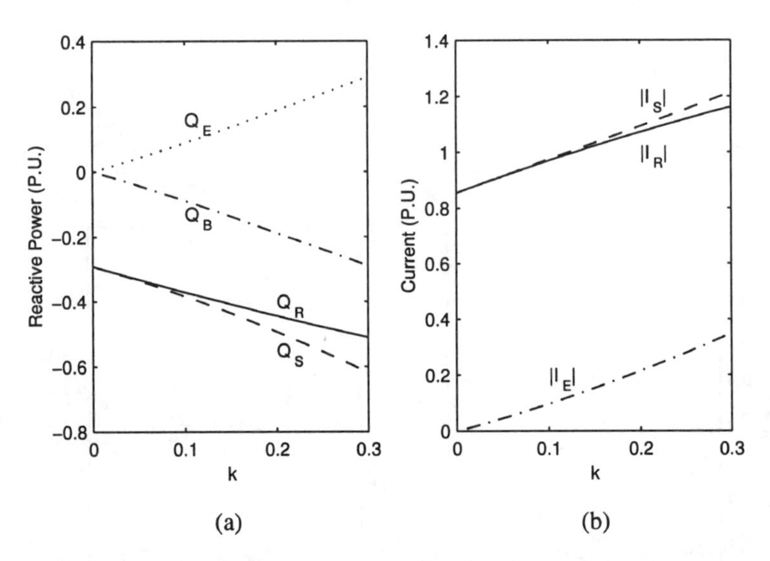

Figure 6.8 Effect of quadrature-phase voltage injection on current and reactive power components ($X_S = X_R = 0.4$ p.u., $\phi = 90°$, $V_S = 1\angle 40°$, $V_R = 1\angle 0°$)

where $k = |V_B|/|V_E|$ and angle ϕ is kept constant at 90°. Figure 6.7 shows variations of electrical quantities obtained from simultaneous solution of equations (6.8) to (6.13) when parameter k is varied between 0 to 0.3. Figure 6.7(a) shows that the injected quadrature-phase voltage V_B varies between 0 to 0.26 per unit as k varies between 0 to 0.3. Correspondingly V_E changes from 0.94 to 0.87 per unit. Voltage injection increases power transfer from 0.80 to 1.05 per unit, Figure 6.7(b). The increase in power transfer is accompanied by a noticeable increase in reactive power demand from the sending end, i.e. -0.29 to -0.62 per unit, while the change in reactive power demand from the receiving end is noticeably less, Fig. 6.8(a). If the sending end cannot provide the required reactive power, the phase shifter cannot increase the real power transfer of the line. Figure 6.8(b) shows that the phase shifting process noticeably increases the line current, particularly the sending end current, and the exciting branch current of the phase shifter. Equations (6.8) to (6.13) also can be used to investigate impact of various system parameters and operating points on the mutual impacts of the system and the phase shifter. For example imposing the condition $X_S+S_R=0.8$ per unit and varying X_R from 0 to 0.8 per unit, can be used to investigate the effect of phase shifter location on power transfer.

6.5 Power circuit configurations for SPS

This section describes various power circuit configurations for static phase shifting. The grouping is primarily based upon the types of power electronic circuits used as converter sections of SPSs. Some of the configurations use two distinct transformers as boosting and exciting transformers. Other configurations eliminate one transformer and combine the functions of the two in one transformer.

In an SPS configuration where a distinct boosting transformer is used, e.g. SPS of Figure 6.1, the primary winding is connected in series with respect to the system. This type of transformer connection is the same as that of an instrument current-transformer (CT). Therefore, the terminals of the secondary winding of the transformer must be either shorted (ideal scenario) or must be connected through a small impedance to provide a path for current flow under all system conditions. Power electronic converters of some SPS configurations, e.g. that of Section 6.5.4, provide the required current path. If the SPS converter circuit is not capable of providing the current path under all conditions, an auxiliary circuit must be included to ensure a current path. The auxiliary circuit is usually a power electronic circuit which operates in coordination with the SPS converter.

The primary winding of an exciting transformer, Figure 6.1, is connected in parallel with respect to the power system. Therefore, a short-circuit of its secondary winding is equivalent to a fault for the system. Various combinations of on/off states of the converter switches must not result in a short-circuit path for the secondary windings of the exciting transformer.

6.5.1 Substitution of mechanical tap-changer by electronic switches

Figure 6.9 shows a schematic diagram of an SPS which is constructed from a conventional phase shifter by direct substitution of mechanical tap-changer (switches) of BT with electronic switches S_{E1} to S_{En} [7]. S_{Ei} and S_B are realized from anti-parallel connection of conventional thyristor switches. S_B is open when either of S_{Ei}s is conducting. S_B provides a short-circuit path for the secondary winding of BT when all S_{Ei}s are open. Control logic of the SPS must prevent simultaneous conduction of two or more S_{Ei}s to avoid partial secondary winding short-circuit of ET.

S_{Ei} is not phase controlled. The SPS control system turns on the desired S_{Ei} at the corresponding current zero-crossing. When S_{E1} is conducting (and all other S_{Ei}s are open), the maximum voltage is injected in the system. The minimum voltage is injected when S_{En} is conducting. Magnitude of the injected voltage is adjusted in either equal or unequal discrete steps between the maximum and the minimum values. The number of steps is determined by the number of switches. Depending on either delta or star connection of the primary windings of ET, the injected voltage is either in quadrature-phase or in-phase with respect to the corresponding system phase voltage. In-phase voltage injection can be used for rapid compensation of voltage sag in sub-transmission and distribution systems.

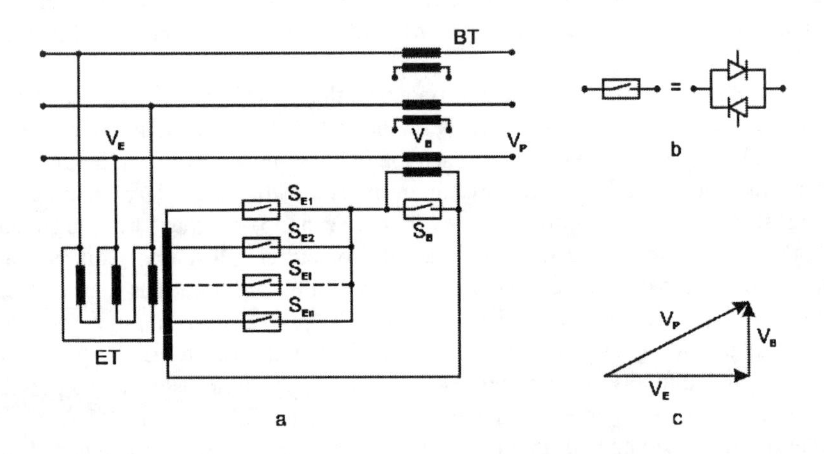

Figure 6.9 (a) Schematic diagram of a SPS based on substitution of mechanical tap-changer by electronic switches; (b) Switch structure; (c) Voltage phasor diagram

Phase-angle regulating transformers are conventionally used for steady-state (slow) power flow regulation [8]. Dynamic (fast) power flow control can be achieved if mechanical tap-changer of a phase-angle regulating transformer is substituted by bidirectional thyristor switches [9]. Application of forced-commutated electronic switches for substituting mechanical tap-changer of distribution level transformers is reported in References 10 and 11.

6.5.2 AC controller

6.5.2.1 Line-frequency ac controller

Figure 6.10 shows schematic diagram of an SPS in which a three-phase ac controller is used as an interface converter between ET and BT [12]. Bidirectional switches S_{E1}, S_{E2} and S_B are composed of antiparallel connected thyristor switches. Delay-angle control strategy is adopted to determine the turn-on instant of each thyristor. Corresponding to each phase, ET provides a leading (lagging) quadrature-phase voltage component $+V_q$ $(-V_q)$ with respect to the system voltage. Magnitude of the injected voltage (V_B) is determined by the delay-angle of S_{E1} (S_{E2}). During the time interval, in each cycle, when S_{E1} (S_{E2}) is not conducting, S_B must conduct to provide a path for the secondary side current of BT. Boosting (bucking) mode of operation is the case when $+V_q$ $(-V_q)$ is injected in the system. If the system requires only voltage boosting (bucking), ET does not need to have a centre-tap connection and S_{E2} (S_{E1}) can be spared [13]. The SPS of Figure 6.10 also can be used for in-phase (180° out-of-phase) voltage injection, if the primary side windings of ET are star-connected [14–15]. A scheme for four-quadrant phase-angle control is reported in Reference 16.Technical drawbacks associated with the SPS of Figure 6.10 are:

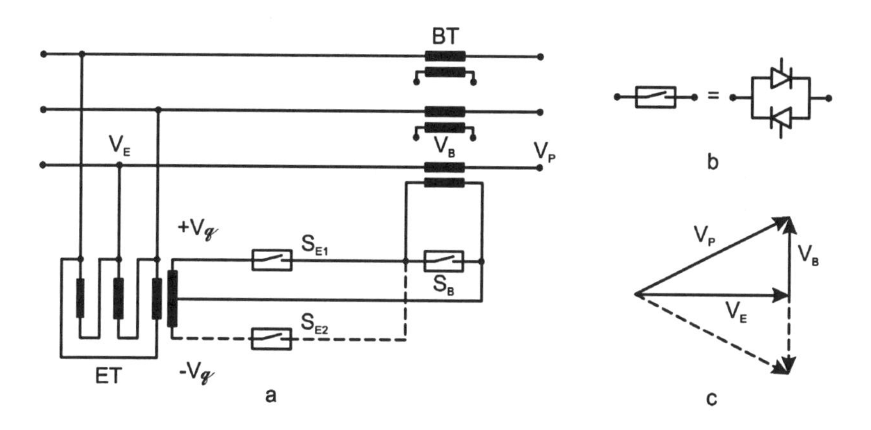

Figure 6.10 (a) Schematic diagram of an SPS based on line-frequency ac controller; (b) Switch structure; (c) Voltage phasor diagram

- Firing-angle control of thyristor switches S_{E1}, S_{E2}, and S_B generates harmonic voltage components in addition to the fundamental frequency voltage component. Injection of harmonic voltage components can be of concern with respect to the power quality of the system.
- Necessary conditions to turn-on each thyristor switch depend on the system operating condition, i.e. phase-angle between corresponding currents and voltages. There exist operating scenarios for which the necessary conditions to turn-on the required thyristor switch are not satisfied. Therefore, the SPS may not be able to provide voltage control under all possible system operating conditions.

6.5.2.2 Pulse-width modulation (PWM) ac controller

Limitations of the SPS of Figure 6.10 are conceptually overcome, if a three-phase PWM ac controller is used as the converter section. Figure 6.11 shows a schematic diagram of an SPS [17] based on a three-phase PWM ac controller. Each of switches S_{Ea}, S_{eb}, and S_{Ec} is composed of antiparallel connection of a diode and a semiconductor switch with on-off control capability (forced-commutation), e.g. a GTO thyristor. A PWM switching pattern is used to turn the switch on and off [18]. When the forced-commutated switch of S_{Ea} (S_{Eb}, S_{Ec}) is on, the corresponding secondary windings of ET and BT are electrically connected. Current flow is either through the forced-commutated switch or its antiparallel diode, depending on the current direction. When S_{Ea} (S_{Eb}, S_{Ec}) is off, forced-commutated switch S_B through a three-phase diode-rectifier provides a free-wheeling path for the secondary side current of BT.

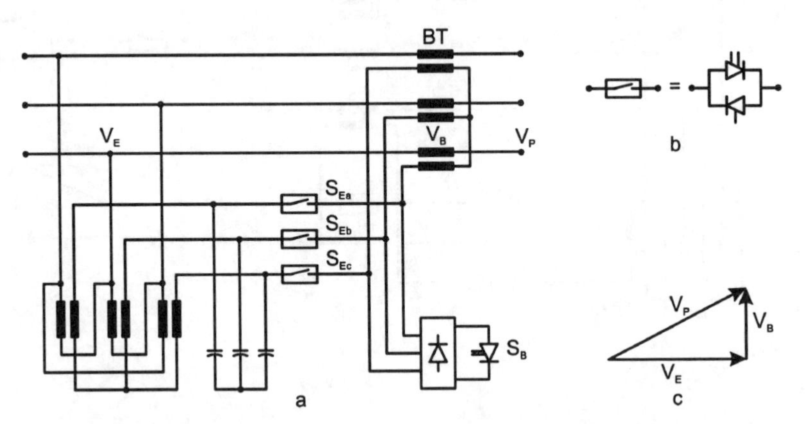

Figure 6.11 (a) Schematic diagram of a PWM ac controller based SPS; (b)Switch structure; (c) Voltage phasor diagram

The magnitude of the injected voltage is controlled by varying the conduction interval (duty-cycle control) of the forced-commutated switches. ET provides in-phase and quadrature-phase voltage if its primary windings are either star- or delta-connected respectively. The technical merits of an SPS based on a PWM ac controller are as follows.

- High frequency PWM switching patterns shift harmonic voltage components to high frequency range. High frequency harmonic components do not propagate widely and are technically easier to be filtered.
- In comparison to a line-frequency ac controller, the required conditions for proper operation of a PWM ac controller are not dependent on the operating point of the system.
- A PWM ac controller provides higher speed of response for an SPS as compared with a line-frequency ac controller. A PWM ac controller results in higher switching losses when compared with a line-frequency ac controller. Switching loss depends on the PWM switching frequency. An alternative configuration for a PWM ac controller based SPS is described in [19].

6.5.3 Single-phase ac–ac bridge converter

6.5.3.1 Delay-angle controlled ac–ac bridge converter

Figure 6.12 shows a schematic diagram of an SPS that uses single-phase ac–ac bridge converter for quadrature-phase voltage injection. Each leg of the converter consists of a bidirectional switch. The bidirectional switch is realized from anti-parallel connection of thyristor switches. The magnitude of the injected voltage is determined by the delay-angle (firing-angle) control of the thyristor switches [20]. The main drawback of delay-angle control is voltage harmonic generation.

6.5.3.2 Discrete-step controlled ac-ac bridge converter

Harmonic generation as a result of delay-angle control of a single-phase ac–ac bridge converter is avoided, if discrete-step control strategy is used. In discrete-step control, each thyristor switch is turned on only at the zero-crossing instant of the corresponding voltage. Figure 6.13 shows a single-line diagram of a discrete-step controlled ac–ac bridge converter and its output voltage (V_{AB}). Figure 6.13 shows that the converter output voltage can have three distinct values of $+V_S$, 0 and $-V_S$ independent of the direction of current *i*. Simultaneous on-state of switches S_1 (S_3) and S_2 (S_4) constitutes a short-circuit across the input voltage source V_S, and must be prevented by the control logic of the converter. If all switches S_1 to S_4 are simultaneously off, then load current i is interrupted. This condition also must be prevented by the control logic.

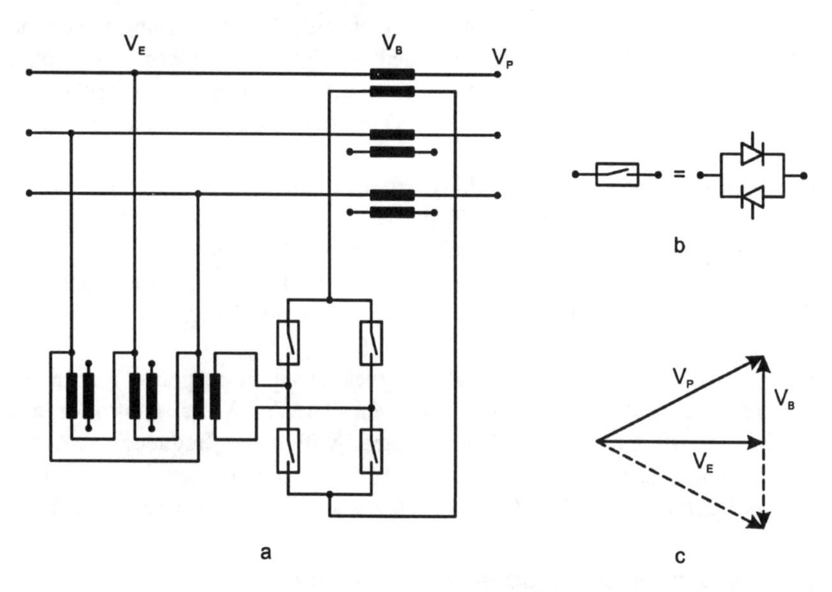

Figure 6.12 Schematic diagram of a SPS based on delay-angle controlled ac–ac converter bridge

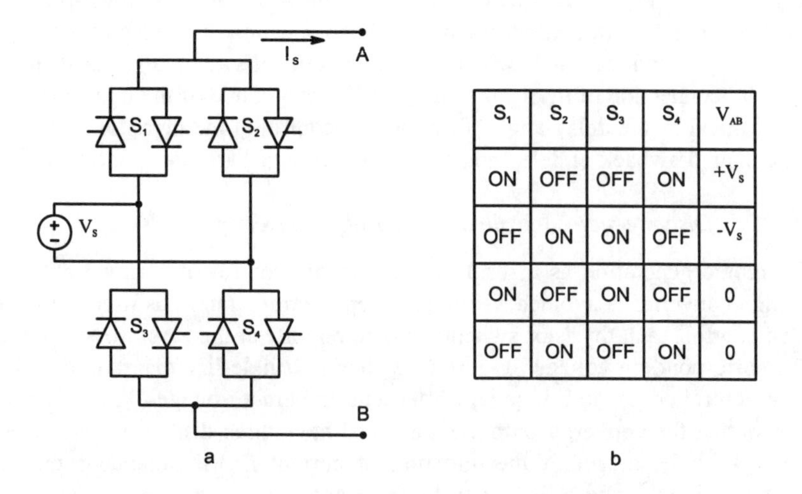

S_1	S_2	S_3	S_4	V_{AB}
ON	OFF	OFF	ON	$+V_s$
OFF	ON	ON	OFF	$-V_s$
ON	OFF	ON	OFF	0
OFF	ON	OFF	ON	0

Figure 6.13 (a) Single-phase ac–ac bridge converter; (b) Output voltage of bridge converter based on discrete-step control

Figure 6.14 shows schematic diagram of an SPS based on multiple discrete-step controlled ac–ac converter bridges [21–22]. Injected voltage is in quadrature-phase with respect to the corresponding system phase voltage. The magnitude of the injected voltage is adjusted in discrete steps by the converter. The number (N) of the secondary windings of ET is equal to the number of bridge converters. The turns ratios of the windings can be either equal or unequal. For the system of Figure 6.14 the turns ratios differ by a factor of three. This arrangement generates (3^N=27) discrete-step voltage levels to be injected in the system [21].

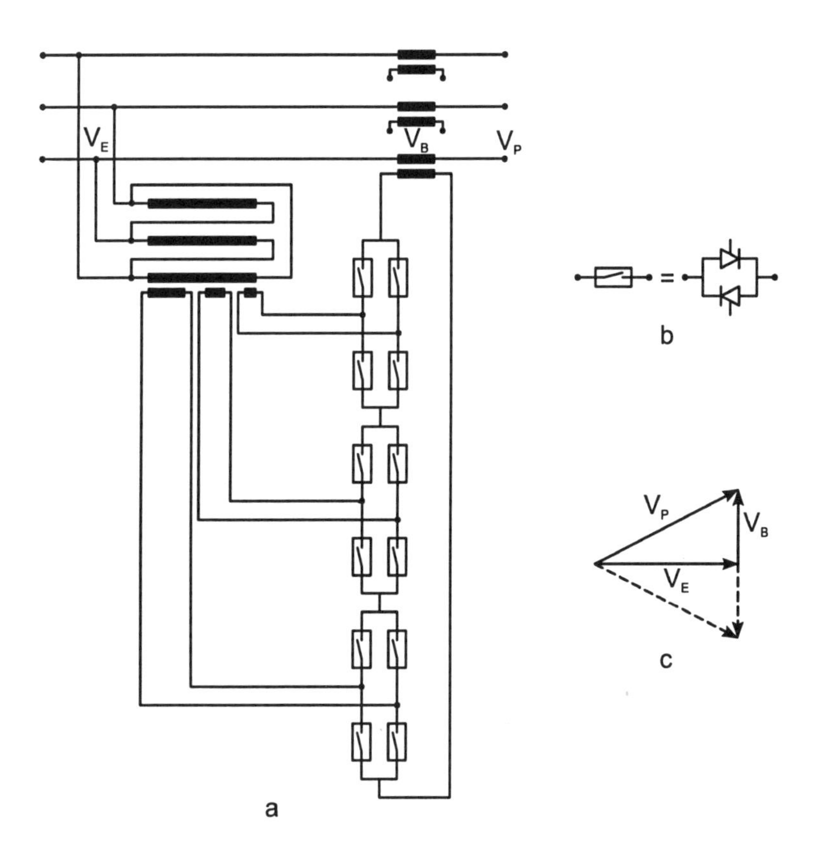

Figure 6.14 (a) Schematic diagram of an SPS based on discrete-step ac–ac bridge converter; (b) Switch structure; (c) Voltage phasor diagram

Reference [23] describes an SPS configuration based on a discrete-step controlled ac–ac bridge converter which provides voltage injection in four quadrants. This SPS configuration needs two exciting transformers. One of the exciting transformers can be spared at the expense of a more complicated winding structure of the remaining transformer, while discrete-step control over both magnitude and phase-angle of the injected voltage is retained [24].

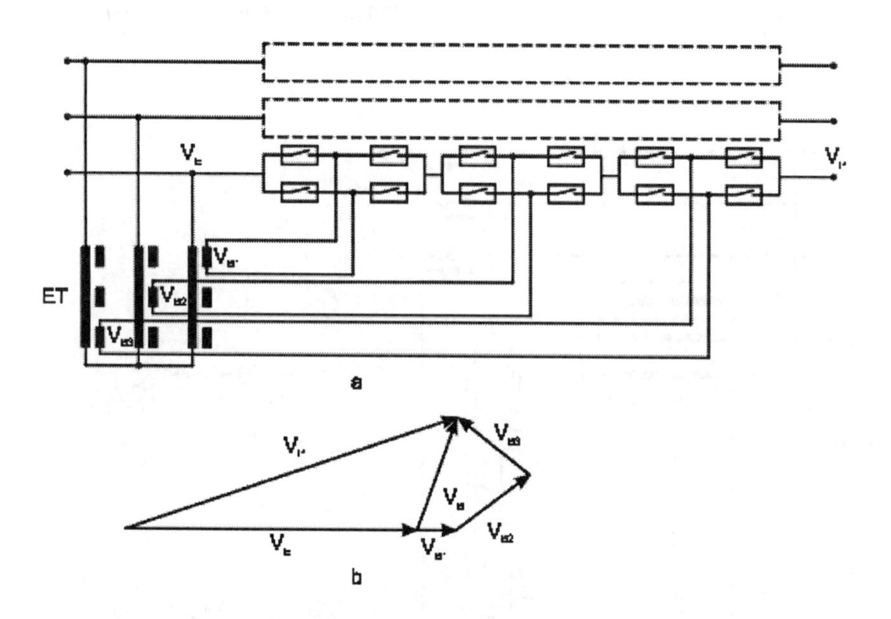

Figure 6.15 (a) Schematic diagram of an SPS that does not use a boosting transformer; (b) Voltage phasor diagram

Figure 6.15 shows an SPS configuration which does not use a boosting transformer and directly inserts ac–ac bridge converters in the transmission line [25]. The structure of each single-phase bridge is the same as that of Figure 5.14 (b). The injected voltage V_B is a combination of three components, V_{B1}, V_{B2}, and V_{B3}. These voltage components are provided from the three phases of ET. ET is either star-connected or delta-connected at the primary side. The phasor diagram of Figure 6.15 illustrates that both magnitude and phase-angle of the injected voltage V_B is controlled in discrete steps.

Figure 6.16 provides a configuration for voltage injection based on the use of tap windings of a step-up (or step-down) power transformer [24]. The configuration can be considered as an SPS which requires neither a boosting transformer nor an exciting transformer. Single-phase ac–ac bridge converters control the injected voltage supplied by the tap windings. Both magnitude and phase-angle of the injected voltage are varied in discrete steps. References [26] and [27] describe two simplified versions of the SPS of Figure 6.16. The SPS of [26] injects a fixed magnitude with an angle of either +60° or -120° with respect to the system phase voltage. The configuration of [27] injects either an in-phase or 180° out-of-phase voltage with respect to the system phase voltage. The configuration of [27] is primarily intended for rapid voltage-magnitude regulation, e.g. voltage sag compensation, and not for phase shifting.

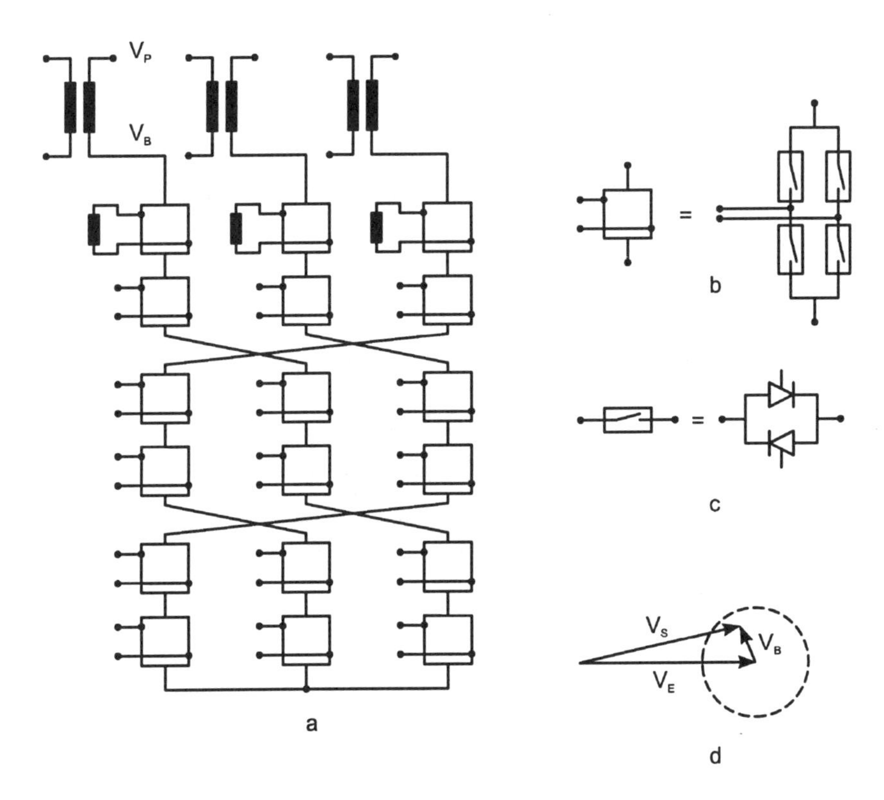

Figure 6.16 (a) Schematic diagram of an SPS which does not require exciting and boosting transformers; (b) Converter structure; (c) Switch structure; (d) Voltage phasor diagram

6.5.4 PWM voltage source converter (VSC)

The SPS of Figure 6.17 uses an ac–dc–ac converter system as the converter interface between exciting and boosting transformers [24 and 28–29]. The converter system is composed of two PWM voltage source converters (VSCs) that share a dc-link capacitor. Each VSC is a three-phase full-bridge configuration. Each arm of the bridge consists of a bidirection switch which is composed of antiparallel connection of a diode and a switch with turn on-off (forced-commutation) capability, e.g. GTO thyristor.

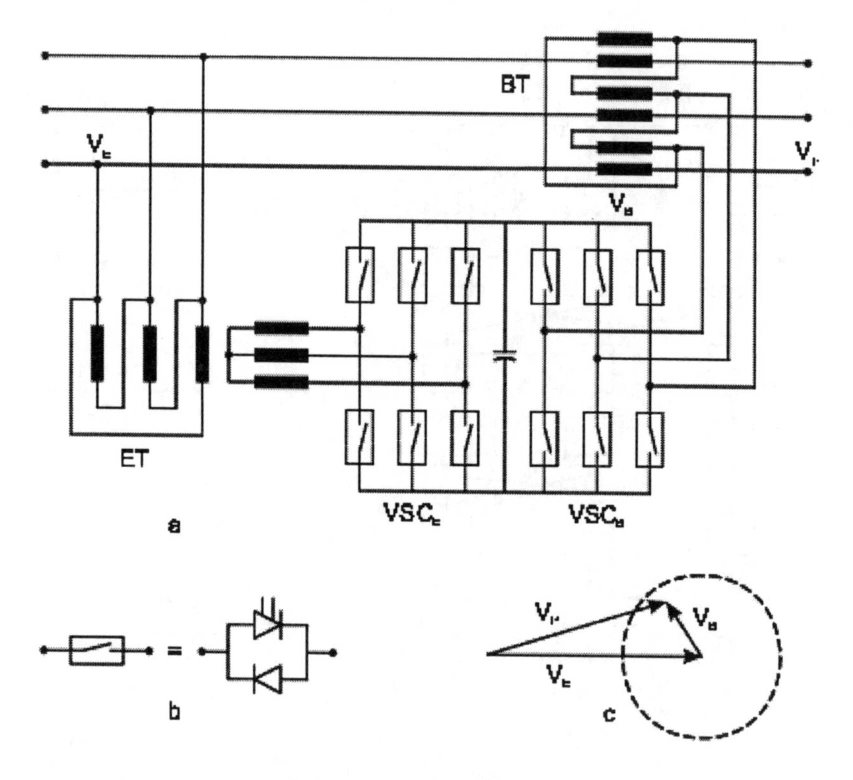

Figure 6.17 (a) Schematic diagram of a SPS based on PWM VSC converter; (b) Switch structure; (c) Voltage phasor diagram

VSC_B provides independent control of magnitude and phase-angle of the injected voltage. VSC_E regulates the dc-link capacitor voltage. VSC_E can also provide reactive power compensation for the system and consequently control the magnitude of V_E. In contrast to the SPS configurations described in previous sections, a VSC based SPS can absorb/inject reactive power. The steady-state equivalent circuit of a VSC based SPS is illustrated by Figure 6.5 (b). A VSC based SPS provides independent, continuous control over real and reactive flow of the line. The circled area of the phasor diagram of Figure 6.17 encompasses the area that the tip of vector V_B can be continuously varied within.

In addition to its function as an SPS, the configuration of Figure 6.17 also can operate as a shunt compensator, series compensator, and shunt/series active power filter. In the technical literature this configuration is more widely known as Unified Power Flow Controller (UPFC). UPFC is the subject of a separate chapter in this book.

6.5.5 PWM current source converter (CSC)

The ac–dc–ac converter section of Figure 6.17 also can be realized from PWM current source converters instead of PWM VSCs. Figure 6.18 shows a schematic diagram of an SPS which is based on PWM CSCs [30]. The converter section is composed of two three-phase PWM CSCs that share a series dc-link reactor. Each CSC is a three-phase full-bridge configuration. Each arm of the bridge consists of a switch with turn-on and turn-off capability, e.g. a GTO thyristor.

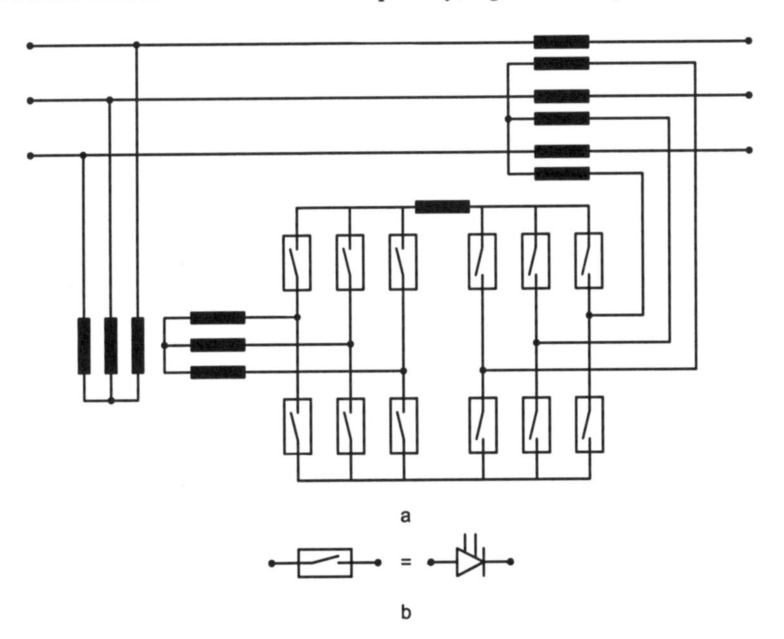

Figure 6.18 (a) Schematic diagram of a PWM CSC based SPS; (b) Switch structure

CSC_B regulates line power flow through control of the line current [30]. This is in contrast to the principles of operation of the VSC based SPS of Figure 6.17 where magnitude and phase-angle of the injected voltage are controlled. The main function of CSC_E is to control the dc link current. A CSC directly controls its ac side current and provides inherent overcurrent protection for semiconductor switches during system fault scenarios. A VSC has no inherent over current limit capability and must be protected, e.g. bypassed, during transient conditions by means of auxiliary devices.

6.5.6 Other SPS circuit configurations

The SPS configurations presented in Figures 6.9 to 6.18 are intended to illustrate the main electrical components and basic converter circuits. Corresponding to the SPS of Figure 6.17, References 31 and 32 describe elaborate arrangements of VSCs to meet practical constraints, e.g. switch ratings, and operational requirements, e.g. harmonics. Reference 24 briefly introduces the concept of augmenting conventional phase shifters with VSCs to provide capability for rapid small-signal phase-angle regulation. Reference 33 and 34 introduce two circuit configurations for in-phase voltage injection for distribution voltage levels.

6.6 SPS applications

6.6.1 Steady-state

Similar to conventional (mechanical) phase shifters, SPSs can be used for steady-state power flow and/or voltage regulation. An SPS regulates steady-state power flow primarily by adjusting the phase-angle of the system voltage by means of voltage injection. Voltage injection by means of an SPS directly impacts reactive power flow through the line and reactive power demand from both ends of the line. The SPS configurations of Figures 6.17 and 6.18 can control real power and reactive power independently. This indicates that for a pre-specified value of real power flow, the SPS can absorb/inject appropriate reactive power and maintain reactive power flow and reactive power demand from both ends within permissible operational constraints.

In contrast to the SPSs of Figures 6.17 and 6.18, the SPS configurations of Figures 6.9 to 6.12 and 6.13 to 6.16 cannot independently control real and reactive power flows. If the SPS adjusts real power flow to a pre-specified value, reactive power flow and reactive power demand from both ends of the line are determined by the overall system parameters and the operating condition. If reactive power flow through the line or reactive power demand from each end of the line violates the operational constraints, real power flow has to be respecified and accordingly adjusted. This may impose severe constraints on the range of real power control by a SPS, particularly in the case of (1) long transmission lines and (2) when either end of a line is a relatively weak system.

Steady-state power flow control by means of a SPS can be used for:

- power flow regulation of a transmission corridor,
- power sharing among parallel lines,
- and loop-flow prevention [35].

Similar to mechanical on-load tap-changing transformers, SPSs can be used for steady-state voltage regulation at distribution voltage levels. In contrast to mechanical tap-changers, SPSs can restore voltage in response to system dynamics, e.g. faults on adjacent distribution feeders and load energization. A steady-state mathematical model of an SPS is described in Section 6.3. Implementation of a SPS steady-state model within a power-flow analysis model is explained in References [36] and [37].

6.6.2 Small-signal dynamics

Due to its inherent slow response, a mechanical phase shifter cannot be used for dynamic voltage injection. In contrast, an SPS can dynamically inject voltage in a system and correspondingly control power flow in response to power system electromechanical oscillations. Electromechanical oscillatory modes of a power system usually have natural frequencies in the sub-synchronous frequency range. Rapid voltage injection by an SPS can be used to dynamically control power flow to enhance damping and mitigate electromechanical oscillatory modes, e.g. inter-area modes, inertial (hunting) modes and torsional modes. Applications of SPSs for mitigation of torsional oscillations and inter-area oscillations have been proposed and examined in References [6] and [26].

Systematic analysis and design of an SPS control system for mitigation of small-signal dynamics requires a linearized dynamic model of the SPS. Depending on the frequency range of oscillations and converter configuration of the SPS, various SPS linear models with different degrees of mathematical details are needed. References [6], [26], [29], and [36] provide linearized small-signal models of SPSs applicable in the sub-synchronous frequency range.

6.6.3 Large-signal dynamics

Large-signal electrical disturbances, e.g. faults, fault clearing, out-of-phase synchronization, line switching, and reclosures can result in large-signal electromechanical transients. Such disturbances can even lead to instability. Rapid phase shifting by means of an SPS can be used to either mitigate transients within the desired time span after the disturbance inception or even to prevent transient instability [38–40].

Analysis of an SPS behaviour, in the context of transient stability, requires a nonlinear model of the SPS and its control system [41]. During system large-signal transients, converter components of an SPS are subjected to abnormal over-current and/or over-voltage stresses. SPS converter design requires investigation and quantification of such stresses. This type of transient study needs detailed SPS model for digital time-domain simulation suitable for the analyses of electromagnetic transients.

6.7 Summary

This chapter briefly describes the concept of phase shifting in electric power systems and highlights the technical merits of semiconductor-controlled (static) phase shifters over conventional (mechanical) phase shifters. Based upon the steady-state principles of operation of a Static Phase Shifter (SPS), a steady-state SPS model is developed. Although simplifying assumptions are used, the model provides adequate insight to explore the impact of various parameters and operating conditions/constraints on the SPS system. Based on currently feasible semiconductor switches and converter topologies, five SPS configurations are identified. The first group is based on substituting a mechanical tap-changer by semiconductor switches. The second group uses ac controllers. The third group is based on a single-phase ac–ac bridge converter topology. The fourth and the fifth group use PWM voltage source converters and PWM current source converters. The basic power circuits, principles of operations and salient features of each group are briefly described. The chapter is concluded by a brief description of SPS applications in power systems.

6.8 References

1 Dobsa, J., 'Transformers for in-phase, phase-angle and quadrature-phase regulation', *Brown Boveri Review*, 8, pp. 376–383, 1972.

2 Ibrahim, M.A., and Stacom, F.P., 'Phase angle regulating transformer protection', *IEEE Transactions on Power Delivery*, 9, (1), pp. 394–402, 1994.

3 Brandes, W., and Haubrich, H.J., 'Benefits of phase shifting transformers to power flow in meshed high voltage power systems', *Proceedings of Fourth International Conference on AC and DC Power Transmission*, London, pp. 447–452, 1982.

4 Wood, P., Bapt, V., and Putkovich, R.P., 'Study of improved load-tap-changing for transformers and phase-angle regulators', *EPRI Report EL-6079*, 1988.

5 IEEE FACTS Terms & Definitions Task Force, 'Proposed terms and definitions for flexible ac transmission systems (FACTS)', *IEEE Transactions on Power Delivery*, 12, (4), pp. 1848-1853, 1997.
6 Nabavi-Niaki, S.A., 'Modelling and applications of unified power flow controller', Ph.D. Thesis, University of Toronto, 1996.
7 Dewan, S.B., and Straughen, A., 'Power semiconductor circuits', John Wiley & Sons, Toronto, pp. 204-211, 1975.
8 Stevenson, W.D., 'Elements of power system analysis', McGraw Hill, New York, 4th edn., pp. 214-216, 1982.
9 Nyati, S., Eitzmann, M., Kappenman, J., Van House, D., Mohan, N., and Edris, A., 'Design issues for a single core transformer thyristor controlled phase-angle regulator', *IEEE Transactions on Power Delivery*, 10, (4), pp. 2013-2019, 1995.
10 Bauer, P., De Haan, S.W.H., and Paap, G.C., 'Electronic tap changer for 10 kV distribution transformer', *Proceedings of European Power Electronics Conference*, Trondheim, Norway, pp. 3.1010-3.1015, 1997.
11 Bauer, P., and De Haan, S.W.H., 'Protective device for electronic tap changer for distribution transformer', *Proceedings of European Power Electronics Conference*, Trondheim, Norway, pp. 4.282-4.285, 1997
12 Arnold, C.P., Duke, R.M., and Arrillaga, J., 'Transient stability improvement using thyristor controlled quadrature voltage injection', *IEEE Transactions on Power Apparatus and Systems*, 100, (3), pp. 1382-1388, 1981.
13 Arrillaga, J., and Duke, R.M., 'Thyristor controlled quadrature boosting', *Proceedings of IEE*, 126, (6), pp. 493-498, 1979.
14 Arrillaga, J., Wood, G., and Duke, R.M., 'Thyristor-controlled in-phase boosting for HVdc converters', *Proceedings of IEE*, 127, (4), pp. 221-227, 1980.
15 Arrillaga, J., and Duke, R.M., 'A static alternative to the transformer on-load tap-changer', *IEEE PES Conference Record*, New York, F-79-171-0, 1979.
16 Arrillaga, J., Duke, R.M., and Arnold, C.P, 'Thyristor-controlled 4-quadrant voltage injection', *IEE AC and DC Conference Record*, London, pp. 65-68, 1985.
17 Lopes, A.C., Joos, G., and Ooi, B.T., 'A PWM quadrature booster phase-shifter for FACTS', *IEEE Transactions on Power Delivery*, 11, (4), pp. 1999-2004,1996.
18 Lopes, A.C., Joos, G., and Ooi, B.T., 'A high-power PWM quadrature booster phase shifter based on a multimodule ac controller', *IEEE Transactions on Power Electronics*, 13, (2), pp. 357-363, 1998.
19 Johnson, B.K., and Venkataramanan, G., 'A hybrid solid state phase shifter using PWM ac controller', *IEEE Power Engineering Society*, Winter Meeting, Tampa, PE-134-PWRD-0-12-1997, 1998.

20 Mohan, N., 'Continuous alternating-voltage regulation using a thyristor bridge', *Proceedings of IEE*, 126, (10) , 1979.
21 Baker, R., Guth, G., Egli, W., and Eglin, P., 'Control algorithm for a static phase shifting transformer to enhance transient and dynamic stability of large power systems', *IEEE Transactions on Power Apparatus and Systems*, 101, (9), pp. 3532-3542, 1982.
22 Guth, G., Baker, R., and Eglin, P., 'Static thyristor controller regulating transformer for ac-transmission', *AC and DC Transmission Conference Record*, London, pp. 69-72, 1982.
23 Iravani, M.R., 'Thyristor-controlled phase shifter for ac transmission', *Power Electronics Specialists Conference Record*, Vancouver, pp. 291-300, 1986.
24 Iravani, M.R., and Maratukulam, D., 'Review of semiconductor-controlled (static) phase shifters for power system applications', *IEEE Transactions on Power Systems*, 9, (4), pp. 1833-1839, 1994.
25 Mohan, N., 'MPTC, An economical alternative to universal power flow controller', *European Power Electronics Conference*, Trondheim, pp. 3.1027-3.1032, 1997.
26 Hammad, A.E., and Dobsa, J., 'Application of a thyristor controlled phase angle regulating transformer for damping subsynchronous oscillations', *Power Electronics Specialists Conference Record*, pp. 111-120, 1983.
27 Demiric, O., Torrey, D.A., Degeneff, R.C., Schaeffer, F.K., and Frazer, R.H., 'A new approach to solid-state on load tap changing transformer', *IEEE Transactions on Power Delivery*, 13, (3), pp. 952-961, 1988.
28 Gyugyi, L., 'Unified power-flow control concept for flexible ac transmission systems', *IEE Proceedings-C*, 139, (4), pp. 323-331, 1992.
29 Ooi, B.T., Dai, S.Z., and Galiana, F.D., 'A solid-state PWM phase shifter', *IEEE Transactions on Power Delivery*, 8, (2), pp. 573-579, 1993.
30 Jager, J., Herold, G., and Hosemann, G., 'Current-source controlled phase shifting transformer as a new FACTS equipment for high dynamical power flow control', *Record of CIGRE Symposium on Power Electronics in Electric Power Systems*, Tokyo, paper 520-03, 1995.
31 Gyugyi, L., Hingorani, N.G., Nannerry, P.R., and Tai, N., 'Advanced static VAR compensators using gate-turn-off thyristors for utility applications', *Record of International Conference on Large High Voltage Electric Systems (CIGRE)*, Paris, paper 23-203, 1990.
32 Schauder, C., Gernhardt, M., Stacey, E., *et al.*, 'Operation of ±100 MVAR TVA STATCOM', *IEEE Transactions on Power Delivery*, 12, (4), pp. 1805-1811, 1997.
33 ABB Technical Guide, On-load tap-changers type UED, ABB Components, 1ZSE-5492-157, 1997.

34 Larson, T., Innanen, R., and Norstrom, G., 'Static electronic tap-changer for fast phase voltage control', *Record of European Power Electronics (EPE) Conference*, Trondheim, pp. 3.956-3.958, 1997.
35 Heyward, D., Miller, J.M., and Balmat, B.M., 'Operating problems with parallel flows', *IEEE Transactions on Power Systems*, 6, (3), pp. 1024-1034, 1991.
36 Nabavi-Niaki, A. and Iravani, M.R., 'Steady-state and dynamic model of unified power-flow controller (UPFC) for power system studies', *IEEE Transactions on Power Systems*, 11, (4), pp. 1937-1943, 1996.
37 Ambriz-Perez, H., Acha, E., Fuerte-Esquivel, C.R., and De La Torre, A., 'Incorporation of a UPFC model in an optimal power flow using Newton's method', *IEE Proceedings on Generation, Transmission and Distribution*, 145, (3), pp. 336-344, 1998.
38 O'Kelly, D., and Musgrave, G., 'Improvement of power-system transient stability by phase-shift insertion', *IEE Proceedings*, 120, 2, pp. 247-252, 1973.
39 Cresap, R.L., Taylor, C.W., and Kreipe, M.J., 'Transient stability enhancement by 120-degree phase rotation', *IEEE Transactions on Power Systems and Apparatus*, 100, (2), pp. 745-753, 1981.
40 Edris, A.A., 'Enhancement of first-swing stability using a high-speed phase shifter', *IEEE Transactions on Power Systems*, 6, (3), pp. 1113-1118, 1991.
41 Arabi, S., and Kundur, P., 'A versatile FACTS model for power flow and stability simulation', *IEEE Transactions on Power Systems*, 11, (4), pp. 1944-1950, 1996.

Chapter 7

The unified power flow controller

Laszlo Gyugyi and Colin D. Schauder

7.1 Introduction

The power transmitted over an ac transmission line is a function of the line impedance, the magnitude of sending-end and receiving-end voltages, and the phase angle between these voltages. Traditional techniques of reactive line compensation and step-like voltage adjustment are generally used to alter these parameters to achieve power transmission control. Fixed and mechanically switched shunt and series reactive compensation are employed to modify the natural impedance characteristics of transmission lines in order to establish the desired effective impedance between the sending- and receiving-ends to meet power transmission requirements. Voltage regulating and phase shifting transformers with mechanical tap-changing gears are used to minimize voltage variation and control power flow. These conventional methods provide adequate control under steady-state and slowly changing system conditions, but are largely ineffective in handling dynamic disturbances. The traditional approach to contain dynamic problems is to establish generous stability margins enabling the system to recover from faults, line and generator outages, and equipment failures. This approach, although reliable, generally results in a significant under utilization of the transmission system.

As a result of recent environmental legislation, rights-of-way issues, construction cost increases, and deregulation policies, there is an increasing recognition of the necessity to utilize existing transmission system assets to the maximum extent possible. To this end, electronically controlled, extremely fast reactive compensators and power flow controllers have been developed within the overall framework of the Flexible AC Transmission System (FACTS) initiative. These compensators and controllers either use conventional reactive components (capacitors and reactors) and tap-changing transformer arrangements with thyristor valves and control electronics or employ switching power converters, as synchronous voltage sources, which can internally generate reactive power for, and also exchange real power with, the ac system.

The Unified Power Flow Controller (UPFC) is a member of this latter family of compensators and power flow controllers which utilize the synchronous voltage source (SVS) concept for providing a uniquely comprehensive capability for transmission system control. Within the framework of traditional power transmission concepts, the UPFC is able to control, simultaneously or selectively, *all* the parameters affecting power flow in the transmission line (i.e., voltage, impedance, and phase angle). Alternatively, it can provide the unique functional capability of independently controlling *both* the real and reactive power flow in the line. These basic capabilities make the Unified Power Flow controller the most powerful device presently available for transmission system control.

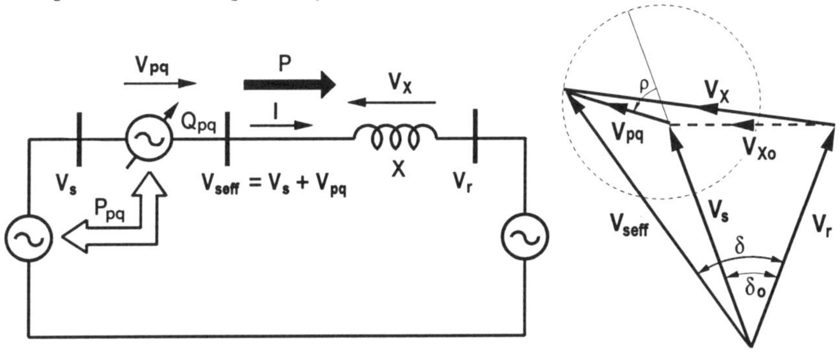

Figure 7.1 Conceptual representation of the UPFC in a two-machine power system

7.2 Basic operating principles and characteristics

The Unified Power Flow Controller (UPFC) was devised for the real-time control and dynamic compensation of ac transmission systems, providing multi-functional flexibility required to solve many of the problems facing the power delivery industry.

From the conceptual viewpoint, the UPFC is a generalized synchronous voltage source (SVS), represented at the fundamental (power system) frequency by voltage phasor V_{pq} with controllable magnitude V_{pq} ($0 \leq V_{pq} \leq V_{pqmax}$) and angle ρ ($0 \leq \rho \leq 2\pi$), in series with the transmission line, as illustrated for an elementary two-machine system (or for two independent systems with a transmission link intertie) in Figure 7.1. In this arrangement the SVS generally exchanges both reactive and real power with the transmission system. Since, by definition (see Section 1.4.2.1), an SVS is able to generate only the reactive power exchanged, the real power must be supplied to it, or absorbed from it, by a suitable power supply or sink. In the UPFC arrangement the real power the SVS exchanges is provided by one of the end buses (e.g., the sending-end bus), as indicated in the figure. (This arrangement conforms to the objective of *controlling* the power flow by the UPFC rather than increasing the generation capacity of the system.)

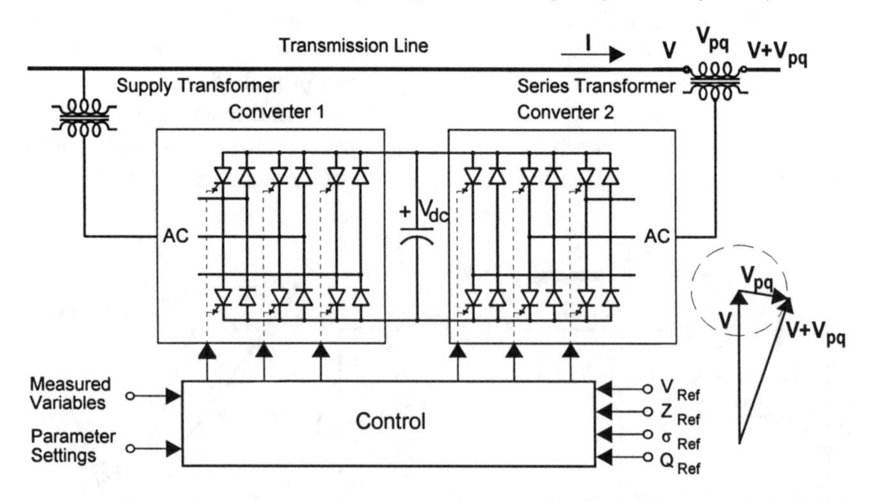

Figure 7.2 Implementation of the UPFC by two back-to-back voltage-sourced converters

In the presently used practical implementation, the UPFC consists of two *voltage-sourced converters* using *gate turn-off (GTO) thyristor* valves, as illustrated in Figure 7.2. These converters, labelled "Converter 1" and "Converter 2" in the figure, are operated from a common dc link provided by a dc storage capacitor. This arrangement functions as an ideal ac to ac power converter in which the *real* power can freely flow in either direction between the ac terminals of the two converters, and each converter can independently generate (or absorb) *reactive* power at its own ac output terminal.

Converter 2 provides the main function of the UPFC by injecting a voltage $\boldsymbol{V}_{pq}$ with controllable magnitude V_{pq} and phase angle ρ in series with the line via an insertion transformer. This injected voltage acts essentially as a synchronous ac voltage source. The transmission line current flows through this voltage source resulting in reactive and real power exchange between it and the ac system. The reactive power exchanged at the ac terminal (i.e., at the terminal of the series insertion transformer) is generated internally by the converter. The real power exchanged at the ac terminal is converted into dc power which appears at the dc link as a positive or negative real power demand.

The basic function of Converter 1 is to supply or absorb the real power demanded by Converter 2 at the common dc link. This dc link power is converted back to ac and coupled to the transmission line via a shunt-connected transformer. Converter 1 can also generate or absorb controllable reactive power, if it is desired, and thereby provide independent shunt reactive compensation for the line. It is important to note that whereas there is a closed "direct" path for the *real power* negotiated by the action of series voltage injection through Converters 1 and 2 back to the line, the corresponding *reactive power* exchanged is supplied or absorbed locally by Converter 2 and therefore does not have to be transmitted by the line. Thus, Converter 1 can be operated at a unity power factor or be controlled to have a reactive power èxchange with the line independent of the reactive power exchanged by Converter 2. This means that there is *no* reactive power flow through the UPFC.

7.2.1 Conventional transmission control capabilities

Viewing the operation of the Unified Power Flow Controller from the standpoint of traditional power transmission based on reactive shunt compensation, series compensation, and phase shifting, the UPFC can fulfill all these functions and thereby meet multiple control objectives by adding the injected voltage V_{pq}, with appropriate amplitude and phase angle, to the (sending-end) terminal voltage V_s. Using phasor representation, the basic UPFC power flow control functions are illustrated in Figure 7.3.

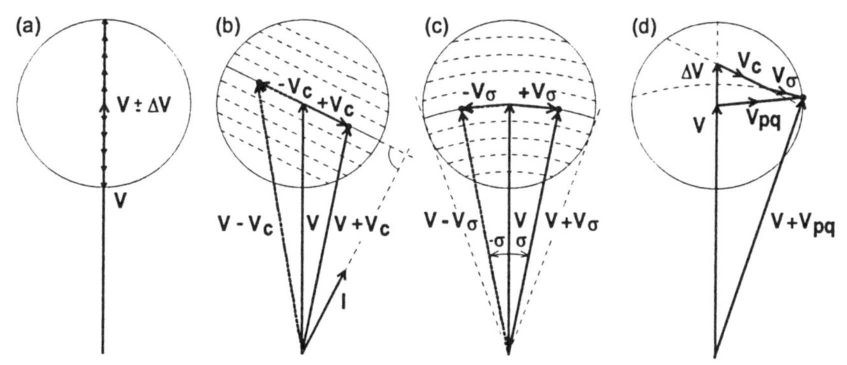

(a) Voltage regulation
(b) Line impedance compensation
(c) Phase shifting
(d) Simultaneous control of voltage, impedance and angle

Figure 7.3 Range of transmittable real power P and receiving-end reactive power demand Q, vs. transmission angle δ of a UPFC controlled transmission line

Voltage regulation with continuously variable in-phase/anti-phase voltage injection, shown at *a* for voltage increments $V_{pq}=\pm\Delta V$ ($\rho=0$). Functionally this is similar to that obtainable with a transformer tap-changer having infinitely small steps.

Series reactive compensation is shown at *b* where $V_{pq}=V_q$ is injected in quadrature with the line current I. Functionally this is similar to, but more general than the controlled series capacitive and inductive line compensation. This is because the UPFC injected series compensating voltage can be kept constant, if desired, independent of line current variation, whereas the voltage across the series compensating (capacitive or inductive) impedance varies with the line current.

Phase shifting (transmission angle regulation) is shown at *c* where $V_{pq}=V_\sigma$ is injected with an angular relationship with respect to V_s that achieves the desired σ phase shift (advance or retard) without any change in magnitude. Thus the UPFC can function as a perfect phase shifter. From the practical viewpoint, it is also important to note that, in contrast to conventional phase shifters, the ac system does not have to supply the reactive power the phase shifting process demands since it is internally generated by the UPFC converter.

Multi-function power flow control, executed by simultaneous terminal voltage regulation, series capacitive line compensation, and phase shifting, is shown at *d* where $V_{pq}=\Delta V+V_q+V_\sigma$. This functional capability is unique to the UPFC. No single conventional equipment has similar multi-functional capability.

The general power flow control capability of the UPFC, from the viewpoint of conventional transmission control, can be illustrated best by the real and reactive power transmission versus transmission angle characteristics of the simple two-machine system shown in Figure 7.1. With reference to this figure, the transmitted power P and the reactive power $-jQ_r$, supplied by the receiving-end, can be expressed as follows:

$$P - jQ_r = V_r\left(\frac{V_s + V_{pq} - V_r}{jX}\right)^* \tag{7.1}$$

where symbol * means the conjugate of a complex number and $j=\sqrt{-1}$. If $V_{pq}=0$, then equation (7.1) describes the uncompensated system, that is,

$$P - jQ_r = V_r\left(\frac{V_s - V_r}{jX}\right)^* \tag{7.2}$$

Thus, with $V_{pq}\neq 0$, the total real and reactive power can be written in the form:

$$P - jQ_{\rm r} = \boldsymbol{V}_{\rm r}\left(\frac{\boldsymbol{V}_{\rm s} - \boldsymbol{V}_{\rm r}}{jX}\right)^* + \frac{\boldsymbol{V}_{\rm r}\boldsymbol{V}_{\rm pq}^*}{-jX} \tag{7.3}$$

Substituting

$$\boldsymbol{V}_{\rm s} = V\mathrm{e}^{j\delta/2} = V\left(\cos\frac{\delta}{2} + j\sin\frac{\delta}{2}\right) \tag{7.4}$$

$$\boldsymbol{V}_{\rm r} = V\mathrm{e}^{-j\delta/2} = V\left(\cos\frac{\delta}{2} - j\sin\frac{\delta}{2}\right) \tag{7.5}$$

and

$$\boldsymbol{V}_{\rm pq} = V_{\rm pq}\mathrm{e}^{j(\delta/2+\rho)} = V_{\rm pq}\left\{\cos\left(\frac{\delta}{2}+\rho\right) + \sin\left(\frac{\delta}{2}+\rho\right)\right\} \tag{7.6}$$

the following expressions are obtained for P and $Q_{\rm r}$:

$$P(\delta,\rho) = P_{\rm o}(\delta) + P_{\rm pq}(\rho) = \frac{V^2}{X}\sin\delta - \frac{VV_{\rm pq}}{X}\cos\left(\frac{\delta}{2}+\rho\right) \tag{7.7}$$

and

$$Q_{\rm r}(\delta,\rho) = Q_{\rm or}(\delta) + Q_{\rm pq}(\rho) = \frac{V^2}{X}(1-\cos\delta) - \frac{VV_{\rm pq}}{X}\sin\left(\frac{\delta}{2}+\rho\right) \tag{7.8}$$

where

$$P_{\rm o}(\delta) = \frac{V^2}{X}\sin\delta \tag{7.9}$$

and

$$Q_{\rm or}(\delta) = -\frac{V^2}{X}(1-\cos\delta) \tag{7.10}$$

are the real and reactive power characterizing the power transmission of the uncompensated system at a given angle δ. Since angle ρ is freely variable between 0 and 2π at any given transmission angle δ $(0\leq\delta\leq\pi)$, it follows that $P_{\rm pq}(\rho)$

and $Q_{pq}(\rho)$ are controllable between $-VV_{pq}/X$ and $+VV_{pq}/X$ independent of angle δ. Therefore, the transmittable real power P is controllable between

$$P_o(\delta) - \frac{VV_{pq}}{X} \leq P_o(\delta) \leq P_o(\delta) + \frac{VV_{pq}}{X} \tag{7.11}$$

and the reactive power Q_r is controllable between

$$Q_o(\delta) - \frac{VV_{pq}}{X} \leq Q_o(\delta) \leq Q_o(\delta) + \frac{VV_{pq}}{X} \tag{7.12}$$

at any transmission angle δ, as illustrated in Figure 7.4.

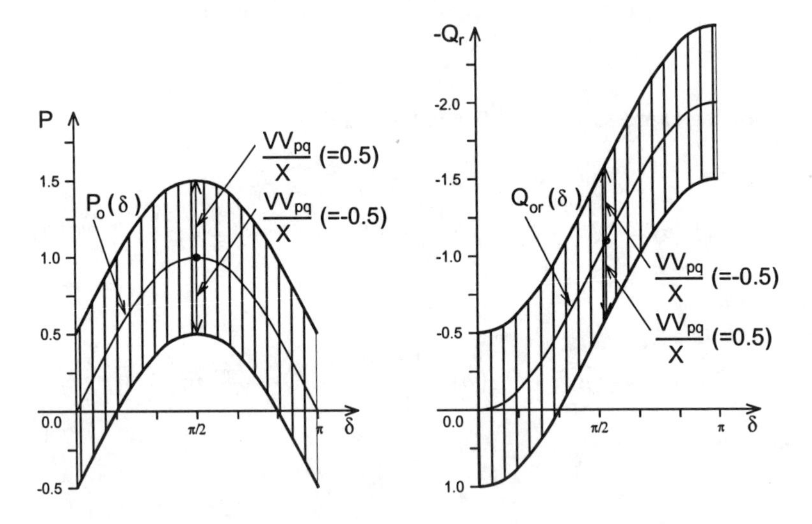

Figure 7.4 Range of transmittable real power P and receiving-end reactive power demand Q_r vs. transmission angle δ of a UPFC controlled transmission line

The powerful, previously unattainable, capabilities of the UPFC summarized above in terms of conventional transmission control concepts, can be integrated into a generalized power flow controller that is able to maintain prescribed, and independently controllable, real power P and reactive power Q in the line. Within this concept, the conventional terms of series compensation, phase shifting, etc., become irrelevant; the UPFC simply controls the magnitude and angular position of the injected voltage in real time so as to maintain or vary the real and reactive power flow in the line to satisfy load demand and system operating conditions.

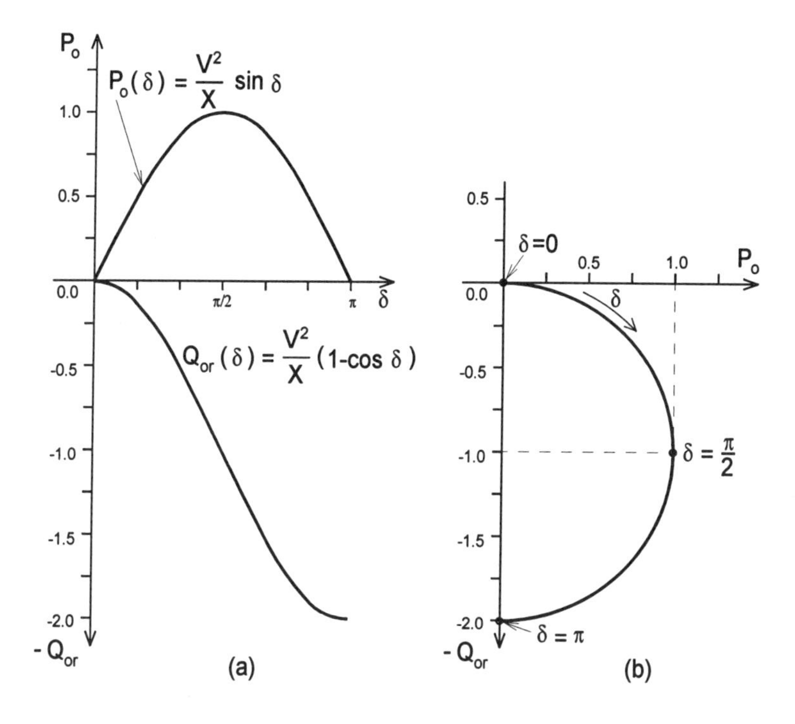

Figure 7.5 (a) Transmittable real power P_o and receiving-end reactive power demand Q_{or}, vs. transmission angle δ of a two machine system (b) and the corresponding Q_{or}, vs. P_o loci

7.2.2 Independent real and reactive power flow control

In order to investigate the capability of the UPFC to control real and reactive power flow in the transmission line, consider again Figure 7.1. Let us first assume that the injected compensating voltage, V_{pq} is zero. Then we obtain an elementary two machine (or two bus ac intertie) system with sending-end voltage V_s, receiving-end voltage V_r, transmission angle δ, and line (or tie) impedance X (assumed, for simplicity, inductive). With these, the normalized transmitted power, $P_o(\delta)=\{V^2/X\}\ \sin\delta = \sin\delta$, and the normalized reactive power, $Q_o(\delta)=Q_{os}(\delta)=-Q_{or}(\delta)=\{V^2/X\}\ \{1-\cos\delta\} = 1-\cos\delta$, supplied at the ends of the line, are shown plotted against angle δ in Figure 7.5(a). The relationship between real power $P_o(\delta)$ and reactive power $Q_{or}(\delta)$ can readily be expressed with $V^2/X=1$ in the following form:

$$Q_{or}(\delta) = -1 - \sqrt{1 - \{P_o(\delta)\}^2} \tag{7.13}$$

or

$$\{Q_{or}(\delta) + 1\}^2 + \{P_o(\delta)\}^2 = 1 \tag{7.14}$$

Equation (7.14) describes a circle with a radius of 1.0 around the center defined by coordinates P=0 and Q_r=-1 in a $\{Q_r, P\}$ plane, as illustrated for positive values of P in Figure 7.5(b). Each point of this circle gives the corresponding P_o and Q_{or} values of the uncompensated system at a specific transmission angle δ. For example, at δ=0, P_o=0 and Q_{or}=0;at δ=30°, P_o=0.5 and Q_{or}=-0.134; at δ=90°, P_o=1.0 and Q_{or}= –1.0;etc.

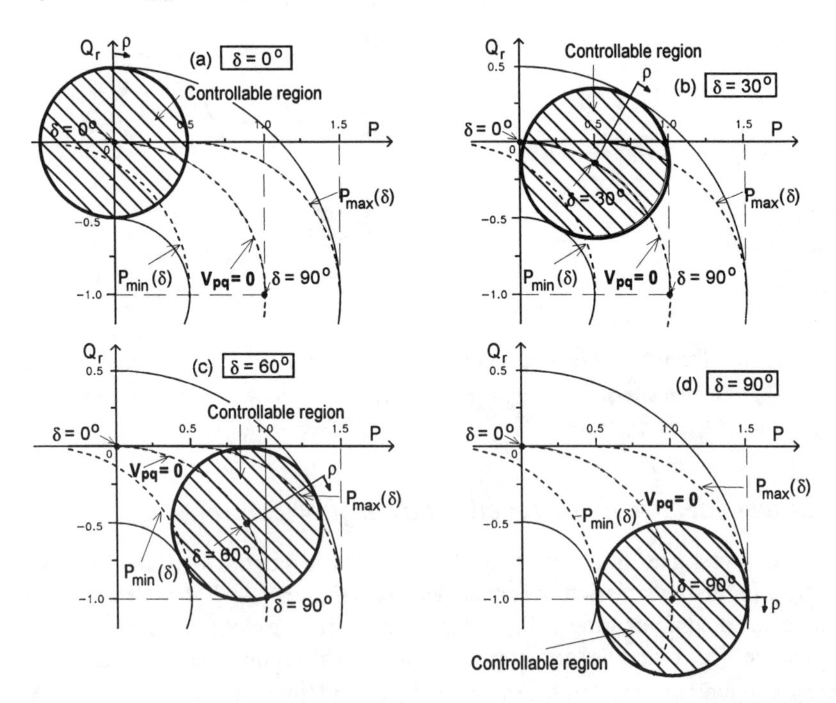

Figure 7.6 Control region of the attainable real power P and receiving-end reactive power demand Q_r with a UPFC-controlled transmission line at (a) δ=0°, (b) δ=30°, (c) δ=60°, and (d) δ=90°

Refer again to Figure 7.1 and assume now that V_{pq}≠0. It follows from equations (7.3), or (7.7) and (7.8), that the active and reactive power change from their uncompensated values, $P_o(\delta)$ and $Q_{or}(\delta)$, as functions of the magnitude V_{pq}

and angle ρ of the injected voltage phasor V_{pq}. Since angle ρ is an unrestricted variable ($0 \leq \rho \leq 2\pi$), the boundary of the attainable control region for $P(\delta,\rho)$ and $Q_r(\delta,\rho)$ is obtained from a complete rotation of phasor V_{pq} with its maximum magnitude V_{pqmax}. It follows from the above equations that this control region is a circle with a center defined by coordinates $P_o(\delta)$ and $Q_{or}(\delta)$ and a radius of V_rV_{pq}/X. With $V_s=V_r=V$), the boundary circle can be described by the following equation:

$$\{P(\delta,\rho)-P_o(\delta)\}^2+\{Q_r(\delta,\rho)-Q_{or}(\delta)\}^2=\left\{\frac{VV_{pqmax}}{X}\right\}^2 \tag{7.15}$$

The circular control regions defined by equation (7.15) are shown in Figures 7.6(a)–(d) for V=1.0, V_{pqmax}=0.5, and X=1.0 (per unit or p.u. values) with their centers on the circular arc characterizing the uncompensated system (equation 7.14) at transmission angles δ=0, 30°, 60°, and 90°. In other words, the centers of the control regions are defined by the corresponding $P_o(\delta),Q_{or}(\delta)$ coordinates at angles δ=0, 30°, 60°, and 90° in the $\{Q_r,P\}$ plane.

Consider first Figure 7.6(a), which illustrates the case when the transmission angle is zero (δ=0). With V_{pq}=0, P, Q_r, (and Q_s) are all zero, i.e., the system is at standstill at the origin of the Q_r, P coordinates. The circle around the origin of the $\{Q_r,P\}$ plane is the loci of the corresponding Q_r and P values, obtained as the voltage phasor V_{pq} is rotated a full revolution ($0\leq\rho\leq360°$) with its maximum magnitude V_{pqmax}. The area within this circle defines all P and Q_r values obtainable by controlling the magnitude V_{pq} and angle ρ of phasor V_{pq}. In other words, the circle in the $\{Q_r,P\}$ plane defines all P and Q_r values attainable with the UPFC of a given rating. It can be observed, for example, that the UPFC with the stipulated voltage rating of 0.5 p.u. is able to establish 0.5 p.u. power flow, in either direction, without imposing *any* reactive power demand on either the sending-end or the receiving-end generator. (This statement tacitly assumes that the sending-end and receiving-end voltages are provided by independent power systems which are able to supply and absorb real power without internal angular change.) Of course, the UPFC, as illustrated, can force the system at one end to supply reactive power for, or absorb that from, the system at the other end. Similar control characteristic for real power P and the reactive power Q_r can be observed at angles δ=30°, 60°, and 90° in Figures 7.6(b), (c), and (d).

In general, at any given transmission angle δ, the transmitted real power P, as well as the reactive power demand at the receiving-end Q_r, can be controlled freely by the UPFC within the boundary circle obtained in the $\{Q_r,P\}$ plane by rotating the injected voltage phasor V_{pq} with its maximum magnitude a full revolution. Furthermore, it should be noted that, although the above presentation focuses on the receiving-end reactive power, Q_r, the reactive component of the line current, and the corresponding reactive power, can actually be controlled with respect to the voltage selected at any point of the line.

Figures 7.6(a)–(d) clearly demonstrate that the UPFC, with its unique capability to control independently the real and reactive power flow at any transmission angle, provides a powerful, hitherto unattainable, new tool for transmission system control.

In order to put the capabilities of the UPFC into proper perspective in relation to other, related power flow controllers, such as the Thyristor-Switched and Thyristor-Controlled Series Capacitor (TSSC and TCSC), the Static Synchronous Series Compensator (SSSC), and the Thyristor-Controlled Phase Shifter (or Phase-Angle Regulator - TCPAR), a basic comparison between the power flow control characteristics of these and the UPFC is presented in the next section. The basis chosen for this comparison is the capability of each type of power flow controller to vary the transmitted real power and the reactive power demand at the *receiving-end.* The receiving-end var demand is usually an important factor because it significantly influences the variation of the line voltage with load demand, the overvoltage at load rejection, and the steady-state system losses. Similar comparison, of course, could easily be made at the sending-end, or at other points of the transmission line, but the results would be quite similar for all practical transmission angles.

7.2.3 Comparison of the UPFC to the controlled series compensators and phase shifters

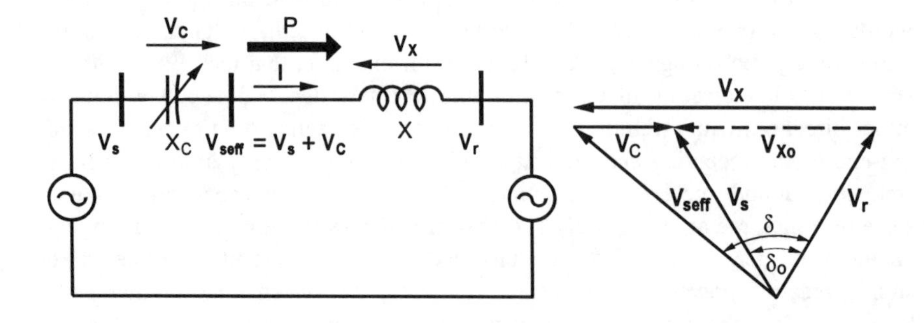

Figure 7.7 Two machine system with controlled series capacitive compensation

Consider again the simple two machine system (two bus intertie) shown in Figure 7.1. This model can be used to establish the basic transmission characteristic of the controlled series compensators (TSSC, TCSC, and SSSC) and the Thyristor-Controlled Phase Angle Regulator by using them in place of the UPFC, as symbolically illustrated in Figure 7.7 for controlled capacitive compensation, to establish their effect on the Q versus P characteristic of the system at different transmission angles.

7.2.3.1 Comparison of the UPFC to controlled series compensators

Thyristor Switched Series Capacitor (TSSC) schemes employ a number of capacitor banks in series with the line, each with a thyristor by-pass switch. This arrangement in effect is equivalent to a series capacitor whose capacitance is adjustable in a step-like manner. Thyristor-Controlled Series Capacitor (TCSC) schemes typically use a thyristor-controlled reactor in parallel with a capacitor to vary the effective series compensating capacitance and thereby the compensating voltage. In practice, several capacitor banks, each with its own thyristor-controlled reactor, may be used to meet specific application requirements. For the purpose of the present investigation, the TCSC, regardless of its practical implementation, can be considered simply as a continuously variable capacitor whose impedance is controllable in the range of $0 \leq X_C \leq X_{Cmax}$. The controllable series capacitive impedance, provided by the TSSC and TCSC, cancels part of the reactive line impedance resulting in a reduced overall transmission impedance (i.e., in an electrically shorter line) and correspondingly increased transmittable power.

The Static Synchronous Series Compensator (SSSC) injects a continuously variable series compensating voltage in quadrature with the line current. In contrast to the series capacitor schemes, the compensating voltage of the SSSC can be controlled independent of the line current (i.e., independent of the transmission angle δ).

Controlled series compensators provide a series compensating voltage that, by definition, is in quadrature with the line current. Consequently, they can affect *only* the magnitude of the current flowing through the transmission line. At any given setting of the capacitive impedance of the TSSC or TCSC, or of the compensating voltage of the SSSC, a particular effective overall line impedance is defined at which the transmitted power is strictly determined by the transmission angle (assuming a constant amplitude for the end voltages). Therefore, the reactive power demand at the end-points of the line is determined by the transmitted real power in the same way as if the line was uncompensated but had a lower line impedance. Consequently, the relationship between the transmitted power P and the reactive-power demand at the receiving-end Q_r can be represented by a single Q_r–P circular locus, similar to that shown for the uncompensated system in Figure 7.5(b), at a given output (compensating impedance or voltage) setting of the series compensator. This means that for a continuously controllable compensator an infinite number of Q_r–P circular loci can be established by using the basic transmission relationships, i.e., $P=\{V^2/(X-X_q)\}\sin\delta$ and $Q_r=\{V^2/(X-X_q)\}\{1-\cos\delta\}$, with X_q varying between 0 and X_{qmax} or 0 and $\pm X_{qmax}$ (where $X_q=X_C$ for the TSSC and TCSC and $X_q=|V_q/I|=V_q/I$ for the SSSC). Evidently, a given transmission angle defines a single point on each Q_r-P locus obtained with a specific value of X_q. Thus, the progressive increase of X_q from zero to X_{qmax} could be viewed as if the point defining the corresponding P and Q_r values at the given transmission angle on the first Q_r–P locus (uncompensated line) moves through an infinite number of Q_r–P loci representing progressively increasing series compensation, until it finally reaches the last Q_r–P locus that represents the system with maximum series compensation. The first Q_r–P locus, representing the uncompensated power transmission, is the *lower* boundary curve for the TSSC and TCSC and identified by $(Q_r–P)_{X_q=0}$. The last Q_r–P locus, representing the power transmission with maximum capacitive impedance compensation, is the *upper* boundary curve for the TSSC and TCSC and identified by $(Q_r–P)_{X_{cmax}}$. The lower and upper boundary curves for the SSSC are different from that of TSSC and TCSC and therefore they are identified by $(Q_r–P)_{+V_{qmax}}$ and $(Q_r–P)_{-V_{qmax}}$. The difference is partially due to the SSSC's capability to inject the compensating voltage both with 90° lagging (capacitive) or 90° leading (inductive) relationship with respect to the line current. Thus, it can both increase and decrease the transmitted power. Also, the SSSC can maintain maximum compensating voltage with decreasing (theoretically, even with zero) line current. For these reasons, the SSSC has a considerably wider control range at low transmission angles than both the TSSC and TCSC. Note that all Q_r–P circular curves are considered only for the normal operating range of the transmission angle ($0\le\delta\le90°$).

The plots in Figures 7.8(a)–(d) present at the four transmission angles (δ=0°, 30°, 60°, and 90°) the range of Q_r–P control for the series reactive compensators, TSSC/TCSC and SSSC, with the same 0.5 p.u. maximum voltage rating stipulated for the UPFC. In each figure the upper and lower boundary curves identified above for the TSSC/TCSC and SSSC are shown by broken- and dotted-lines for reference. The previously derived circular control region of the UPFC is also shown (by heavy broken-lines) for comparison.

Consider Figure 7.8(a) which illustrates the case when the transmission angle, δ, is zero. (For this special case, the two machine system is again assumed to represent two independent power systems with an ac line intertie.) Since both the TSSC and TCSC are an actively-controlled, but functionally *passive* impedance, the current through the line with the compensated line impedance $(X$-$X_C)$ remains invariably zero, regardless of the actual value of X_C. Thus both P and Q_r are zero, the system is at standstill, and therefore cannot be changed by reactive impedance compensation. An ideal SSSC, represented by a theoretical ac voltage source could establish power flow between the two sources. However, a practical SSSC with internal losses could only do that if an external power supply (connected to the dc terminals of the converter) replenished these losses and kept the SSSC in operation at zero (and at small) line current. By contrast, since the UPFC is a self-sufficient voltage source (whose losses are supplied by the shunt converter), it can force up to 0.5 p.u. real power flow in either direction and also control reactive power exchange between the sending-end and receiving-end buses within the circular control region shown in the figure.

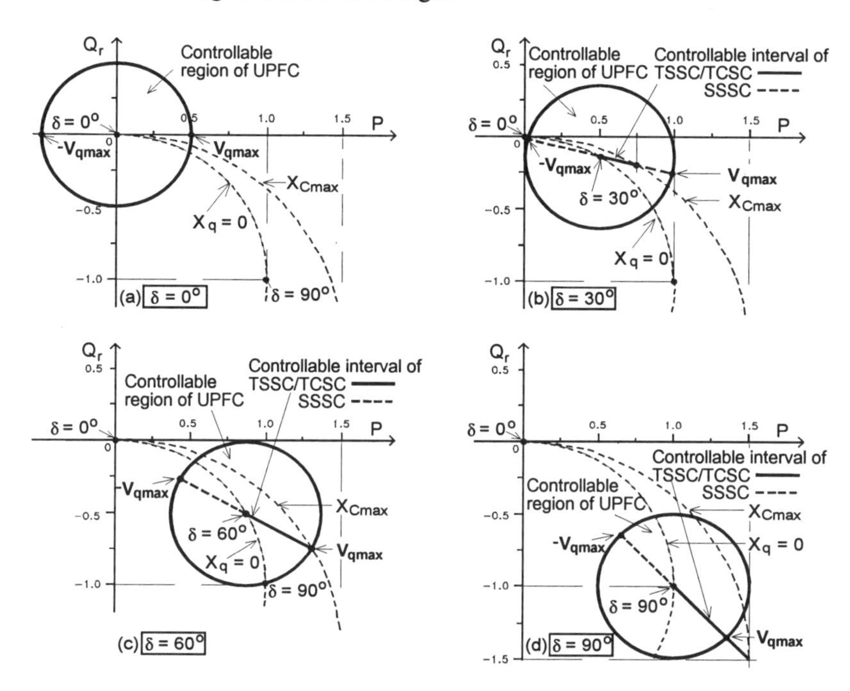

Figure 7.8 Attainable Q_r and P values with controlled capacitive compensation (points on the heavy straight line inside the circle) and those with the UPFC (any point inside the circle) at (a)δ=0°, (b)δ=30°, (c)δ=60°, and (d)δ=90°

The Q_r–P characteristic of the TSSC and TCSC at δ=30° is shown in Figure 7.8*b*. As seen, the relationship between Q_r and P, for the total range of series compensation, $0 \le \delta \le 90°$, is defined by a straight line connecting the two related points on the lower and upper boundary curves, $(Q_r–P)_{Xq=0}$ and $(Q_r–P)_{Xcmax}$, which represent the power transmission at δ=30° with zero and, respectively, maximum series compensation. It can be observed that, as expected, the TSSC and TCSC control the real power by changing the degree of series compensation (and thereby the magnitude of the line current). However, the reactive power demand of the line *cannot* be controlled independently; it remains a direct function of the transmitted real power obtained at δ=30° with varying X_C ($0 \le X_C \le X_{Cmax}$). It is also to be noted that the achievable maximum increase in transmittable power attainable with the TSSC and TCSC is a *constant percentage* (defined by the maximum degree of series compensation) of the power transmitted with the uncompensated line at the given transmission angle. In other words, the maximum increase attainable in *actual* transmitted power is much less at small transmission angles than at large ones. This is due to the fact that the TSSC and TCSC are a series impedance and thus the compensating voltage they produce is proportional to the line current, which is a function of angle δ. The SSSC, being a *reactive* voltage source (with respect to the line current) can provide maximum compensating voltage, theoretically down to zero line current. (The limit, as previously mentioned, is the ability of the line to replenish the losses of the SSSC.) It can also reverse the polarity of the compensating voltage. These characteristics mean that the SSSC can provide compensation over a wider range, between the boundaries of $(Q_r\text{–}P)_{+V_{qmax}}$ and $(Q_r\text{-}P)_{-V_{qmax}}$, than can the TSSC and TCSC. However, due to its strictly reactive compensation capability, the SSSC cannot control the reactive line power and thus the reactive power, Q_r, remains proportional to the real power P. In other words, the SSSC simply lengthens the control range of the TSSC and TCSC, without changing their Q_r versus P characteristic. By contrast, since the UPFC is a self-sufficient voltage source with the capability of exchanging *both* reactive and real power, the magnitude and angle of the compensating voltage it produces is *independent* of the line current (and of angle δ). Therefore the maximum change (increase or decrease) in transmittable power, as well as in receiving-end reactive power, achievable by the UPFC is not a function of angle δ and is determined solely by the maximum voltage the UPFC is rated to inject in series with the line (assumed 0.5 p.u. in this comparison). This characteristic can be observed in the figures where the radius of the circular loci defining the control region of the UPFC remains the same for all four transmission angles considered.

Figures 7.8(c) and (d), showing the Q_r–P characteristics of the TSSC, TCSC and the UPFC at δ=60° and δ=90°, respectively, further confirms the above observations. The range of the TSSC and TCSC for real power control remains a constant *percentage* of the power transmitted by the uncompensated line at all transmission angles ($0 \le \delta \le 90°$), but the maximum actual change in transmitted real power progressively increases with increasing δ and reaches that of the SSSC and UPFC at δ=90°. However, it should be noted that whereas the maximum transmitted power of 1.5 p.u., obtained with all the reactive compensators, TSSC, TCSC, and SSSC, at full compensation is associated with 1.5 p.u. reactive power demand at the receiving-end, the same 1.5 p.u. power transmission is achieved only with 1.0 p.u. reactive power demand when the line is compensated by a UPFC.

From Figures 7.8(a)–(d), it can be concluded that the UPFC has superior power flow control characteristics compared to the TSSC, TCSC, and SSSC: it can control independently both real and reactive power over a broad range, its control range in terms of actual real and reactive power is independent of the transmission angle, and it can control both real and reactive power flow in either direction at zero (or at a small) transmission angle.

7.2.3.2 Comparison of the UPFC to the thyristor-controlled phase shifter

Ideal phase shifters provide series voltage injection in series with the line so that the voltage applied at their input terminals appears with the same magnitude but with $\pm\sigma$ phase difference at their output terminals. Most practical phase shifters (also called Phase Angle Regulators - PARs- and "Quadrature Boosters") provide controllable *quadrature* voltage injection in series with the line. The commercially available PARs (for brevity, this term will hereon be used) employ a shunt-connected *excitation* transformer with appropriate taps on the secondary windings, a mechanical tap-changing gear, and a series *insertion* transformer. The excitation transformer has wye to delta (or delta to wye) windings and thus the phase to phase secondary voltages are *in quadrature* with respect to the corresponding primary phase to neutral voltages. These voltages, via the tap-changing gear and the series insertion transformer, are injected in series with the appropriate phases of the line. As a result, the two sets of three-phase voltages, obtained at the two ends of the series insertion transformer, are phase shifted with respect to each other by an angle σ, where $\sigma = \tan^{-1}(V_\sigma/V)$; V_σ is the magnitude of the series injected voltage and V is the magnitude of the line voltage (phase to neutral). Since the voltage injection can be of either polarity, σ may represent either phase advance or retard.

The Thyristor-Controlled Phase Angle Regulators (TCPARs) are functionally similar to the conventional PARs, except for the mechanical tap-changer which is replaced by an appropriate thyristor switch arrangement. Like their mechanically-controlled counterparts, they also provide controllable, bi-directional, quadrature voltage injection. However, their control can be continuous or step-like, but the continuous control is usually associated with some harmonic generation. For the purpose of the present investigation, the practical implementation of the TCPAR is unimportant. For simplicity, the TCPAR is considered as an *ideal* phase angle regulator, which is able to vary continuously the phase angle between the voltages at the two ends of the insertion transformer in the control range of $-\sigma_{max} \leq \sigma \leq \sigma_{max}$, without changing the magnitude of the phase shifted voltage from that of the original line voltage that would be done by quadrature voltage injection employed in most practical phase shifter implementations.

A basic, and in the present investigation important, attribute of all conventional (mechanical and thyristor-controlled) phase angle regulators, including the ideal one stipulated above, is that the *total VA* (i.e., *both* the real power *and* the vars) exchanged by the series insertion transformer appears at the primary of the excitation transformer as a load demand on the power system. Thus, both the real *and* the reactive power the phase angle regulator supplies to, or absorbs from, the line when it injects the quadrature voltage must be absorbed from it, or supplied to it by the ac system. By contrast, the UPFC *itself* generates the reactive power part of the total VA it exchanges as a result of the series voltage injection and it presents *only* the real power part to the ac system as a load demand.

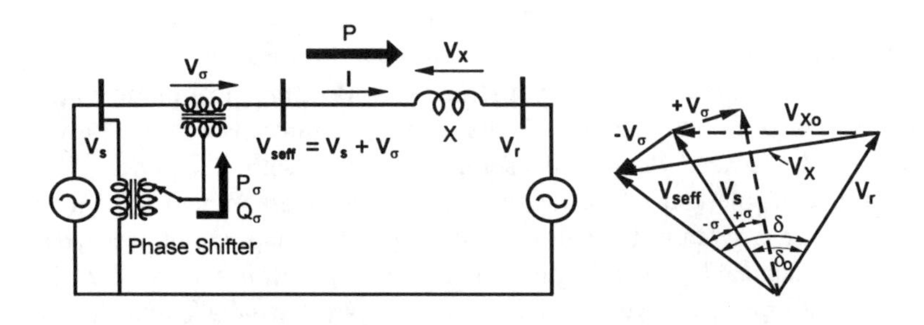

Figure 7.9 Two machine system with controlled phase shifter compensation

Consider Figure 7.9, where the previously considered two machine system is shown again with a TCPAR assumed to function as an ideal phase angle regulator. The transmitted power and the reactive power demands at the sending-end and receiving-end can be described by relationships analogous to those characterizing the uncompensated system: $P=\{V^2/X\}\sin\delta'$ and $Q_r=Q_s=\{V^2/X\}\{1-\cos\delta'\}$, where $\delta'=\delta-\sigma$. Thus, it is clear that the TCPAR cannot increase the maximum transmittable power, $P=V^2/X$, or change Q_r at a fixed P. Consequently, the Q_r–P relationship, with transmission angle δ' ($0\leq\delta'\leq 90°$) controlling the actual power transmission, is identical to that of the uncompensated system shown in Figure 7.5. The function of the TCPAR is simply to establish the *actual* transmission angle, δ', required for the transmission of the desired power P, by adjusting the phase-shift angle σ so as to satisfy the equation $\delta'=\delta-\sigma$ at a given δ, the angle existing between the sending-end and receiving-end voltages. In other words, the TCPAR can vary the transmitted power at a fixed δ, or maintain the actual transmission angle δ' constant in the face of a varying δ, but it *cannot* increase the maximum transmittable power or control the reactive power flow independently of the real power.

The plots in Figures 7.10(a)–(d) present the Q_r–P relationship for the TCPAR in comparison to that of the UPFC at δ=0°, 30°, 60°, and 90°. It can be observed in these figures that the control range centres around the point defined by the given value of angle $\delta=\delta_0$(=0°, 30°, 60°, 90°) on the Q_r–P locus characterizing the uncompensated power transmission (V_σ=0 for the TCPAR and V_{pq}=0 for the UPFC). As the phase-shift angle σ is varied in the range of $-30° \leq \sigma \leq 30°$ (which corresponds to the maximum inserted voltage of V_σ=0.5 p.u.), this point moves on the uncompensated Q_r–P locus in the same way as if the fixed angle δ_0 was varied in the range of $\delta_0-30° \leq \delta_0 \leq \delta_0+30°$. Consequently, the TCPAR, in contrast to the series reactive compensators TSSC, TCSC, and SSSC, can control effectively control the real power flow, as illustrated in Figure 7.10(a), when the angle between the sending-end and receiving-end voltages is zero (δ=0).

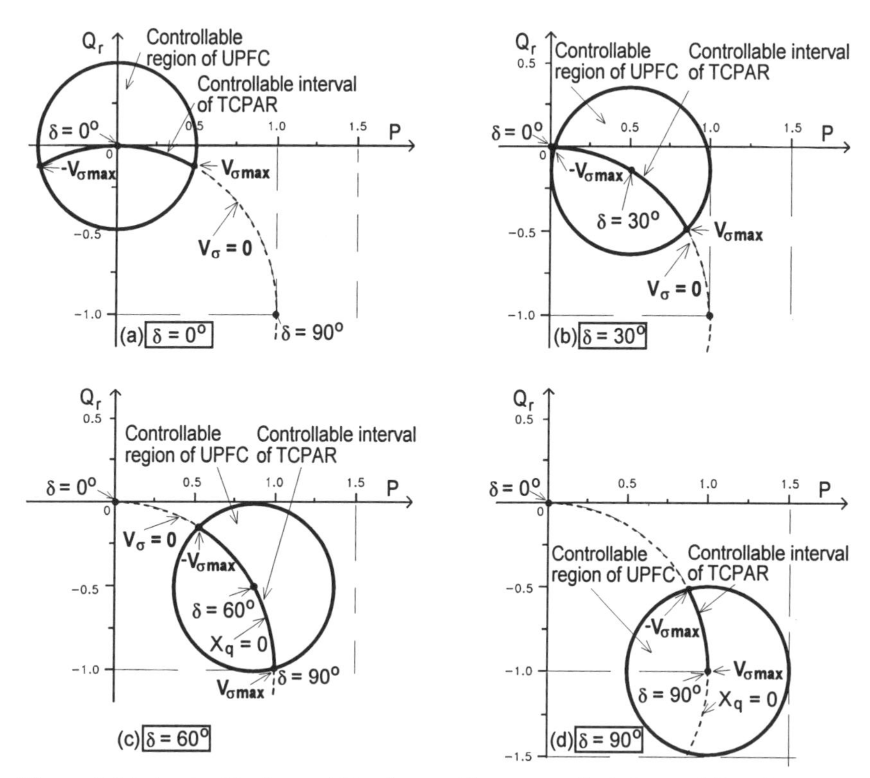

Figure 7.10 Attainable Q_r and P values with a controlled phase shifter (points on the heavy arc inside the circle) and those with the UPFC (any point inside the circle) at (a) δ=0°, (b) δ=30°, (c) δ=60°, and (d) δ=90°

The Q_r versus P plots in Figure 7.10 clearly show the superiority of the UPFC over the TCPAR for power flow control: the UPFC has a wider range for real power control *and* facilitates the independent control of the receiving-end reactive power demand over a broad range. For example, it is seen that the UPFC can facilitate up to 1.0 p.u. real power transmission with unity power factor at the receiving-end (Q_r=0), whereas the reactive power demand at the receiving-end would increase with increasing real power (Q_r reaching 1.0 p.u. at P=1.0 p.u.), in the manner of an uncompensated line, when the power flow is controlled by the TCPAR.

7.3 Control and dynamic performance

The superior operating characteristics of the UPFC are due to its unique ability to inject an ac compensating voltage *vector* with arbitrary magnitude and angle in series with the line upon command, subject only to equipment rating limits. With suitable electronic controls, the UPFC can cause the series-injected voltage vector to vary rapidly and continuously in magnitude and/or angle as desired. Thus, it is not only able to establish an operating point within a wide range of possible *P, Q* conditions on the line, but also has the inherent capability to transition rapidly from one such achievable operating point to any other.

The term *vector*, instead of phasor, is used in this section to represent a set of three *instantaneous* phase variables, voltages, or currents, that sum to zero. The symbols $\tilde{v}$ and $\tilde{\imath}$ are used for voltage and current vectors. The reader will recall that these vectors are not stationary, but move around a fixed point in the plane as the values of the phase variables change, describing various trajectories, which become circles when the phase variables represent a balanced, steady-state condition. For the purpose of power control it is useful to view these vectors in an orthogonal coordinate system with p and q axes such that the p axis is always coincident with the instantaneous voltage vector $\tilde{v}$ and the q axis is in quadrature with it. In this coordinate system the p-axis current component, i_p, accounts for the instantaneous real power and the q-axis current component, i_q, for the reactive power. Under balanced steady-state conditions, the p-axis and q-axis components of the voltage and current vector are constant quantities. This characteristic of the described vector representation makes it highly suitable for the control of the UPFC by facilitating the de-coupled control of the real and reactive current components.

The UPFC control system may be divided functionally into *internal* (or converter) control and *functional operation* control. The internal controls operate the two converters so as to produce the commanded series injected voltage and, simultaneously, draw the desired shunt reactive current. The internal controls provide gating signals to the converter valves so that the converter output voltages will properly respond to the internal reference variables, i_{pRef}, i_{qRef} and $\tilde{v}_{pqRef}$, in accordance with the basic control structure shown in Figure 7.11. As can be observed, the series converter responds directly and independently to the demand for series voltage vector injection. Changes in series voltage vector, $\tilde{v}_{pq}$, can therefore be effected virtually instantaneously. In contrast, the shunt converter operates under a closed-loop current control structure whereby the shunt real and reactive power components are independently controlled. The shunt reactive power (if this option is used, for example, for terminal voltage control) responds directly to an input demand. However, the shunt real power is dictated by another control loop that acts to maintain a preset voltage level on the dc link, thereby providing the real power supply or sink needed for the support of the series voltage injection. In other words, the control loop for the shunt real power ensures the required real power balance between the two converters. As mentioned previously, the converters do not (and could not) exchange reactive power through the link.

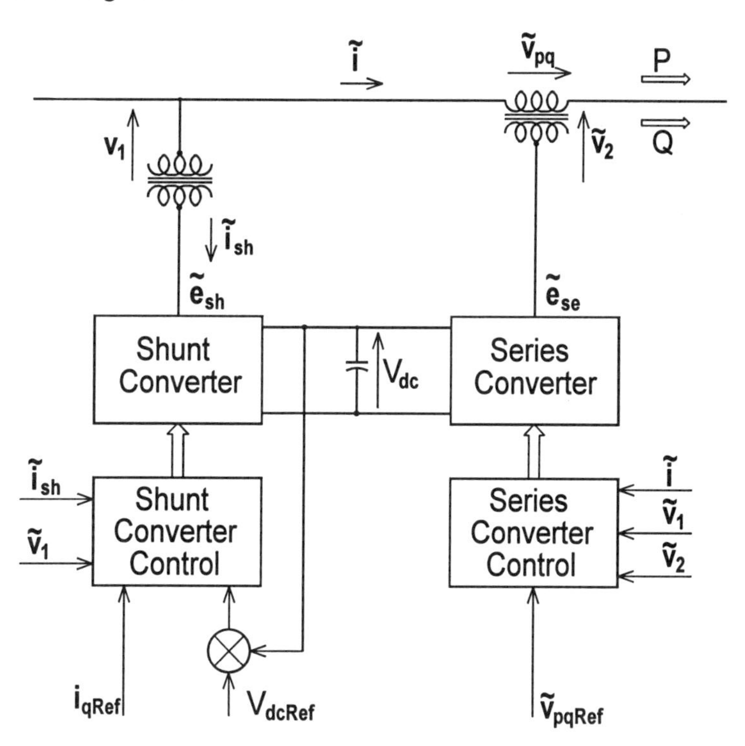

Figure 7.11 Basic UPFC control scheme

The *functional operation* control defines the functional operating mode of the UPFC and is responsible for generating the *internal* references, $\tilde{v}_{pqRef}$ and i_{qRef}, for the series and shunt compensation to meet the prevailing demands of the transmission system. The functional operating modes and compensation demands, represented by *external* (or system) reference inputs, can be set manually (via a computer keyboard) by the operator or dictated by an automatic *system optimization* control to meet specific operating and contingency requirements. An overall control structure, showing the *internal,* the *functional operation*, and *system optimization* controls with the *internal* and *external* references is presented in Figure 7.12.

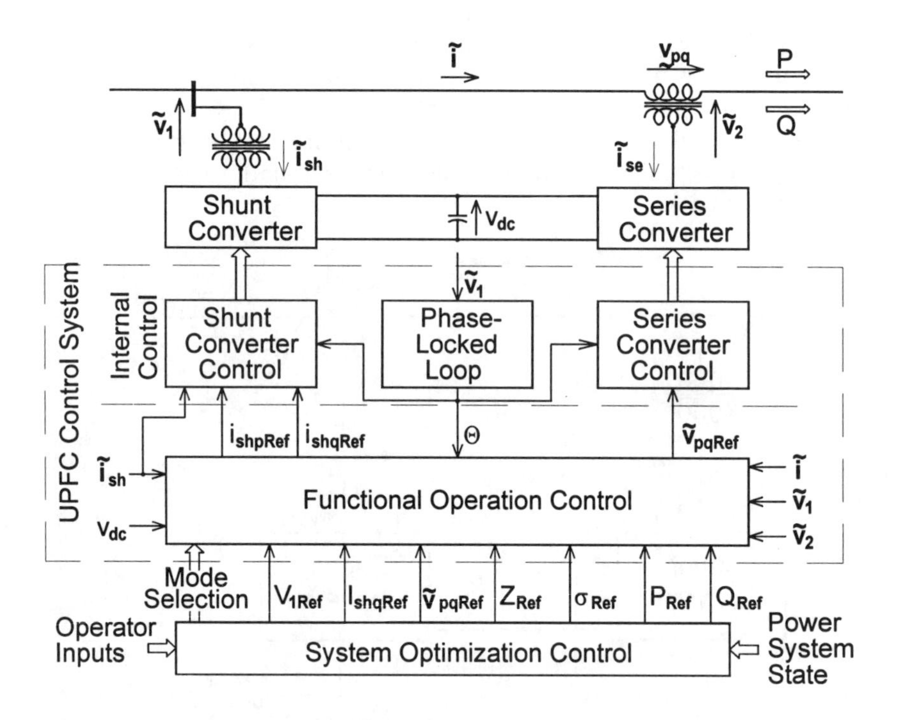

Figure 7.12 Overall UPFC control structure

7.3.1 Functional operating and control modes

The capability of unrestricted series voltage injection together with independently controllable reactive power exchange offered by the circuit structure of two dc-coupled converters, facilitate several operating and control modes for the UPFC. These include the option of reactive shunt compensation and the free control of series voltage injection according to a prescribed functional approach selected for power flow control. The UPFC circuit structure also allows the total de-coupling of the two converters (i.e., separating the dc terminals of the two converters) to provide independent reactive shunt compensation (STATCOM) and reactive series compensation (SSSC) without any real power exchange.

7.3.1.1 Functional control of the shunt converter

The shunt converter is operated so as to draw a controlled current, i_{sh}, from the line. One component of this current, i_{shp}, is automatically determined by the requirement to balance the real power of the series converter. The other current component, i_{shq}, is reactive and can be set to any desired reference level (inductive or capacitive) within the capability of the converter. The reactive compensation control modes of the shunt converter are very similar to those commonly employed on conventional static var compensators.

Reactive power (VAR) control mode. In reactive power control mode the reference input is an inductive or capacitive var request. The shunt converter control translates the var reference into a corresponding shunt current request and adjusts the gating of the converter to establish the desired current. The control in a closed-loop arrangement uses current feedback signals obtained from the output current of the shunt converter to enforce the current reference. A feedback signal representing the dc bus voltage, v_{dc}, is also used ensure the necessary dc link voltage.

Automatic voltage control mode. In voltage control mode (which is normally used in practical applications), the shunt converter reactive current is automatically regulated to maintain the transmission line voltage to a reference value at the point of connection, with a defined droop characteristic. The droop factor defines the per unit voltage error per unit of converter reactive current within the current range of the converter. The automatic voltage control uses voltage feedback signals, usually representing the magnitude of the positive sequence component of bus voltage $\tilde{\boldsymbol{v}}_1$.

7.3.1.2 Functional control of the series converter

The series converter controls the magnitude and angle of the voltage vector $\tilde{\boldsymbol{v}}_{pq}$ injected in series with the line. This voltage injection is, directly or indirectly, always intended to influence the flow of power on the line. However, $\tilde{\boldsymbol{v}}_{pq}$ is dependent on the operating mode selected for the UPFC to control power flow. The possible operating modes include:

Direct voltage injection mode. The series converter simply generates the voltage vector, $\tilde{\boldsymbol{v}}_{pq}$, with the magnitude and phase angle requested by the reference input. This operating mode may be advantageous when a separate system optimization control coordinates the operation of the UPFC and other FACTS controllers employed in the transmission system. A special case of direct voltage injection is when the injected voltage vector, $\tilde{\boldsymbol{v}}_{pq}$, is kept in quadrature with the line current vector, $\tilde{\boldsymbol{i}}$, to provide purely reactive series compensation.

Line impedance compensation mode. The magnitude of the injected voltage vector, $\tilde{\boldsymbol{v}}_{pq}$, is controlled in proportion to the magnitude of the line current, $\tilde{\boldsymbol{i}}$, so that the series insertion emulates a reactive impedance when viewed from the line.

The desired impedance is specified by reference input and in general it may be a complex impedance with resistive and reactive components of either polarity. A special case of impedance compensation is when the injected voltage is kept in quadrature with respect to the line current to emulate purely reactive (capacitive or inductive) compensation. This operating mode may be selected to match existing series line compensations in the system.

Phase angle shifter mode. The injected voltage vector $\tilde{\mathbf{v}}_{\mathbf{pq}}$ is controlled with respect to the "input" bus voltage vector $\tilde{\mathbf{v}}_1$ so that the "output" bus voltage vector $\tilde{\mathbf{v}}_2$ is phase shifted relative to $\tilde{\mathbf{v}}_1$ by an angle specified by the reference input. A special case of phase shifting is when $\tilde{\mathbf{v}}_{\mathbf{pq}}$ is kept in quadrature with $\tilde{\mathbf{v}}_1$ to emulate the "quadrature booster".

Automatic power flow control mode. The magnitude and angle of the injected voltage vector, $\tilde{\mathbf{v}}_{\mathbf{pq}}$, is controlled so as to force such a line current vector, $\tilde{\mathbf{i}}$, that results in the desired real and reactive power flow in the line. In automatic power flow control mode, the series injected voltage is determined automatically and continuously by a closed-loop control system to ensure that the desired P and Q are maintained despite system changes. The transmission line containing the UPFC thus appears to the rest of the power system as a high impedance power source or sink. This operating mode, which is not achievable with conventional line compensating equipment, has far reaching possibilities for power flow scheduling and management. It can also be applied effectively to handle dynamic system disturbances (e.g., to damp power oscillations).

7.3.1.3 Stand alone shunt and series compensation

The UPFC circuit structure offers the possibility of operating the shunt and series converters independently of each other by disconnecting their common dc terminals and splitting the capacitor bank. In this case, the shunt converter operates as a stand-alone STATCOM, and the series converter as a stand-alone SSSC. This feature may be included in the UPFC structure in order to handle contingencies (e.g., one converter failure) and be more adaptable to future system changes (e.g., the use of both converters for shunt only or series only compensation). In the stand-alone mode, of course, neither converter is capable of absorbing or generating real power so that operation only in the reactive power domain is possible. In the case of the series converter this means severe limitations in the available control modes. Since the injected voltage must be in quadrature with the line current only controlled reactive voltage compensation or reactive impedance emulation are possible for power flow control.

7.3.2 Basic control system for P and Q control

As illustrated in Figure 7.12, the UPFC has many possible operating modes. However, in order to keep focused on the subject of this chapter, only the automatic power flow control mode, providing independent control for real and reactive power flow in the line, will be considered further. This control mode utilizes most of the unique capabilities of the UPFC and it is expected to be used in the majority of practical applications, just as the shunt compensation is used normally for automatic voltage control. Accordingly, block diagrams giving greater details of the control schemes are shown for the series converter in Figure 7.13(a) and for the shunt converter in Figures 7.13(b) and 7.13(c) for operating in these modes. For clarity, only the most significant features are shown in these figures while less important signal processing and limiting functions have been omitted.

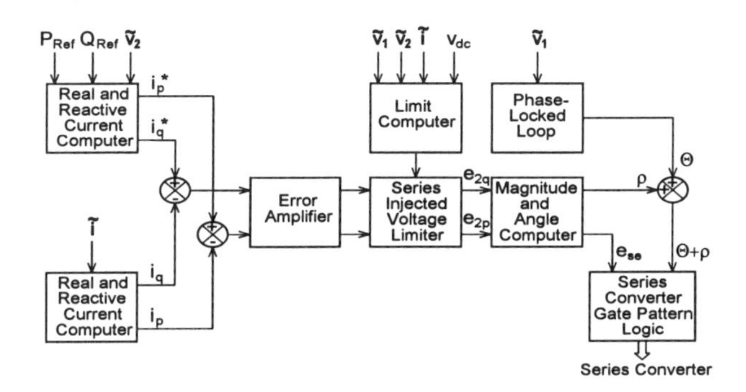

Figure 7.13(a) Functional block diagram of the series converter control

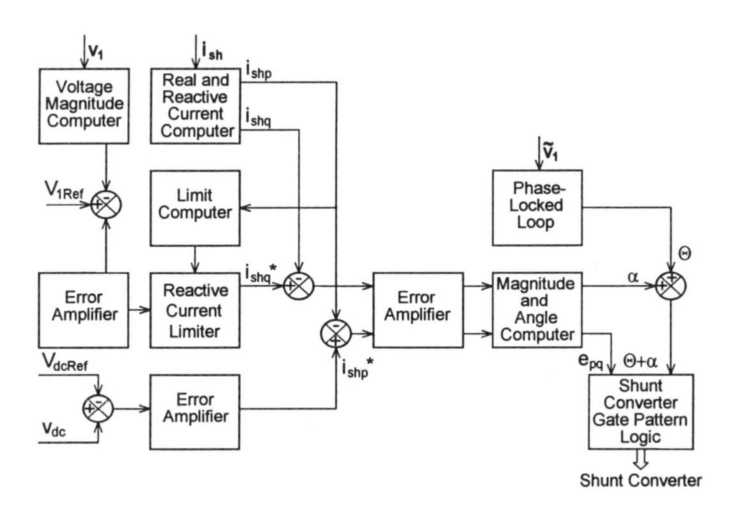

Figure 7.13(b) Functional block diagram of the shunt converter control for operation with constant dc link voltage

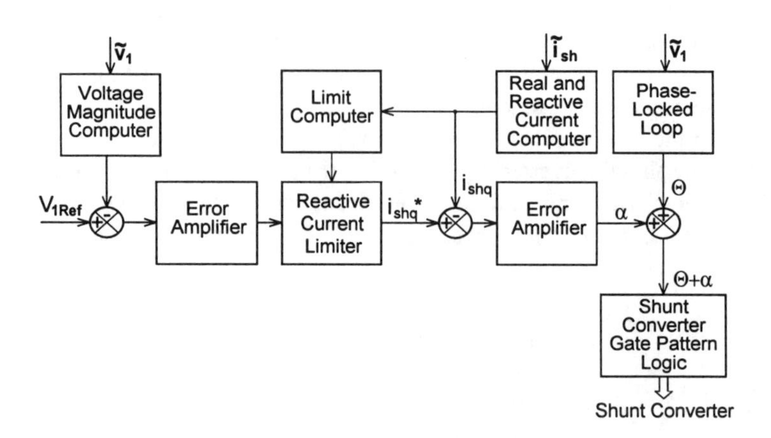

Figure 7.13(c) Functional block diagram of the shunt converter control for operation with varying dc link voltage

The control scheme shown in Figure 7.13(a) assume that the series converter can generate output voltage with controllable magnitude and angle at a given dc bus voltage. The control scheme for the shunt converter shown in Figure 7.13(b) also assumes that the converter can generate output voltage with controllable magnitude and angle. However, this may not always be the case, since the converter losses and harmonics can be reduced by allowing the dc voltage to vary according to the prevailing shunt compensation demand. Although the variation of the dc voltage inevitably reduces the attainable magnitude of the injected series voltage when the shunt converter is operated with high reactive power absorption, in many applications this may be an acceptable trade-off. In the control scheme for the shunt converter shown in Figure 7.13(c) the magnitude of the output voltage is directly proportional to the dc voltage and only its angle is controllable. With this control scheme the dc capacitor voltage is changed (typically in the ±12% range) by momentary angle adjustment that forces the converter to exchange real power with the ac system to meet the shunt reactive compensation requirements.

As shown in Figure 7.13(a) the automatic power flow control for the series converter is achieved by means of a vector control scheme that regulates the transmission line current, using a synchronous reference frame (established with and appropriate phase-locked loop producing reference angle θ) in which the control quantities appear as dc signals in the steady state. The appropriate reactive and real current components, i_q^* and i_p^*, are determined for a desired P_{Ref} and Q_{Ref}. These are compared with the measured line currents, i_q and i_p, and used to drive the magnitude and angle of the series converter voltage, $e_{se,pq}$ and ρ, respectively.

Note that a voltage limiter in the forward path is employed to enforce practical limits, resulting from system restrictions (e.g., voltage and current limits) or equipment and component ratings, on the series voltage injected.

A vector control scheme is also used for the shunt converter, as illustrated by the block diagrams of Figures 7.13(b) and (c). In this case the controlled quantity is the current $\tilde{i}_{sh}$ drawn from the line by the shunt converter. The real and reactive components of this current, however, have a different significance. For the scheme of Figures 7.13(b), the reference for the reactive current, $i_{shq}{}^{*}$, is generated by an outer voltage control loop, responsible for regulating the ac bus voltage, and the reference for the real-power bearing current, $i_{shp}{}^{*}$, is generated by a second voltage control loop that regulates the dc bus voltage. In particular, the real power negotiated by the shunt converter is regulated to balance the dc power from the series converter and maintain a desired bus voltage. The dc voltage reference, V_{dcRef}, may be kept substantially constant. In the scheme shown in Figure 7.13(c), the outer voltage loop regulates the ac bus voltage *and* also controls the dc capacitor voltage. This outer loop changes the angle α of the converter voltage with respect to the ac bus voltage until the dc capacitor voltage reaches the value necessary for the reactive compensation demanded. The closed-loop controlling the output of the series converter is responsible for maintaining the magnitude of the injected voltage, v_{pq}, in spite of the variable dc voltage.

The most important limit for the shunt converter is imposed on the shunt reactive current which is a function of the real power being passed through the dc bus to support the real power demand of the series converter. This prevents the shunt converter current reference from exceeding its maximum rated value.

The control block diagrams shown in Figure 7.13 represent only a selected part of the numerous control algorithms needed if additional operating modes of the UPFC are also to be implemented. The block diagrams also omit control functions related to converter protection, as well as sequencing routines during operating mode changes and start-up and shut-down procedures. The overall control system typically incorporates several sophisticated computers and the extensive use of electronics.

7.3.3 Dynamic performance

The dynamic performance of the UPFC is illustrated by real-time voltage and current waveforms obtained in a representative TNA (Transient Network Analyzer) hardware model shown schematically by a simplified single-line diagram in Figure 7.14. The simple, two-bus power system modeled includes the sending-end and receiving-end generators with two parallel transmission lines which are represented by lumped reactive impedances as shown in the figure. One of the lines is controlled by a model UPFC. The converters and the magnetic structure of the UPFC model accurately represent a 48-pulse structure used in an actual transmission application (see Section 7.4). The UPFC power circuit model is operated by the actual control used in the full scale system. The performance of the UPFC is demonstrated for power flow control, for operation under power system oscillation and transmission line faults.

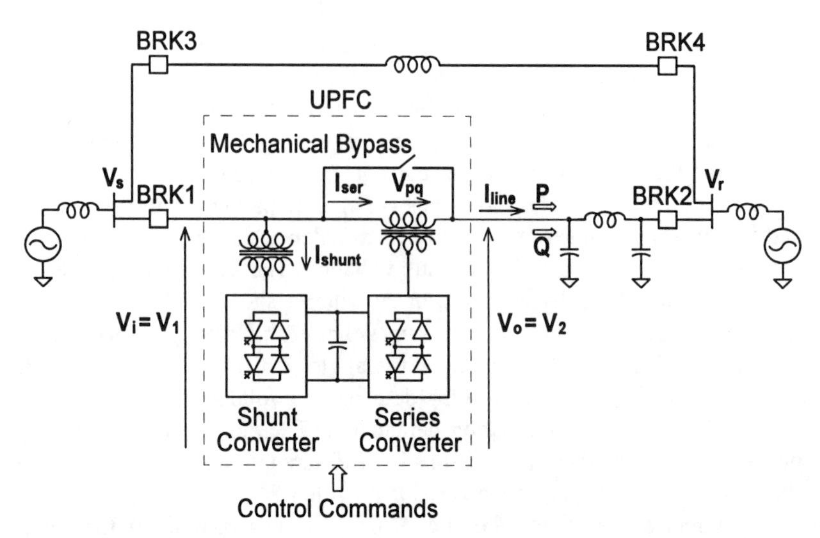

Figure 7.14 Simplified schematic of the power system and UPFC model used for TNA evaluation

7.3.3.1 Power flow control

The performance of the UPFC for real and reactive power flow control is demonstrated under the conditions of keeping the sending- and receiving-end bus voltages constant (the same 1.0 p.u. magnitude, and a fixed transmission angle) and operating the UPFC in the automatic power flow control mode. As established previously, in this operating mode the UPFC regulates the real and reactive line power to given reference values. As illustrated in Figure 7.15, the UPFC, via appropriate P_{Ref} and Q_{Ref} inputs to its control, is instructed to perform a series of step changes in rapid succession. Firstly P is increased, then Q, followed by a series of decreases, ending with a negative value of Q. The waveforms, showing the injected voltage $v_{pq}(t)$, the system voltage at the two ends of the insertion transformer, $v_i(t)=v_1(t)$ and $v_o(t)=v_2(t)=v_1(t)+v_{pq}(t)$, and the line current, illustrate clearly the operation of the UPFC. It can be observed that the closed-loop controlled UPFC easily follows the P_{Ref} and Q_{Ref} references, changing the power flow in approximately a quarter cycle. (Note that in a real system the power flow would normally be changed much more gradually in order to avoid possible dynamic disturbances.)

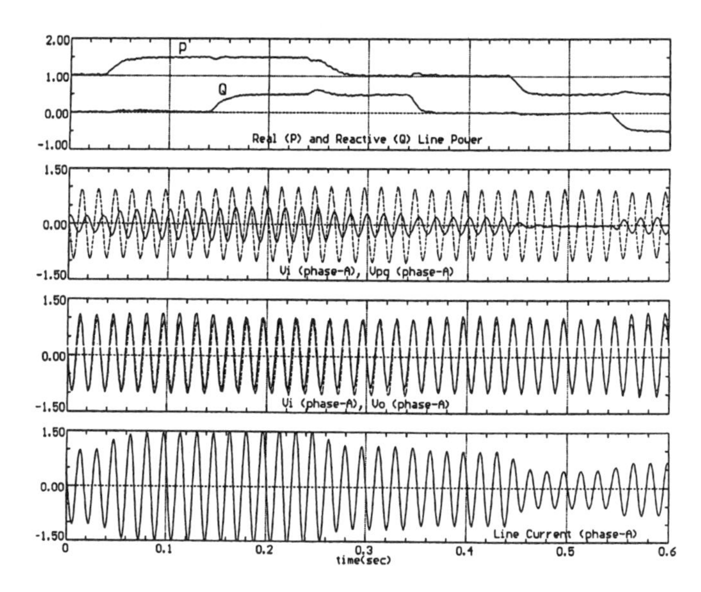

Figure 7.15 TNA simulation results for step-like changes in P and Q demands

7.3.3.2 Operation under power system oscillation

The unique versatility of the UPFC for power flow control can well be demonstrated when the power system is subjected to dynamic disturbances resulting in power oscillations. Since the UPFC actually controls the effective sending-end voltage, it is capable of forcing a desired power flow on the transmission line under dynamic system conditions as well as in the steady-state. This capability can be used in several different ways to meet system requirements. If constant power flow is to be maintained, the UPFC will act to provide this, even if the conditions on the sending and receiving end buses are varying. In essence, the UPFC will dynamically decouple the two (sending-end and receiving-end) buses. If preferred, the UPFC can be commanded to force an appropriately varying power level on the line that will effectively damp the prevailing power oscillation.

To demonstrate the potential of the UPFC under dynamic (oscillatory) conditions, the TNA system model was provided with a receiving-end (V_r) bus programmed to have a damped second-order phase-angle response characteristic of a generator with a large inertia. The simple algorithm governing this mechanism assumes a defined source of mechanical power to the generator supporting the V_r bus and matches this against the electrical power being delivered from the bus to the two transmission lines. Excess mechanical power causes an acceleration with a resultant phase angle advance, and vice versa. In the steady state, the bus angle assumes a value in which the electrical and mechanical powers are exactly equal. The modified system model is shown in Figure 7.16.

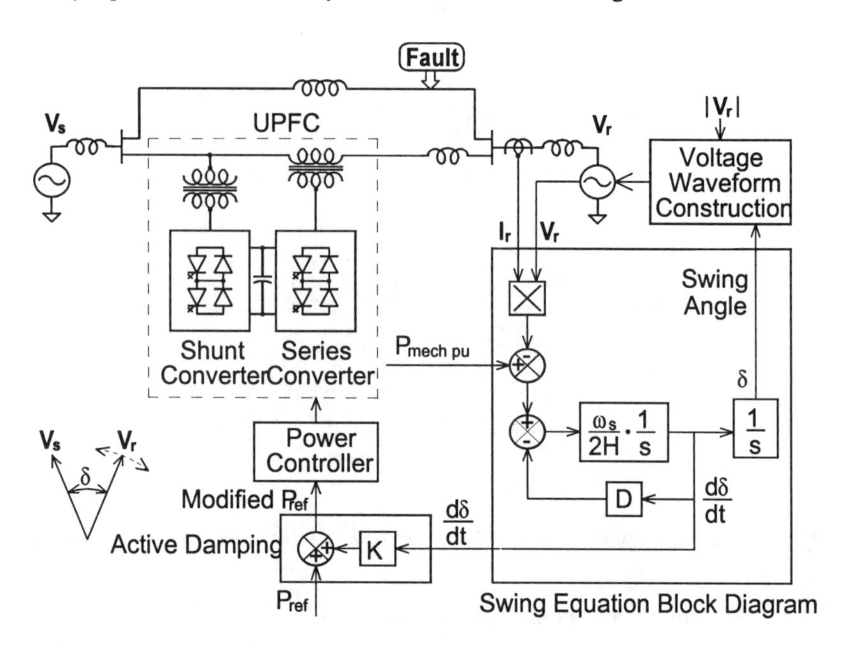

Figure 7.16 Block diagram showing the "swing bus" and control algorithms for power oscillation damping

To initiate a power oscillation in this simple system model, a fault was applied for a duration of several cycles through an impedance to ground at the V_r bus as shown in Figure 7.16, simulating a distant fault condition. Three cases are presented in Figures 7.17 through 7.19 to illustrate dynamic response of the UPFC under various operating modes. The initial conditions for all three cases are identical, with the mechanical power request for generator V_r programmed to produce a 1.0 p.u. real power flow from V_s to V_r. Line impedances are such that with zero compensation ($v_{pq}=0$), the UPFC line carries 0.75 p.u. real power while the remaining 0.25 p.u. real power flows through the parallel line. The UPFC is then operated to obtain 1.0 p.u. real power flow on its line, so that no power is transferred through the parallel line. In each of the three cases, the UPFC achieves this initial power flow using a different operating mode.

The results of Figure 7.17 show the UPFC in *direct voltage injection mode* where the magnitude and angle of the injected voltage are adjusted by the operator (or by the *system optimization control*) until the desired power flow is achieved. In this operating mode, the UPFC has no effect on the dynamic performance of the system due to the applied fault, appearing only as a voltage source of fixed magnitude and angle, added to the sending end voltage.

The results of Figure 7.18 show the UPFC in *automatic power flow control mode with a constant reference*. Here the UPFC holds the power flow on its line constant while the oscillating power required to synchronize the generators is carried entirely by the parallel line. Note the change in swing frequency since the impedance of the UPFC's line no longer dynamically couples the two generators.

In the final case, shown in Figure 7.19, the UPFC is in *automatic power flow control mode with active damping control*. The oscillation damping control algorithm is shown in Figure 7.16, where the rate of change of the differential phase angle between the sending-end and receiving-end buses ($d\delta/dt$) is sensed and fed into the real power command, P_{Ref}, for the UPFC with the correct polarity and an appropriate gain. Clearly, it may be difficult to obtain the feedback for this algorithm in a real system, but the purpose here is to show the powerful capability of the UPFC to damp system oscillations. However, it seems plausible that any of those system variables available at, or transmittable to a given location of the power system, which were found to be effective inputs to other type of power flow controllers (e.g., TCSC) for power oscillation damping, would be applicable to the UPFC as well.

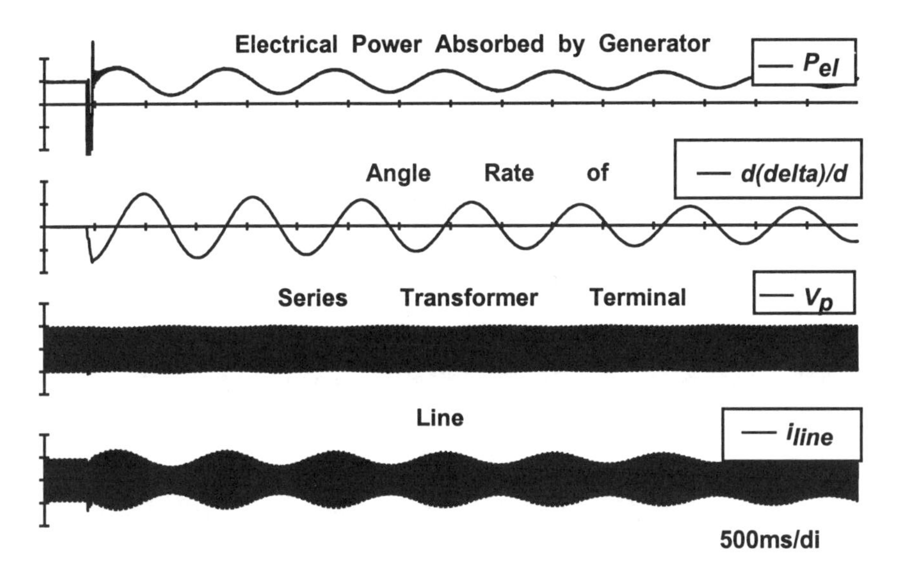

Figure 7.17 Power flow control during power oscillation with the UPFC in direct voltage injection mode (UPFC remains neutral to damping)

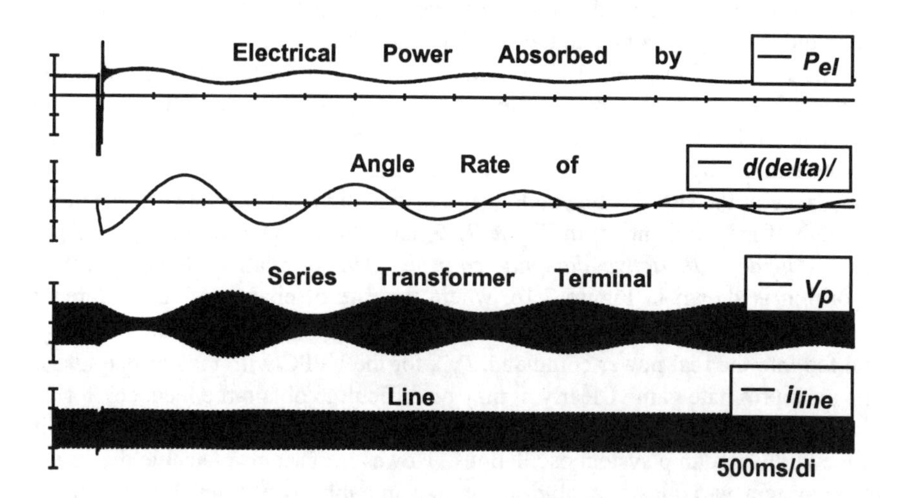

Figure 7.18 Power flow control during power oscillation with the UPFC in automatic power flow control with constant reference (UPFC maintains constant power flow in the controlled line)

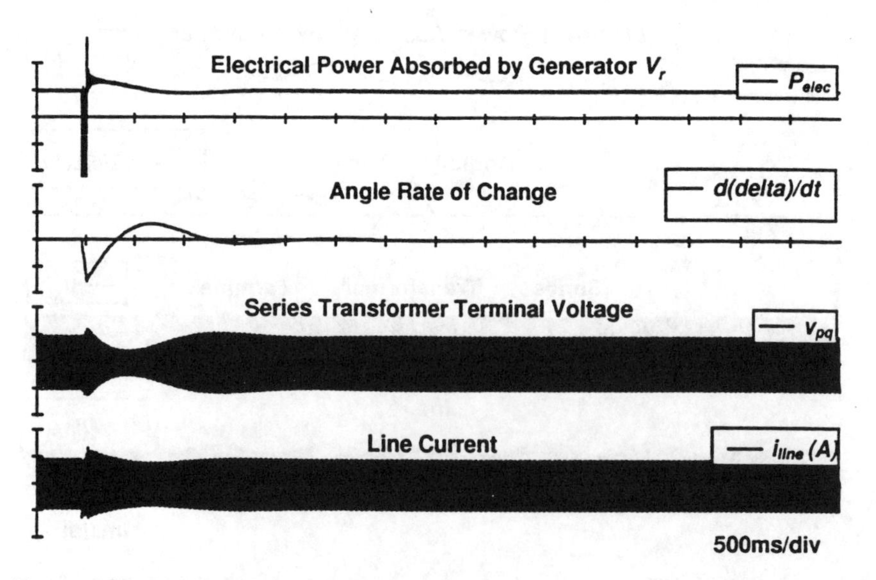

Figure 7.19 Power flow control during power oscillation with an active damping control is added to the automatic power flow control of the UPFC (UPFC acts to damp the power oscillation)

7.3.3.3 Operation under line faults

The current of the compensated line flows through the series converter of the UPFC. Depending on the impedance of the line and the location of the system fault, the line current during faults may reach a magnitude which would far exceed the converter rating. Under this condition the UPFC would typically assume a *bypass* operating mode. In this mode the injected voltage would be reduced to zero and the line current, depending on its magnitude, would be bypassed through either the converter valves, electronically reconfigured for terminal shorting, or through a separate high current thyristor valve. For the contingency of delayed fault clearing, a mechanical bypass breaker would also be normally employed.

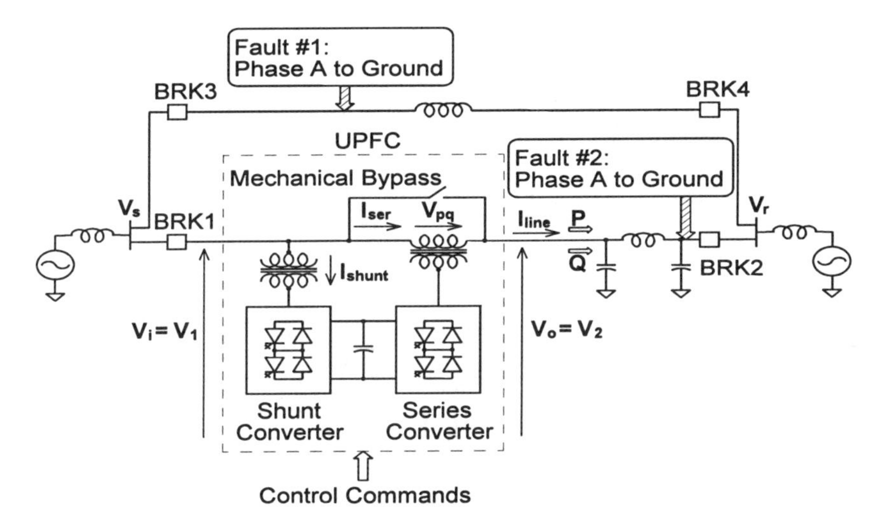

Figure 7.20 Simplified schematic of the power system and UPFC TNA model showing the locations of external and internal system faults

Two cases of line faults are considered in this section. The first is an external fault (fault is on the non-UPFC line) with normal clearing time. As illustrated in Figure 7.20, phase A of the parallel line faulted to ground at point **Fault #1**, effectively at the stiff sending end bus. The fault path has zero impedance, and prior to the onset, the UPFC is in automatic power flow control mode, controlling the power on its line at P=1.0 p.u. and Q=0.02 p.u. Six cycles after the fault, breakers BRK3 and BRK4 open, clearing the fault and restoring voltage to the UPFC. The breakers reclose nine cycles after opening.

Resulting waveforms for this case are given in Figure 7.21. When the UPFC senses the overcurrent on faulted phase A, it immediately (at point 1 in the figure) activates the electronic bypass to protect the series converter. During the fault, the shunt converter may, if desired, remain operational to supply reactive compensation. However, the gross voltage unbalance caused by the single line-to-ground fault, may cause considerable distortion on the compensating currents. These waveforms show normal fault clearing conditions, where the fault current is conducted by the electronic bypass switch. Should the fault clearing be delayed beyond the thermal capacity of the electronic switch, a mechanical bypass would be initiated. If the series transformer is mechanically bypassed, a specific reinsertion sequence for the UPFC would be required, as shown in the next fault case.

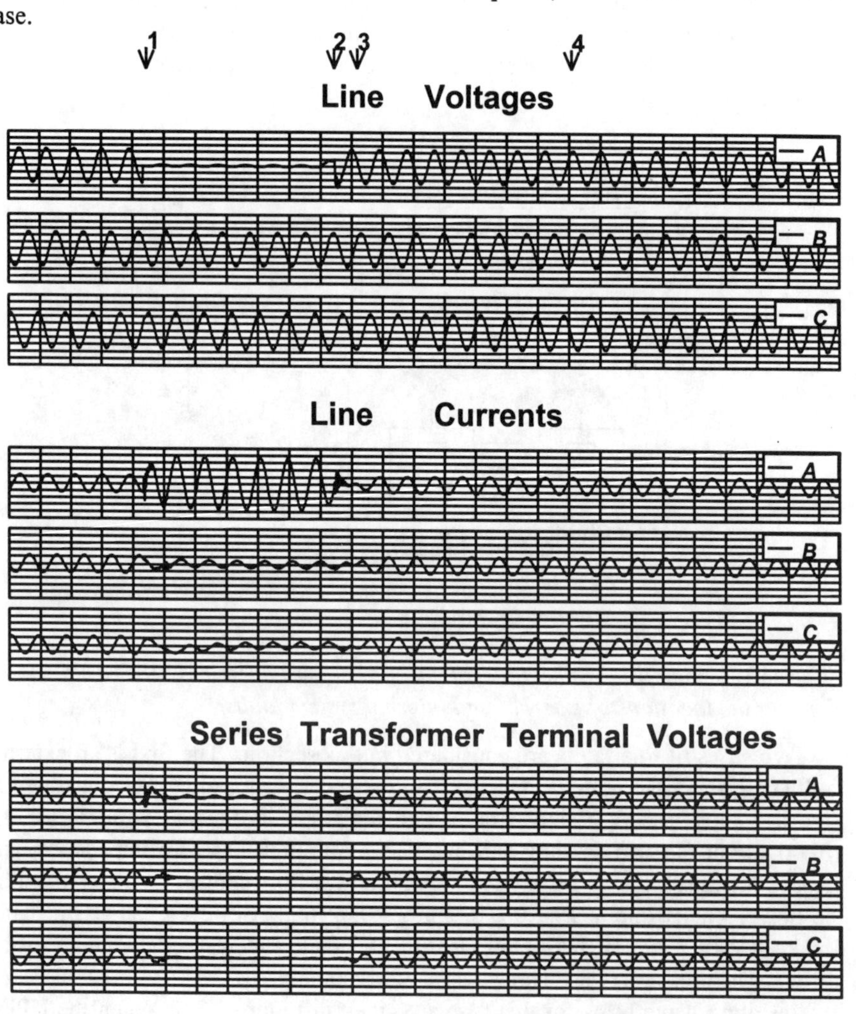

Figure 7.21 UPFC response to an external phase A to ground fault with normal clearing

Six cycles after initiation the fault is cleared (at point 2) when BRK3 and BRK4 open, restoring balanced voltage to the line with the UPFC. Line conditions quickly return to normal and the UPFC responds by removing the electronic bypass and immediately returning to the pre-fault power flow control mode (point 3). Breakers BRK3 and BRK4 reclose with no noticeable effect (point 4).

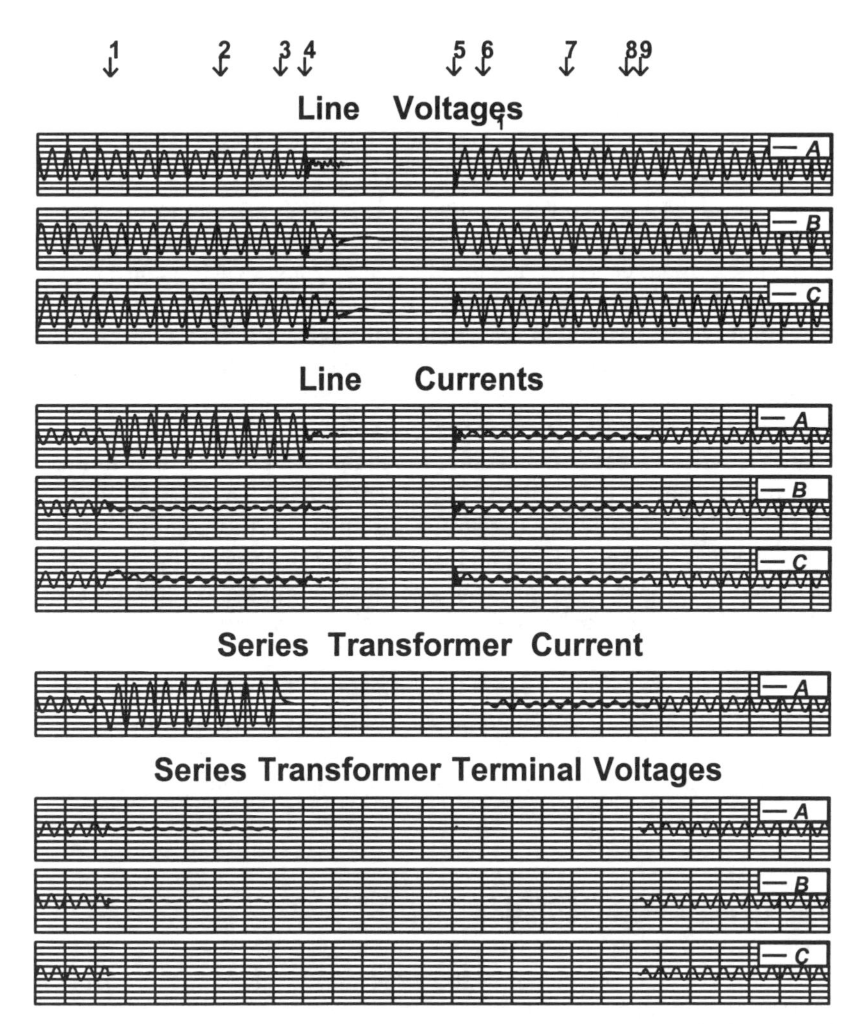

Figure 7.22 UPFC response to an internal phase A to ground fault with delayed clearing

The second fault case represents an internal fault (i.e., fault is on the UPFC line) with delayed fault clearing. This case consists of a single line to ground internal fault, with phase A faulted to ground at point **Fault #2**, the receiving-end bus in the single line diagram of Figure 7.20. The fault path has zero impedance and, as before, at the onset the UPFC is operating in automatic power flow control mode with P=1.0 p.u. and Q=0.02 p.u. Fault clearing by opening breakers BRK1 and BRK2 is delayed until 12 cycles after fault initiation. The breakers reclose nine cycles after opening.

Resulting waveforms for this case are given in Figure 7.22. As indicated previously, when the UPFC senses the overcurrent on faulted phase A, it immediately (at point 1 in the figure) activates the electronic bypass to protect the series converter. However, these waveforms illustrate the case where the fault clearing is assumed delayed beyond the thermal capacity of the electronic switch and a mechanical bypass is initiated (point 2). Approximately four cycles later the mechanical bypass across the primary of the insertion transformer closes and carries the line current, thus relieving the electronic bypass switch (point 3).

A short time later (point 4), breakers BRK1 and BRK2 open to clear the fault. This action leaves the UPFC in its mechanically bypassed state on an isolated line. Once the fault has been cleared, the breakers reclose, restoring voltage to the UPFC (point 5). Line conditions quickly return to normal and the UPFC is able to initiate reinsertion of the series transformer (point 6). During the reinsertion procedure the series converter forces current (i_{ser}) through the series transformer to match to the line current (i_{line}), driving the current through the mechanical bypass to zero in order to achieve a transient-free reinsertion. At this point the value of the series injected voltage is fixed and a signal is sent to open the mechanical bypass (point 7). Several cycles later the mechanical bypass, carrying zero current, opens (point 8). The "bumpless" reinsertion of the series transformer into the line is complete.

Once the series transformer is in the line again, the UPFC can remain in voltage insertion mode or transition to any other desired post-fault operating mode. To illustrate this capability, this case shows the UPFC returning to the pre-fault operating mode (automatic power flow control) and reference values (point 9).

7.4 The first UPFC installation

The world's first UPFC installation with two identical converters, each rated for ±160 MVA, was commissioned in June 1998 at the Inez substation of the American Electric Power (AEP) in Kentucky, USA. The project was a joint development effort by the Westinghouse Electric Corporation, the Electric Power Research Institute (EPRI) and AEP.

7.4.1 Application background

The functional versatility and comprehensive compensation capability of the UPFC eminently suited the reinforcement requirements of the power system in the Inez area. This load area with a power demand of approximately 2000 MW is served by long and heavily loaded 138 kV transmission lines. Even under normal system conditions, the heavy loading may result in voltage levels as low as 95 per cent, which is considered the lowest acceptable level for reliable power supply. Single contingency conditions would cause thermal overloads and severe voltage drops, and double contingencies could result in area wide blackouts.

The reinforcement of the power system in the area required both increased power transmission capability and voltage support. The planning studies indicated that the UPFC would, under normal conditions, regulate the voltage within one percent and with the addition of a 138 kV high capacity transmission line would, under contingency conditions, be able to transmit up to 950 MWs into the area.

In order to increase the system reliability and provide flexibility for future system changes, the UPFC installation was designed to allow self-sufficient operation of the shunt converter as an idependent STATCOM and the series converter as an independent Static Synchronous Series Compensator (SSSC). It is also possible to couple both converters together to provide either shunt only or series only compensation.

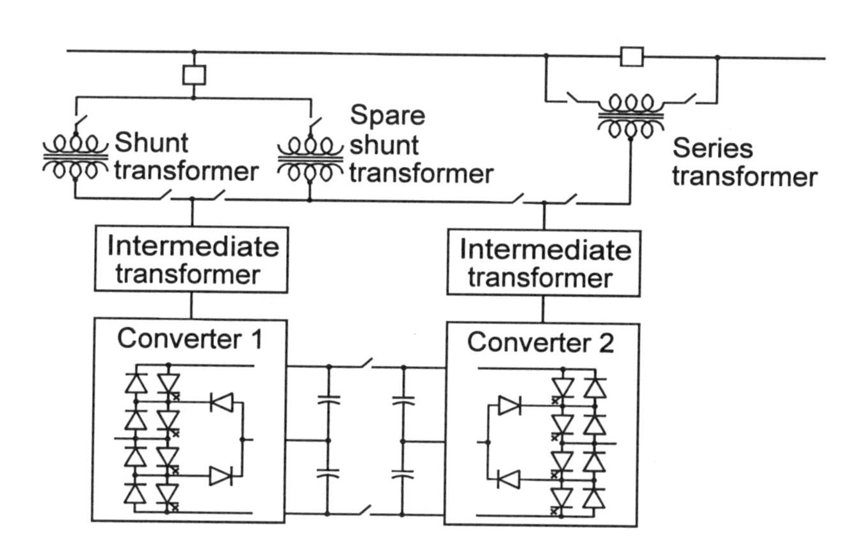

Figure 7.23 Simplified diagram of the UPFC installed at Inez

7.4.2 Power circuit structure

The UPFC equipment comprises two identical GTO-thyristor-based converters, each rated at ±160 MVA. Each converter includes multiple high-power GTO valve structures feeding an intermediate (low voltage) transformer. The converter output is a three-phase voltage set of nearly-sinusoidal (48-pulse) quality that is coupled to the transmission line by a conventional (3-winding to 3-winding) main coupling transformer. The converter-side voltage of the main transformer is 37 kV line-line (for both shunt and series transformers.) The shunt-connected transformer has a 138 kV delta-connected primary, and the series transformer has three separate primary windings each rated at 16 percent of the phase voltage.

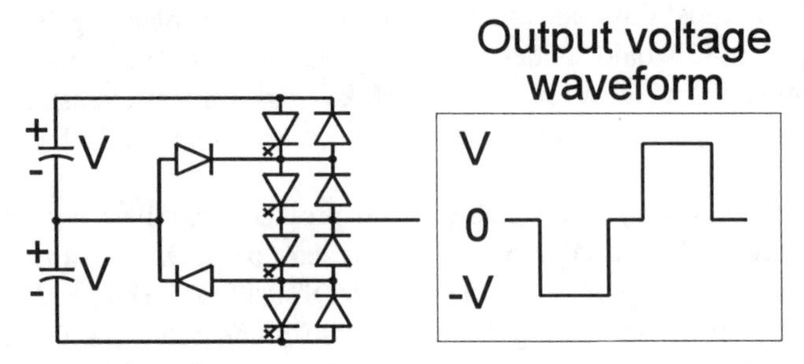

Figure 7.24 Basic three-level converter pole employed in the Inez UPFC and associated output voltage waveform

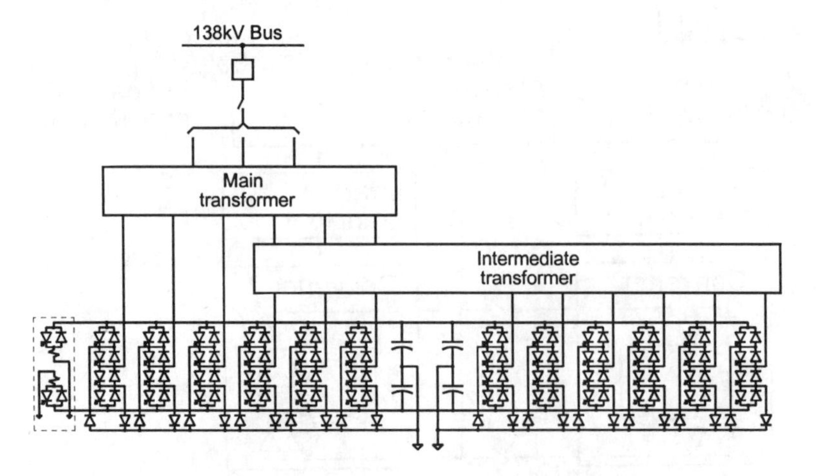

Figure 7.25 Simplified power circuit schematic of the 48-pulse converter used in the Inez (a) UPFC and (b)corresponding output voltage waveform

Figure 7.26 UPFC converter hall (Courtesy of Siemens Power T&D, Orlando)

Figure 7.27 Aerial view of the Inez Substation (Courtesy of AEP)

To maximize the versatility of the installation, two identical main shunt transformers and a single main series transformer have been provided as illustrated in Figure 7.23. With this arrangement, a number of power circuit configurations are possible. Converter 1 can operate as a STATCOM with either one of the two main shunt transformers, while Converter 2 operates as an SSSC. Alternatively, Converter 2 can be connected to the spare main shunt transformer and can operate as an additional STATCOM. With the latter configuration a shunt reactive capability of ±320 MVA becomes available. The power circuit arrangement indicates the priority of shunt compensation at this location.

The converters are constructed from three-level poles, each composed of four valves. The basic circuit for a three-level pole, together with a typical output voltage waveform, is illustrated symbolically in Figure 7.24. The three-level pole offers the additional flexibility of a step in the output voltage that can be controlled in duration, either to vary the fundamental output voltage, or to assist in waveform construction. Each converter uses 48 valves in 12 three-level poles with a nominal dc voltage of 24 kV (+12kV and -12kV with respect to the mid-point). The valves are composed of a number (eight in the outer and nine in the inner valves) of 4500V, 4000A GTOs, each with its associated anti-parallel diode, and snubber components. They are operated in each pole at 60 Hz switching rate and the phase of the switching is strategically controlled from one pole to the next to facilitate harmonic elimination. The pole ac outputs are summed via the intermediate transformer at the secondary windings of the main coupling transformer. The voltage appearing across each winding is a nearly sinusoidal 48-pulse waveform as illustrated in Figure 7.25. The total MVA rating of the intermediate transformer is approximately 50 percent of the main transformer rating. Figure 7.26 shows a view of the actual converter valve hall and Figure 7.27 shows an overall aerial view of the UPFC installation at the Inez Substation.

7.4.3 Control system

Both converters comprising the UPFC are controlled from a single central control system housed in three cabinets in the control room. Two of the cabinets house the relay interface and signal conditioning, while a single cabinet contains the control electronics. The conceptual structure of the control system is shown in Figure 7.28.

The actual control algorithms that govern the instantaneous operation of the two converters are performed in the real-time control electronics which employs multiple digital signal processors. The real-time control communicates with the pole electronics mounted on each pole via the valve interface that is linked to the poles by fiber optic cables. The status processor is connected to every part of the system, including the cooling system and all of the poles, by serial communications. During runtime it continually monitors the operation of all subsystems, collecting and analyzing status information. It is responsible for all startup and shutdown sequences and for organizing and annunciation of all alarm conditions. The status processor is serially connected to a graphical display terminal which provides the local operator interface. A hierarchical arrangement of graphical display screens gives the operator access to all system settings and parameters, and provides extensive diagnostic information right down to the individual GTO modules.

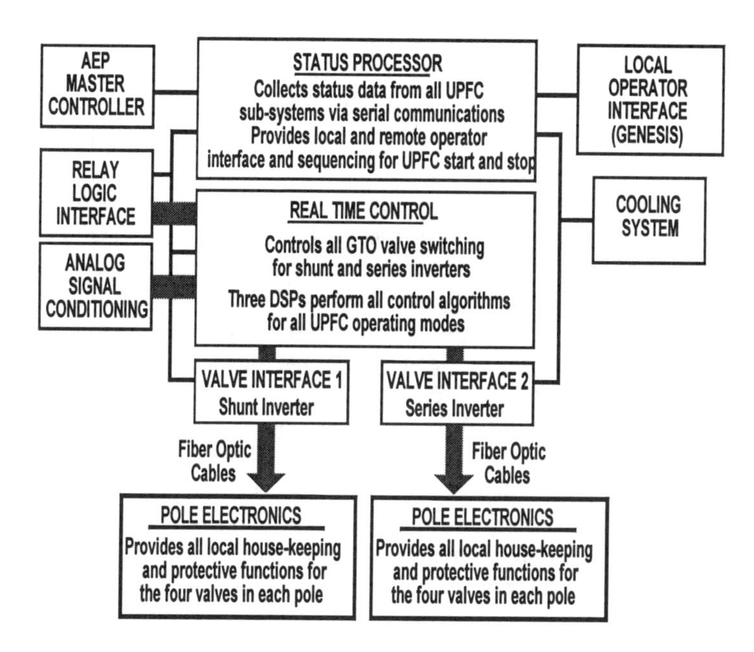

Figure 7.28 Conceptual control structure of the Inez UPFC

7.4.4 Commissioning test results

7.4.4.1 General discussion

In the course of commissioning the UPFC, tests were performed to verify its predicted capability. For this purpose, the UPFC was directed to produce large controlled swings of real and reactive power on the Big Sandy line, and sizeable swings of voltage at the Inez station, while measurements were recorded. It must be emphasized that these swings do not in any sense represent "normal" duty for the UPFC in this application. Arbitrary variations of this kind can be disruptive to the power system because they force real power flows to be redistributed in the network and affect voltage regulation at other stations. AEP system operators defined acceptable boundary limits for the tests and, in addition, slow ramping functions were applied to the control references for the UPFC automatic power flow controller. Measured data was available from the UPFC control system and was recorded at one second intervals. Only signal quantities with little alternating content were selected for recording (such as power, current, and voltage phasor magnitudes/angles, etc.).

Five representative cases have been selected for the purpose of this presentation. The first three cases show the UPFC independently controlling line P, line Q, and Inez bus voltage respectively. The fourth case is for the UPFC maintaining unity power factor on the line, and the final case is a demonstration of the series inverter operating as an SSSC. In all cases (except the SSSC), the UPFC is operating with the shunt inverter in automatic voltage control mode and the series inverter in automatic line power flow control mode. Each set of results is annotated using the sign convention as defined in Figure 7.29 Note in particular that P and Q for the line are measured at the line-side terminals of the series insertion transformer. This is the actual power at the end of the line, and is defined as positive *towards* Big Sandy. The real and reactive power for each of the two converters is also recorded. Note that the converters independently generate reactive power but that their real power is substantially equal and opposite. In each case a few stationary points (between major transitions) have been chosen and a phasor diagram drawn to represent the operating condition at that point. This is considered to be helpful, because it is difficult to interpret each situation from the time plots alone. The interpretation is made even more difficult by the fact that the power network "adjusts" following each major transition, especially with regard to the voltage phase angles at Inez and Big Sandy. The following five sections will take the form of a specific commentary for each of the cases.

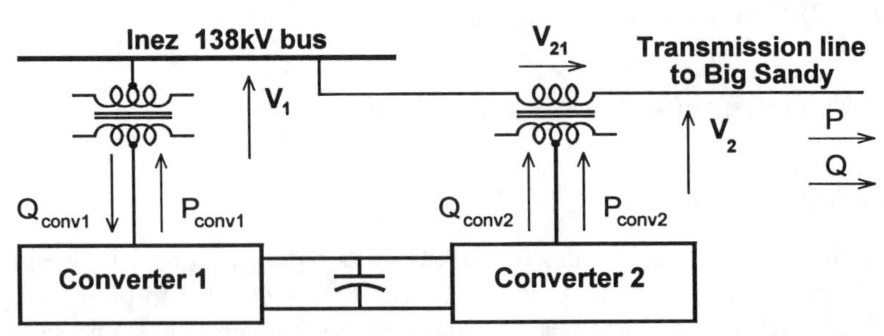

Figure 7.29 Definition of polarity conventions used in the test results for power and voltage measurements

7.4.2.2 *Case 1: UPFC changing real power (P)*

Refer to Figure 7.30 (p.312). This case starts with the UPFC idling near zero injected voltage and the real power flow on the line near the "natural" level of 150 MW *from* Big Sandy. The shunt converter is regulating the Inez bus to 1.0 p.u. by generating about 60 Mvar capacitive, and about 36 Mvar are being delivered into the line. The objective for this test is to maintain the Inez bus voltage and the line Q unaltered, while making big step changes in line P.

The UPFC is first commanded to raise the line power to 240 MW. It does this by injecting a voltage of about 0.16 p.u. roughly in quadrature (lagging) with the Inez bus voltage. To satisfy the required conditions, the shunt converter drops its capacitive output to about 20 Mvar, and the series converter delivers about 40 Mvar capacitive to the line. Real power exchange between the converters is about 8 MW.

The second transition commanded is a 170 MW drop in the line power to 70 MW. This is accomplished with little change in the magnitude of the injected voltage, but about 180 degrees phase shift, so that the injected voltage is now still roughly in quadrature with the Inez bus voltage, but *leading*. The shunt converter produces about 85 Mvars capacitive, the series converter reverses its output to 10 Mvar inductive, and about 8 MW flows between the converters through the dc bus. The final transition returns the system to the initial operating point.

Throughout this test the Inez bus voltage is tightly regulated at 1.0 p.u. and Q on the line stays constant. Note however that the voltage, V_2, applied to the transmission line, is lowered by a few percent relative to V_1 to achieve the first swing and raised by a few percent for the second. It should also be noted that the large changes in real power arriving at Inez must, of course, be balanced by an equal and opposite total change in the power on the other lines leaving the station. For this to happen a change in the phase angle of the Inez bus voltage, V_1, is unavoidable.

At the time when these tests were performed the natural power flow on the line was too high for the UPFC to demonstrate its unique ability to reverse power flow. On other occasions, however, when the line has been lightly loaded, this has been demonstrated successfully and the UPFC has driven real power back towards Big Sandy.

7.4.2.3 *Case 2: UPFC changing reactive power (Q)*

Refer to Figure 7.31 (p.313). The initial conditions for this test are similar to Case 1. The objective of the test is to regulate the Inez voltage at 1.0 p.u. and keep the line real power, P, constant while causing large steps in the line reactive power, Q.

For the first swing, the UPFC reference for Q is changed from +30 Mvar to –30 Mvar. After the change 30 Mvar is being received *from* the line, compared with the 30 Mvar delivered *to* the line initially. The UPFC forces the change by injecting about 0.05 p.u. voltage roughly in anti-phase with V_1. The line voltage, V_2, is consequently reduced in magnitude by about five percent. For the second step, the Q reference is taken to +100 Mvar (i.e. 100 Mvar to the line). This time the injected voltage is in phase with V_1 so that V_2 is increased by about five percent. The final step reduces the Q reference to zero. The line is now fed at unity power factor with V_2 reduced by about 2.5 percent relative to V_1.

It is very interesting to note that the changes in Q at the line terminals are balanced almost entirely by equal and opposite changes in the reactive output of the shunt converter, which acts to maintain the Inez voltage. This brings to light a fascinating capability of the UPFC. In essence, it can "manufacture" inductive or capacitive Mvars using the shunt converter and "export" this reactive power into a particular transmission line (i.e. the one with the series insertion transformer connected), without changing the local bus voltage *and* without changing the reactive power on any of the other lines leaving the substation. It can therefore regulate the station bus voltages at both the sending and receiving ends of a transmission line, while still freely controlling the real power flow, P, on the line.

7.4.2.4 Case 3: UPFC changing local bus voltage

Refer to Figure 7.32 (p.314). The objective for this test is to produce a large voltage change at Inez on command, while maintaining an unaltered level of P and Q on the line. For the test the voltage reference is stepped from an initial value of 0.985 p.u., to 1.02 p.u., to 0.95 p.u., and back to 0.985 p.u. The UPFC successfully holds the line P and Q constant (using very small changes in injected voltage), while the shunt converter goes from its initial output of 40 Mvar capacitive, to 100 Mvar capacitive, to 0 Mvar, and back to 40 Mvar capacitive.

In a sense, this test is the opposite of the previous one. In the previous case, the Inez bus and the other transmission lines were "insulated" from changes in the reactive loading of the Big Sandy line. In this case, the line is insulated from the large voltage swings at Inez, while the other lines leaving the station all experience changes in reactive loading.

7.4.2.5 Case 4: UPFC holding unity power factor

Refer to Figure 7.33 (p.315). The objective for the test is to maintain unity power factor looking into the transmission line, while producing large swings in real power, P, on the line, and also maintaining the Inez bus at 1.0 p.u. This case is really a combination of Cases 1 and 2. It is of particular interest because driving a line at unity power factor should, in principle, make it possible to deliver the largest amount of real power into the line for the lowest current. This should result in the most efficient use of the line from a thermal point of view. Naturally, the reactive power consumed by the line itself must now be supplied at the other end of the line, resulting in higher voltages at that end.

7.4.2.6 Case 5: Series converter operating in SSSC mode

Refer to Figure 7.34 (p.316). For this test, the shunt converter is disconnected from the dc terminals of the series converter and is completely out of service. Consequently the Inez bus voltage is not regulated. The series converter injects voltage into the line essentially in quadrature with the prevailing line current. The injection angle deviates from true quadrature to draw real power from the line for converter losses and to charge and discharge the dc bus capacitor banks. By means of the quadrature voltage injection, the SSSC is able to raise or lower the line current, but cannot *independently* alter P and Q. In principle, the SSSC can reverse the direction of power flow on a line, but the control becomes difficult as the line current is reduced through zero, at which point real power cannot be drawn from the line. SSSC operation is an important subset of full UPFC operation, since it can be used when the shunt converter is not available. A single series-connected converter installation may be the most cost effective solution for applications where a simpler form of power flow control is sufficient.

The objective of the test is simply to show the SSSC raising and lowering the power on the Big Sandy line. The SSSC is operated to control the magnitude and the polarity of the injected voltage (i.e. the line power is not automatically controlled). Voltage injections are selected to give a sequence of approximately 100 MW, 180 MW, 250 MW, and finally 200 MW on the line. Note the corresponding changes in Q. From the phasor diagrams it can be clearly seen how $\boldsymbol{I}_{\text{line}}$ maintains a constant phase relationship to $\boldsymbol{V}_1$ and is always in quadrature with $\boldsymbol{V}_{21}$. The natural flow on the line is between 100 MW and 180 MW. Consequently the polarity of $\boldsymbol{V}_{21}$ is reversed from lagging $\boldsymbol{I}_{\text{line}}$ to leading in this transition. The polarity reversal is accomplished by taking the dc bus voltage first to zero, then raising it again with 180 degree phase shift in the converter output voltage.

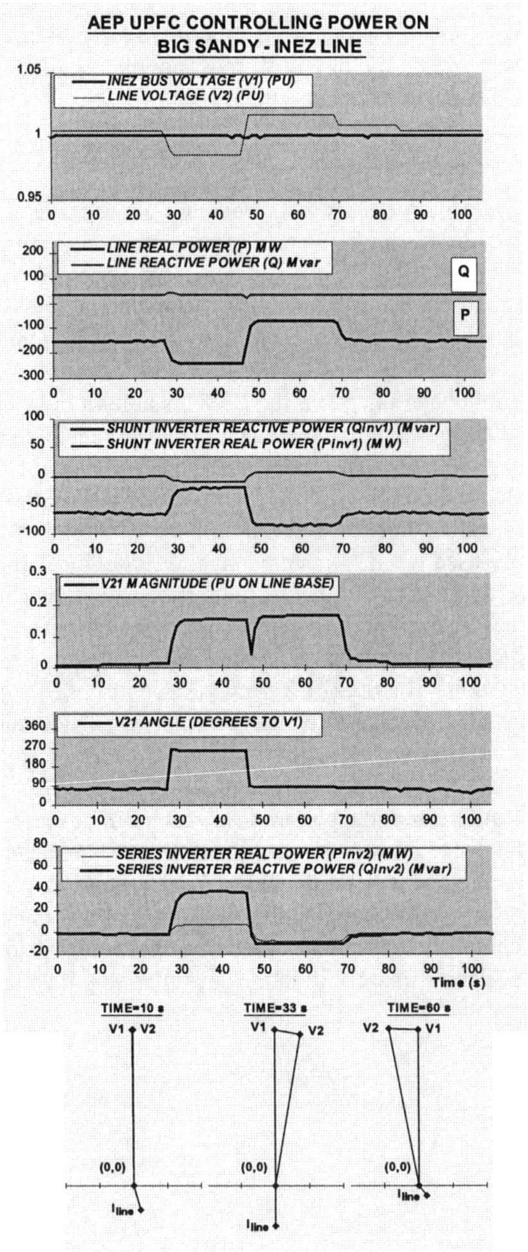
AEP UPFC CONTROLLING POWER ON
BIG SANDY - INEZ LINE
INEZ BUS VOLTAGE (V1) (PU)
LINE VOLTAGE (V2) (PU)
LINE REAL POWER (P) MW
LINE REACTIVE POWER (Q) Mvar
Q
P
SHUNT INVERTER REACTIVE POWER (Qinv1) (Mvar)
SHUNT INVERTER REAL POWER (Pinv1) (MW)
V21 MAGNITUDE (PU ON LINE BASE)
V21 ANGLE (DEGREES TO V1)
SERIES INVERTER REAL POWER (Pinv2) (MW)
SERIES INVERTER REACTIVE POWER (Qinv2) (Mvar)
Time (s)
TIME=10 s
TIME=33 s
TIME=60 s
V1
V2
(0,0)
Iline

Figure 7.30 Test Case 1: UPFC changing real power (P)

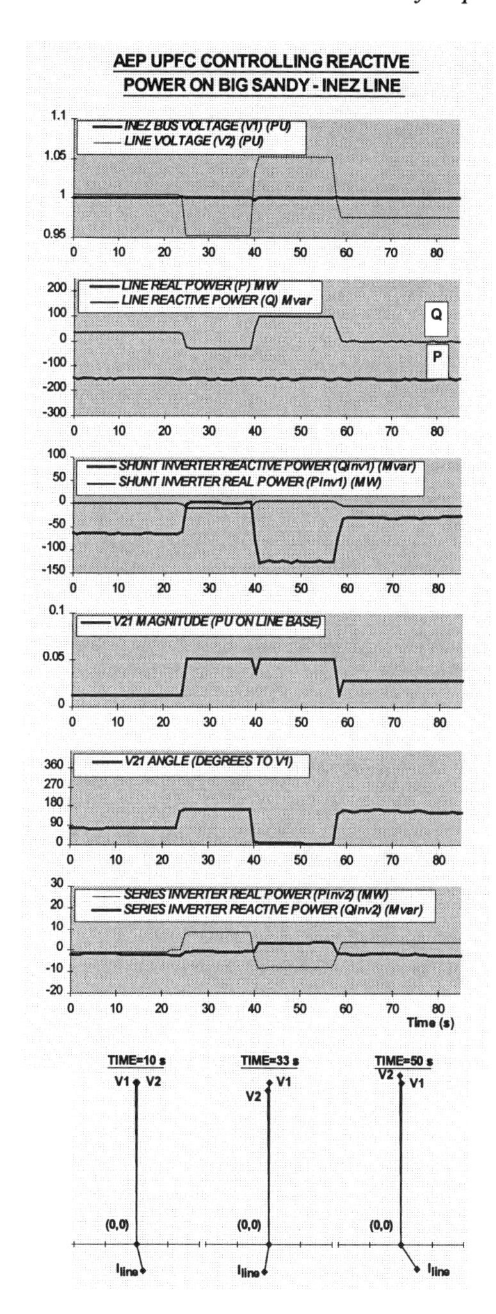

Figure 7.31 Test Case 2: UPFC changing reactive power (Q)

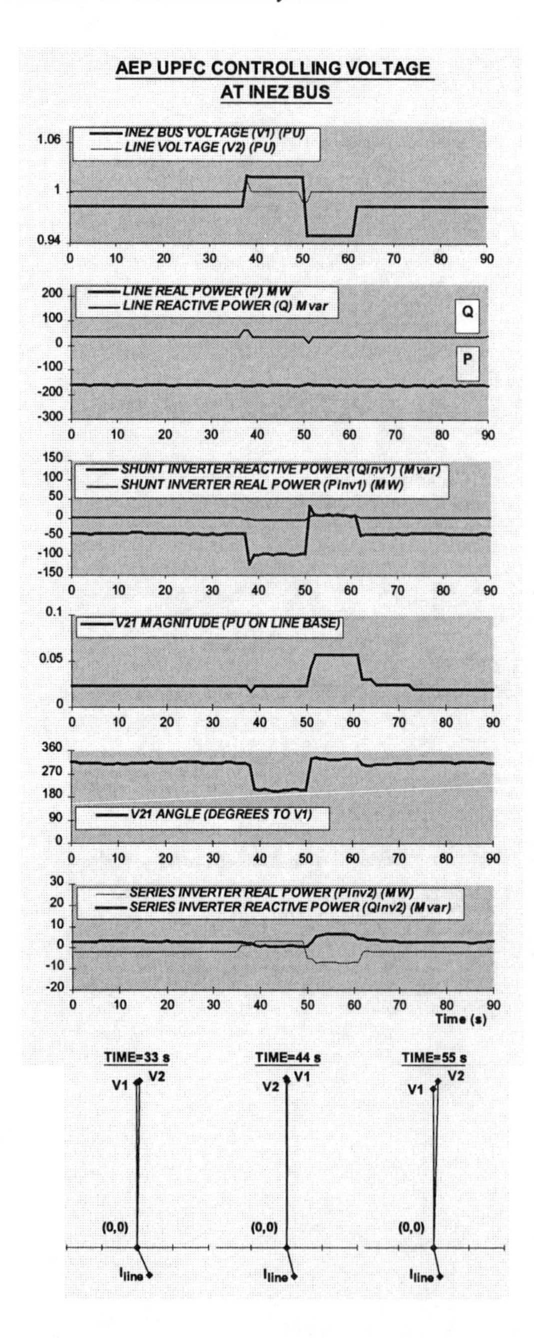

Figure 7.32 Test Case 3: UPFC changing Inez bus voltage

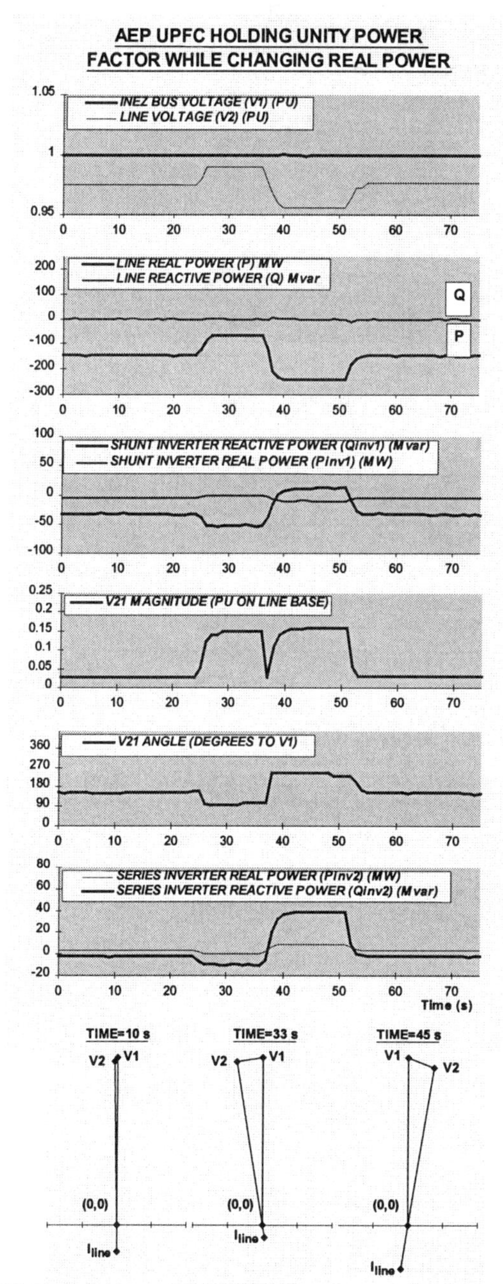

Figure 7.33 Test Case 4: UPFC holding unity power factor

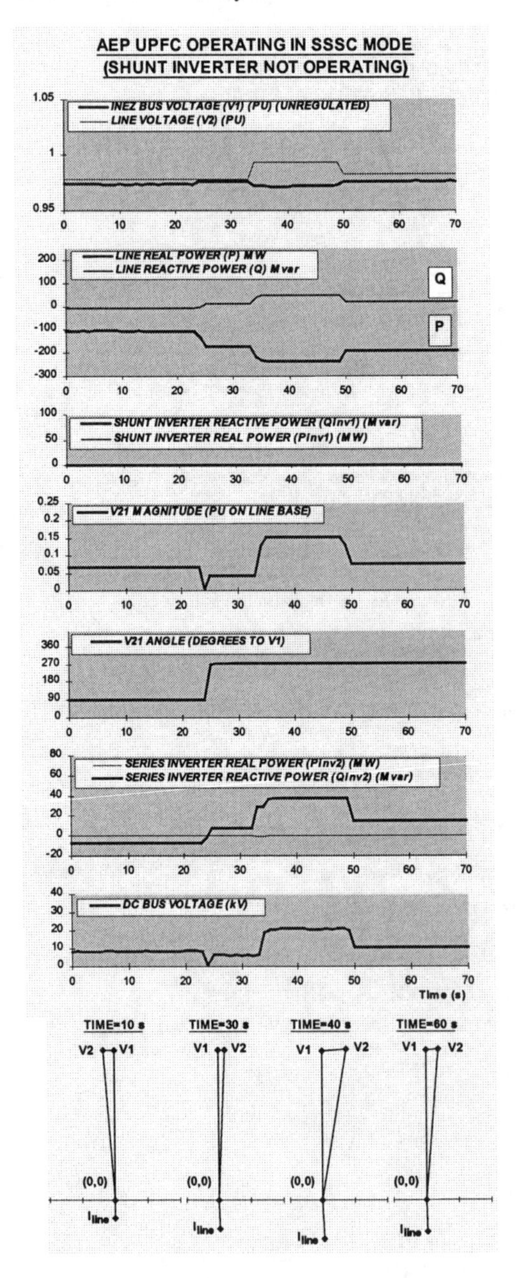

Figure 7.34 Test Case 5: Series converter operating in SSSC mode

7.5 Summary

From the viewpoint of conventional transmission compensation and control, the Unified Power Flow Controller can provide simultaneous, real-time control of all or any combination of the basic power system parameters (transmission voltage, line impedance and phase angle) which determine the transmittable power. Thus, the UPFC can fulfill all the functions of reactive shunt compensation, series compensation, and phase shifting, and thereby meet multiple power flow control objectives. However, the UPFC can also be viewed as a generalized power flow controller that is able to maintain prescribed and *independently* controllable real power and reactive power in the line. Within this concept, the conventional terms of series compensation, phase shifting, etc., become irrelevant; the UPFC controls the magnitude and angular position of the injected series voltage in real time so as to maintain or vary the real and reactive power flow in the line to satisfy load demands under normal, *as well as* temporary system operating conditions and contingencies. With sufficient rating, the UPFC can provide this generalized P and Q control function over a broad operating range, and at any practical transmission angle.

When compared to other related power flow controllers, such as the thyristor-controlled series capacitor (TCSC) and the thyristor-controlled phase angle regulator, it becomes evident that the UPFC is unique in its capability to control both real and reactive power. The capability of the UPFC to force and maintain strategically chosen values of P and Q at a given point (terminal) of the line by continuous closed-loop feedback control opens up a range of new possibilities for power system control. Apart from the obvious advantages for power scheduling and automatic power limiting, the fast dynamic response of the UPFC will automatically oppose transient disturbances and can prevent power oscillations from existing on the line.

TNA model and actual field tests have verified the powerful capabilities of the UPFC in rapidly controlling P and Q, while maintaining highly regulated bus voltage. These tests have also demonstrated the functional versatility of the UPFC and its ability to counteract dynamic system disturbances in a predefined manner prevent power oscillations. The UPFC installation at Inez represents a milestone in establishing the superiority of modern transmission system control with the use of high power switching converters.

7.6 References

1 Gyugyi, L., "A Unified Power Flow Control Concept for Flexible AC Transmission Systems," *IEE PROCEEDINGS-C*, 139, (4), 1992.

2 Schauder, C.D. and Mehta, H., "Vector Analysis and Control of Advanced Static Var Compensators," *IEE PROCEEDINGS-C*, 140, (4), 1993.
3 Gyugyi, L., "Dynamic Compensation of AC Transmission Lines by Solid-State Synchronous Voltage Sources," *IEEE Transactions on Power Delivery*, 9, (2), 1994.
4 Gyugyi, L. et al., "The Unified Power Flow Controller: A New Approach to Power Transmission Control," *IEEE Transactions on Power Delivery*, 10, (2), 1995.
5 Gyugyi, L. et al., "Static Synchronous Series Compensator: A Solid-State Approach to the Series Compensation of Transmission Lines," *IEEE Transactions on Power Delivery*, 12, (1), 1997.
6 Rahman, M., et al., "UPFC Application on the AEP System: Planning Considerations," *IEEE Transactions on Power Systems*, 12, (4), 1997.
7 Mihalic, R., et al., "Improvement on Transient Stability Using Unified Power Flow Controller," *IEEE Transactions on Power Systems*, 11, (1), 1997.
8 Schauder, C.D., et al., "Operation of the Unified Power Flow Controller (UPFC) Under Practical Constrains," *IEEE Transactions on Power Delivery*, 13, (2), 1998.
9 Schauder, C.D., et al., "AEP UPFC Project: Installation, Commissioning, and Operation of the ±160 MVA STATCOM (Phase 1)", PE-515-PWRD-0-12-1997.
10 Edris, A., et al., "Controlling the Flow of Real and Reactive Power," *IEEE Computer Applications in Power*, 11, (1), 1998.
11 Renz, B.A., et al., "World's First Unified Power Flow Controller on the AEP System," CIGRE Paper No. 14-107, 1998.
12 Gyugyi, L., et al., "The Interline Power Flow Controller Concept: A New Approach to Power Flow management in Transmission Systems," *IEEE Transactions on Power Delivery,* 14, (3), 1999.
13 Renz, B.A., et al., "AEP Unified Power Flow Controller Performance" IEEE/PES PE-042-PWRD-0-12-1998.

Chapter 8

Electromagnetic transient simulation studies

J.Y. Liu and Y.H. Song

8.1 Introduction

Electromagnetic transient simulation studies are necessary for assessing the interactions between the power electronics subsystem of FACTS and the power system network, and for control design and evaluation. SPICE and EMTP are two digital simulation tools widely used in the simulations of power electronics devices. SPICE stands for Simulation Program with Integrated Circuit Emphasis. It is better suited for use in power electronics device models, while EMTP is a very powerful program for modelling power electronics applications in power systems because of the built-in models for various power system components. In addition, specialized programs such as NETOMAC and EMTDC have also been used. These programs have been used to digitally simulate various FACTS devices [1-6], including SVC, CSC, STATCOM, and UPFC. In this chapter, detailed description on UPFC [7] will be given to illustrate the procedures involved.

For simulating the UPFC under electromagnetic transient states, the UPFC model should be set up including its power electronics subsystem. Figure 8.1 shows the detailed topology of the two voltage source type bridge inverters, which are distinguished by the following features: (i) The capacitor is across the dc link; (ii) The systems are on the ac side; (iii) Each valve has an antiparallel diode across it. It is assumed that the dc link voltage is always present and sufficiently high with respect to the ac line voltage so that the antiparallel diodes are normally reverse biased. The valves are triggered on and off by logical signals to their gates from the firing control block. This type of circuit configuration has been widely used in various industrial applications such as transportation transaction, machine drives and UPS. The UPFC using above typical GTO based voltage-source inverters is illustrated in Figure 8.2, in which each inverter leg is composed of a GTO valve and a diode valve in antiparallel connection to permit bidirectional current flow.

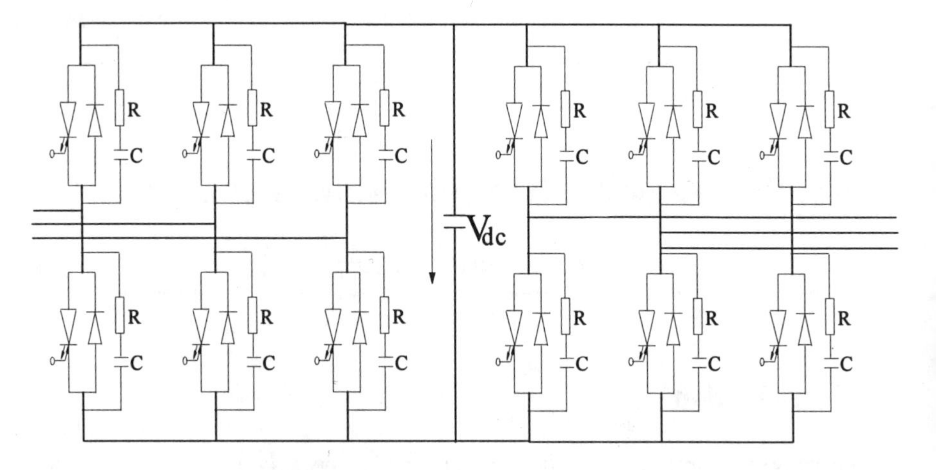

Figure 8.1 The internal relationship

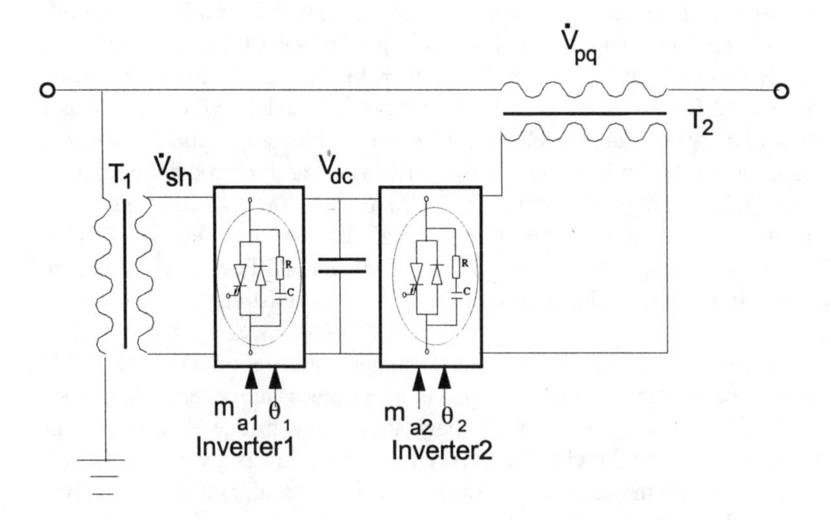

Figure 8.2 UPFC arrangement

In general, two methods [8] are employed to trigger the turn-on and turn-off signals of converters as the internal controllers: (i) Square wave method; (ii) Pulse width modulated method (PWM). The former is widely used in rectifiers, inverters and is also considered in UPFC design. Reference [9] demonstrates the performance of a UPFC employing 48-pulse inverters controlled by the square wave method. Although the square wave method shows high quality performance, low harmonic generation, minimum operating losses, it requires a more complex magnetic structure. In recent years, the PWM has been considered to develop new

FACTS devices, such as the PWM-STATCON, PWM-HVDC, series-type PWM compensator, and PWM phase-shifter [10–13]. This is because PWM converters have some important characteristics such as: (i) near sinusoidal current waveforms; (ii) 0–360^0 angle operation; (iii) bidirectional power transfer capability through reversal in the direction of flow of the dc link current; and (iv) direct and continuous control of the source voltages on both sides without change of dc-link voltage. If the PWM method is used in the UPFC, the UPFC will generate low harmonics, require simple magnetic structure, and be relatively inexpensive. Thus the PWM design approach was initially chosen for simulation studies of the UPFC. This chapter will investigate the simple double six pulse converters with the objective of better understanding the basic relationship between the control and functions of the PWM UPFC.

So far, many PWM methods have been developed with their own advantages in different fields. Generally speaking, PWM schemes are divided into two groups: (i) carrier-based PWM; (ii) carrierless PWM. Their characteristics and application in various areas can be found in Reference [14]. The carrier-based PWM includes: the sinusoidal PWM scheme (SPWM), the modified sinusoidal PWM technique, the third-harmonic injection PWM technique, and the harmonic injection PWM technique. In particular, SPWM has a number of advantages [14]. This chapter presents the performance and capability of the UPFC using the SPWM technique, and its aim is to provide information as to how the converters are controlled to realize the functions of the UPFC.

8.2 Principles of the UPFC based on SPWM inverters

The UPFC proposed for control of active and reactive power in ac systems is typically involved with the use of a forced-voltage source six-pulse inverter bridge which is illustrated in Figure 8.2. When the inverter operates under the control of the SPWM, it gives the relationship between the fundamental component of the ac voltage V_{sh} and the direct voltage V_{dc}:

$$V_{sh} = K(m_{a1}) V_{dc} \tag{8.1}$$

for $m_{a1} \leq 1.0$, $K(m_{a1})=0.612 m_{a1}$, 3-phase

A single phase diagram corresponding to a VSI connected to the utility system through a transformer is given in Figure 8.3. In this case, the general expression for the apparent power flowing between the ac mains side V_0 and the ac side V_{sh} of the VSI is as follows:

$$S_0 = \frac{V_0 V_{sh}}{X} \sin\theta_1 - j(\frac{V_0 V_{sh}}{X}\cos\theta_1 - \frac{V_0^2}{X}) \tag{8.2}$$

Where θ_1 is the phase displacement between V_0 and V_{sh}.

When the shunt part and series part of the UPFC operate under the SPWM, they have different functions. According to the concepts of the UPFC, the functions of the series part are achieved by adding an appropriate voltage phasor V_{pq} to the terminal phasor V_0 as shown in Figure 8.4. Because V_{pq} can be regulated by amplitude and angle, it is important to analyze how V_{pq} is regulated by SPWM. It is assumed that the dc link voltage V_{dc} in the UPFC circuit is kept constant by inverter 1, which can be readily realized by changing the phase angle θ_1 between V_0 and V_{sh}. Therefore, the series voltage output of ac side terminal of inverter 2 can be obtained:

$$\dot{V}_{pq} = T_2 m_{a2} V_{dc}(\cos\theta_2 + j\sin\theta_2)/2 \tag{8.3}$$

where T_2 is the ratio of the series transformer, m_{a2} modulation ratio of inverter 2, and θ_2 is the phase shift angle between V_{pq} and V_0. Thus V_{pq} is defined by m_{a2} and θ_2 and can be proportionally controlled by different m_{a2} and θ_2 according to the concepts of the UPFC which is shown in Figure 8.4. In this way, the UPFC can partially fulfill the functions of voltage regulation, series compensation, phase angle regulation, and multi-function power flow control through regulating inverter 2 based on the SPWM method.

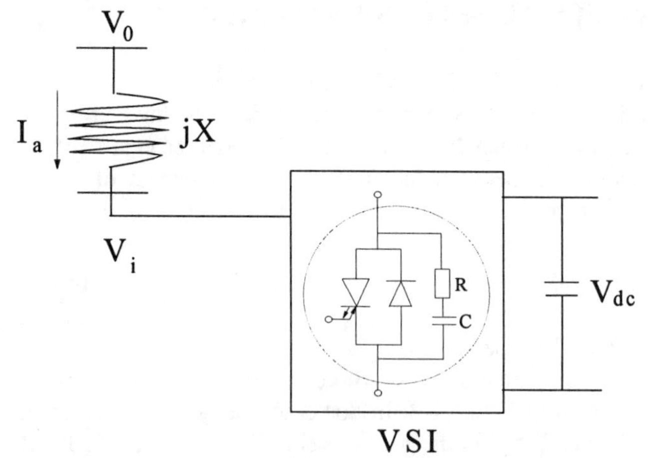

Figure 8.3 A single diagram corresponding to a VSI connected to the utility system through a transformer

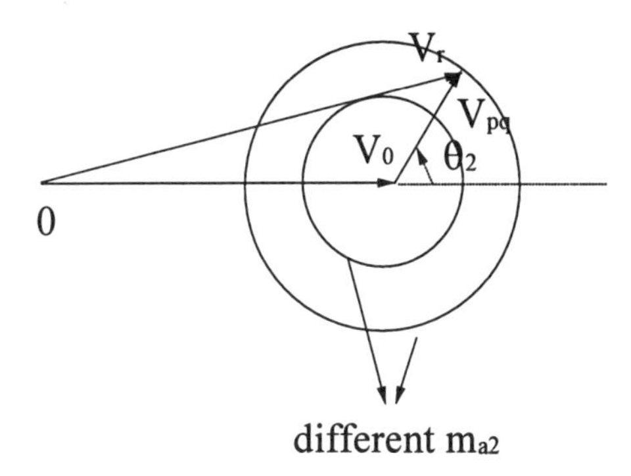

Figure 8.4 Phasor diagram of multiple control scheme

For the shunt part of the UPFC, it not only provides the active power to charge or discharge the direct link capacitor and keep V_{dc} constant, but also has the function of a synchronous solid-state var compensator (SVC), which can control the voltage V_0 through regulating V_{sh}. It is in principle straightforward to meet the requirements of regulating V_{dc} and V_0 simultaneously with SPWM through control of θ_1 and m_{a1} respectively. However, from equation (8.2), changing V_{sh} not only results in changing the reactive power flow but also changing the active power flow, and can thus lead to changes in V_{dc}. At this time, θ_1 should be regulated in order to keep V_{dc} constant. Under the above assumptions, the inverter 1 has the following operating characteristics:

(1) Active power flow is bilateral. It goes from V_0 bus to V_{sh} bus for lagging θ_1 and vice versa for leading θ_1;
(2) Assuming θ_1 is used to keep V_{dc} constant, the shunt part of the UPFC absorbs reactive power when $V_0 > V_{sh}$, which can be realized through decreasing m_{a1};
(3) Assuming θ_1 is used to keep V_{dc} constant, the shunt part of the UPFC supplies reactive power when $V_0 < V_{sh}$, which can be realized through increasing m_{a1}.

The SPWM UPFC can thus control transmission line terminal voltage and power along the line by regulating m_{a1}, θ_1, m_{a2}, and θ_2 of inverters 1 and 2.

8.3 EMTP/ATP simulation

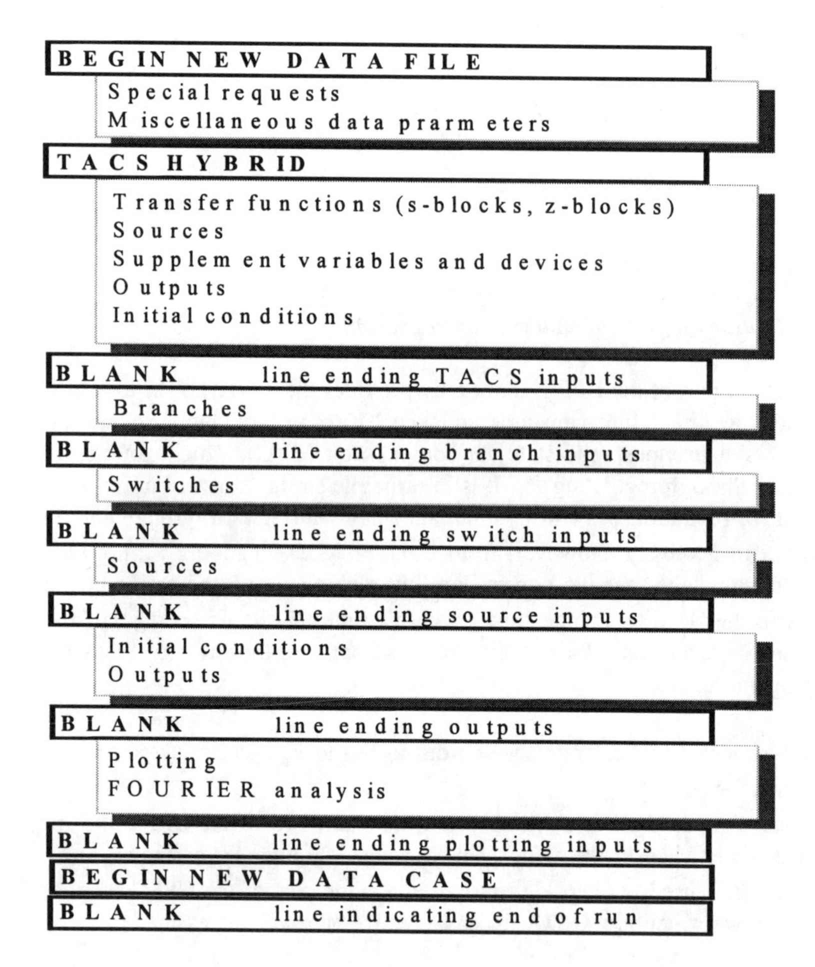

Figure 8.5 Date file structure for EMTP/TACS HYBRID case

8.3.1 The EMTP/ATP program

EMTP – Electromagnetic Transients Program is a full – featured transient analysis program, initially developed for electrical power systems. It is also capable of simulating controls, power electronics, and hybrid situations. The program features an extremely wide variety of modelling capabilities encompassing electromagnetic and electromechanical oscillations ranging in duration from microseconds to seconds. Some main features of it are described as follows:

8.3.1.1 Types of EMTP studies

EMTP is used for a wide variety of studies. Some of the important applications associated with power electronics and FACTS applications include:

- HVDC operation and controls
- Various FACTS operations
- General control system analysis
- Protection systems
- Harmonic propagation analysis

8.3.1.2 Structure of data format of EMTP

Figure 8.5 gives a date file structure combined general data format of EMTP with that of Transient Analysis of Control Systems (TACS).

8.3.1.3 TACS and selected representation of the GTO device model

In EMTP, power systems transients and control systems could be modelled simultaneously to study their dynamic interaction. "Sensors" pick up signals from the power system (often briefly called the Network) for input to the control system (called TACS). Commands are forwarded from the control system to the power system as shown in Figure 8.6.

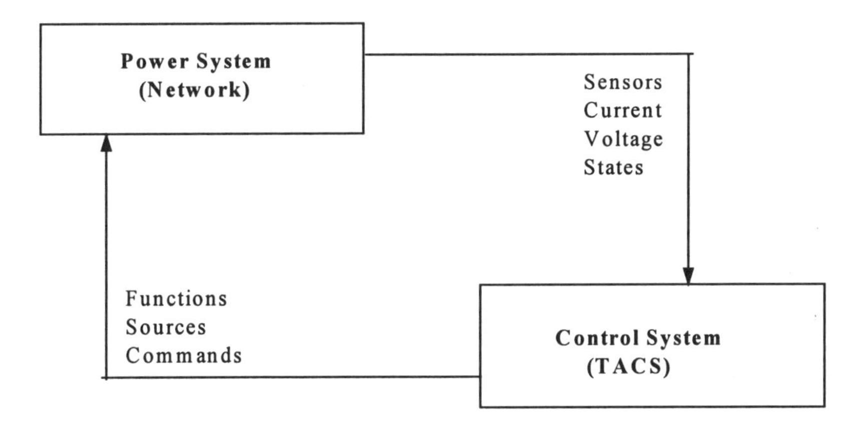

Figure 8.6 Interaction between power system (network) and control system (TACS)

The procedure for setting up models of FACTS and the UPFC by EMTP/TACS is to derive mathematical models from the real-world device configurations. For example, the GTO is a key device in the UPFC whose original prototype and characteristics are shown in Figure 8.7 (a) and (b). However, it can be simplified to Figure 8.7 (c) so as to match the device type provided by EMTP, shown in Figure 8.8. In this case, the GTO becomes an idealised switch, which can be controlled by signal feedback from the network according to TACS working. These signals should be carefully chosen and calculated from the scheme of control strategies. Therefore, from this point of view, modelling, simulation and control of the UPFC is an integrated process.

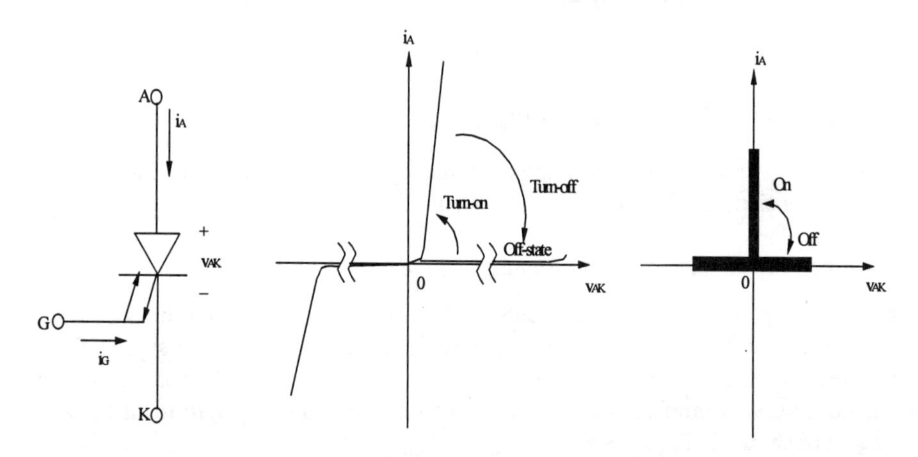

Figure 8.7 A GTO: (a) symbol, (b) i-v characteristics, (c) idealised characteristics

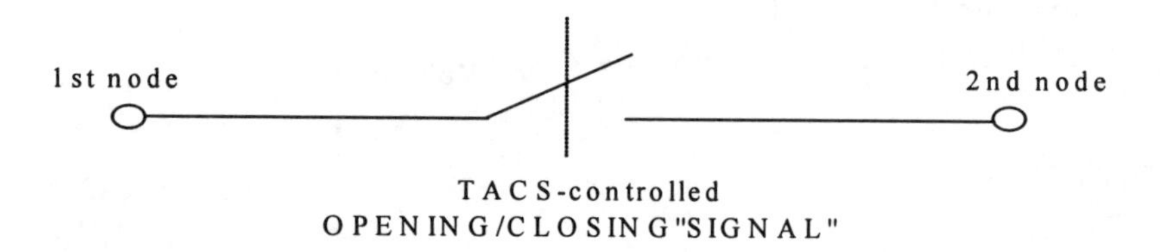

Figure 8.8 Type-11 switch for diode and valve

8.3.2 SPWM scheme generated by EMTP/ATP TACS

SPWM switching is a scheme where a control signal $V_{control}$ (constant sinusoidal wave or varying in time according to control mode) is compared with a repetitive switching frequency triangular waveform, in order to generate switching signals. Controlling the switch duty ratios in this way allows the average ac voltage output to be controlled. With reference to Figure 8.9 (the diagram is generated by TACS of EMTP according to SPWM theory which will be described in detail in the next section), the frequency of the triangular waveform establishes the inverter switching frequency f_s and is generally kept constant along with its amplitude V_{tri}. The control signal $V_{control}$ is used to modulate the switch duty ratio and has a frequency f_1, which is the desired fundamental frequency of the inverter voltage output, recognizing that the inverter output voltage will not be a perfect sine wave and will contain voltage components at harmonic frequencies of f_1. Thus, the amplitude modulation ratio m_a is defined as

$$m_a = \frac{V_{control}}{V_{tri}} \tag{8.4}$$

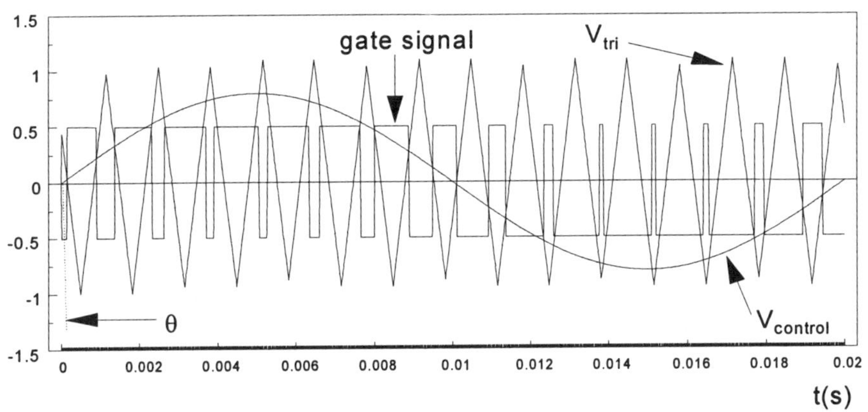

Figure 8.9 PWM signal generated by TACS

The frequency modulation ratio m_f is defined as

$$m_f = \frac{f_s}{f_1} \tag{8.5}$$

Therefore, three parameters are adjusted to adapt the simulation of the UPFC system interaction:

(1) In order to keep the SPWM operating in the linear modulation range, m_a should be from 0 to 1. In this case, $V_{control}$ determines the amplitude of the sinusoidal modulating waveform and therefore modifies the position of its intersections with the constant amplitude triangular carrier waveform, and hence m_a. $V_{control}$ is generally derived from the system control objective, whose amplitude and phase depend on their different requirements.

(2) Because the power system keeps a constant frequency of 50Hz, m_f is chosen as a large number, 15 in this case (i.e a repetition rate of 750 Hz), which means that

the triangular waveform signal and the control signal are synchronized to each other and the amplitudes of harmonics are small. Therefore, both the frequencies of the $V_{control}$ and the V_{tri} are kept constant during the UPFC operation.

(3) The phase displacement between $V_{control}$ and V_{tri} could be regulated according to the demands of the magnitude and direction of power flow. Therefore, two regulating parameters m_a and θ can be employed as internal controls to manipulate the turn-on and turn-off signals of two back-to-back inverters of the UPFC.

8.3.3 EMTP model development for systems with UPFC

The simulation system suitable for electromagnetic study is often chosen to be a simple system with the objective of understanding the UPFC internal regulation concept and interaction with the system. The system as shown in Figure 8.10 includes two three-phase 400kV sources which have a 10^0 phase separation, a double circuit 62 km 400 kV transmission line, and the UPFC which is connected to the system at the sending end of the transmission line. This type of 400 kV double circuit is the major type used in British EHV transmission systems. The whole system contains the necessary components for simulating UPFC, which is implemented in EMTP/TACS data format. Each circuit component has its own associated data input format specified by the EMTP manual. Besides these system circuit components, some factors must be taken into consideration in order to operate the system properly, for instance, the snubber circuit needs to avoid numerical instability in the EMTP simulation, power meters are used as the output simulation results and input of closed-loop controllers. The whole list of components for the EMTP model for this system is as follows:

- Sources and transmission lines
- Transformers: shunt and series
- UPFC and snubber circuit
- SPWM generator for gate firing signals of GTO
- Synchronization
- Power meters and sampling
- Harmonics filters
- Closed-loop control circuits

This section mainly describes how to set up the EMTP model for the first six components. The others will be introduced in the following sections.

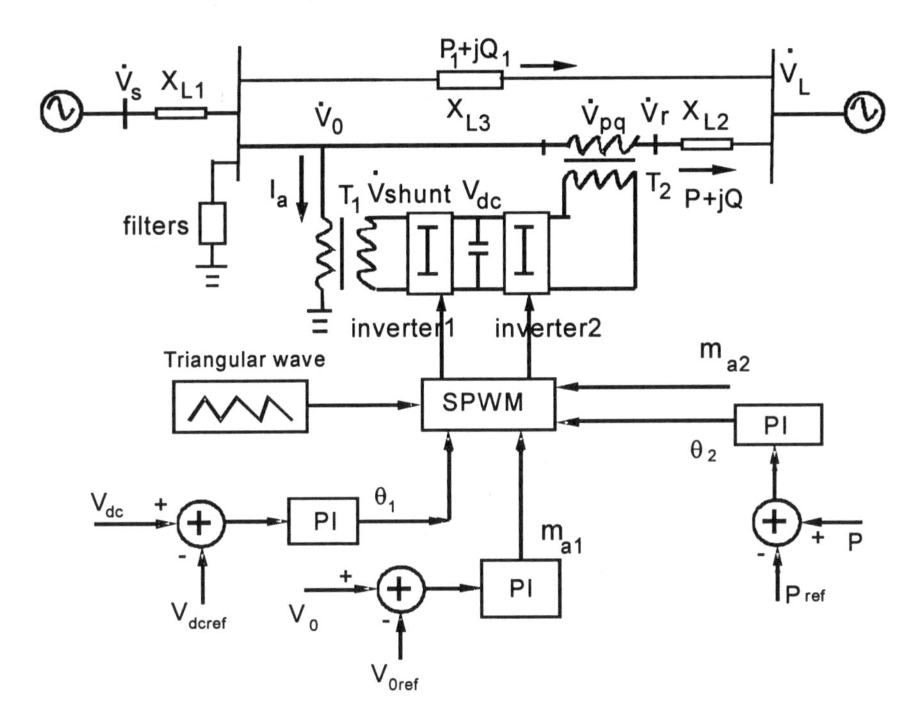

Figure 8.10 EMTP UPFC model in a simple system with control of UPFC

8.3.3.1 Sources and transmission lines

(i) EMTP provides many types of sources to be chosen for different studies, such as voltage source, current source and series voltage sources. These sources can be defined as step, ramp, slope ramp, normal sinusoidal function, with given starting and stopping times and phases. In the UPFC study, the two terminal sources are specified as type 14 voltage sources of normal sinusoidal function, whose parameters include: connection bus name, peak value of the amplitude and phase displacement per phase. There is a defined phase displacement between two sources to allow power transmission.

(ii) The transmission lines are adopted as the resistance, inductance and capacitance matrix (both symmetrical components and phase components) for the given configuration of overhead conductors through the 'LINE CONSTANTS' function provided by EMTP. This matrix, the representation of transmission line parameters in more detail than that of conventional nominal PI circuit, can easily be inserted into the EMTP data file to provide a more accurate model of the lines. The input of 'LINE CONSTANTS' includes: tower configuration, skin effect parameters and conductor physical parameters.

8.3.3.2 Transformer type: the shunt and series transformers of UPFC

There are two types of transformers used in the UPFC EMTP data file. One is the three phase transformer for the shunt transformer of the UPFC, whose configuration is wye-wye. Another is the single phase transformer for the series one of the UPFC. Both types of transformers can take account of saturation effects. However, no saturation effects are considered for simplicity. The ratio for the primary winding with respect to secondary winding and leakage impedance of the windings are the parameters needed for input of the transformer type.

8.3.3.3 UPFC and snubber circuit

The voltage source inverter is the heart of the UPFC. The three, full-wave inversion bridge is built using three identical GTO inverter legs. A dc capacitor is the link to the voltage source inverters of the shunt part and the series part of the UPFC. They are represented in the EMTP model by a type-11 diode and a type-11 TACS controlled switch respectively. The type-11 switch can be used to simulate a switch which may be simultaneously controlled by any given TACS variable while following the simple opening/closing rules of a standard diode.

(i) The type-11 switch acts as a diode when the open/close signal is not applied (no TACS control).
(ii) When the open/close signal is specified (the switch acts neither as a diode nor as a valve), the type-11 switch is purely TACS-controlled and can be used successfully to model GTOs. A positive signal will result in an immediate closing of the switch and this will remain closed as long as this positive signal is active. It can conduct arbitrarily large currents with zero voltage drops. A negative signal will result in an immediate opening of the switch (irrespective of the instantaneous current flow) and this will remain open as long as the negative signal is active. It can block arbitrarily large forward and reverse voltages with zero current flow. The change of signal from positive/negative to negative/positive is assumed to be instantaneous when triggered. The trigger power requirement is assumed negligible. In this case, the type-11 switch acts as an ideal GTO, whose signal comes from internal control blocks of the SPWM.

It is worth mentioning the role of the snubbers in the EMTP model. Physically, voltage snubbers are required solely to prevent the switching device from seeing an excessive rate of voltage change. However, the snubbers shown in Figure 8.11 of this EMTP representation are also for controlling the numerical oscillation associated with the Trapezoidal solution method. As a result, the parameters required for the snubbers in the EMTP model are dependent on the circuit being simulated and the step size selected for the solution. Generally, the minimum RC time constant should be greater than 2-3 times the step size. For this study, simple RC snubber circuits are used, while assuring the numerical stability of the simulations.

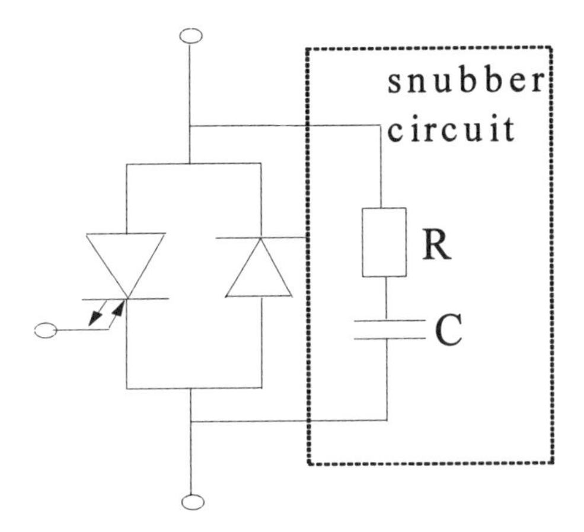

Figure 8.11 Diagram of circuit on every bridge of VSI-inverter

8.3.3.4 SPWM generator for firing signals of GTO

To effectively control the UPFC and especially to control GTO switches, it is necessary to model the generation of SPWM signals which are used to trigger the turn-on or turn-off of the GTO. In this respect, the Transient Analysis of Control System (TACS) (or its equivalent MODELS) of the EMTP provides the way to set up the SPWM switch scheme for controlling the GTO thyristor valves of the inverter [15]. Through using various functions such as the transfer function blocks and FORTRAN-like logical statements provided by TACS, the SPWM control signal where the amplitude and frequency modulation ratios can be changed is shown in Figures 8.9 and 8.12. The SPWM control accepts an analogue sinusoidal modulating waveform signal from each of the 3-phases and an analogue triangular carrier signal, and based on detecting the intersection points it generates gating signals to the GTOs of the bridge. The block labelled $m_a\sin(\omega t+\theta)$ is the key to realize the regulation of SWPM. When m_a, ω, and θ are specified as some values, the UPFC operates under the control of the open-loop controller. When m_a, ω, θ are derived from the system operation and set-points, the UPFC acts under the control of the closed-loop controller. Both cases have their different purposes and will be simulated in the following sections. From the expression, it can be seen that the controlled variable can have very large changes in its amplitude and phase in full circle under the control of SPWM.

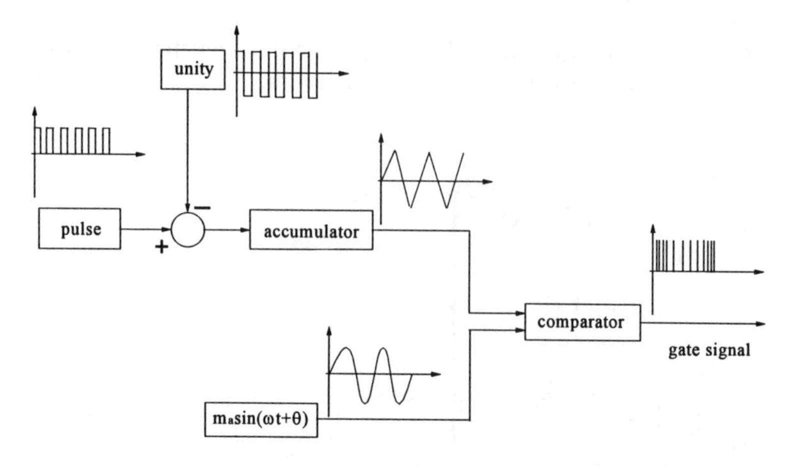

Figure 8.12 Transfer function of TACS to generate gate signal of the inverter

8.3.3.5 Synchronization

In the generation of SPWM signals, the phase displacement θ is the ideal phase of the controlled variable with reference to the phase of the connected bus which is assumed to be zero degrees. This assumes the existence of a perfect three-phase voltage source (i.e. infinite bus) right at the point of common coupling (PCC) between the AC and DC system. In technical terms, it can be said that the system interface has infinite strength, or that it has an infinite effective short circuit ratio (ESCR). However, this is not true in practice because of voltage distortion at the PCC and voltage synchronization problems. For example, the phase of V_0 in Figure 8.10 is not zero and its wave does not always have a pure sinusoidal waveform. The absence of perfect voltage sources at PCC implies the following:

(i) AC voltages at the AC/DC interface will always have some degree of distortion, even with filters installed.
(ii) Three-phase bus voltages at the interface may be unbalanced, say in the event of a single-line-to-ground fault on the AC side.
(iii) As the loading of the system changes, so do the bus voltages. The changes in bus voltages at PCC occur in magnitude as well as phase angle.

In order to successfully obtain the required information on the phase angle Θ of the fundamental AC busbar voltage, a voltage synchronization system is introduced as follows:

Assume AC bus voltage as:

$$V_0 = V_m \cos(\omega t + \Theta) \tag{8.6}$$

Through the use of Fourier analysis:

$$C_1 \equiv \frac{2}{T}\int_{t-T}^{t} v(t)\cos\omega t dt$$
$$= \frac{2}{T}\left\{\int_{0}^{t} v(t)\cos\omega t dt - \int_{t-T}^{0} v(t)\cos\omega t dt\right\} \quad (8.7)$$

$$S_1 \equiv \frac{2}{T}\int_{t-T}^{t} v(t)\sin\omega t dt$$
$$= \frac{2}{T}\left\{\int_{0}^{t} v(t)\sin\omega t dt - \int_{t-T}^{0} v(t)\sin\omega t dt\right\} \quad (8.8)$$

Note that the above equations can be implemented in TACS with the same concepts as the power meter (whose realization will be described in 'Power meters and sampling').

V_m can now be obtained from

$$V_m = \sqrt{C_1^2 + S_1^2} \quad (8.9)$$

The determination of Θ needs more elaboration. The difficulties arise from the lack of structured IF_THEN_ELSE statements in EMTP. To circumvent the problem, the following expression is suggested:

$$\Theta = a\tan\left(\frac{-S_1}{C_1}\right) + \{C_1.LT.0\} * \pi \quad (8.10)$$

The above expression will correctly return values of Θ ranging from -90^0 to +270^0, which generally covers the operation range of the phase for any given bus voltages.

Once Θ is obtained, $m_a\sin(\omega t+\theta)$ can be changed into $m_a\sin(\omega t+\theta+\Theta)$, in which the controlled variable will synchronize with the AC connected bus.

8.3.3.6 Power meters and sampling

In EMTP, some function monitoring of variables has been provided, such as voltages, currents and instantaneous powers. However, of more practical interest is the average power over a period of time *T* because this is often treated as the input to the proposed closed-loop controller of the UPFC. The rate of sampling is also important to the performances of controllers, which is directly related to the numerical stability of the controller.

(i) Power meters

The instantaneous power in any electrical component is given by

$$p(t) = v(t) * i(t) \tag{8.11}$$

The average power over a period of time T is formulated by:

$$P = \frac{1}{T}\int_{t-T}^{t} p(t)\mathrm{d}t \tag{8.12}$$

The period of time T is usually one period of the fundamental frequency f. For 50 Hz, the period becomes:

$$T = \frac{1}{f} = \frac{1}{50} = 0.02\ \mathrm{s} \tag{8.13}$$

The average power P is constant in steady-state. But the average power varies with time during transients. Then the average power $P(t)$ is of interest during the last half cycle on a continuous basis. It can be found by continuously computing $p(t)$ and splitting the integral into two components.

$$\begin{aligned} P(t) &= \frac{1}{T}\left[\int_{0}^{t} p(t)\mathrm{d}t - \int_{0}^{t-T} p(t)\mathrm{d}t\right] \\ &= \frac{1}{T}\left[\overline{P}(t) - \overline{P}(t-T)\right] \end{aligned} \tag{8.14}$$

The second integral has the value of the first integral delayed by T seconds. This suggests an implementation in TACS. The delay of the value of $\overline{P}(t)$ by time T is readily obtained by using TACS Device Code 53: Transport Delay. Its output value is equal to the input value delayed by time T.

(ii) Sampling and control references

In this EMTP UPFC model, closed-loop control is accomplished by monitoring the variables to the transmission line and UPFC and then generating gate signals for the inverters to create voltages that will regulate system variables. Sampling of these variables must be at a high enough rate to accurately characterize all variables to be controlled. Then the sampled variables are further processed and sent to the controllers as the practical input. To ensure the accuracy of the overall system modelling and to avoid possible numerical oscillation, simulation of this entire system involving fast GTO switching actions requires a much smaller time step. If the actual sampling rate is ignored and every calculated point is used for control reference derivation, any interaction between the sampling and the firing control would not be correctly simulated. Since the control response times are so short for the UPFC application, these interactions can be important. Discrete sampling is represented in the model by using a type-58 TACS device to simulate the sampling.

Finally, a diagram including the system, UPFC, and its internal control as well as system state definitions is shown in Figure 8.10.

8.4 Open-loop simulation

In this section simulation results from EMTP using detailed modeling of the UPFC based on SPWM inverters are presented. The SPWM switch scheme for controlling the GTO thyristor valves of the inverters has been set up using TACS with EMTP, in which the SPWM control signal can be generated according to open-loop simulation or closed-loop control studies. Two sets of some typical open-loop simulation results follow:

8.4.1 Simulation of SPWM UPFC regulation performance

(i) The first group of results only consider the series part of the UPFC and the V_{dc} is kept to 65kV. When V_{dc} is kept constant, V_{pq} is regulated by changing m_{a2} and θ_2, which is shown in Figures 8.12 – 8.15. When the system operates only with the control of the series injected voltage, it can be clearly seen from the figures when θ_2 is changed from 0^0 - 360^0, V_{pq} rotates according to the analysis of Figure 8.4. And when m_{a2} decreases V_{pq} also reduces which is shown in Figure 8.16.

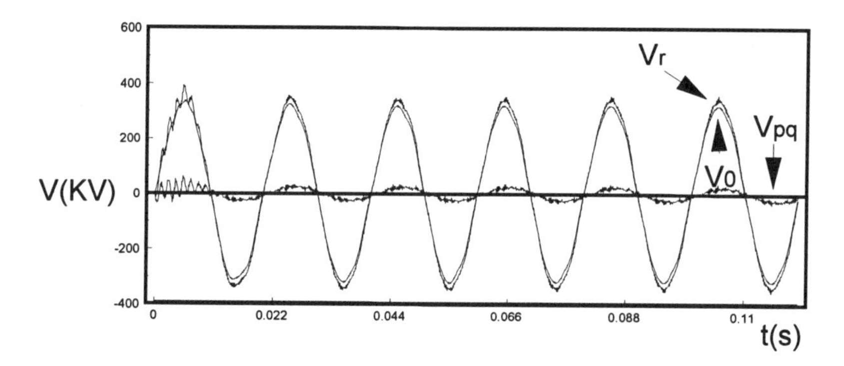

Figure 8.12 Simulation results of V_0, V_{pq}, and V_r when V_{dc}=65kV, m_{a2}=0.8, and θ_2=0°

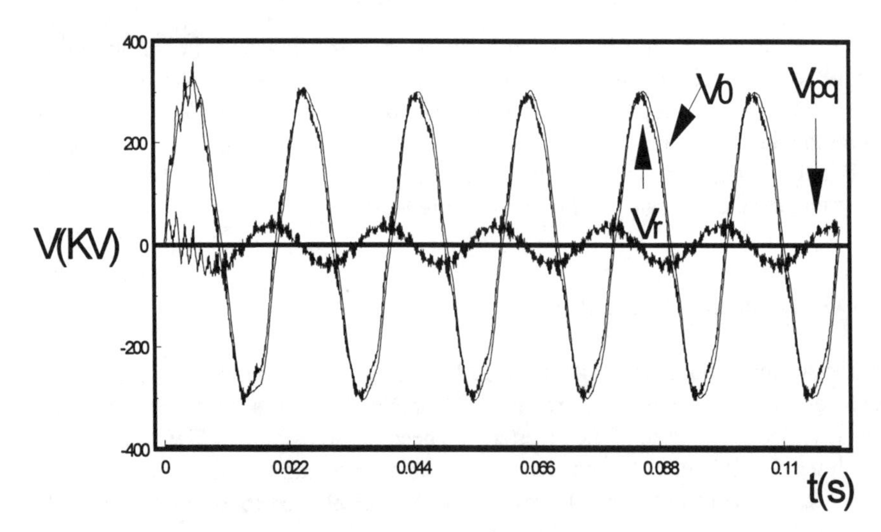

Figure 8.13 Simulation results of V_0, V_{pq}, and V_r when V_{dc}=65kV, m_{a2}=0.8, and θ_2=90°

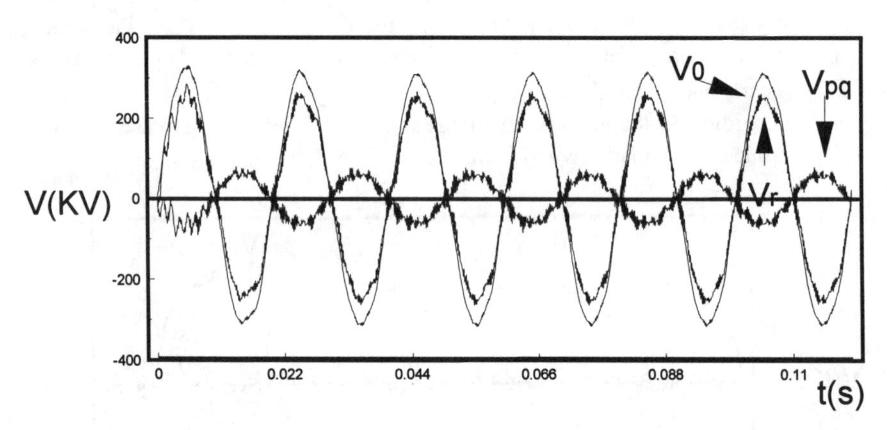

Figure 8.14 Simulation results of V_0, V_{pq}, and V_r when V_{dc}=65kV, m_{a2}=0.8, and θ_2=180°

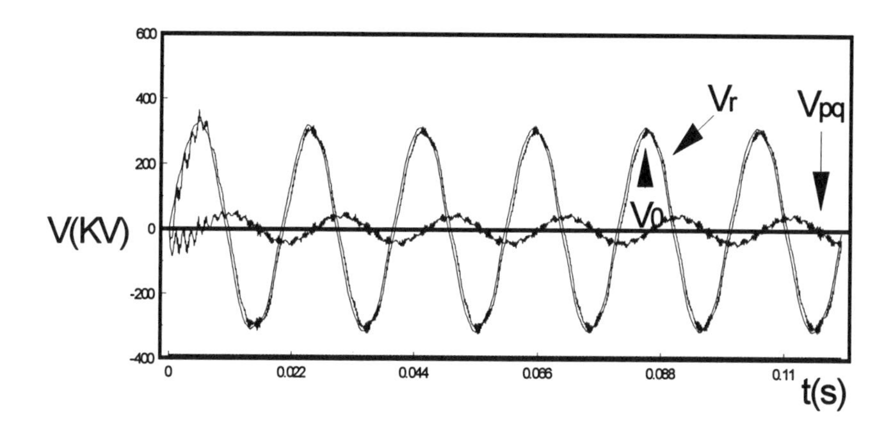

Figure 8.15 Simulation results of V_0, V_{pq}, and V_r when V_{dc}=65kV, m_{a2}=0.8, and θ_2=270°

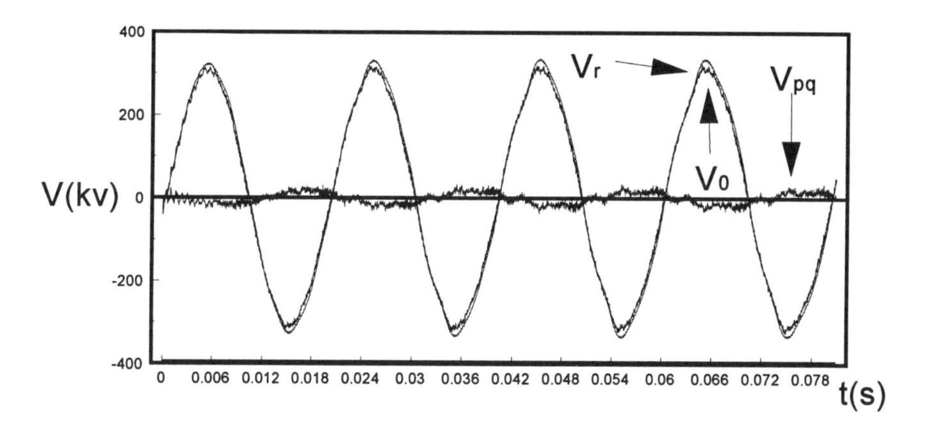

Figure 8.16 Simulation results of V_0, V_{pq}, and V_r when V_{dc}=65kV, m_{a2}=0.4, and θ_2=90°

(ii) The second group of results mainly concern the shunt part of the UPFC. When V_{dc} is kept constant at 65kV in this case through regulation of θ_1, then V_0 is controlled through regulation of m_{a1}, which is shown in Figures 8.17–8.19. When the UPFC connects to the V_0 bus, the shunt part of the UPFC can compensate the reactive power needed by the system in order to increase V_0 or absorb the reactive power to decrease V_0. From these figures, different conditions which show leading or the lagging compensation are presented. When m_{a1}=0.8, there is no exchange of reactive power and V_0 and I_a are thus in the same phase. When m_{a1}=1.0, the compensation is maximized and I_a thus leads V_0. When m_{a1}=0.6, the inverter absorbs reactive power and I_a thus lags V_0.

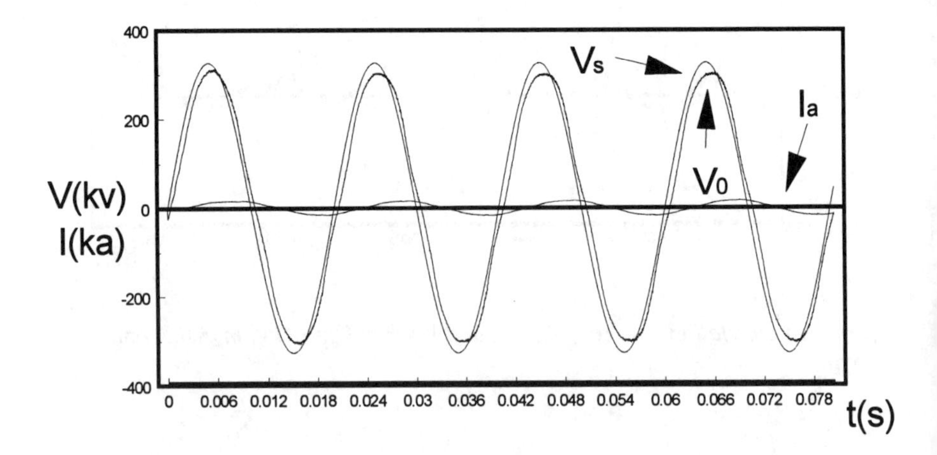

Figure 8.17 Simulation results of V_0, V_s, and I_a when m_{a1}=0.6, V_{dc}=65kV, m_{a2}=0.8, and θ_2=90°

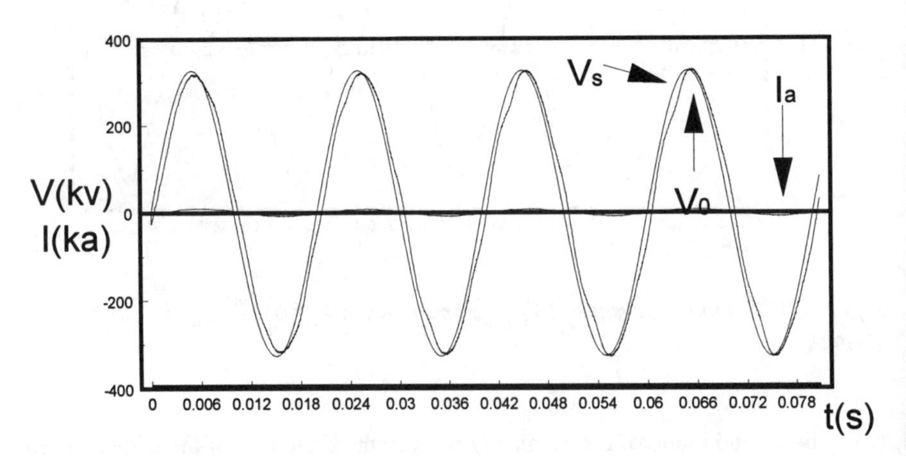

Figure 8.18 Simulation results of V_0, V_s, and I_a when m_{a1}=0.8, V_{dc}=65kV, m_{a2}=0.8, and θ_2=90°

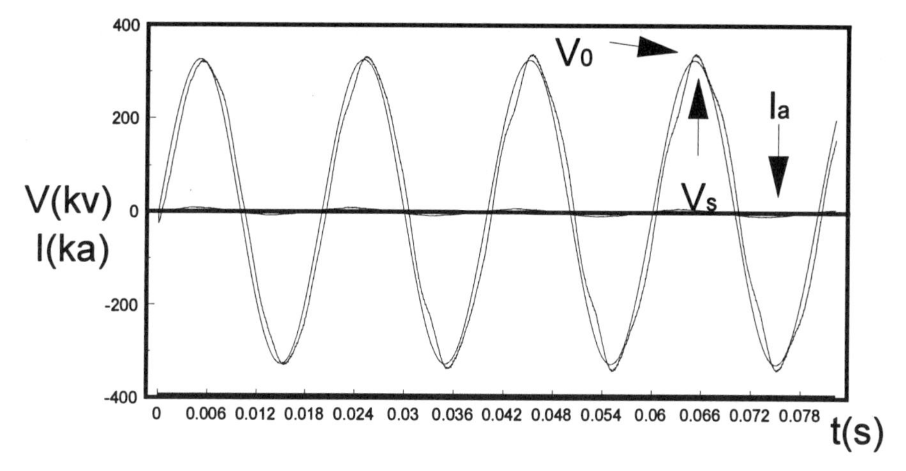

Figure 8.19 Simulation results of V_0, V_s, and I_a when m_{a1}=1.0, V_{dc}=65kV, m_{a2}=0.8, and θ_2=90°

Table 8.1 List of regulating parameters and power under different cases (the series part of the UPFC)

case	m_{a2}	θ_2 (degree)	P (MW)	Q (MVar)
1	0.8	0	459	411
2	0.8	90	990	-250
3	0.8	180	0	-200
4	0.8	270	-350	390
5	1.0	90	1169	-337
6	0.6	90	761	-127
7	0.4	90	552	-55

8.4.2 Results of the power flow and voltage support under control of SPWM UPFC

When V_{dc} is kept constant, the amplitude and phase of V_{pq} can be regulated to control the line power flow. In these cases corresponding to Figures 8.12 – 8.25, the double circuit transmission line power flow is effectively controlled, which is shown in Table 8.1. Examining the phasor of V_{pq} in relation to others shows that the angle between V_r and V_L as well as the magnitude of V_r qualitatively reaches a maximum at θ_2=90^0 which transfers maximum P; but when θ_2=180^0 there is no change; and the magnitudes of both V_r and P decrease for values of m_{a2} less then 1.0. When V_0 is controlled by the shunt part of the UPFC, the UPFC acts as a STATCON to give voltage support. Its impacts on power flow can also be found in Table 8.2 which corresponds to Figures 8.16–8.19.

Table 8.2 List of regulating parameters and power under different cases

case	m_{a1}	θ_1 (degree)	m_{a2}	θ_2 (degree)	P (MW)	Q (Mvar)
1	0.6	-15	0.8	90	860	-290
2	0.8	-18	0.8	90	950	-250
3	1.0	-20	0.8	90	990	-250

8.4.3 Operating envelope of UPFC

Based on a large number of simulation results and analysis, the operating envelope of the UPFC at a specified condition can be deduced. Generally speaking, it is difficult to define the operating envelope of the UPFC because all four parameters m_{a1}, θ_1, m_{a2} and θ_2 will couple together to affect the regulation of the SPWM UPFC. As to the normal operation of GTO valves, it often keeps V_{dc} at a minimum voltage under which the valves will extinct. In this case, it is easy to obtain the operating envelope of power vs. m_{a2} and θ_2 for the series part and the full operating envelope of the UPFC at V_{dc}=65kV; this is shown in Figures 8.20 and 8.21. From Figure 8.20, it can be seen that P and Q can be regulated within a wide range just by control of m_{a2} and θ_2. Figure 8.21 shows that the operating envelope of power will be extended when inverter 1 also takes part in regulating.

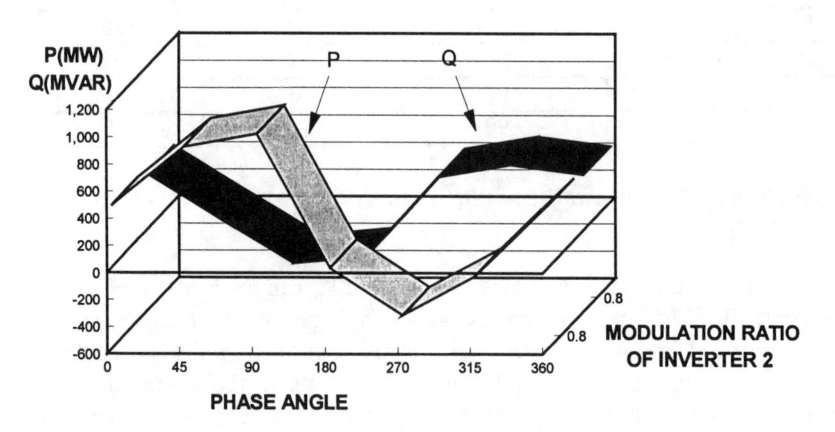

Figure 8.20 Operating envelope of the series part

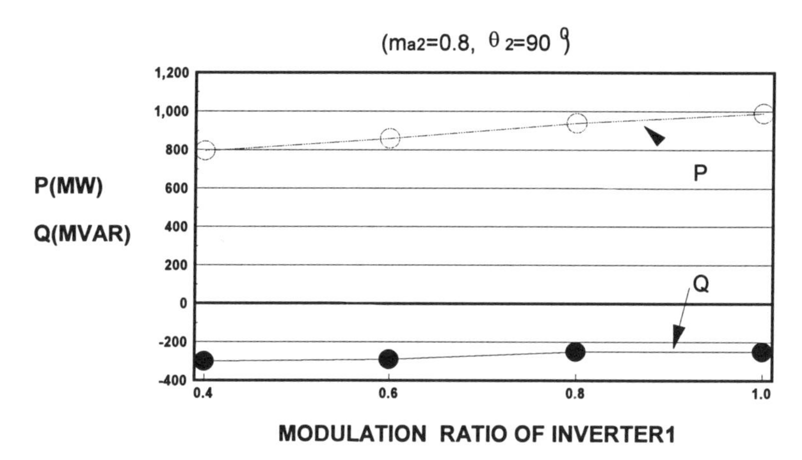

Figure 8.21 Operating envelope of the series part of the UPFC using PWM regulation

8.5 Close-loop simulation

An internal controller for the UPFC has been designed for preliminary evaluation studies. In this respect, the UPFC controller is designed as a proportional-integral feedback type controller which is shown in Figure 8.22. In Figure 8.22, X_{input} represents the variable from the system to be controlled, such as active transmission line power or voltage of dc-link capacitor. $X_{\text{reference}}$ is the desired value. A PI-type function is the conventional structure. The limiter forces the SPWM to operate in a linear range. Y_{output} is one of input control signals to the SPWM, that is m_a, or θ or ω in Figure 8.11. The UPFC internal controller in Figure 8.11 has the same structure as in Figure 8.22.

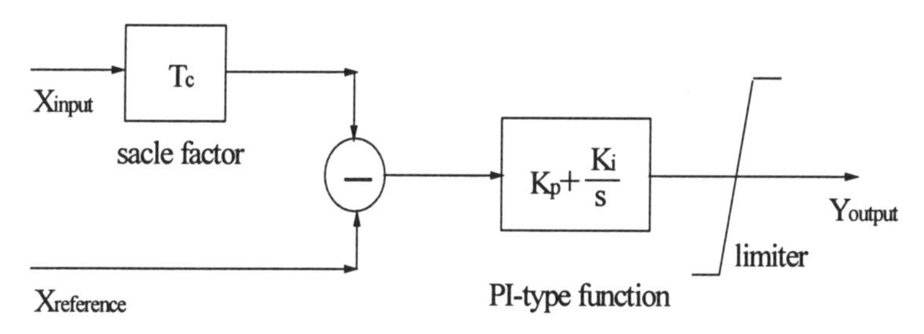

Figure 8.22 PI-type function block

There are three control parts in the UPFC:

(1) The PI controller for the series part of the UPFC controls θ_2 to follow the variation of P under the condition that m_{a2} is kept constant. P from the system is measured through a scaling factor and compared with the reference value. Through PI-type function, the error between the reference and the measured is regulated to null. Generally, the modulation index m_{a2} can also be used to regulate the active power P of the transmission line. However, the regulating range of m_{a2} on P is narrower than that of θ_2 (from Figures 8.20 and 8.21). So θ_2 is chosen to control P only. In this case the capability of the series part is maximum if m_{a2}=1.0. It is important to notice that reactive power Q on the transmission line can not be controlled independently from P through either m_{a2} or θ_2, which means that P can be controlled to the desired value through θ_2, while Q is forced to follow the path defined by Figure 8.20.

(2) The PI controller of the shunt part keeps V_0 constant by regulating m_{a1} while PI control of θ_1 regulates V_{dc}. In this case, the shunt part plays two roles in the operation of the UPFC: one is as dc voltage regulator using θ_1, which must convert the same amount of active power to replace the power that has been drained by the series part placed across the dc link; another is reactive power generator using m_{a1}, which can generate or absorb reactive power from full leading to full lagging in order to increase the feasible range of the connected-bus voltage.

(3) The capacitor on the dc side must remain properly charged in order for both parts of the UPFC to operate according to control objectives. V_{dc} and V_{deref} are the monitored and the pre-set dc voltage, respectively. The two signals are compared and the difference is magnified through a PI controller. The output θ_1 of the PI controller is modulated with a sinusoidal signal synchronised with the system and hence regulates the dc voltage through charging or discharging the dc capacitor.

Three PI controllers need to be designed to give a wide range of feasible solutions according to the operating envelope shown in Figures 8.20 and 8.21. However, it should be pointed out that such PI controllers are designed only to demonstrate the functions of the UPFC. For example, the PI-type function of P–θ_2 should be designed to achieve any desired P within 0^0–360^0 according to Figure 8.22. But the PI-type function in our design can only operate within 0^0–180^0 of θ_2. So other design methods are needed to realize the required functions.

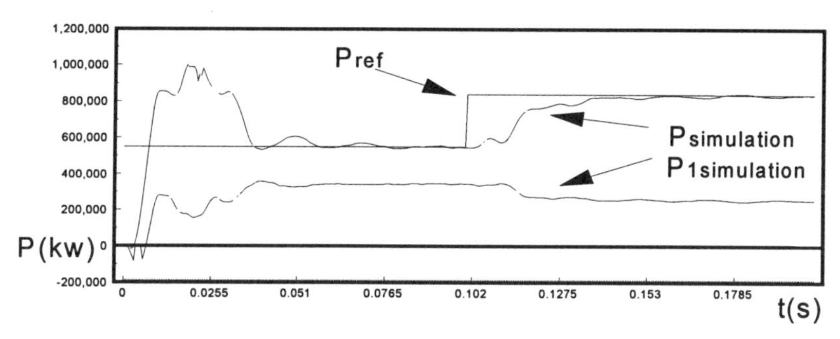

Figure 8.23(a) The simulation results of the internal controller's performance corresponding to variations of the reference

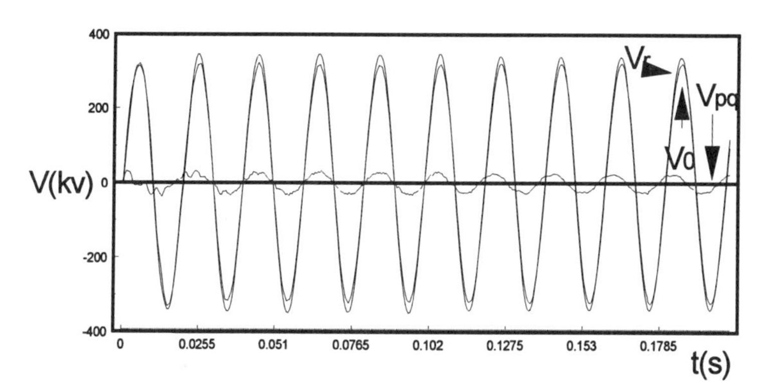

Figure 8.23(b) The simulation results of the internal controller's performance corresponding to variations of the reference

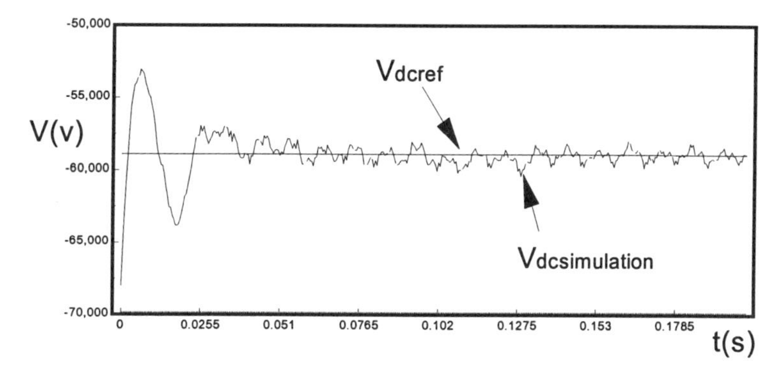

Figure 8.23(c) The simulation results of the internal controller's performance corresponding to variations of the reference

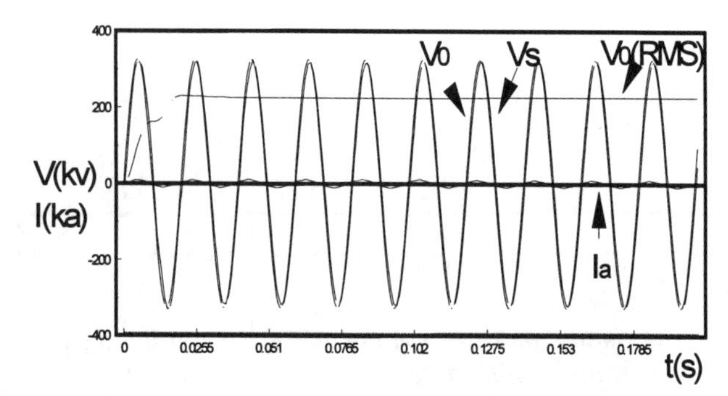

Figure 8.23(d) The simulation results of the internal controller's performance corresponding to variations of the reference

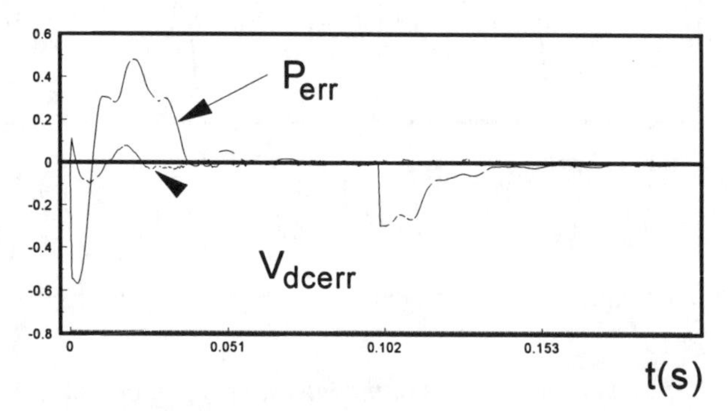

Figure 8.23(e) The simulation results of the internal controller's performance corresponding to variations of the reference

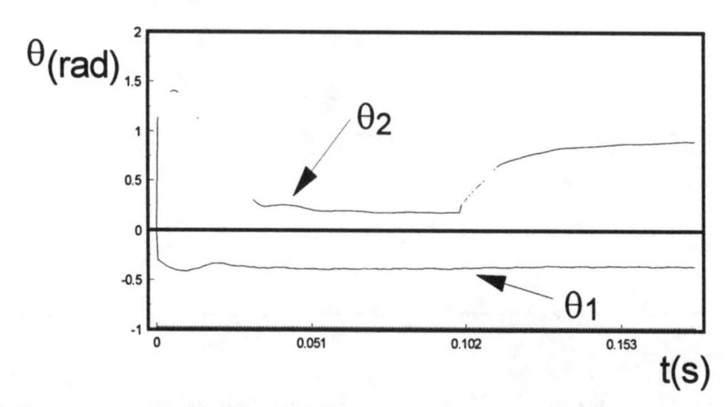

Figure 8.23(f) The simulation results of the internal controller's performance corresponding to variations of the reference

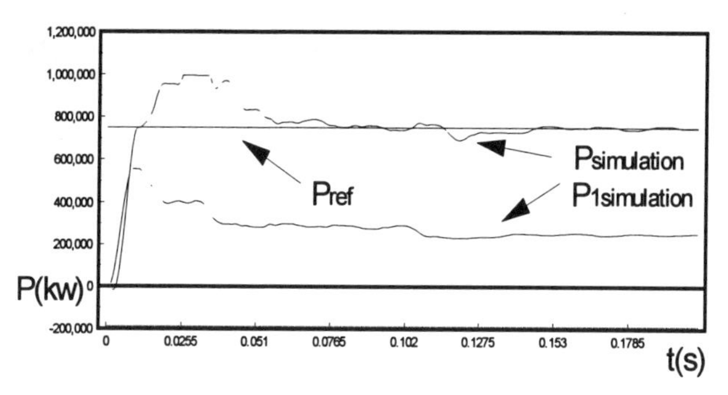

Figure 8.24(a) The simulation results of the internal controller's performance corresponding to variations of the reference

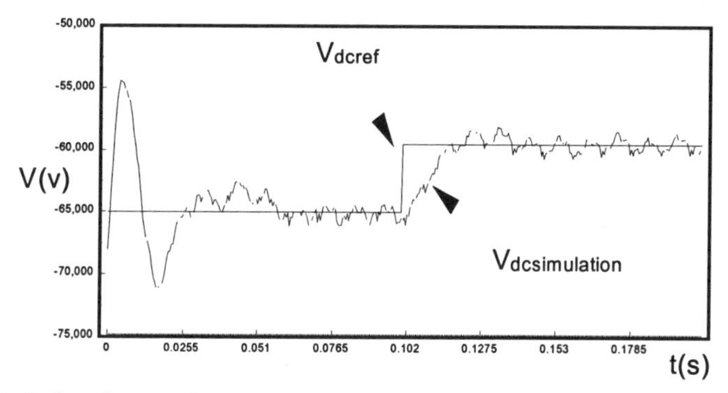

Figure 8.24(b) The simulation results of the internal controller's performance corresponding to variations of the reference

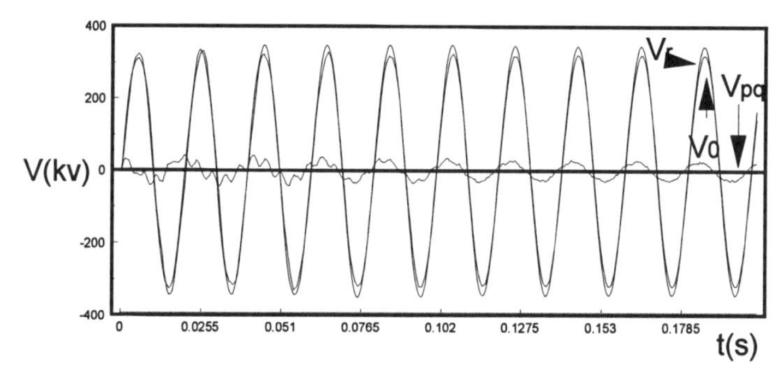

Figure 8.24(c) The simulation results of the internal controller's performance corresponding to variations of the reference

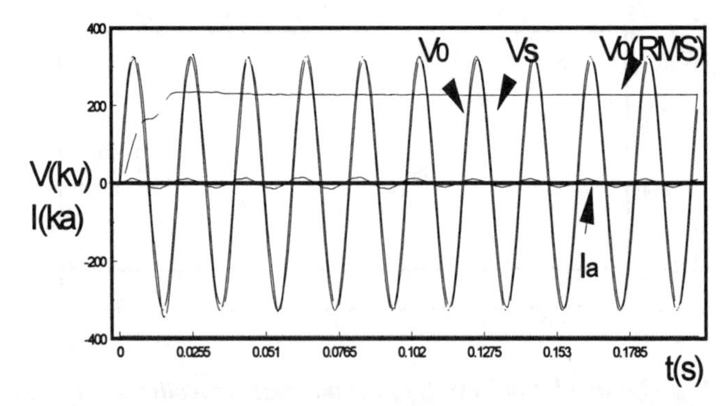

Figure 8.24(d) The simulation results of the internal controller's performance corresponding to variations of the reference

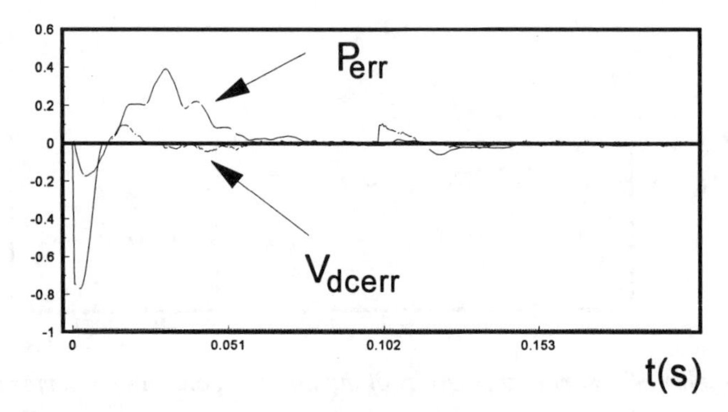

Figure 8.24(e) The simulation results of the internal controller's performance corresponding to variations of the reference

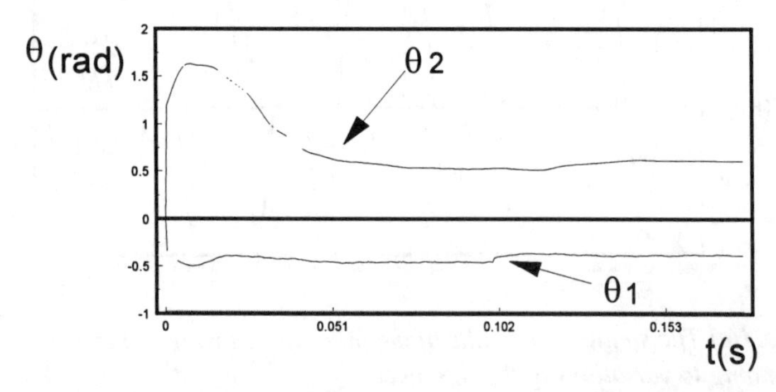

Figure 8.24(f) The simulation results of the internal controller's performance corresponding to variations of the reference

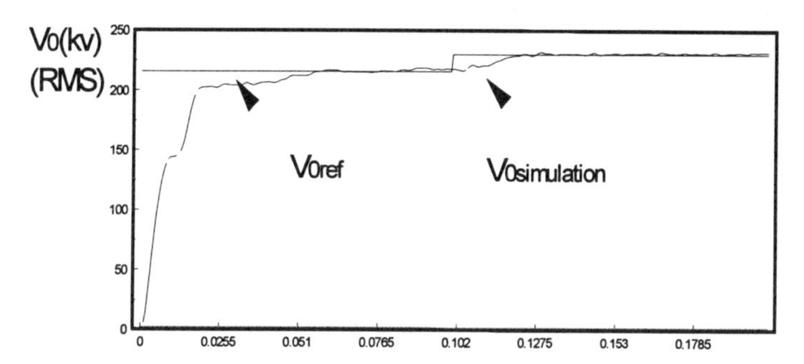

Figure 8.25(a) The simulation results of the internal controller's performance corresponding to variations of the reference

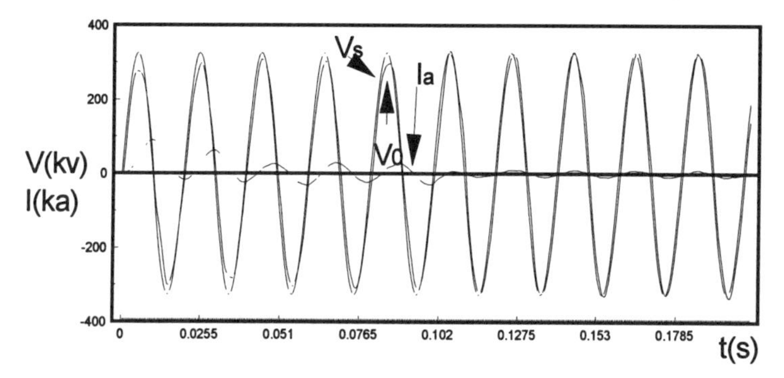

Figure 8.25(b) The simulation results of the internal controller's performance corresponding to variations of the reference

In order to demonstrate the performance of the controller, three simulation results are shown in Figures 8.23, 8.24, and 8.25 where $P_{\text{simulation}}$, $Q_{\text{simulation}}$, P_{ref}, $V_{0\text{simulation}}$, and $V_{0\text{ref}}$ represent phasor quantities, θ_1 and θ_2 are outputs of PI functions, and P_{err} and V_{dcerr} are the error to be controlled to null. The performances of control function and system impacts are classified into three parts:

(1) In Figure 8.23, V_{dc} and V_0 references are kept constant during the simulation, and the references to P change from one steady-state to another at time 0.1s. By control of θ_2 tracing the change of P, P_{err} is quickly controlled to null during the control period although P jumps to 820,000 kW from 585,000 kW at 0.1s. In this case, θ_1 used to maintain dc voltage constant is nearly unchanged.

(2) In Figure 8.24, the reference voltage V_{dcref} for V_{dc} varies at 0.1s while V_0 and P are kept constant. Using PI-type control, V_{dc} can easily and quickly be modulated even with a large change in the reference value.
(3) In Figure 8.25, only V_0 changes with time, and I_a comes from lagging in phase with V_0.

In conclusion, it can be clearly seen that under these different conditions the UPFC is effectively controlled to trace the variations of the system, and the transient process from one steady state to another state is very short, within several cycles. In the closed-loop control demonstrations of PI-type function, the problems of power flow control and voltage support are also solved simultaneously. The whole control process is stable and the controller is robust and operates under different scenarios.

8.6 Conclusions

This chapter has described some research results of EMTP-based digital studies of SPWM UPFC. Useful insights into UPFC performance have been attained through detailed analysis and simulation of the UPFC internal structure and regulation method. Voltage and power control simulation results have demonstrated the functions of the UPFC. Studies have also indicated that PWM is a potentially promising method for the effective regulation of the UPFC.

The whole process of setting up an EMTP based UPFC model has been discussed in detail. Firstly, SPWM scheme generation and the main factors affecting the functions and all aspects concerned with the model have been mathematically explained. Secondly, how the series part and shunt part of the SPWM UPFC are controlled and coordinated has been analyzed. Simulation results of SPWM UPFC regulation performance and open-loop control have been given to validate the described theory. The simulator has further been used to investigate the operating envelope and closed-loop control. This EMTP-based simulation of SPWM UPFC implementation has provided a useful tool to assist in the development and validation of more detailed and practical models of the UPFC for further studies.

8.7 Acknowledgment

The authors gratefully acknowledge the support of NGC and useful discussions with Dr A M Foss, Mr M Allison and other NGC engineers.

8.8 References

1 Arrillaga, J. and Smith, B., *AC-DC power system analysis*, IEE, 1998

2 Vasconelos, A.N., Ramos, A.J.P., Monteiro, J.S., Lima, M.V.B.V., Silva, H.D., and Lins, L.R., "Detailed modeling of an actual static var compensator for electromagnetic transient studies", *IEEE Trans Power Systems*, 7, (1), pp.11-19, 1992.

3 Lefebvre, S. and Gerin-Lajoie, L., "A static compensator model for the EMTP", *IEEE Trans Power Systems*, 7, (2), pp.477-484, 1992.

4 Tenorio, A.R.M., Ekanayake, J.B., and Jenkins, N., "Modelling of FACTS Devices", *The sixth international conference on AC and DC transmission, IEE*, UK, pp.340-345, 29 April-3 May 1996.

5 Povh, D. and Mihalic, R., "Simulation of power electronic equipment", EPRI Workshop, 1994

6 Song, Y.H. and Liu, J.Y., "Electromagnetic transient simulation studies of sinusoidal PWM unified power flow controllers", *European Transactions on Electrical Power*, 8, (3), pp.187-192, 1998.

7 Gyugyi, L., "Unified power flow control concept for flexible ac transmission systems", *Proc IEE, Pt.C*, 139, (4), pp.323-331, 1992.

8 Working Group 37.16 of CIGRE, "Interaction of system planning in the development of the transmission network of the 21st century", 210-02, CIGRE Symposium, Power Electronics in Electric Power Systems, Tokyo, 1995.

9 Mohan, N., Undeland, T.M., and Robbins, W.P., *Power electronics: converters, applications, and design*, John Wiley & Sons, 1992

10 Gyugyi, L., Schauder, C.D., Williams, S.L., Rietman, T.R., Torgerson, D.R., and Edris, A., "The unified power flow controller: a new approach to power transmission control", *IEEE Trans. on Power Delivery*, (2), pp.1085-1097, 1995.

11 Moran, L.T., Ziogas, P.D., and Joos, G., "Analysis and design of a three phase synchronous solid state var compensator", *IEEE Trans. on IA*, 25, (4), pp.598-608, 1989.

12 Ooi, B.T., Wang, X., "Voltage angle lock loop control of the boost type PWM converter for HVDC application", *IEEE Trans. on PE*, 55, (2), pp.229-235, , 1990.

13 Hatziadoniu, C.J. and Funk, A.T., "Development of a control scheme for a series-connected solid-state synchronous voltage source", IEEE summer meeting 405-1, 1995

14 Ooi, B.T., Dai, S.-Z., and Galiaua, F.D., "A solid state PWM phase shifter", *IEEE Trans. on Power Delivery*, 8, (2), pp.573-579, 1993.

15 Holtz, J., "Modulation Pulsewidth – A Survey", *IEEE Trans. on Industrial Electronics*, 39, (5), pp.410-420, 1992.

16 *EMTP/ATP Rule Book*, version 1.0.

Chapter 9

Steady-state analysis and control

Y.H. Song and J.Y. Liu

9.1 Introduction

Advances in high-power solid-state devices and control technologies have led to the development of a new range of flexible ac transmission systems (FACTS) devices. Thyristor-controlled static var compensators (SVC) are already in use, and phase shifters (PS) and unified power flow controllers (UPFC) are currently under development. The SVC is a thyristor switched parallel capacitor/inductor circuit used to supply leading and lagging reactive power compensation for voltage control. The PS (with many different types) can be used to control power flow by adding a voltage which is out of phase with the line voltage. The UPFC uses solid state synchronous voltage sources for both shunt and series compensation and for phase angle control. The UPFC thus provides simultaneous control of voltage and power flow, along with reactive power compensation. In addition, these devices can be used for dynamic and transient stability improvement.

In order to effectively investigate the impact of FACTS devices on power systems, their modelling and implementation in power system software is essential. Generally speaking, there are three types of modelling: (i) electromagnetic models for detailed equipment level investigation [1–3], as described in Chapter 8; (ii) steady-state models for system steady state operation evaluation [4–14], which is the topic of this chapter; and (iii) dynamic models for stability studies [12–15], which will be discussed in Chapters 10 and 11. Although various studies have been made on steady-state analysis and control aspects, it is widely recognised that conventional power flow algorithms are convergence failure-prone when applied to power systems with embedded FACTS devices. In particular, this chapter will use the UPFC as an example device to illustrate four aspects of steady-state analysis in FACTS studies: (1) steady-state modelling of FACTS devices; (2) power flow and controlled power flow; (3) constraints handling and (4) comparison studies of different FACTS devices.

Most of the steady-state models of the UPFC treat the UPFC either as one series voltage source and one shunt current source, or as two voltage sources (both the series and the shunt are represented by voltage sources). A generalized power flow controller (GPFC) has also been proposed to represent various power flow control means with a unified mathematical model and a unified data format [11]. The first part of this chapter focuses on development of a steady-state UPFC model and its implementation in a power flow algorithm. Firstly, power injection transformation of a two voltage source UPFC model is derived in rectangular form. When formulated in rectangular form, the power flow equations are quadratic. Some numerical advantages can be obtained from the form. The rectangular form also leads naturally to the idea of an "optimal multiplier". After detailed considerations of issues in implementation of UPFC in power flow by various power flow algorithms, an improved method based on optimal multiplier [20, 21] for ill-conditioned systems is adopted. The proposed UPFC model and power flow algorithm have been programmed and vigorously tested in a number of systems.

With the application of such controllable devices, the need arises to ascertain the most appropriate power system control strategy both for an individual UPFC, and for their global co-ordination across a system. Controllable shunt compensation schemes, such as the SVC, have been extensively used on transmission systems for the past decade, and the appropriate power system control strategies are now well established. Control strategies for controllable series devices, however, are significantly less developed, particularly for meshed power systems. Such devices are intrinsically different from shunt devices in that they potentially have a much greater effect on power flows, and must be able to accommodate the loss of the circuit into which the device is connected. In this respect, [17] presents an optimal power flow method to derive UPFC control to regulate line flow and at the same time to minimise power losses. A Newton-type algorithm for the control of power flow in electrical power networks has been developed in [5, 6]. Furthermore, one of the practical questions is "what kind of control should a UPFC take when its internal limits are violated?". Reference [16] presents a strategy to alleviate some constraints of UPFC with the objective of achieving maximum power transfer. The method of [16] uses θ_{pq} to alleviate the violations of these limits, which does not consider the participation of V_{pq} and V_{sh}. Thus, it loses some degrees of control freedom. When some limit is violated by a variable, it is necessary to set the violation variable at the limit value in order to achieve the highest efficiency of UPFC. In this situation, the original control performance may deteriorate and the original control objective should transform into another type, that is, the limit values of these violation variables are specified as the control objectives. Thus based on the proposed optimal multiplier based power flow for UPFC studies and a power injection model, the second part of the chapter proposes the use of a power injection model to derive control parameters for UPFC to achieve the required line active power control and bus voltage support. The proposed method offers a number of advantages: it does not change the symmetrical structures of the Jacobian matrix, avoids the initialisations of control parameters and can cover a wide control range of UPFC due to the characteristics of optimal multiplier power flow algorithms employed. Furthermore, the impact of device limit constraints is discussed in detail, including series injection voltage magnitude, line current through the series inverter, real power transfer between the shunt inverter and series inverter, shunt side current and shunt injection voltage magnitude.

To evaluate the effectiveness and benefits of various FACTS devices, it is essential to perform some comparison studies [2], particularly at the planning stage. The concept of security regions and optimal power flow based scalar measures are used in reference [3] to compare the impact of various FACTS devices on the behaviour of power systems. The chapter also compares the ratings and control flexibility of three major FACTS devices. Finally, the chapter describes some numerical results on two practical systems, which clearly illustrate the effectiveness of the proposed algorithm.

9.2 Steady-state UPFC model for power flow studies

9.2.1 Principles of UPFC

The UPFC consists of two switching converters, as shown in Figure 9.1(a). These converters are operated from a common dc link provided by a dc storage capacitor. Inverter 2 provides the power flow control of the UPFC by injecting an ac voltage V_{pq} with controllable magnitude and phase angle in series with the transmission line via a series transformer. Inverter 1 is to supply or absorb the real power demand by inverter 2 at the common dc link. It can also generate or absorb controllable reactive power and provide shunt reactive power compensation.

9.2.2 Steady-state UPFC representation

Studies show that the UPFC can be represented in steady-state by the two voltage sources with appropriate impedances as shown in Figure 9.1(b). The voltage sources can then be represented by the relationship between the voltages and amplitude modulation ratios, and phase shifts of UPFC. In this model, the shunt transformer impedance and the transmission line impedance including the series transformer impedance are assumed to be constant. No power loss is considered within the UPFC. However, the proposed model and algorithm can easily include these when required.

9.2.3 Power injection model of UPFC

The two voltage source model of UPFC is converted into two power injections in rectangular form for power flow studies. The advantage of power injection representation is that it does not destroy the symmetric characteristics of the admittance matrix. When formulated in rectangular form, the power flow equations are quadratic. Some numerical advantages can be obtained from the form. The rectangular form also leads naturally to the idea of an "optimal multiplier", which will be discussed in Section 9.3.

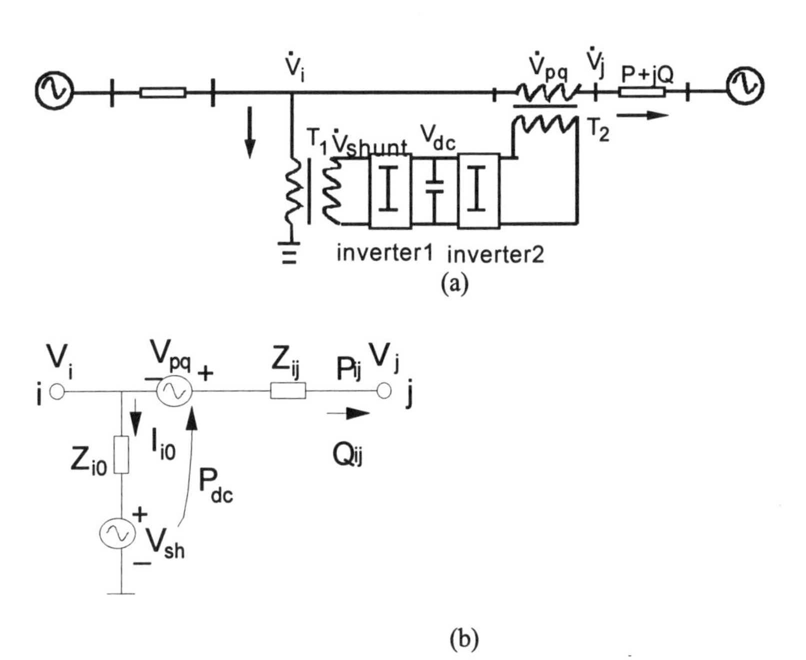

Figure 9.1 The UPFC structure and its two voltage source model

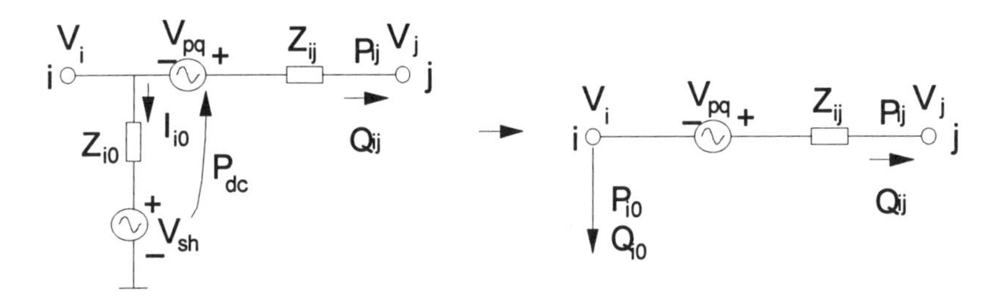

Figure 9.2 The shunt side of UPFC is converted into power injection at busbar i only

From Figure 9.2, the first step is to transform the shunt side of UPFC into a power injection at busbar *i* only. Thus,

$$S_{i0} = P_{i0} + jQ_{i0} = \dot{V}_i\left(\frac{\dot{V}_i - \dot{V}_{sh}}{Z_{i0}}\right)^* \tag{9.1}$$

$$P_{i0} = G_{i0}(V_i^2 - e_i e_{sh} - f_i f_{sh}) + B_{i0}(e_i f_{sh} - f_i e_{sh}) \tag{9.2}$$

$$Q_{i0} = G_{i0}(e_i f_{sh} - f_i e_{sh}) + B_{i0}(e_i e_{sh} + f_i f_{sh} - V_{sh}^2) \tag{9.3}$$

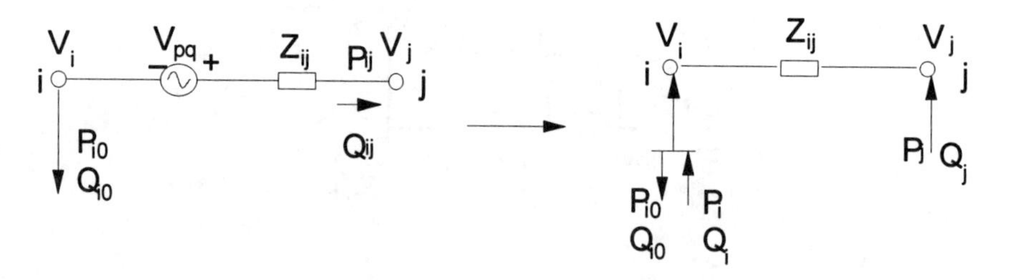

Figure 9.3 The series side of UPFC is converted into two power injections at buses i and j

The second step is to convert series source of UPFC into two power injections at both busbars i and j, which is shown in Figure 9.3. Therefore, we have

$$S_i = P_i + jQ_i = \dot{V}_i\left(-\frac{\dot{V}_{pq}}{Z_{ij}}\right)^* \tag{9.4}$$

$$P_i = G_{ij}(-e_i e_{pq} - f_i f_{pq}) + B_{ij}(e_i f_{pq} - f_i e_{pq}) \tag{9.5}$$

$$Q_i = G_{ij}(e_i f_{pq} - f_i e_{pq}) + B_{ij}(e_i e_{pq} + f_i f_{pq}) \tag{9.6}$$

$$S_j = P_j + jQ_j = \dot{V}_j\left(\frac{\dot{V}_{pq}}{Z_{ij}}\right)^* \tag{9.7}$$

$$P_j = G_{ij}(e_j e_{pq} + f_j f_{pq}) + B_{ij}(f_j e_{pq} - e_j f_{pq}) \tag{9.8}$$

$$Q_j = G_{ij}(f_j e_{pq} - e_j f_{pq}) - B_{ij}(e_j e_{pq} + f_j f_{pq}) \tag{9.9}$$

$$P_{dc} = \mathrm{Re}\left[\dot{V}_{pq}\left(\frac{\dot{V}_i + \dot{V}_{pq} - \dot{V}_j}{Z_{ij}}\right)^*\right]$$

$$= G_{ij}(V_{pq}^2 + e_i e_{pq} - e_j e_{pq} + f_i f_{pq} - f_j f_{pq}) + B_{ij}(e_i f_{pq} - e_j f_{pq} - f_i e_{pq} + f_j e_{pq}) \tag{9.10}$$

where S_{i0} is the power injection of shunt side of the UPFC at the i busbar, P_{dc} is the power transfer from shunt side to series side, S_i and S_j are two power injections transformed from the series voltage at i and j busbars. When power loss inside the UPFC is neglected, i.e

$$P_{i0} = P_{dc} \tag{9.11}$$

then

$$S_{i(inj)} = S_i - S_{i0} \tag{9.12}$$

$$P_{i(inj)} = G_{ij}(-V_{pq}^2 - 2e_i e_{pq} - 2f_i f_{pq} + e_j e_{pq} + f_j f_{pq}) + B_{ij}(e_j f_{pq} - f_j e_{pq}) \tag{9.13}$$

$$\begin{aligned} Q_{i(inj)} &= G_{ij}(e_i f_{pq} - f_i e_{pq}) + B_{ij}(e_i e_{pq} + f_i f_{pq}) \\ &\quad - G_{i0}(-f_i e_{sh} + e_i f_{sh}) + B_{i0}(V_i^2 - e_i e_{sh} - f_i f_{sh}) \end{aligned} \tag{9.14}$$

$$S_{j(inj)} = S_j \tag{9.15}$$

$$P_{j(inj)} = G_{ij}(e_j e_{pq} + f_j f_{pq}) + B_{ij}(f_j e_{pq} - e_j f_{pq}) \tag{9.16}$$

$$Q_{j(inj)} = G_{ij}(f_j e_{pq} - e_j f_{pq}) - B_{ij}(e_j e_{pq} + f_j f_{pq}) \tag{9.17}$$

Thus, two power injections $(P_{i(inj)}, Q_{i(inj)})$ and $(P_{j(inj)}, Q_{j(inj)})$ represent all features of the steady-state UPFC model.

Normally the UPFC control parameters are given in polar form which is more intuitive because the state variables are voltage magnitudes and angles, and have physical meaning. They can be easily transformed from one form to another by the following relationship:

$$V_{pq} = \sqrt{e_{pq}^2 + f_{pq}^2} \tag{9.18}$$

$$\theta_{pq} = a\tan\left(\frac{f_{pq}}{e_{pq}}\right) \tag{9.19}$$

$$V_{sh} = \sqrt{e_{sh}^2 + f_{sh}^2} \tag{9.20}$$

$$\theta_{sh} = a\tan\left(\frac{f_{sh}}{e_{sh}}\right) \tag{9.21}$$

9.3 Representation of UPFC for power flow

9.3.1 UPFC modified Jacobian matrix elements

In power flow, the two power injections $(P_{i(inj)}, Q_{i(inj)})$ and $(P_{j(inj)}, Q_{j(inj)})$ of a UPFC can be treated as generators. However, because they vary with the connected busbar voltage amplitudes and phases, the relevant elements of the Jacobian matrix will be modified at each iteration. The formation of Jacobian matrix is $J = \begin{bmatrix} H & N \\ M & L \end{bmatrix}$ for PQ bus. Based on the equations (9.13), (9.14), (9.16), and (9.17), the following additional elements of the Jacobian matrix owing to the injections of the UPFC at the i and j busbars can be derived:

for i busbar, $i \neq j$ then

$$\Delta H_{ij} = \frac{\partial P_{i(inj)}}{\partial f_j} = G_{ij} f_{pq} - B_{ij} e_{pq} \tag{9.22}$$

$$\Delta N_{ij} = \frac{\partial P_{i(inj)}}{\partial e_j} = G_{ij} e_{pq} + B_{ij} f_{pq} \tag{9.23}$$

$$\Delta M_{ij} = \frac{\partial Q_{i(inj)}}{\partial f_j} = 0 \tag{9.24}$$

$$\Delta L_{ij} = \frac{\partial Q_{i(inj)}}{\partial e_j} = 0 \tag{9.25}$$

when $i=j$

$$\Delta H_{ii} = \frac{\partial P_{i(inj)}}{\partial f_i} = -2 G_{ij} f_{pq} \tag{9.26}$$

$$\Delta N_{ii} = \frac{\partial P_{i(inj)}}{\partial e_i} = -2 G_{ij} e_{pq} \tag{9.27}$$

$$\Delta M_{ii} = \frac{\partial Q_{i(inj)}}{\partial f_i} = -G_{ij} e_{pq} + B_{ij} f_{pq} + G_{i0} e_{sh} + 2 B_{i0} f_i - B_{i0} f_{sh} \tag{9.28}$$

$$\Delta L_{ii} = \frac{\partial Q_{i(inj)}}{\partial e_i} = G_{ij} f_{pq} + B_{ij} e_{pq} - G_{i0} f_{sh} + 2B_{i0} e_i - B_{i0} e_{sh} \tag{9.29}$$

for j busbar, $j \neq i$

$$\Delta H_{ji} = \frac{\partial P_{j(inj)}}{\partial f_i} = 0 \tag{9.30}$$

$$\Delta N_{ji} = \frac{\partial P_{j(inj)}}{\partial e_i} = 0 \tag{9.31}$$

$$\Delta M_{ji} = \frac{\partial Q_{j(inj)}}{\partial f_i} = 0 \tag{9.32}$$

$$\Delta L_{ji} = \frac{\partial Q_{j(inj)}}{\partial e_i} = 0 \tag{9.33}$$

when $j=i$

$$\Delta H_{jj} = \frac{\partial P_{j(inj)}}{\partial f_j} = G_{ij} f_{pq} + B_{ij} e_{pq} \tag{9.34}$$

$$\Delta N_{jj} = \frac{\partial P_{j(inj)}}{\partial e_j} = G_{ij} e_{pq} - B_{ij} f_{pq} \tag{9.35}$$

$$\Delta M_{jj} = \frac{\partial Q_{j(inj)}}{\partial f_j} = G_{ij} e_{pq} - B_{ij} f_{pq} \tag{9.36}$$

$$\Delta L_{jj} = \frac{\partial Q_{j(inj)}}{\partial e_j} = -G_{ij} f_{pq} - B_{ij} e_{pq} \tag{9.37}$$

9.3.2 Normal (open-loop) and controlled (close-loop) power flow with UPFC

There are two aspects when handling the UPFC in steady state analysis: (i) When the control parameters of UPFC are given (i.e. $e_{\mathrm{sh}} f_{\mathrm{sh}}$ $e_{\mathrm{pq}} f_{\mathrm{pq}}$ are given), a power flow programme is used to evaluate the impact of the given UPFC on the system under various system conditions. In this case, UPFC is operated in an open-loop form. The corresponding power flow is treated as normal power flow. This will be described in Section 9.4. (ii) As UPFC can be used to control line flow and bus voltage, control techniques are needed to derive the UPFC control parameters to achieve the required objective. In this case, the UPFC is operated in a close loop form. The corresponding power flow is called controlled power flow, which will be described in Section 9.5.

9.4 Implementation of UPFC in power flow studies

9.4.1 Difficulties with implementation of UPFC in power flow

The implementation of UPFC models in power flow is essentially a controlled power flow problem. There are four basic ways to handle this problem. They are: (i) Error feedback adjustment, which involves modifications of a control variable to maintain another functionally dependent variable at a specified value. (ii) Automatic adjustment, in which the control parameters are directly considered as independent variables so that the resulting Jacobian matrix will be enlarged. (iii) Sensitivity-based adjustment, which is derived from Taylor series expansion of the perturbed system equations around the initial operating point and (iv) Recently, user defined model approach, which has been developed in a number of commercial packages, provides the facilities for the user to build their own models. However, there are difficulties in handling UPFCs because of the following two reasons.

On one hand, this is due to the following peculiarities existing in the UPFC: (i) power transfer between two inverters; (ii) wide regulation range of control parameters; (iii) non-linear multivariable control and (iv) stressed conditions of the system with UPFC. All these make the power flow structure very complex and can lead to an increase of iterations or even divergence of power flow.

On the other hand, this is because: (i) Sensitivity-based adjustment can only work in a very narrow linear range around the initial operating point. (ii) User defined model has no way of dealing with complicated relationship of two converters of UPFC. UPFC modelling requires the change of relevant elements of Jacobian matrix. However, user-defined power flow softwares do not allow users to directly modify the Jacobian matrix and only provide facilities for the iteration between the main program and the user defined model. This iteration sometimes diverges, especially when the system is heavily loaded or ill-conditioned. (iii) Although an automatic adjustment method has a good convergence, it needs much work to reconstruct the Jacobian matrix. Some useful properties such as symmetry and sparsity disappear. It also takes more steps to achieve convergence.

9.4.2 Optimal multiplier power flow algorithm

In this section, optimal multiplier power flow algorithm [20, 21] is adopted as it offers a number of advantages in handling ill-conditioned power flow, including: (i) The optimal multiplier acts as an adaptive gain, which reduces if the Jacobian becomes ill-conditioned. In this way it can give the maximum rate of convergence at each iteration. (ii) The optimal multiplier power flow method can be used to detect the distance between the desired operating point and the closest unfeasible point. Thus it provides a measure of degree of controllability. (iii) It can provide computational efficiency without destroying the advantages of conventional power flow when used together with error-feedback adjustment to implement UPFC model. In this section, a brief description is given.

Consider the power flow equations for an N busbar system:

$$S = F(X) \tag{9.38}$$

where S is a vector of the power injections at all buses except the slack bus.

$$S = [P_1,\ldots,P_M,P_{M+1},\ldots,P_{N-1},Q_1,\ldots,Q_M,V_{M+1}^2,\ldots,V_{N-1}^2] \tag{9.39}$$

where M represents the number of PQ bus, N–M–1 is the number of PV bus.

When X is expressed using rectangular coordinates of bus voltages, we have

$$X = [e_1,\ldots,e_{N-1},f_1,\ldots,f_{N-1}] \tag{9.40}$$

and f is the function of the bus power balance constraints.

$$F = [f_{p1},\ldots,f_{pN-1},f_{q1},\ldots,f_{qM},f_{qM+1},\ldots,f_{qN-1}] \tag{9.41}$$

The tth iteration of the power flow solves the equations:

$$\Delta X^t = -J(X^t)^{-1}[F(X^t) - S] \tag{9.42}$$

$$X^{t+1} = X^t + \mu\Delta X^t \tag{9.43}$$

where $J(X^t)$ is the Jacobian matrix at the tth iteration.

The μ in equation (9.43) is a scalar multiplier used to control the updating of variables at each iteration. In the traditional Newton-Raphson algorithm, μ=1 at each iteration. With Optimal Multiplier Power flow Algorithm, a scalar 'optimal multiplier' μ is chosen to minimize a cost function, shown as equation (9.43), in the direction given by ΔX^t so that the updates of variables at each iteration converge in a optimal way to the solution point. In this respect, the cost function is set to equal one half the norm of the power flow mismatch equations:

$$Goal(X^{t+1}] = \frac{1}{2}[f(X^t + \mu\Delta X^t) - S]^T [f(X^t + \mu\Delta X^t) - S] \tag{9.44}$$

Then μ is determined by the following steps:

Let

$$a = [a_1, \ldots, a_N]^T = S - F[X_t] \tag{9.45}$$
$$b = [b_1, \ldots, b_N]^T = -J\Delta X \tag{9.46}$$
$$c = [c_1, \ldots, c_N]^T = -F(\Delta X) \tag{9.47}$$

Then the equation (9.44) becomes:

$$Goal = \sum_{i=1}^{N}(a_i + \mu b_i + \mu^2 c_i)^2 = Goal(\mu) \tag{9.48}$$

In order to solve the minimal value of the objective function Goal, one solves:

$$\frac{\mathrm{d}Goal}{\mathrm{d}\mu} = 0 \tag{9.49}$$

Therefore,

$$g_0 + g_1\mu + g_2\mu^2 + g_3\mu^3 = 0 \tag{9.50}$$

where

$$g_0 = \sum_{i=1}^{N} (a_i b_i)$$

$$g_1 = \sum_{i=1}^{N} (b_i^2 + 2 a_i c_i)$$

$$g_2 = 3\sum_{i=1}^{N} (b_i c_i)$$

$$g_3 = 2\sum_{i=1}^{N} c_i^2$$

Thus the equation (9.50) can be solved easily by the Cardan's formula analytically or Newton's method numerically.

9.4.3 Power flow procedure with UPFC

The overall procedures for the proposed algorithm can be summarised as:

Step 1: Input data needed by conventional power flow; Order busbar optimally; Form admittance matrix; Input UPFC series and shunt voltage amplitudes and phases;

Step 2: Form conventional Jacobian matrix; Modify the Jacobian matrix using UPFC injection elements to become the enhanced Jacobian matrix according to equations (9.22) - (9.37);

Step 3: Use the enhanced Jacobian matrix to solve busbar voltage until the convergence of all power injections is achieved. In this step, the optimal multiplier μ is calculated using Newton method to solve equation (9.41). When the mismatch at every busbar is less than prescribed error, the power flow converges. Otherwise go to Step 2;

Step 4: Output system voltages and line flows; Display various information about the UPFC.

9.5 Power injection based power flow control method

9.5.1 General concepts

For relieving thermal and voltage transmission constraints (i.e. power flow control and voltage support), an overall UPFC control strategy is envisaged which involves (i) a central controller issuing setpoint voltage and circuit real power flow settings, and (ii) local UPFC controllers to enable the UPFC to achieve the required setpoints. This section describes a novel method for UPFC local control – a power injection model based control co-ordination algorithm (PIM).

In steady modelling of UPFC, the UPFC represented by two voltage sources of series part and shunt part is often transformed into a pair of power injections ($P_{i(inj)}$, $Q_{i(inj)}$) ($P_{j(inj)}$, $Q_{j(inj)}$), as shown in Figures 9.2 and 9.3. From the viewpoint of effects of these power injections on the system, $Q_{i(inj)}$ can be independently regulated to support busbar voltage connected at the shunt part, $P_{i(inj)}$ and $P_{j(inj)}$ are used to manipulate line active power with equal magnitude but at reverse direction and $Q_{j(inj)}$ (when UPFC loss is neglected) can control both the j busbar voltage and line reactive power. Under lossless conditions, the UPFC operating condition can be specified by the following three quantities shown in Figures 9.2 and 9.3: Q_{sh} – reactive power supplied through shunt current; Q_{ser} – reactive power supplied through series voltage injection and P_{dc} – real power supplied across the dc-link from shunt to series converters.

In most applications, the central controller places on the UPFC a requirement to regulate two power system parameters, namely busbar voltage and circuit real power flow. For a UPFC with three degrees of freedom, a spare degree of freedom is available which may be used in an advantageous manner, for instance, to minimise device rating requirements. An initial algorithm, referred to as the PIM algorithm, applies the constraint $Q_{j(inj)}$ =0, where $Q_{j(inj)}$ relates to the power injection model of the UPFC. This enables the PIM algorithm to be formulated based on: (a) use of $Q_{i(inj)}$ to control bus voltage V_i and (b) use of $P_{i(inj)}$ and $P_{j(inj)}$ to control the circuit real power.

9.5.2 Decoupled rectangular co-ordinate power flow equations

Firstly, the model of UPFC as well as its power injection transformation are derived in rectangular form. When formulated in rectangular form, the power flow equations are quadratic. Some numerical advantages can be obtained from the form. The rectangular form also leads naturally to the idea of an "optimal multiplier". In PIM, the foundation is the decoupled rectangular co-ordinate power flow equations (9.51) and (9.52):

$$[\Delta P / e] = [B'][\Delta f] \tag{9.51}$$

$$[\Delta Q / e] = [B''][\Delta e] \tag{9.52}$$

Where, the elements of the matrices $[B']$ and $[B'']$ are the negative of the elements of the imaginary part of bus admittance matrix $[-B]$.

9.5.3 Closed-loop voltage control strategy by reactive power injection

Assuming that the controlled busbar voltage magnitude is V_i^{spe}, then the control strategy is as follows:
Convergence condition:

$$Max\{V_i^{spe} - V_i^k\} \leq \varepsilon \tag{9.53}$$

Iteration process:

$$\Delta Q_{i(inj)}^k = \Delta Q_{i(inj)}^{k-1} - B_{ii} \Delta e_i^k e_i^k \tag{9.54}$$

$$\Delta e_i^k = e_i^k - e_i^{k-1} \tag{9.55}$$

$$e_i^k = V_i^{spe} \cos\theta_i^{k-1} \tag{9.56}$$

$$\theta_i^{k-1} = a\tan\left(\frac{f_i^{k-1}}{e_i^{k-1}}\right) \tag{9.57}$$

where i is the controlled busbar; k is the kth power flow iteration; $\Delta Q_{i(inj)}^k$ is the incremental reactive power injection needed to control busbar voltage at the busbar i; B_{ii} is the busbar admittance of busbar i and ε is the tolerance of controlled voltage .

9.5.4 Closed-loop line transfer active power control strategy by active power injections

Assuming that the control objective of line transfer active power is P_l^{spe}, then the proposed controller is presented as:
Convergence condition:

$$Max\{P_l^{spe} - P_l^k\} \leq \sigma \tag{9.58}$$

Iteration process

$$\Delta P_{i(inj)}^k = \Delta P_{i(inj)}^{k-1} - (P_l^{spe} - P_l^{k-1})/(w_{li}/e_i^{k-1} - w_{lj}/e_j^{k-1}) \tag{9.59}$$

$$\Delta P_{j(inj)}^k = -\Delta P_{i(inj)}^k \tag{9.60}$$

Where l represents the controlled line; $\Delta P_{i(inj)}^k$ is the incremental busbar active power injection at busbar i needed to control line transfer power; $\Delta P_{j(inj)}^k$ is the incremental busbar active power injection at busbar j needed to control line transfer power ; w_{li}, w_{lj} are weight factors for line l transfer active power to busbar active power injections at busbar i and j and σ is the tolerance of controlled line transfer power. The weight factors w_{li}, w_{lj} are formulated from the linear relationship of line active power and busbar active power injections. The incremental value of line l active power as ΔP_l can be expressed in terms of changes in busbar active power injections as follows:

$$\Delta P_l = \frac{\partial P_l}{\partial P_1}\Delta P_1 + \cdots + \frac{\partial P_l}{\partial P_i}\Delta P_i + \cdots + \frac{\partial P_l}{\partial P_n}\Delta P_n \tag{9.61}$$

While the partial derivatives of the line active powers with respect to injections can be expressed:

$$\frac{\partial P_l}{\partial P_i} = \frac{\partial P_l}{\partial \delta_1}\frac{\partial \delta_1}{\partial P_i} + \cdots + \frac{\partial P_l}{\partial \delta_i}\frac{\partial \delta_i}{\partial P_i} + \cdots + \frac{\partial P_l}{\partial \delta_n}\frac{\partial \delta_n}{\partial P_i} \tag{9.62}$$

9.5.5 Solution of UPFC parameters

Based on the results, the pair of power injections $(P_{i(inj)}, Q_{i(inj)})$ and $(P_{j(inj)}, Q_{j(inj)})$ derived from the closed-loop controllers can be used to obtain the values of two voltage sources of the UPFC. During this stage of calculation, these power injections and UPFC control parameters are linked through the equations (9.13, 9.14, 9.16, 9.17) obtained from the transformation process of UPFC models from voltage sources to power injections.

9.6 Control of UPFC constrained by internal limits

9.6.1 The internal limits of UPFC device

In practice, there are a number of constraints imposed by device limits which may affect the capability of the UPFC. In this section, the following five typical limits will be considered:

(i) the series injection voltage magnitude (V_{pq})
(ii) the line current through the series inverter (I_{se})
(iii) the active power transfer between the shunt inverter and series inverter (P_{dc})
(iv) the shunt side current (I_{sh})
(v) the shunt injection voltage magnitude (V_{sh})

Mathematically, these constraints can be expressed as:

$$V_{pq} \leq V_{pq}^{L} \tag{9.63}$$

$$I_{se} = \left| \frac{\dot{V}_i + \dot{V}_{pq} - \dot{V}_j}{Z_{ij}} \right| \leq I_{se}^{L} \tag{9.64}$$

$$P_{dc} = \mathrm{Re}\left[\dot{V}_{pq} \left(\frac{\dot{V}_i + \dot{V}_{pq} - \dot{V}_j}{Z_{ij}} \right)^{*} \right] \leq P_{dc}^{L} \tag{9.65}$$

$$I_{sh} = \left| \frac{\dot{V}_i - \dot{V}_{sh}}{Z_{i0}} \right| \leq I_{sh}^{L} \tag{9.66}$$

$$V_{sh} \leq V_{sh}^{L} \tag{9.67}$$

where superscript L in each equation represents the maximum limit value of the variable.

9.6.2 *Considerations of internal limits in power flow control methods*

In general, the series injection voltage and line constraints are enforced by adjusting the scheduled levels of series P and Q. The maximum current through the series inverter is the line thermal current. While the real power transfer between the shunt inverter and series inverter is strictly an equipment rating. The shunt voltage current and voltage are constrained by the rating of the shunt inverter, which must be at least as large as the real power transfer between the two inverters. Additional capability will be required to provide the reactive current needed to regulate bus voltage. Hence, the shunt inverter current is limited by relaxing the scheduled bus voltage.

All the above limits of UPFC should be enforced in power flow calculations. A simple alternative is that these constraints checked at each iterative step in conventional power flow may be used in the UPFC constraints. If there are violations of these limits, it may slow down convergence, or more seriously, cause the solution to oscillate or even diverge. In order to alleviate these limits, the common practice is either to decrease the scheduled control objective levels or to adjust control parameters. However, this method can not be adopted in the proposed PIM because PIM can only obtain control parameters until the whole procedure of PIM converges at the final iterative step. Therefore, developing another way of handling UPFC limits in PIM is necessary.

When PIM gives the final results of power flow using UPFC control objectives, all constraints of UPFC can be checked. If violations of these constraints occur, the control parameters of UPFC can be adjusted to avoid these limits and thus the original scheduled control objectives are impossible to achieve. The effects of the adjusted parameters of UPFC owing to their constraint violations on the power flow are evaluated through the modified power injection model of UPFC. The way of dealing with these constraints in PIM at the final step has more advantages than that of the conventional way because it avoids the interference of constraints in the process of control objectives and also avoids computation oscillation or divergence.

When PIM ends with power flow results including UPFC parameters, it is quick to check whether these constraints have been violated. If not, PIM outputs with satisfactory scheduled control objectives and UPFC parameters within the ranges of device ratings. Otherwise, PIM will modify UPFC parameters and the associated control objectives.

9.6.3 Strategies for handling the constraints

The following rules are proposed to design UPFC control when the internal limits of UPFC are violated:

(i) If V_{pq} violates the constraint, fix it at its maximum value in the remaining computation process;
(ii) If one of I_{se} and P_{dc} is out of the range of limit, fix it as the limit value and then solve θ_{pq} and/or V_{pq} using the associated equations;
(iii) If both I_{se} and P_{dc} limits are violated, solve V_{pq} and/or θ_{pq} respectively at first, then choose the one which ensures both I_{se} and P_{dc} are within the limits;
(iv) If I_{sh} is beyond the limit while V_{sh} is within the range, relax V_i to alleviate I_{sh} and use V_i to obtain the associated reactive power injection so as to achieve the new specified V_i;
(v) If violation of V_{sh} occurs, set it as the constraint value and then use it to examine whether I_{sh} is out of range or not. If it leads to violation of I_{sh}, repeat step (iv);
(vi) If I_{sh} and V_{sh} both are out of range, use their maximum constraints to solve the new specified V_i.

The above rules can be mathematically implemented by the following control strategies:

(i) Set I_{se}^{L} as control objective

Owing to the limit violation of I_{se}, the modification of θ_{pq} can be formulated from I_{se}^{L} in equation (9.68):

$$\cos(\theta_{pq}-\alpha_1)\le\frac{\cos\alpha_1}{2B_{ij}^2V_{pq}(e_i-e_j)}*\left[(I_{se}^{L})^2-B_{ij}^2V_{pq}^2-B_{ij}^2(e_i-e_j)^2-B_{ij}^2(f_i-f_j)^2\right] \tag{9.68}$$

$$\alpha_1 = a\tan\left(\frac{f_i - f_j}{e_i - e_j}\right) \tag{9.69}$$

Based on equations (9.68) and (9.69), one method to alleviate I_{se} from its limit is to get a new θ_{pq} while keeping V_{pq} constant, this has been demonstrated in [10]. If only θ_{pq} is used to regulate I_{se}, it may have many solutions that satisfy equation (9.68). However, in this situation, UPFC does not operate at I_{se}^L and thus loses its advantage. In order to change the way of θ_{pq} which has limited capability of regulating I_{se}, V_{pq} is taken for granted to regulate I_{se} along with θ_{pq}. Therefore, the more general formulae of modifying V_{pq} and θ_{pq} are derived as follows to alleviate I_{se}:

$$\cos(\theta_{pq} - \alpha_1) = C_{se} \tag{9.70}$$

$$-B_{ij}^2 V_{pq}^2 - C_{se}\left|\frac{2B_{ij}^2(e_i - e_j)}{\cos\alpha_1}\right| V_{pq} + \left[(I_{se}^L)^2 - B_{ij}^2(e_i - e_j)^2 - B_{ij}^2(f_i - f_j)^2\right] = 0 \tag{9.71}$$

$$-1 \le C_{se} \le 1 \tag{9.72}$$

Here, a factor C_{se} is introduced to secure the solution of θ_{pq}, whose value can be specified within the range of [-1, 1]. C_{se} introduced here enables the UPFC to operate at I_{se}^L and makes full use of its rating through both V_{pq} and θ_{pq}. So, once I_{se} is found out of the range, the control objective becomes I_{se}^L. Using equations (9.70)-(9.72), the modified V_{pq} and θ_{pq} change the power injections ($P_{i(inj)}$, $P_{j(inj)}$, $Q_{j(inj)}$) and thus control I_{se} towards I_{se}^L.

(ii) Set P_{dc}^L as control objective

Modification of θ_{pq} can be obtained from P_{dc}^L in equation (9.65) due to the violation of P_{dc}:

$$P_{dc}^L = \left|B_{ij} V_{pq}(e_i - e_j)\frac{\sin(\theta_{pq} + \alpha_2)}{\cos\alpha_2}\right| \tag{9.73}$$

$$\alpha_2 = a\tan\left(\frac{f_j - f_i}{e_i - e_j}\right) \tag{9.74}$$

Similar to C_{se}, another factor C_{dc} is given to obtain both V_{pq} and θ_{pq} so as to alleviate P_{dc}.

$$\sin(\theta_{pq} + \alpha_2) = C_{dc} \tag{9.75}$$

$$V_{pq} = \left| \frac{P_{dc}^L \cos\alpha_2}{C_{dc} B_{ij}(e_i - e_j)} \right| \tag{9.76}$$

$$-1 \le C_{dc} \le 1, \quad C_{dc} \ne 0.0 \tag{9.77}$$

P_{dc}^L will be regarded as the new control objective once P_{dc} violates its limit.

(iii) Set I_{sh}^L as control objective

If I_{sh} violates its limit, we can relax V_i to alleviate it. For example, if it violates its upper limit, we have from equation (9.66):

$$V_i^2 - 2f_{sh}f_i - 2e_{sh}e_i + V_{sh}^2 = \left(\frac{I_{sh}^L}{B_{i0}} \right)^2 \tag{9.78}$$

Then V_i can be obtained from the following quadratic equation:

$$V_i^2 + (-2f_{sh}\sin\theta_i - 2e_{sh}\cos\theta_i)V_i + (V_{sh}^2 - \left(\frac{I_{sh}^L}{B_{i0}} \right)^2) = 0 \tag{9.79}$$

$$\Delta Q_{i(inj)}^k = \Delta Q_{i(inj)}^{k-1} - B_{ii}\Delta e_i^k e_i^k \tag{9.80}$$

Therefore, $Q_{i(inj)}$ is modified by the solved V_i while other injected powers ($P_{i(inj)}$, $P_{j(inj)}$, $Q_{j(inj)}$) are assumed unchanged.

9.7 Test results

9.7.1 Power flow

In order to investigate the feasibility of the proposed technique, a large number of power systems of different sizes and under different system conditions have been tested. All the results indicate good convergence and high accuracy achieved by the proposed method. In this section, a 306-bus practical system has been presented to numerically demonstrate its performance. It consists of 306 buses, 521 lines, 38 generators, 147 transformers (including 35 on-line tap changes) and 171 load buses. The total generation is 11000.0 MW. The system is divided into seven areas. The aim of four UPFC installations in the main voltage sensitive points and the associated transmission lines is to balance power transmissions and improve voltage profile. Many operating conditions have been investigated to achieve the optimal operation of the system with UPFCs. Table 9.1 gives one of the results with specified UPFC parameters compared with the power flow results are without UPFC. Figures 9.4 and 9.5 show the μ and the real power mismatch with four UPFCs. The program converges at 4 iterations.

Table 9.1 Comparison between power flow results with and without UPFCs

UPFC location and parameters	11-23: V_{pq}=0.032∠90.0° V_{sh}=0.974∠-4.21° 36-54: V_{pq}=0.02∠45° V_{sh}=0.99∠3.8° 60-62: V_{pq}=0.015∠180° V_{sh}=1.00∠3.2° 67-68: V_{pq}=0.01∠235.0° V_{sh}=1.012∠7.2°			
power flows without UPFC	V_{11}=1.0215 P_{11-23}=0.8448	V_{36}=0.96302 P_{36-54}=-2.1519	V_{60}=1.03336 P_{60-62}=-0.2902	V_{67}=1.05197 P_{67-68}=-0.4897
Power flows with UPFCs	V_{11}=1.000 P_{11-23}=-7.0819	V_{36}=0.9800 P_{36-54}=-2.3015	V_{60}=1.010 P_{60-62}=-0.2960	V_{67}=1.020 P_{67-68}=-0.1296

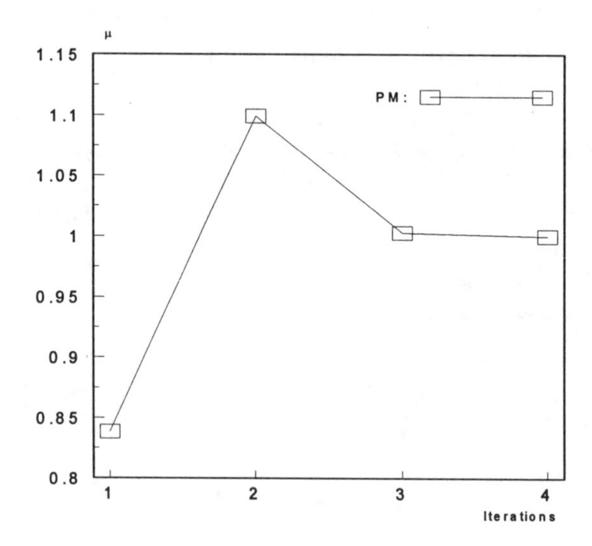

Figure 9.4 Optimal multiplier μ versus iteration number

9.7.2 Controlled power flow

9.7.2.1 Test system and un-reinforced studies

A single-line diagram of the HV circuits of the test system is shown in Figure 9. 6 [23]. It comprises: (i) 13 HV buses with attached load and fixed shunt compensation; (ii) 42 circuits connecting the HV buses, arranged as a combination of single and double circuit lines. All HV circuits have a designated thermal rating and an X/R ratio of 10. All but three of lines can be outaged under contingency conditions; (iii) 10 LV buses with attached generation with transformer connection to HV buses; and (iv) controlled shunt compensation applied to five of the HV buses. There are twelve key double-circuits associated with the boundary A. In order to allow for alternative generation patterns, the generation at each location can be increased by up to 20% from the baseline generation pattern.

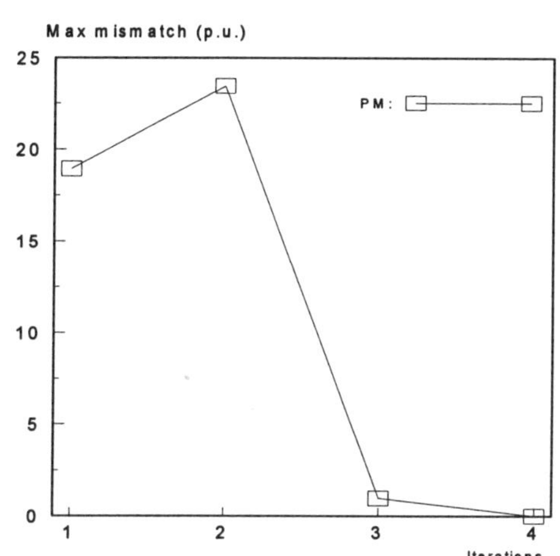

Figure 9.5 Maximum mismatch versus iteration number

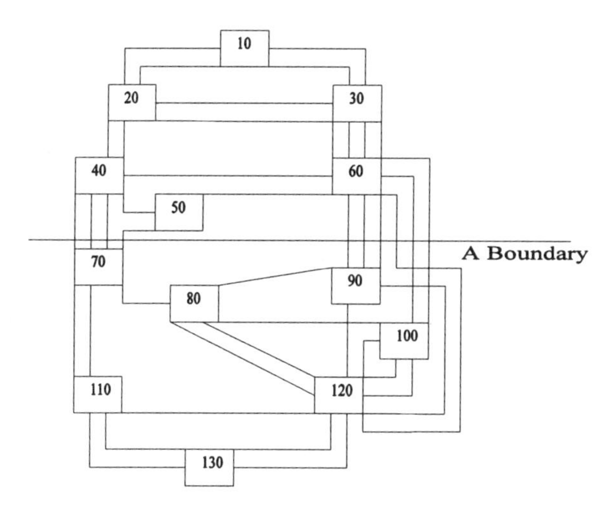

Figure 9.6 The meshed test network

In order to establish a common base for comparison of reinforcement strategies, the following design criteria are applied: (i) All circuit current flows to be within their thermal capabilities under intact, single-circuit outage and double-circuit outage conditions and (ii) All HV bus voltages to be within the range 0.975-1.025pu under intact system conditions. All HV bus voltages to be within the range 0.95-1.025pu under single-circuit and double-circuit outage conditions, and in particular, before any adjustment to generator step-up transformer tap ratios. The power flow results of two operation conditions and outages on the test system are listed in Tables 9.2 and 9.3.

Table 9.2 Power flow results with contingence without UPFC

Case	Outage line	Transfer capability across A boundary	Lowest voltage (p.u)	Violations of line thermal limits and voltage limits
Case 1 (Default)	60-100 60-120	10056MW	V_{90}=0.94126	line 50-60 106.5% line 60—90 104.0%
Case 2 (Scenario A)	40-70	13154MW	V_{50}=0.95764	line 50-60 108.4% line 60-90 108.49%

Table 9.3 Unbalanced power sharing among transmission lines across the boundary A

	Case (without UPFCs)			
	Case 1		Case 2	
Line	Transfer active power (MW)	Percentage of thermal limit	Transfer active power (MW)	Percentage of thermal limit
60—90	592.9	61.60%	634.5	63.91%
60—90	307.7	63.93%	329.3	66.89%
60—90	2002.0	104.00%	2142.6	108.49%
50—60	-1010.3	106.51%	-996.9	108.40%

Default and Scenario A are two typical operating conditions, the former is normal generation-load pattern, the latter is the result of 20% increase in generation of the upper part of the network based on Default which leads to a big increase of power transfer across boundary A. From Tables 9.2 and 9.3, it is seen that the transfer capability across boundary A is limited by violations of line thermal and busbar voltage constraints and the power sharing among the line across boundary A is unbalanced and also there is enough space left for these lines to transfer the given power without exceeding thermal limits. If the system is designed to transfer the given power across boundary A under these operating points, there are two basic ways which can be used: one is to build new transmission lines or replace the line with a new and higher thermal limit transmission line; another is to install control devices to share loads among the boundary lines, in this case, the total transfer capability of this method is determined by the sum of thermal limits of these lines and thus it enhances the system transfer capability. Therefore, UPFCs can expect to play an important role in enhancing the system in this aspect.

All the simulation results of this section are studied based on Cases 1 and 2. The definitions of Cases 1 and 2 are: Case 1 is 'Default' operation condition with outage of two lines 60-100 and 60-120; Case 2 is 'Scenario A' operation condition with outage of line 40-70.

9.7.3 Convergence analysis of controlled power flow

The convergence of the proposed controlled power flow has been tested in a large number of cases under different system conditions and with various numbers of UPFCs. All these test results clearly show the quadratic convergence of the proposed PIM. This is largely because that PIM does not change the features such as optimal bus ordering, sparisity and symmetric properties of Jacobian matrix and quick forward and backward substitutions. Also as discussed in section 7.7.1, the power flow method adopted is based on optimal multiplier algorithms which offers a number of advantages.

As an example, test results illustrate the convergence of the proposed PIM with a UPFC installed along line 90-60 under case 2. The control objective is V_{90}=1.01 p.u. and $P_{90\text{-}60} = -17.0$ p.u.. The actual results of controlled states with a UPFC obtained by PIM are V_{90}=1.00984 p.u. and $P_{90\text{-}60} = -17.034$ p.u. The PIM converges at 12 iterations. Although the initial mismatches of bus power injections are big owing to participation of additional power injections, they tend to converge quickly.

9.7.4 Control performance analysis

In order to verify the proposed PIM, various schemes for controlling busbar voltage and line transfer power have been studied. Some results are shown in Tables 9.4 and 9.5. In both tables, PIM not only shows its satisfactory ability to trace control objectives, but also derives UPFC control parameters directly without any initial assumptions at local modes. A single UPFC installed along line 90-60 under Case 2 is studied to present the performances of PIM using different control objectives and the mappings of power injections of PIM and real parameters of UPFC, which can provide information about rating of UPFC. The different control objectives are designed to increase or decrease the specified value around the operating points. Under Case 2, V_{90}=0.96884 p.u., $P_{90\text{-}60}$=-20.786 p.u. Four control cases are demonstrated in Table 9.4, which gives the actual results of controlled states and its UPFC parameters. From the table, it is seen that the PIM can handle a wide range of control objectives without losing accuracy. For instance, the amplitude of V_{pq} can vary from 0.0498 to 0.1291 p.u. and its phase changes from 81.30^0 to 261.45^0, which is difficult to handle by conventional methods. Furthermore, Table 9.5 presents the mapping of power injections of the proposed method and the parameters of UPFC, which justifies effectiveness of the PIM. Although the power injections are obtained through decoupled control of busbar voltage and line active power, PIM links these power injections using the UPFC internal relations and thus simulates the effects of the UPFC. These mappings provide the foundation of determining the rating of UPFC.

Table 9.4 Bus voltage and line power flow performances controlled by UPFC using the proposed PIM (The angles of UPFC sources are with respect to the angle of slack bus voltage of the system)

Control case	Given control values (p.u.)	Actual control values (p.u.)	Parameters of UPFC (p.u.)
1	V_{90}=0.99 line 90-60: P=-18.78	V_{90}=0.9898 line 90-60: P=-18.79	UPFC for 90-60: $V_{pq}=0.0672\angle 261.39^0$ $V_{sh}=1.0106\angle -23.52^0$
2	V_{90}=1.01 line 90-60: P=-17.0	V_{90}=1.0098 line 90-60: P=-17.03	UPFC for 90-60: $V_{pq}=0.1291\angle 261.45^0$ $V_{sh}=1.0576\angle -21.72^0$
3	V_{90}=0.99 line 90-60: P=-23.0	V_{90}=0.9895 line 90-60: P=-22.96	UPFC for 90-60: $V_{pq}=0.0613\angle 81.3^0$ $V_{sh}=1.0068\angle -27.42^0$
4	V_{90}=0.95 line 90-60: P=-18.78	V_{90}=0.9498 line 90-60: P=-18.79	UPFC for 90-60: $V_{pq}=0.0498\angle 261.30^0$ $V_{sh}=0.9177\angle -24.15^0$

Furthermore, multi-UPFCs have also been investigated. For example, Table 9.6 presents the control performance of PIM under Cases 1 and 2 with two UPFCs installed along lines 50-60 and 90-60, respectively. From both tables, it can be clearly seen that PIM can achieve different specified control objectives without any initial assumptions of UPFC parameters. The PIM co-ordinates the UPFCs in the process of designing controllers in terms of power injections.

Table 9.5 The impact of PIM on the UPFC operating condition (all in p.u., the series reactance of the UPFC is assumed to be 0.01p.u.; the shunt reactance is 0.005)

Control	Power injection model conditions			UPFC conditions		
Case	$Q_{i(inj)}$	$P_{j(inj)}$	$Q_{j(inj)}$	Q_{sh}	P_{dc}	Q_{ser}
1	2.8293	-4.9838	0.0	-4.0784	-0.3255	1.5766
2	7.4493	-9.6091	0.0	-9.6222	-0.4107	3.3812
3	4.8211	4.5311	0.0	-3.4176	0.3654	-1.1312
4	-7.0189	-3.6739	0.0	6.0981	-0.3693	1.1008

Table 9.6 The impact of PIM on the UPFC operating condition (two UPFCs installation)

Case	UPFC locations	Given control values (p.u.)	Actual control values (p.u.)
1	Line 50-60 Line 90-60	V_{50}=1.0 line 50-60: P=-9.0 V_{90}=0.99 line 90-60: P=-17.8	V_{50}=1.0023 line 50-60: P=-8.967 V_{90}=0.99473 line 90-60: P=-17.65
2	Line 50-60 Line 90-60	V_{50}=1.0 line 50-60: P=-8.5 V_{90}=0.98 line 90-60: P=-17.5	V_{50}=1.00238 line 50-60: P=-8.476 V_{90}=0.98205 line 90-60: P=-17.405

In order to verify the proposed control algorithm under contingency, some outage operating conditions are used to evaluate the performance of the UPFC control on the test network. Based on Cases 1 and 2 of Tables 9.2 and 9.3, Table 9.7 gives the system performances with the installations of UPFCs. Investigation shows that if the violations of their voltage limits and line thermal limits of both cases are totally alleviated, it needs two UPFCs installed along lines 50-60 and 90-60. The results are summarised in Table 9.6, in which, both line thermal limit violations have been effectively alleviated and power sharing among lines tends to balance thus increasing power transfer capability across boundary A without constraint violations. This also shows that co-ordination of the UPFCs at different locations for solving power sharing and increasing power transfer has been achieved by PIM.

Table 9.7 Summary of power flow results with UPFC

Case	UPFC location line	Lowest voltage (p.u.)	Line	Active power (MW)	Line thermal limit (%)
Case 1	50-60 60-90	V_{70}= 0.982	60-90 60-90 60-90 50-60	940.4 488.0 1832.1 -896.7	95.51 99.12 93.81 91.87
Case 2	50-60 60-90	V_{90}= 0.981	60-90 60-90 60-90 50-60	917.3 476.0 1785.5 -847.6	92.08 96.36 90.93 88.74

Table 9.8 Results of UPFC with constraint limit check of I_{se} (C_{se}=0.1929)

	Results of I_{se} limit violation	Results of alleviating I_{se} violation
Original power flow results	V_{90}= 0.96884 P_{90-60}= -20.786	
Control objectives	V_{90}= 0.99 P_{90-60}= -23.00	
Actual power flow results	V_{90}= 0.9895 P_{90-60}= -22.966	V_{90}= 0.98649 P_{90-60}= -22.20
Power injection values	$P_{i(inj)}$= -4.5311 $Q_{i(inj)}$ = 4.8221 $P_{j(inj)}$ = 4.5311 $Q_{j(inj)}$ = -4.0	$P_{i(inj)}$ = -2.5712 $Q_{i(inj)}$ = 4.8211 $P_{j(inj)}$ = 2.5712 $Q_{j(inj)}$ = -5.4696
V_{pq} θ_{pq}	V_{pq}= 0.0816 θ_{pq}= 123.04^0	V_{pq} = 0.0823 θ_{pq} = 146.07^0
V_{sh} θ_{sh}	V_{sh}= 0.9909 θ_{sh}= -28.94^0	V_{sh}= 0.9814 θ_{sh}= -27.47^0
I_{se}	I_{se}=27.4885	I_{se}=24.83
I_{se}^L	I_{se}^L=25.0	
Iteration	3	
Modified θ_{pq}	θ_{pq}= 146.06^0	

9.7.5 Alleviation of constraint limit violations using the proposed control strategy

Extensive numerical studies have been carried out to evaluate the proposed strategies. As examples, Tables 9.8, 9.9, and 9.10 present several typical results. They all start with ideal control objectives and PIM can really do so. However, these objectives lead to UPFC limit violations, for example I_{se} of Tables 9.8 and 9.9 and I_{sh} of Table 9.11. In these situations, making use of relations of limit values with UPFC parameters and power injections, PIM ignores the control objectives and modifies power flow control towards the satisfactory parameters within their limits. All tables are investigated under 'Scenario A' with one UPFC installed along line 90-60 and with different initial control objectives, which yield violations of I_{se} and I_{sh}.

Table 9.9 Results of UPFC with constraint limit check of I_{se} (C_{se}=0.92)

	Results of I_{se} limit violation	Results of alleviating I_{se} violation
Original power flow results	V_{90}= 0.96884 $P_{90\text{-}60}$= -20.786	
Control objectives	V_{90}= 0.99 $P_{90\text{-}60}$= -23.00	
Actual power flow results	V_{90}= 0.9895 $P_{90\text{-}60}$= -22.966	V_{90}= 0.98929 $P_{90\text{-}60}$= -22.191
Power injection values	$P_{i(inj)}$= -4.5311 $Q_{i(inj)}$ = 4.8221 $P_{j(inj)}$ = 4.5311 $Q_{j(inj)}$ = -4.0	$P_{i(inj)}$ = -2.5481 $Q_{i(inj)}$ = 4.8211 $P_{j(inj)}$ = 2.5481 $Q_{j(inj)}$ = -0.4570
V_{pq} θ_{pq}	V_{pq}= 0.0816 θ_{pq}= 123.04^0	V_{pq} = 0.0349 θ_{pq} = 90.024^0
V_{sh} θ_{sh}	V_{sh}= 0.9909 θ_{sh}= -28.94^0	V_{sh}= 1.008 θ_{sh}= -26.759^0
I_{se}	I_{se}=27.4885	I_{se}=24.97
I_{se}^{L}	I_{se}^{L}=25.0	
C_{se}	0.92	
Iteration	4	
Modified V_{pq}	V_{pq}=0.0349	
Modified θ_{pq}	θ_{pq}= 90.157^0	

Table 9.10 Results of UPFC with constraint limit check of I_{sh}

	Results of I_{sh} limit violation	Results of alleviating I_{sh} violation
Original power flow results	V_{90}= 0.96884 $P_{90\text{-}60}$= -20.786	
Control objectives	V_{90}= 1.01 $P_{90\text{-}60}$= -23.00	
Actual power flow results	V_{90}= 1.0096 $P_{90\text{-}60}$= -22.974	V_{90}= 1.00245 $P_{90\text{-}60}$= -22.970
Power injection values	$P_{i(inj)}$= -3.8474 $Q_{i(inj)}$ = 10.6426 $P_{j(inj)}$ = 3.8474 $Q_{j(inj)}$ = -4.0	$P_{i(inj)}$ = -3.8474 $Q_{i(inj)}$ = 9.1764 $P_{j(inj)}$ = 3.8474 $Q_{j(inj)}$ = -4.0
V_{pq} θ_{pq}	V_{pq}= 0.0752 θ_{pq}= 127.43^0	V_{pq} = 0.0753 θ_{pq} = 127.34^0
V_{sh} θ_{sh}	V_{sh}= 1.0377 θ_{sh}= -27.49^0	V_{sh}= 1.0236 θ_{sh}= -27.67^0
I_{sh}	I_{sh}=5.8119	I_{sh}=4.482
I_{sh}^L	I_{sh}^L =4.5	
Iteration	3	
Modified V_{90}	V_{90}=1.0032	
Modified $Q_{i(inj)}$	$Q_{i(inj)}$ =9.1764	

In Tables 9.8 and 9.9, I_{se} violates its limit according to the control objective and thus PIM fixes I_{se} at the I_{se}^L and obtains the modified V_{pq} and/or θ_{pq} and the associated power injection values. After PIM gets new power flow results using these modified power injection control, it can be seen that violation of I_{se} can be efficiently alleviated with 3–4 iterative solutions. In fact, the final V_{pq} and θ_{pq} have been treated as the control values to modify power injections and thus it achieves alleviation of the I_{se} limit, which can be seen that the modified θ_{pq} = 146.06^0 is the same as the final result of phase of V_{pq} of Table 9.8. In Table 9.8, only θ_{pq} is used to modify power injections while V_{pq} remains nearly unchanged. In this typical case, C_{se}=0.1929. However, under the same case, when C_{se} is chosen to be 0.92, V_{pq} and θ_{pq} both co-operate to modify power injections to achieve the same goal, which is shown in Table 9.9. Compared to Table 9.8, results of Table 9.9 have one more control degree of freedom to achieve the same objective. This shows that the proposed general control equations (70)–(72) have more advantages applicable to alleviation of I_{se} within I_{se}^L. In Table 9.4, V_{90} is relaxed from 1.0096 p.u. of I_{sh}

violating limit to 1.0032 p.u. without violation. Here, V_{90} is first obtained from equation (9.79) specified I_{sh}^{L}, and then used to modify $Q_{i(inj)}$ while keeping power injections $(P_{i(inj)}, P_{j(inj)}, Q_{j(inj)})$ unchanged. The modified $Q_{i(inj)}$ is obtained through busbar voltage – reactive power injection controller equation (9.80). At last, I_{sh} violation has been alleviated through relaxing V_{90}.

9.7.6 Comparison of UPFC, SVC, and PS

Many operations of SVC, PS, and UPFC have been studied. Here, one operation condition, i.e. the second condition and line 50-70 outage, is presented to demonstrate the comparison. According to Table 9.2, two line thermal limits and one bus voltage limit are violated in this condition. Generally speaking, they can be alleviated either by two UPFCs or two PSs and one SVC. In this section, only one UPFC is installed along line 90-60 for comparison with SVC and PS. In this case, one SVC and one PS are assumed to be installed at bus 90 and line 90-60 respectively so as to achieve the same UPFC performance.

In this study, two cases are simulated:

(1) Only one PS is installed along 90-60 for comparison with UPFC.

(2) One SVC connected at bus 90 is set as 3.0p.u. for simplicity and the angle of PS is changed from -10.0^0 to $+10.0^0$.

In both cases, power flow results from these open-loop controls, i.e. bus 90 voltage and line 90-60 active power are recorded and regarded as the specified control objective values. Then, the power injections, ratings and control parameters of UPFC can be obtained.

Although the UPFC and the PS have the same injected power models, their representations of these injected powers are different owing to their different regulation mechanisms. In Figures 9.7 and 9.8, line power percentage represents the regulating power of line 50-60 by the PS and UPFC respectively where the UPFC keeps the bus 50 voltage the same as the conditions of the PS. From both figures in view of power injections, it is known that the capabilities of both sides of the UPFC are less than that of the PS in the first case.

When SVC and PS are combined to achieve performance identical to the UPFC, their ratings also have a difference, which is shown in Figure 9.9. Here, two assumptions are made: (i) the rating of PS plus SVC is obtained from the sum of PS in MVA and SVC in Mvar although strictly it is not so simple as directly adding the MVA to Mvar; (ii) the shunt part and the series part of UPFC are treated as having the same rating. However, it must be pointed out that the conclusions here are based on a particular network configuration and the models used in this study are simplified.

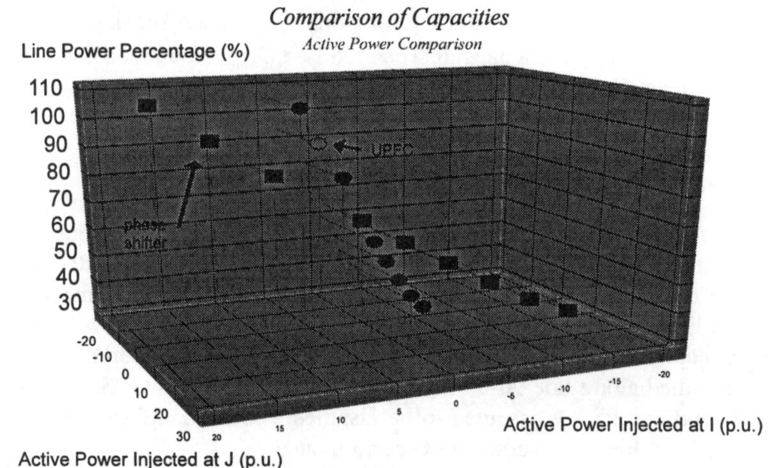

Figure 9.7 Comparison of PS and UPFC in terms of active power injections at both ends of transmission line

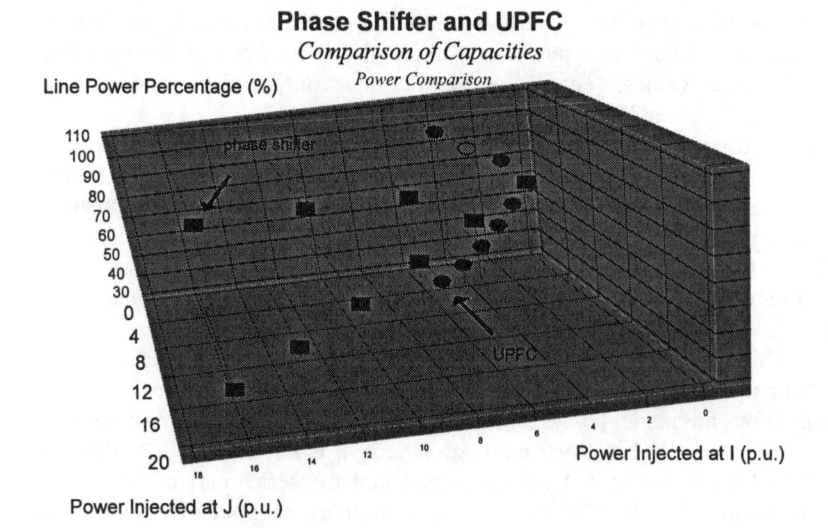

Figure 9.8 Comparison of PS and UPFC in terms of complex power injections at both ends of transmission line

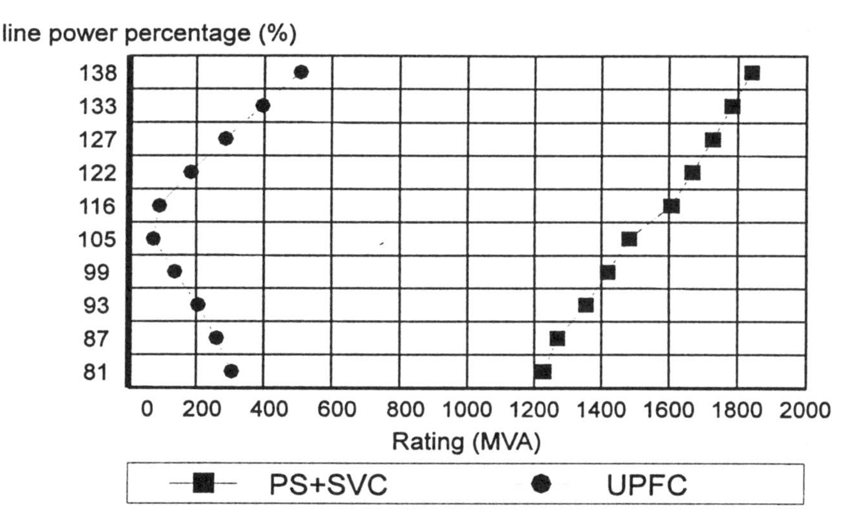

Figure 9.9 Comparison of PS+SVC and UPFC

9.8 Conclusions

This chapter focuses on the steady state analysis and control of systems with FACTS devices with particular reference to the UPFC. A steady-state UPFC model is proposed and its power injection transformation is derived in rectangular form. The optimal multiplier power flow method for ill-conditioned system is applied to implement the UPFC model.

Then a novel method for the steady state control of UPFC for power flow control and voltage support is presented. Essentially, it is based on a power injection model and optimal multiplier power flow. The proposed power injection power flow control can be effectively used to derive UPFC control parameters to achieve the required control objectives. A number of internal limits imposed on the UPFC have been considered, including series injection voltage magnitude, line current through the series inverter, real power transfer between the shunt inverter and series inverter, shunt side current and shunt injection voltage magnitude. The proposed constrained control strategies can co-ordinate the available control freedom to achieve an efficient usage of the UPFC when constrained by the internal limits.

The comparison of rating and control flexibility of three major FACTS devices is also included in the numerical examples. The proposed UPFC model and power flow control algorithms have been vigorously tested in a number of systems. The results on two practical systems clearly illustrate the effectiveness of the proposed algorithms.

9.9 Acknowledgment

The authors gratefully acknowledge the support of NGC and useful discussions with Dr A. M. Foss, Dr D. T. Y. Cheng and Mr M. Allison and other NGC engineers.

9.10 References

1 Daniel, D., Le Du, A., Poumarede, C., Therond, P.G., Langlet, B., Taisne, J.P., and Collet-Billon, V., "Power electronics: An effective tool for network development? An electricite de France answer based on the development of a prototype unified power flow controller", *Proc of CIGRE Power Electronics in Electric Power Systems*, Tokyo, paper 520-02, May, 1995.
2 Song, Y.H. and Liu, J.Y., "Electromagnetic transient simulation studies of sinusoidal PWM unified power flow controllers", *European Transactions on Electrical Power*, 8, (3), pp.187-192, 1998.
3 Lombard, X., and Ptherond, P.G., "Control of unified power flow controller: comparison of methods on the basis of a detailed numerical model", *IEEE Trans. on Power Systems*, 12, (2), pp.824-830, May 1997.
4 Bian, J., Lemak, T.A., Nelson, R.J., and Ramey, D.G., "Power flow controller models for power system simulations", *Proc IEEE/CSEE International Conference on Power System Technology*, Beijing, China, pp.687-691, 1994.
5 Fuerte-Esquivel, C.R. and Acha, E., "Newton-Raphson algorithm for the reliable solution of large power networks with embedded FACTS devices", *Proc IEE, Pt.C*, 143, (5), pp.447-454, 1996.
6 Fuerte-Esquivel, C.R. and Acha, E., "A Newton-type algorithm for the control of power flow in electrical power networks", IEEE, PE-159-PWRS-0-12-1997.
7 Han, Z.X., "Phase shifter and power flow control", *IEEE Trans. on PAS*, 101, (10), pp.3790-3795, 1982.
8 Peterson, N.M., and Meyer, W.S., "Automatic adjustment of transformer and phase shifter taps in the Newton power flow", *IEEE Trans. on PAS*, 90, (1), pp.104-106, 1971.
9 Youssef, R.D., "Phase-shifting transformers in load flow and short circuit analysis: modelling and control", *Proc IEE, Pt.C*, 140, (4), pp.331-335, 1993.
10 IEEE Special Stability Controls working Group, "Static Var Compensator Models for Power Flow and Dynamic Performance Simulation", *IEEE Trans. on Power systems*, 9, (1), pp.229-239, February 1994.

11 Yu, J., "Unified analysis of generalized power flow controller (GPFC) in power system load flow calculation", *Electric Power System and Automation*, 18, (12), pp.56-66, 1994.
12 Gotham, D.J., and Heydt, G.T., "FACTS device models for power flow studies", *The Proceedings of the 27th Annual North American Power Symposium*, Bozeman, Montana, pp.514-519, 1995.
13 Arabi, S. and Kundur, P., "A versatile FACTS device model for power flow and stability simulations", *IEEE 96 Winter Meeting*, paper 96 WM 258-4 PWRS, 1996,
14 Nabavi-Niaki, A., and Iravani, M.R., "Steady-state and dynamic models of unified power flow controller (UPFC) for power system studies", *IEEE 96 Winter Meeting*, paper 96 WM 257-6 PWRS, 1996.
15 Mihallc, R., Zunko, P., and Povh, D., "Modelling of unified power flow controller and its impacts on power oscillation damping", *Proc of CIGRE Power Electronics in Electric Power Systems*, Tokyo, paper 320-01, May, 1995.
16 Bian, J., Ramey, D.G., Nelson, R.J., and Edris, A., "A study of equipment sizes and constraints for a unified power flow controller", *Proc IEEE T&D Conference*, 1996.
17 Noroozian, M., Angquist, L., Ghandhari, M., and Andersson, G., "Use of UPFC for optimal power flow control", *Proc IEEE/KTH Stockholm Power Tech Conference*, Stockholm, Sweden, pp.506-511, June, 1995.
18 Noroozian, M. and Andersson, G., "Power flow control by use of controllable series components", *IEEE 92 Summer Meeting*, paper 92 SM466-3 PWRD, 1992.
19 Chang, S.K., and Brandwajn, V., "Adjusted solution in fast decoupled load flow", *IEEE Trans. on PS*, 3, (2), pp.726-733, May 1988.
20 Hiskens, I.A., "Analysis tools for power systems – contending with nonlinearities", *Proceedings of the IEEE*, 83, (11), pp.1573-1587, November 1995.
21 Iwamoto, S., and Tamura, Y., "A load flow calculation method for ill-conditioned power systems", *IEEE PAS*, PAS -100, (4, pp. 1736-1743), 1981.
22 Glover, J.D., and Digby, G., *Power system analysis and design*, PWS Publishing Company, Boston, 1994.
23 Liu, J.Y., Song, Y.H., and Foss, A.M., "Power flow control and voltage support in a meshed power system using unified power flow controllers", *Proc UPFC'97*, UMIST, UK, 1997
24 Liu, J.Y., and Song, Y.H., "Determining maximum regulating capability of UPFC based on predicting feasibility limit of power systems", *Electric Machines and Power Systems*, 26, (8), pp.789-800, 1998.
25 Laesen, E., Miller, N., Nilsson, S., and Lindgren, S., "Benefits of GTO-based compensation systems for electric utility applications", *IEEE Trans. on Power Delivery*, 7, (4), pp 2056-2062, October 1992.
26 Galiana, F.D., Almeida, K., Toussaint, M., Griffin, J., Altanackovic, D., Ooi, B.O., and Mcgillis, D.T., "Assessment and control of the impact of FACTS

devices on power system performance", *IEEE 1996 Winter Meeting*, Paper No. 96 WM 257-8 PWRS,1996.
27 Liu, J.Y., Song, Y.H., "Comparison studies of unified power flow controllers with static var compensators and phase shifters", *Electric Machines and Power Systems*, 27, (3), 1999.

9.11 Appendix: steady-state modelling of SVC and phase shifter

9.11.1 SVC modelling and implementation

According to the report 'Static Var Compensator Models for Power Flow and Dynamic Performance Simulation' by IEEE Special Stability Controls Working Group [6], SVCs can be modelled as a slope representation using a conventional PV (generator) bus with reactive power limits. Figure 9.10 is the SVC model in which X_{sl} is the slope reactance equal to the per unit slope. The slope is typically 1-5%. At the capacitive limit, the SVC becomes a shunt capacitor. At the inductive limit, the SVC becomes a shunt reactor.

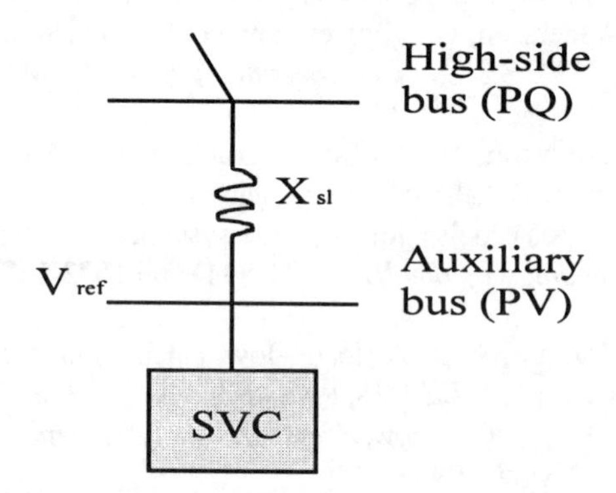

Figure 9.10 A SVC model implemented in power flow software

9.11.2 PS modelling and implementation

A PS is a mechanically-controlled type device. Its mechanism and model used in power flow and transients analysis, has been thoroughly investigated. In our study, a power injection model [7] representing PS in a rectangular form more suitable for its implementation in optimal multiplier power flow is adopted.

The model of a PS is converted to an injection model which is shown in Figure 9.11.

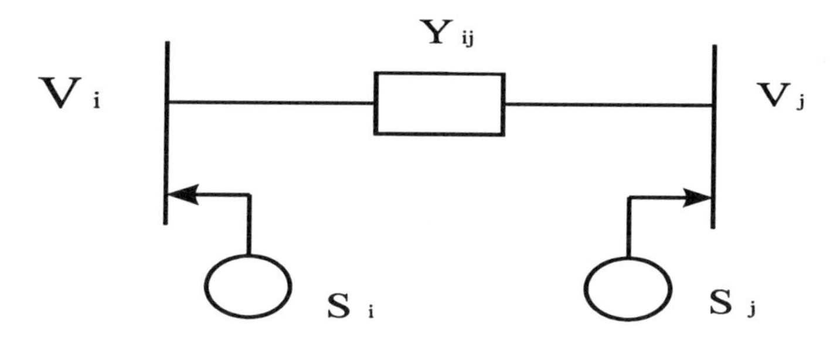

Figure 9.11 The power injection model of a phase shifter

The following are the formulae of such a model:

$$P_{is} = b_{ij} \tan\Phi\left(e_i e_j + f_i f_j\right) \tag{9.A1}$$

$$P_{js} = -b_{ij} \tan\Phi\left(e_i e_j + f_i f_j\right) \tag{9.A2}$$

$$Q_{is} = b_{ij} \tan^2\Phi\left(e_i^2 + f_i^2\right) + b_{ij} \tan\Phi\left(e_j f_i - e_i f_j\right) \tag{9.A3}$$

$$Q_{js} = b_{ij} \tan\Phi\left(e_j f_i - e_i f_j\right) \tag{9.A4}$$

Φ is the phase shift angle.

In order to implement this model into our optimal power flow software, the associated Jacobain matrix elements are modified in each iteration.

Chapter 10

Oscillation stability analysis and control

H. F. Wang

10.1 Introduction

Power system oscillation stability refers to the damping of electromechanical oscillations occurring in power systems with oscillation frequency in the range from 0.2Hz to 2.0Hz. These low-frequency oscillations are the consequence of the development of interconnection of large power systems. Once started, the oscillations would continue for a while and then disappear, or continue to grow, causing system separation.

Low-frequency oscillations were first observed in the Northern American power network in the early 1960s during a trial interconnection of the Northwest Power Pool and the Southwest Power Pool. Later, they have been reported in many countries. A low-frequency oscillation in a power system constrains the capability of power transmission, threatens system security and damages the efficient operation of the power system. Therefore, this problem has caused wide concern and attracted the interests of many researchers. Since the 1960s, a special research and application area has been gradually formed in tackling the problem of low-frequency oscillations. Generally it is classified as analysis and improvement of power system oscillation stability.

To increase power system oscillation stability, the installation of supplementary excitation control, Power System Stabilizer (PSS), is a simple, effective and economical method. To date, most major electric power plants in many countries are equipped with PSS. For the analysis of power system oscillation stability, the linearized Phillips-Heffron model was proposed [1][2], based on which, method of damping torque analysis and phase compensation method were applied to design PSS [2][3]. In the late 1980s, various techniques have been proposed for the co-ordination of multiple PSSs installed in multi-machine power systems to suppress multi-mode oscillations. Oscillation stability analysis and control has been an important and active subject in power system research and applications.

With the advent of Flexible AC Transmission Systems (FACTS) [4], the potential of FACTS damping function, which is referred to as FACTS-based stabilizer in this chapter, in suppressing power system oscillations has attracted interests from both academia and industry, although the damping duty of a FACTS controller often is not its primary function. The capability of FACTS-based stabilizers to increase power system oscillation stability has been explored in many aspects [4–16], by which many constructive results are presented and should be carefully appraised. However, in order to maintain the consistency of the content within the required space, in this chapter, mainly the work conducted by the author and published in *IEEE Transactions* and *IEE Proceedings* on the oscillation stability analysis and control of power systems installed with FACTS-based stabilizers will be presented [18–26].

The chapter is organized as follows:

In Section 10.2, the linearized Phillips-Heffron model of power systems installed with FACTS-based stabilizers is established, which presents an example and format on modelling FACTS-based stabilizers into power systems.

In Section 10.3, the performance of FACTS-based stabilizers installed in power systems is analysed. The results facilitate the proposed selection of robust operating conditions of power systems in which robust FACTS-based stabilizers can be designed.

In Section 10.4, the selection of installing locations and feedback signals of FACTS-based stabilizers is discussed. The connections among the various indices proposed so far are revealed which provides insight into these different indices and a basis for their comparison. The concept of robustness of the installing locations and feedback signals of FACTS-based stabilizers is proposed. A method to achieve an effective and robust selection is introduced.

The chapter is concluded by a brief summary where some problems to be investigated in future are raised.

10.2 Linearized model of power systems installed with FACTS-based stabilizers

Power system oscillation stability and control can be studied using a linearized model of a power system. The general form of the linearized model is

$$\begin{bmatrix} \Delta\dot{\delta} \\ \Delta\dot{\omega} \\ \dot{x} \end{bmatrix} = \begin{bmatrix} 0 & \omega_0 I & 0 \\ A_{21} & A_{22} & A_{23} \\ A_{31} & A_{32} & A_{33} \end{bmatrix} \begin{bmatrix} \Delta\delta \\ \Delta\omega \\ x \end{bmatrix} + \sum_{k=1}^{M} \begin{bmatrix} 0 \\ B_{2k} \\ B_{3k} \end{bmatrix} \Delta u_k$$

$$y_k = [C_{1k}{}^T, C_{2k}{}^T, C_{3k}{}^T] \begin{bmatrix} \Delta\delta \\ \Delta\omega \\ x \end{bmatrix}, K = 1,2,\cdots,M \tag{10.1}$$

where Δu_k is the output control signal of the *k*th stabilizer (PSS or FACTS-based stabilizer) installed in the power system and y_k is the feedback signal of the stabilizer. In this section, the establishment of an augmented Phillips-Heffron model of the power system installed with FACTS-based stabilizers is demonstrated which is of a similar form to that of the traditional Phillips-Heffron model [1–2]. Besides the advantages of the augmented Phillips-Heffron model (a systematic configuration and simple expression), the control function of the FACTS-based stabilizers is clearly demonstrated, the study of power system oscillation stability, such as damping torque analysis and phase compensation methods, and the analysis and design of FACTS-based stabilizers is made more convenient.

10.2.1 Phillips-Heffron model of single-machine infinite-bus power systems installed with SVC, TCSC, and TCPS

The non-linear differential equations from which the traditional Phillips-Heffron linear model of a single-machine infinite-bus power system is derived are:

$$\begin{aligned} \dot{\delta} &= \omega_0 \Delta\omega \\ \Delta\dot{\omega} &= (P_m - P_e - D\Delta\omega)/2H \\ \dot{E}_q' &= (-E_q + E_{qe})/T_{do}' \\ \dot{E}_{qe} &= \mathrm{Reg}(s)(V_{to} - V_t) = K_A(V_{to} - V_t)/(1 + sT_A) \end{aligned} \tag{10.2}$$

where

$$\begin{aligned} P_e &= E_q' V_b \sin\delta / X_{d\Sigma}' - V_b^2 (X_q - X_d') \sin 2\delta / 2X_{d\Sigma}' X_{q\Sigma} \\ E_q &= X_{d\Sigma} E_q' / X_{d\Sigma}' - (X_d - X_d') V_b \cos\delta / X_{d\Sigma}' \end{aligned} \tag{10.3a}$$

$$V_{td} = X_q V_b \sin\delta / X_{q\Sigma},$$
$$V_{tq} = X_L E_q' / X_{d\Sigma}' + V_b X_d' \cos\delta / X_{d\Sigma}' \tag{10.3b}$$

$$X_{d\Sigma}' = X_d' + X_L, X_{q\Sigma} = X_q + X_L, X_{d\Sigma} = X_d + X_L$$

and X_L is the impedance of the transmission line. Without loss of generality, the transfer function of AVR, Reg(s), is assumed to be of the simplest form $\frac{K_A}{1+sT_A}$.

When a FACTS device is installed in the system, equation (10.3) must be modified to include the influence of the FACTS device on the system performance. For an SVC (Static Var Compensator), we can simply assume the SVC's model is

$$\overline{Y}_{svc} = -\Delta f_{SVC} / jX_{SVC} \tag{10.4}$$

From Figure10.1, we have

$$I_{sb} = I_{ts} - Y_{svc} V_{svc}$$
$$V_{svc} = jX_{sb} I_{sb} + V_b = jX_{sb} I_{ts} - jX_{sb} Y_{svc} V_{svc} + V_b$$

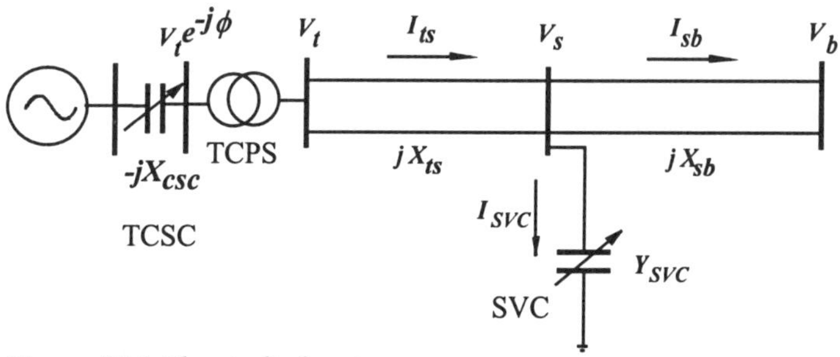

Figure 10.1 The studied system

Thus we can obtain

$$V_t = jX_{sb} I_{ts} + V_{svc} = jX_{TL} I_{ts} + V_b / C \tag{10.5}$$

where

$$C = 1 + jX_{sb} Y_{svc} = 1 - \Delta f_{SVC} X_{sb} / X_{SVC}$$

$$X_{TL} = X_{ts} + X_{sb} / C$$

Equation (10.5) indicates that when the power system is installed with an SVC as shown by Figure10.1, the equivalent system is of a line impedance X_{TL} connected to an infinite bus with a voltage V_b/C. Therefore, the non-linear equations of the power system installed with the SVC have the same form as equation (10.3) except that X_L and V_b in equation (10.3) are replaced by X_{TL} and V_b/C respectively.

For a TCSC (Thyristor-Controlled Series Compensator) applied in the power system, only X_L in equation (10.3) should be replaced by X_{TL} where

$$X_{TL} = X_L - \Delta f_{CSC} X_{CSC} \tag{10.6}$$

For a TCPS (Thyristor-Controlled Phase Shifter), only the variable δ in equation (10.3) should be substituted by

$$\delta + \Delta f_{PS} F_0 \tag{10.7}$$

Therefore, the non-linear equations of equation (10.3) of the power system installed with SVC, TCSC and TCPS can be modified to be

$$P_e = \frac{E_q' V_b \sin(\delta + \Delta f_{PS} F_0)}{CX_{d\Sigma}'} - \frac{V_b^2 (X_q - X_d') \sin 2(\delta + \Delta f_{PS} F_0)}{2C^2 X_{d\Sigma}' X_{q\Sigma}}$$

$$E_q = \frac{X_{d\Sigma} E_q'}{X_{d\Sigma}'} - \frac{(X_d - X_d') V_b \cos(\delta + \Delta f_{PS} F_0)}{CX_{d\Sigma}'} \tag{10.8a}$$

$$V_{td} = \frac{X_q V_b \sin(\delta + \Delta f_{PS} F_0)}{CX_{q\Sigma}},$$

$$V_{tq} = \frac{X_{TL} E_q'}{X_{d\Sigma}'} + \frac{V_b X_d' \cos(\delta + \Delta f_{PS} F_0)}{CX_{d\Sigma}'} \tag{10.8b}$$

where

$$X_{d\Sigma}' = X_d' + X_{TL}, X_{q\Sigma} = X_q + X_{TL}, X_{d\Sigma} = X_d + X_{TL} \tag{10.8c}$$

and X_{TL} is expressed by equations (10.5) and (10.6). By linearizing equation (10.8) at an operating condition of the power system, the Phillips-Heffron model of the power system with SVC, TCSC and TCPS can be obtained as follows:

$$\begin{aligned}
&\Delta\dot{\delta} = \omega_o \Delta\omega \\
&\Delta\dot{\omega} = [-K_1\Delta\delta - K_2\Delta E_q' - K_P\Delta f_* - D\Delta\omega\,]/2H \\
&\Delta\dot{E}_q' = (-K_4\Delta\delta - K_3\Delta E_q' - K_q\Delta f_* + \Delta E_{qe}\,)/T_{d0}' \\
&\Delta\dot{E}_{qe} = [-\Delta E_{qe} - K_A(K_5\Delta\delta + K_6\Delta E_q' + K_V\Delta f_*)]/T_A
\end{aligned} \tag{10.9a}$$

where

$$\begin{aligned}
&K_1 = \partial P_e/\partial\delta, K_2 = \partial P_e/\partial E_q', K_P = \partial P_e/\partial\Delta f_* \\
&K_4 = \partial E_q/\partial\delta, K_3 = \partial E_q/\partial E_q', K_q = \partial E_q/\partial\Delta f_* \\
&K_5 = \partial V_t/\partial\delta, K_6 = \partial V_t/\partial E_q', K_V = \partial V_t/\partial\Delta f_*
\end{aligned} \tag{10.9b}$$

The model can be shown by Figure 10.2.

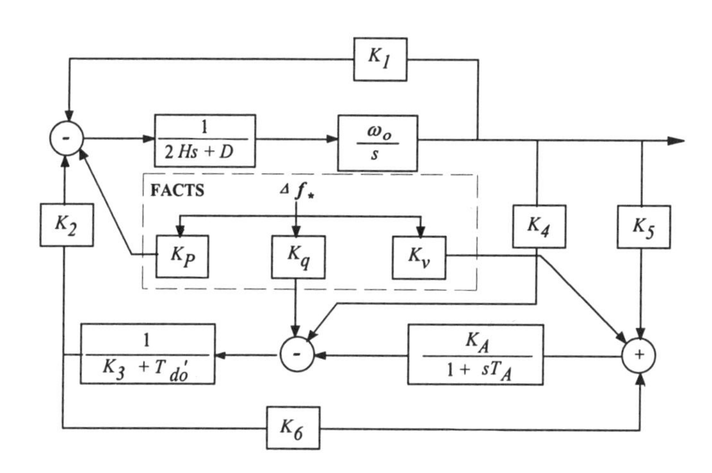

Figure 10.2 The Phillips-Heffron model of a single-machine infinite-bus power system installed with a FACTS-based stabilizer

10.2.2 Phillips-Heffron model of single-machine infinite-bus power system installed with UPFC

Figure 10.3 is a single-machine infinite-bus power system installed with a UPFC (Unified Power System Controller) which consists of an excitation transformer (ET), a boosting transformer (BT), two three-phase GTO based voltage source converters (VSCs) and a DC link capacitor. In Figure 10.3, m_E, m_B and δ_E, δ_B are the amplitude modulation ratio and phase angle of the control signal of each VSC respectively, which are the input control signals to the UPFC. If the general Pulse Width Modulation (PWM) (or optimized pulse patterns or space-vector modulation approach) is adopted for the GTO based VSC, the three-phase dynamic differential equations of the UPFC are

$$\begin{bmatrix} \frac{di_{Ea}}{dt} \\ \frac{di_{Eb}}{dt} \\ \frac{di_{Ec}}{dt} \end{bmatrix} = \begin{bmatrix} -\frac{r_E}{l_E} & 0 & 0 \\ 0 & -\frac{r_E}{l_E} & 0 \\ 0 & 0 & -\frac{r_E}{l_E} \end{bmatrix} \begin{bmatrix} i_{Ea} \\ i_{Eb} \\ i_{Ec} \end{bmatrix} - \frac{m_E v_{dc}}{2l_E} \begin{bmatrix} \cos(\omega t + \delta_E) \\ \cos(\omega t + \delta_E - 120^0) \\ \cos(\omega t + \delta_E + 120^0) \end{bmatrix} + \begin{bmatrix} \frac{1}{l_E} & 0 & 0 \\ 0 & \frac{1}{l_E} & 0 \\ 0 & 0 & \frac{1}{l_E} \end{bmatrix} \begin{bmatrix} v_{Eta} \\ v_{Etb} \\ v_{Etc} \end{bmatrix}$$

$$\begin{bmatrix} \frac{di_{Ba}}{dt} \\ \frac{di_{Bb}}{dt} \\ \frac{di_{Bc}}{dt} \end{bmatrix} = \begin{bmatrix} -\frac{r_B}{l_B} & 0 & 0 \\ 0 & -\frac{r_B}{l_B} & 0 \\ 0 & 0 & -\frac{r_B}{l_B} \end{bmatrix} \begin{bmatrix} i_{Ba} \\ i_{Bb} \\ i_{Bc} \end{bmatrix} - \frac{m_B v_{dc}}{2l_B} \begin{bmatrix} \cos(\omega t + \delta_B) \\ \cos(\omega t + \delta_B - 120^0) \\ \cos(\omega t + \delta_B + 120^0) \end{bmatrix} + \begin{bmatrix} \frac{1}{l_E} & 0 & 0 \\ 0 & \frac{1}{l_E} & 0 \\ 0 & 0 & \frac{1}{l_E} \end{bmatrix} \begin{bmatrix} v_{Bta} \\ v_{Btb} \\ v_{Btc} \end{bmatrix}$$

$$\frac{dv_{dc}}{dt} = \frac{m_E}{2C_{dc}} \begin{bmatrix} \cos(\omega t + \delta_E) & \cos(\omega t + \delta_E - 120^0) & \cos(\omega t + \delta_E + 120^0) \end{bmatrix} \begin{bmatrix} i_{Ea} \\ i_{Eb} \\ i_{Ec} \end{bmatrix}$$

$$+ \frac{m_B}{2C_{dc}} \begin{bmatrix} \cos(\omega t + \delta_B) & \cos(\omega t + \delta_B - 120^0) & \cos(\omega t + \delta_B + 120^0) \end{bmatrix} \begin{bmatrix} i_{Ba} \\ i_{Bb} \\ i_{Bc} \end{bmatrix}$$

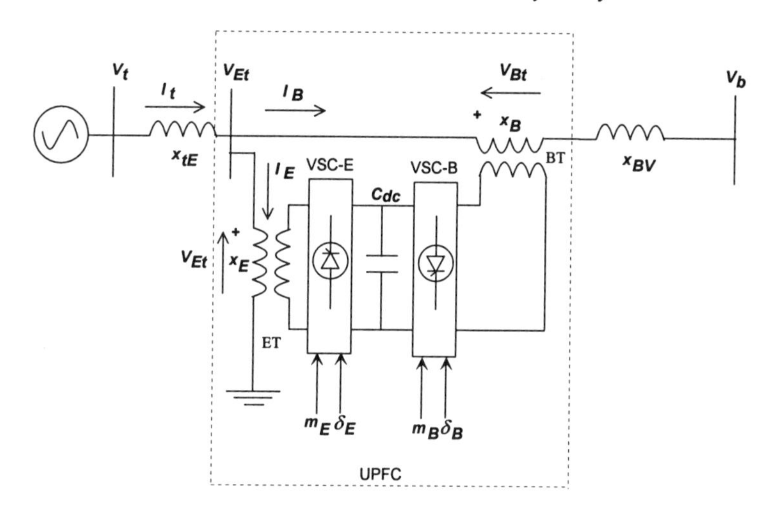

Figure 10.3 A UPFC installed in a single-machine infinite-bus power system

By applying Park's transformation

$$\mathbf{P}\begin{bmatrix} \frac{di_{Ea}}{dt} \\ \frac{di_{Eb}}{dt} \\ \frac{di_{Ec}}{dt} \end{bmatrix} = P\begin{bmatrix} -\frac{r_E}{l_E} & 0 & 0 \\ 0 & -\frac{r_E}{l_E} & 0 \\ 0 & 0 & -\frac{r_E}{l_E} \end{bmatrix}\mathbf{P}^{-1}\mathbf{P}\begin{bmatrix} i_{Ea} \\ i_{Eb} \\ i_{Ec} \end{bmatrix} - \frac{m_E v_{dc}}{2l_E}\mathbf{P}\begin{bmatrix} \cos(\omega t+\delta_E) \\ \cos(\omega t+\delta_E-120^0) \\ \cos(\omega t+\delta_E+120^0) \end{bmatrix} + \mathbf{P}\begin{bmatrix} \frac{1}{l_E} & 0 & 0 \\ 0 & \frac{1}{l_E} & 0 \\ 0 & 0 & \frac{1}{l_E} \end{bmatrix}\mathbf{P}^{-1}\mathbf{P}\begin{bmatrix} v_{Eta} \\ v_{Etb} \\ v_{Etc} \end{bmatrix}$$

$$\mathbf{P}\begin{bmatrix} \frac{di_{Ba}}{dt} \\ \frac{di_{Bb}}{dt} \\ \frac{di_{Bc}}{dt} \end{bmatrix} = \begin{bmatrix} -\frac{r_B}{l_B} & 0 & 0 \\ 0 & -\frac{r_B}{l_B} & 0 \\ 0 & 0 & -\frac{r_B}{l_B} \end{bmatrix}\mathbf{P}^{-1}\mathbf{P}\begin{bmatrix} i_{Ba} \\ i_{Bb} \\ i_{Bc} \end{bmatrix} - \frac{m_B v_{dc}}{2l_B}P\begin{bmatrix} \cos(\omega t+\delta_B) \\ \cos(\omega t+\delta_B-120^0) \\ \cos(\omega t+\delta_B+120^0) \end{bmatrix} + \mathbf{P}\begin{bmatrix} \frac{1}{l_E} & 0 & 0 \\ 0 & \frac{1}{l_E} & 0 \\ 0 & 0 & \frac{1}{l_E} \end{bmatrix}\mathbf{P}^{-1}\mathbf{P}\begin{bmatrix} v_{Bta} \\ v_{Btb} \\ v_{Btc} \end{bmatrix}$$

$$\frac{dv_{dc}}{dt} = \frac{m_E}{2C_{dc}}\left[\cos(\omega t+\delta_E) \quad \cos(\omega t+\delta_E-120^0) \quad \cos(\omega t+\delta_E+120^0)\right]\mathbf{P}^{-1}\mathbf{P}\begin{bmatrix} i_{Ea} \\ i_{Eb} \\ i_{Ec} \end{bmatrix}$$

$$+\frac{m_B}{2C_{dc}}\left[\cos(\omega t+\delta_B) \quad \cos(\omega t+\delta_B-120^0) \quad \cos(\omega t+\delta_B+120^0)\right]\mathbf{P}^{-1}\mathbf{P}\begin{bmatrix} i_{Ba} \\ i_{Bb} \\ i_{Bc} \end{bmatrix}$$

where $\mathbf{P} = \frac{2}{3}\begin{bmatrix} \cos\omega t & \cos(\omega t - 120^0) & \cos(\omega t + 120^0) \\ -\sin\omega t & -\sin(\omega t - 120^0) & -\sin(\omega t + 120^0) \\ \frac{1}{2} & \frac{1}{2} & \frac{1}{2} \end{bmatrix}$

we can obtain

$$\begin{bmatrix} \frac{di_{Ed}}{dt} \\ \frac{di_{Eq}}{dt} \\ \frac{di_{E0}}{dt} \end{bmatrix} = \begin{bmatrix} 0 & \omega & 0 \\ -\omega & 0 & 0 \\ 0 & 0 & 0 \end{bmatrix}\begin{bmatrix} i_{Ed} \\ i_{Eq} \\ i_{E0} \end{bmatrix} + \begin{bmatrix} -\frac{r_E}{l_E} & 0 & 0 \\ 0 & -\frac{r_E}{l_E} & 0 \\ 0 & 0 & -\frac{r_E}{l_E} \end{bmatrix}\begin{bmatrix} i_{Ed} \\ i_{Eq} \\ i_{E0} \end{bmatrix} - \frac{m_E v_{dc}}{2l_E}\begin{bmatrix} \cos\delta_E \\ \sin\delta_E \\ 0 \end{bmatrix} + \begin{bmatrix} \frac{1}{l_E} & 0 & 0 \\ 0 & \frac{1}{l_E} & 0 \\ 0 & 0 & \frac{1}{l_E} \end{bmatrix}\begin{bmatrix} v_{Etd} \\ v_{Etq} \\ v_{Et0} \end{bmatrix}$$

$$\begin{bmatrix} \frac{di_{Bd}}{dt} \\ \frac{di_{Bq}}{dt} \\ \frac{di_{B0}}{dt} \end{bmatrix} = \begin{bmatrix} 0 & \omega & 0 \\ -\omega & 0 & 0 \\ 0 & 0 & 0 \end{bmatrix}\begin{bmatrix} i_{Bd} \\ i_{Bq} \\ i_{B0} \end{bmatrix} + \begin{bmatrix} -\frac{r_B}{l_B} & 0 & 0 \\ 0 & -\frac{r_B}{l_B} & 0 \\ 0 & 0 & -\frac{r_B}{l_B} \end{bmatrix}\begin{bmatrix} i_{Bd} \\ i_{Bq} \\ i_{B0} \end{bmatrix} - \frac{m_B v_{dc}}{2l_B}\begin{bmatrix} \cos\delta_B \\ \sin\delta_B \\ 0 \end{bmatrix} + \begin{bmatrix} \frac{1}{l_E} & 0 & 0 \\ 0 & \frac{1}{l_E} & 0 \\ 0 & 0 & \frac{1}{l_E} \end{bmatrix}\begin{bmatrix} v_{Btd} \\ v_{Btq} \\ v_{Bt0} \end{bmatrix}$$

$$\frac{dv_{dc}}{dt} = \frac{3m_E}{4C_{dc}}\begin{bmatrix} \cos\delta_E & \sin\delta_E & 0 \end{bmatrix}\begin{bmatrix} i_{Ed} \\ i_{Eq} \\ i_{E0} \end{bmatrix} + \frac{3m_B}{4C_{dc}}\begin{bmatrix} \cos\delta_B & \sin\delta_B & 0 \end{bmatrix}\begin{bmatrix} i_{Bd} \\ i_{Bq} \\ i_{B0} \end{bmatrix}$$

For the study of power system oscillation stability, we can ignore the resistance and transients of the transformers of the UPFC, and the equations above become

$$\begin{bmatrix} v_{Etd} \\ v_{Etq} \end{bmatrix} = \begin{bmatrix} 0 & -x_E \\ x_E & 0 \end{bmatrix}\begin{bmatrix} i_{Ed} \\ i_{Eq} \end{bmatrix} + \begin{bmatrix} \frac{m_E\cos\delta_E v_{dc}}{2} \\ \frac{m_E\sin\delta_E v_{dc}}{2} \end{bmatrix}$$

$$\begin{bmatrix} v_{Btd} \\ v_{Btq} \end{bmatrix} = \begin{bmatrix} 0 & -x_B \\ x_B & 0 \end{bmatrix}\begin{bmatrix} i_{Bd} \\ i_{Bq} \end{bmatrix} + \begin{bmatrix} \frac{m_B\cos\delta_B v_{dc}}{2} \\ \frac{m_B\sin\delta_B v_{dc}}{2} \end{bmatrix} \qquad (10.10)$$

$$\frac{dv_{dc}}{dt} = \frac{3m_E}{4C_{dc}}\begin{bmatrix} \cos\delta_E & \sin\delta_E \end{bmatrix}\begin{bmatrix} i_{Ed} \\ i_{Eq} \end{bmatrix} + \frac{3m_B}{4C_{dc}}\begin{bmatrix} \cos\delta_B & \sin\delta_B \end{bmatrix}\begin{bmatrix} i_{Bd} \\ i_{Bq} \end{bmatrix}$$

From Figure10.3 we can have

$$V_t = jx_{tE}I_t + V_{Et}$$

$$V_{Et} = V_{Bt} + jx_{BV}I_B + V_b \tag{10.11}$$

which can be expressed on the d-q co-ordinate as

$$v_{dt} + jv_{qt} = jx_{tE}(i_{Ed} + i_{Bd} + ji_{Eq} + ji_{Bq}) + v_{Etd} + jv_{Etq}$$
$$= x_q(i_{Eq} + i_{Bq}) + j[E_q' - x_d'(i_{Ed} + i_{Bd})]$$

$$v_{Etd} + jv_{Etq} = v_{Btd} + jv_{Btq} + jx_{BV}i_{Bd} - x_{BV}i_{Bq} + V_b \sin\delta + jV_b \cos\delta \tag{10.12}$$

from which we can obtain

$$\begin{bmatrix} i_{Eq} \\ i_{Bq} \end{bmatrix} = \begin{bmatrix} x_q + x_{tE} + x_E & x_q + x_{tE} \\ x_E & -x_B - x_{BV} \end{bmatrix}^{-1} \begin{bmatrix} \dfrac{m_E \cos\delta_E v_{dc}}{2} \\ \dfrac{m_E \cos\delta_E v_{dc}}{2} - \dfrac{m_B \cos\delta_B v_{dc}}{2} - V_b \sin\delta \end{bmatrix}$$

$$\begin{bmatrix} i_{Ed} \\ i_{Bd} \end{bmatrix} = \begin{bmatrix} x_d' + x_{tE} + x_E & x_d' + x_{tE} \\ x_E & -x_B - x_{BV} \end{bmatrix}^{-1} \begin{bmatrix} E_q' - \dfrac{m_E \sin\delta_E v_{dc}}{2} \\ \dfrac{m_B \sin\delta_B v_{dc}}{2} + V_b \cos\delta - \dfrac{m_E \sin\delta_E v_{dc}}{2} \end{bmatrix} \tag{10.13}$$

For the dynamic model of the power system of equation (10.2) we have

$$T_e = P_e = V_q I_q + V_d I_d, \quad E_q = E_q' + (x_d - x_d')i_{dt},$$
$$\begin{matrix} v_{qt} = E_q' - x_d' i_{dt} \\ v_{dt} = x_q i_{qt} \end{matrix}, \; v_t = \sqrt{v_{dt}^2 + v_{qt}^2}, \quad \begin{matrix} i_{dt} = i_{Ed} + i_{Bd} \\ i_{qt} = i_{Eq} + i_{Bq} \end{matrix} \tag{10.14}$$

By linearizing equations (10.2), (10.13) and (10.14) we can obtain the state variable equations of the power system installed with the UPFC to be

$$\begin{bmatrix}\Delta\dot{\delta}\\ \Delta\dot{\omega}\\ \Delta\dot{E}_q'\\ \Delta\dot{E}_{fd}\end{bmatrix}=\begin{bmatrix}0 & \omega_o & 0 & 0\\ -\frac{K_1}{M} & -\frac{D}{M} & -\frac{K_2}{M} & 0\\ -\frac{K_4}{T_{do}'} & 0 & -\frac{K_3}{T_{do}'} & \frac{1}{T_{do}'}\\ -\frac{K_AK_5}{T_A} & 0 & -\frac{K_AK_6}{T_A} & -\frac{1}{T_A}\end{bmatrix}\begin{bmatrix}\Delta\delta\\ \Delta\omega\\ \Delta E_q'\\ \Delta E_{fd}\end{bmatrix}+\begin{bmatrix}0\\ -\frac{K_{pd}}{M}\\ -\frac{K_{qd}}{T_{do}'}\\ -\frac{K_AK_{vd}}{T_A}\end{bmatrix}\Delta v_{dc}$$

$$+\begin{bmatrix}0 & 0 & 0 & 0\\ -\frac{K_{pe}}{M} & -\frac{K_{p\delta e}}{M} & -\frac{K_{pb}}{M} & -\frac{K_{p\delta b}}{M}\\ -\frac{K_{qe}}{T_{do}'} & -\frac{K_{q\delta e}}{T_{do}'} & -\frac{K_{qb}}{T_{do}'} & -\frac{K_{q\delta b}}{T_{do}'}\\ -\frac{K_AK_{ve}}{T_A} & -\frac{K_AK_{v\delta e}}{T_A} & -\frac{K_AK_{vb}}{T_A} & -\frac{K_AK_{v\delta b}}{T_A}\end{bmatrix}\begin{bmatrix}\Delta m_E\\ \Delta\delta_E\\ \Delta m_B\\ \Delta\delta_B\end{bmatrix} \tag{10.15}$$

where $\Delta m_E, \Delta m_B$, $\Delta\delta_E, \Delta\delta_B$ are the deviation of input control signals to the UPFC and Δv_{dc} is that of DC bus voltage between two VSCs. By linearizing the last equation in equation (10.10) we can have

$$\Delta v_{dc}=\frac{1}{K_9+s}(K_7\Delta\delta+K_8\Delta E_q'+K_{ce}\Delta m_E+K_{c\delta e}\Delta\delta_E+K_{cb}\Delta m_B+K_{c\delta b}\Delta\delta_B) \tag{10.16}$$

The linearized model of the power system installed with the UPFC can also be shown by Figure10.2 with

$$\Delta f_* = [\Delta v_{dc} \quad \Delta u_k]$$

$$K_P=\begin{bmatrix}\frac{K_{pd}}{M}\\ \frac{K_{pu_k}}{M}\end{bmatrix}, K_q=\begin{bmatrix}\frac{K_{qd}}{T_{d0}'}\\ \frac{K_{qu_k}}{T_{d0}'}\end{bmatrix}, K_V=\begin{bmatrix}\frac{K_AK_{vd}}{T_AT_{d0}'}\\ \frac{K_AK_{vu_k}}{T_AT_{d0}'}\end{bmatrix} \tag{10.17}$$

if the input control signal to the UPFC, which is selected to be superimposed by the control output of the UPFC-based stabilizer, is Δu_k to be *e* ($\Delta u_k = \Delta m_E$), *b* ($\Delta u_k = \Delta m_B$), δ*e* ($\Delta u_k = \Delta\delta_E$) or δ*b* ($\Delta u_k = \Delta\delta_B$).

10.2.3 Phillips-Heffron model of multi-machine power systems installed with SVC, TCSC, and TCPS

The linearized equations from which the Phillips-Heffron model of an n-machine power system without FACTS-based stabilizers is derived are

$$
\begin{aligned}
&\Delta\dot{\delta} = \omega_0 \Delta\omega \\
&\Delta\dot{\omega} = \boldsymbol{M}^{-1}(-\Delta \boldsymbol{T}_E - \boldsymbol{D}\Delta\omega) \\
&\Delta\dot{\boldsymbol{E}}_q{}' = \boldsymbol{T}_{D0}{}'^{-1}[-\Delta \boldsymbol{E}_q{}' - (\boldsymbol{X}_D - \boldsymbol{X}_D{}')\Delta \boldsymbol{I}_D + \Delta \boldsymbol{E}_{FD}] \\
&\Delta\dot{\boldsymbol{E}}_{FD} = (-\Delta \boldsymbol{E}_{FD} - \boldsymbol{K}_A \Delta \boldsymbol{V}_T)\boldsymbol{T}_A{}^{-1} \\
&\Delta \boldsymbol{T}_E = \Delta \boldsymbol{I}_Q \boldsymbol{E}_{q0}{}' + \boldsymbol{I}_{Q0}\Delta \boldsymbol{E}_q{}' + \Delta \boldsymbol{I}_Q(\boldsymbol{X}_Q - \boldsymbol{X}_D{}')\boldsymbol{I}_{D0} + \boldsymbol{I}_{Q0}(\boldsymbol{X}_Q - \boldsymbol{X}_D{}')\Delta \boldsymbol{I}_D \\
&\Delta \boldsymbol{V}_{TD} = \boldsymbol{X}_Q \Delta \boldsymbol{I}_Q, \quad \Delta \boldsymbol{V}_{TQ} = \Delta \boldsymbol{E}_q{}' - \boldsymbol{X}_D{}' \Delta \boldsymbol{I}_D
\end{aligned}
\tag{10.18}
$$

where the variables with a prefix Δ are all n-order vectors. Others are n-order diagonal matrices. The output current of the ith generator can be expressed on d_i - q_i axes as :

$$
I_i = I_{di} + jI_{qi} = \sum_{k=1}^{n} Y_{ik}[E_{qi}{}' e^{j(90^0+\delta_{ik})} + (X_{qk} - X_{dk}{}')I_{qk} e^{j\delta_{ik}}] \tag{10.19}
$$

where $\boldsymbol{Y} = [Y_{ij}]$ is the system admittance matrix of the power system when only the n generator nodes are kept and $\delta_{ij} = \delta_j - \delta_i$.

Without loss of generality, we assume that in the n-machine power system, a FACTS-based stabilizer will be installed at node 1 (for SVC) or between nodes 1 and 2 (for TCSC or TCPS). In order to obtain a systematic expression for Y_{ij} which includes the influence of the FACTS-based stabilizers, we assume that the first step in forming $\boldsymbol{Y}$ is to obtain an initial system admittance matrix, $\boldsymbol{Y}_{FACTS}$, with the n generator nodes and in addition the nodes where the FACTS-based stabilizers will be installed. Then by deleting the extra nodes associated with the FACTS-based stabilizers, we can finally obtain the system admittance matrix $\boldsymbol{Y}$. So $\boldsymbol{Y}_{FACTS}$ can be arranged as

$$
\boldsymbol{Y}_{FACTS} = \begin{bmatrix} \boldsymbol{Y}_{11} & \boldsymbol{Y}_{12} \\ \boldsymbol{Y}_{21} & \boldsymbol{Y}_{22} \end{bmatrix} \begin{matrix} \leftarrow \text{nodes associated with FACTS} \\ \leftarrow \text{the } n \text{ generator nodes} \end{matrix}
$$

from which we can obtain

$$\boldsymbol{Y} = \overline{\boldsymbol{Y}}'_{22} - \boldsymbol{Y}_{21}\boldsymbol{Y}_{11}^{-1}\boldsymbol{Y}_{12} \tag{10.20}$$

For the SVC-based stabilizer,

$$\boldsymbol{Y}_{11}^{-1} = (y_{11} + j\Delta B_{SVC})^{-1}$$

where ΔB_{SVC} is the output signal of the SVC-based stabilizer and y_{11} is the self admittance at node 1.

For the TCSC-based stabilizer, we have

$$\boldsymbol{Y}_{11}^{-1} = \begin{bmatrix} y_{11} + \dfrac{j\Delta X_{CSC}}{z_{12}(z_{12} - j\Delta X_{CSC})} & y_{12} + \dfrac{-j\Delta X_{CSC}}{z_{12}(z_{12} - j\Delta X_{CSC})} \\ y_{21} + \dfrac{-j\Delta X_{CSC}}{z_{12}(z_{12} - j\Delta X_{CSC})} & y_{22} + \dfrac{j\Delta X_{CSC}}{z_{12}(z_{12} - j\Delta X_{CSC})} \end{bmatrix}^{-1}$$

since the admittance between nodes 1 and 2 changes from $\dfrac{1}{z_{12}}$ to $\dfrac{1}{z_{12} - j\Delta X_{CSC}}$ due to the addition of the output signal of the TCSC-based stabilizer, ΔX_{CSC}.

Similarly, for the TCPS-based stabilizer with ratio $k = k(\Delta\phi)e^{-j\Delta\phi}$,

$$\boldsymbol{Y}_{11}^{-1} = \begin{bmatrix} y_{11} + (\dfrac{1}{k(\Delta\phi)^2} - 1)\dfrac{1}{z_{12}} & y_{12} + (1 - \dfrac{e^{j\Delta\phi}}{k(\Delta\phi)})\dfrac{1}{z_{12}} \\ y_{21} + (1 - \dfrac{e^{-j\Delta\phi}}{k(\Delta\phi)})\dfrac{1}{z_{12}} & y_{22} \end{bmatrix}^{-1}$$

In the expression above, y_{11} and y_{22} are the self admittance's at nodes 1 and 2. y_{12} and y_{21} are the mutual admittance's between them.

So Y_{ij} in equation (10.19) can be expressed as

$$\overline{Y}_{ij} = Y'_{ij} - Y_{ij}(\Delta F) \tag{10.21}$$

where ΔF is the output signal of the FACTS-based stabilizer, which can be ΔB_{SVC}, ΔX_{CSC} or $\Delta\phi$. Therefore from equations (10.19) and (10.21) we can have

$$\Delta I_i = \Delta I_{di} + j\Delta I_{qi} = \sum_{k=1}^{n} \{Y_{ik0}\Delta[E_{qk}'e^{j(90^0+\delta_{ik})} + (X_{qk} - X_{dk}')I_{qk}e^{j\delta_{ik}}] - [E_{qk}'e^{j(90^0+\delta_{ik})} + (X_{qk} - X_{dk}')I_{qk}e^{j\delta_{ik}}]_0 \frac{\partial Y_{ik}(\Delta F)}{\partial \Delta F}\Delta F\} \tag{10.22}$$

where subscript 0 denotes the value at initial state (or steady state) of the variables. Arranging equation (10.22) in the form of matrices we have

$$\Delta \boldsymbol{I}_D = \boldsymbol{Q}_d \Delta \boldsymbol{E}_q' + \boldsymbol{P}_d \Delta\delta + \boldsymbol{M}_d \Delta \boldsymbol{I}_Q + \boldsymbol{A}_d \mathrm{I}_n \Delta F$$
$$\boldsymbol{L}_q \Delta \boldsymbol{I}_Q = \boldsymbol{Q}_q \Delta \boldsymbol{E}_q' + \boldsymbol{P}_q \Delta\delta + \boldsymbol{A}_q \mathrm{I}_n \Delta F$$

Thus we can obtain

$$\begin{aligned} \Delta \boldsymbol{I}_D &= \boldsymbol{Y}_D \Delta \boldsymbol{E}_q' + \boldsymbol{F}_D \Delta\delta + \boldsymbol{B}_D \mathrm{I}_n \Delta F \\ \Delta \boldsymbol{I}_Q &= \boldsymbol{Y}_Q \Delta \boldsymbol{E}_q' + \boldsymbol{F}_Q \Delta\delta + \boldsymbol{B}_Q \mathrm{I}_n \Delta F \end{aligned} \tag{10.23}$$

From equations (10.18) and (10.23) we obtain

$$\begin{aligned} s\Delta\delta &= \omega_0 \Delta\omega \\ s\Delta\omega &= \boldsymbol{M}^{-1}(-\boldsymbol{K}_1 \Delta\delta - \boldsymbol{D}\Delta\omega - \boldsymbol{K}_2 \Delta \boldsymbol{E}_q' - \boldsymbol{K}_P \mathrm{I}_n \Delta F) \\ (\boldsymbol{K}_3 + s\boldsymbol{T}_D)\Delta \boldsymbol{E}_q' &= -\boldsymbol{K}_4 \Delta\delta + \Delta \boldsymbol{E}_{FD} - \boldsymbol{K}_q \mathrm{I}_n \Delta F \\ (\boldsymbol{I} + s\boldsymbol{T}_A)\Delta \boldsymbol{E}_{FD} &= -\boldsymbol{K}_A(\boldsymbol{K}_5 \Delta\delta + \boldsymbol{K}_6 \Delta \boldsymbol{E}_q' + \Delta \boldsymbol{U}_{PSS} + \boldsymbol{K}_V \mathrm{I}_n \Delta F) \end{aligned} \tag{10.24}$$

where $\boldsymbol{I}_n = [1 \quad 1 \quad \cdots \quad 1]^T$. Equation (10.24) is the Phillips-Heffron model of the n-machine power system installed with SVC, TCSC or TCPS-based stabilizers, as shown by Figure 10.4, where $\Delta \boldsymbol{f}_* = \boldsymbol{I}_n \Delta F$.

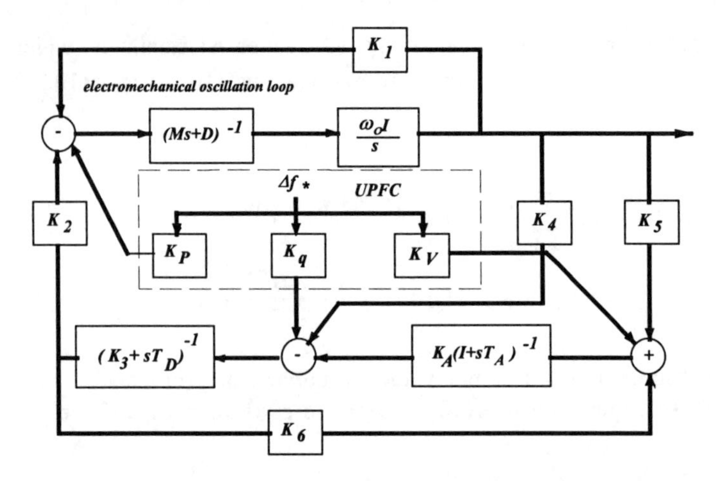

Figure10.4 The Phillips-Heffron model of an n-machine power system installed with a FACTS-based stabilizer

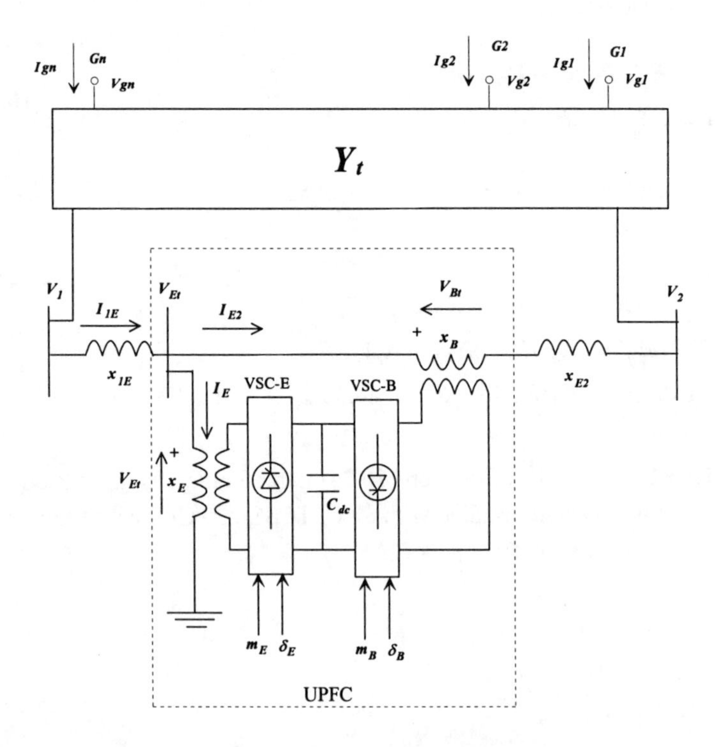

Figure 10.5 An n-machine power system installed with a UPFC

10.2.4 Phillips-Heffron model of multi-machine power systems installed with UPFC

Without loss of generality, we assume that in the n-machine power system, the UPFC will be installed between nodes 1 and 2 in the network as shown in Figure 10.5. In order to include the function of the UPFC in the network admittance matrix $\boldsymbol{Y}$, where only n generator nodes are kept, we assume that the first step in forming $\boldsymbol{Y}$ is to obtain an initial system admittance matrix, $\boldsymbol{Y}_t$, where node 1 and 2 are also kept. That is

$$\begin{bmatrix} 0 \\ 0 \\ \boldsymbol{I}_g \end{bmatrix} = \begin{bmatrix} Y_{11} & Y_{12} & \boldsymbol{Y}_{13} \\ Y_{21} & Y_{22} & \boldsymbol{Y}_{23} \\ \boldsymbol{Y}_{31} & \boldsymbol{Y}_{32} & \boldsymbol{Y}_{33} \end{bmatrix} \begin{bmatrix} V_1 \\ V_2 \\ \boldsymbol{V}_g \end{bmatrix} = \boldsymbol{Y}_t \begin{bmatrix} V_1 \\ V_2 \\ \boldsymbol{V}_g \end{bmatrix} \tag{10.25}$$

where

$$\boldsymbol{I}_g = \begin{bmatrix} I_{g1} & I_{g2} & \cdots & I_{gn} \end{bmatrix}^T, \boldsymbol{V}_g = \begin{bmatrix} V_{g1} & V_{g2} & \cdots & V_{gn} \end{bmatrix}^T$$

With the installation of the UPFC between nodes 1 and 2, the network equation of equation (10.25) becomes

$$\begin{aligned} &Y_{11}'V_1 + I_{1E} + \boldsymbol{Y}_{13}\boldsymbol{V}_g = 0 \\ &Y_{22}'V_2 - I_{E2} + \boldsymbol{Y}_{23}\boldsymbol{V}_g = 0 \\ &\boldsymbol{Y}_{31}V_1 + \boldsymbol{Y}_{32}V_2 + \boldsymbol{Y}_{33}\boldsymbol{V}_g = \boldsymbol{I}_g \end{aligned} \tag{10.26}$$

where Y_{11}' and Y_{22}' are obtained from Y_{11} and Y_{22} by excluding $x_{12} = x_{1E} + x_{E2}$. From Figure 10.5 we can have

$$\begin{aligned} &V_1 = jx_{1E}I_{1E} + V_{Et}, I_E = I_{1E} - I_{E2} \\ &V_{Et} = jx_{E2}I_{E2} + V_{Bt} + V_2 \end{aligned} \tag{10.27}$$

Equation (10.10) for the UPFC can be written as

$$V_{Et} = jx_E I_E + V_E, V_{Bt} = jx_B I_{E2} + V_B \tag{10.28}$$

where

$$V_E = \frac{m_E v_{dc}}{2}(\cos\delta_E + j\sin\delta_E) = \frac{m_E v_{dc}}{2}e^{j\delta_E}$$

$$V_B = \frac{m_B v_{dc}}{2}(\cos\delta_B + j\sin\delta_B) = \frac{m_B v_{dc}}{2}e^{j\delta_B}$$

From equations (10.27) and (10.28) we can obtain

$$\begin{bmatrix} I_{1E} \\ I_{E2} \end{bmatrix} = \frac{1}{x_\Sigma}\begin{bmatrix} -j(x_E + x_{E2} + x_B) & jx_E \\ -jx_E & j(x_{1E} + x_E) \end{bmatrix}\begin{bmatrix} V_1 \\ V_2 \end{bmatrix} + \frac{1}{x_\Sigma}\begin{bmatrix} j(x_{E2} + x_B) & jx_E \\ -jx_E & j(x_{1E} + x_E \end{bmatrix} \tag{10.29}$$

where $x_\Sigma = (x_{1E} + x_E)(x_E + x_{E2} + x_B) - x_E^{\ 2}$. Then by substituting equation (10.29) into equation (10.26) we can have

$$\boldsymbol{I_g} = \boldsymbol{CV_g} + \boldsymbol{F_E}V_E + V_B \tag{10.30}$$

where

$$\boldsymbol{C} = \boldsymbol{Y_{33}} - [\boldsymbol{Y_{31}} \quad \boldsymbol{Y_{31}}]\boldsymbol{Y_t}'^{-1}\begin{bmatrix} \boldsymbol{Y_{13}} \\ \boldsymbol{Y_{23}} \end{bmatrix}, \qquad \boldsymbol{F_E} = -[\boldsymbol{Y_{31}} \quad \boldsymbol{Y_{31}}]\boldsymbol{Y_t}'^{-1}\begin{bmatrix} \dfrac{j(x_{E2} + x_B)}{x_\Sigma} \\ \dfrac{jx_{E1}}{x_\Sigma} \end{bmatrix}$$

$$\boldsymbol{F_B} = -[\boldsymbol{Y_{31}} \quad \boldsymbol{Y_{31}}]\boldsymbol{Y_t}'^{-1}\begin{bmatrix} \dfrac{jx_E}{x_\Sigma} \\ \dfrac{-j(x_{1E} + x_E)}{x_\Sigma} \end{bmatrix},$$

$$\boldsymbol{Y_t}' = \begin{bmatrix} Y_{11}' - \dfrac{j(x_E + x_{E2} + x_B)}{x_\Sigma} & \dfrac{jx_E}{x_\Sigma} \\ \dfrac{jx_E}{x_\Sigma} & Y_{22}' - \dfrac{j(x_{1E} + x_E)}{x_\Sigma} \end{bmatrix}$$

For the n-machine power system, the terminal voltage of the generators can also be expressed in the common co-ordinate as

$$\boldsymbol{V_g} = \boldsymbol{E_q}' - j\boldsymbol{x_d}'\boldsymbol{I_g} - j(\boldsymbol{x_q} - \boldsymbol{x_d}')\boldsymbol{I_q} \tag{10.31}$$

where

$$\boldsymbol{E_q}' = [E_{q1}' \quad E_{q2}' \quad \cdots \quad E_{qn}']^T, \qquad \boldsymbol{I_q} = [I_{q1} \quad I_{q2} \quad \cdots \quad I_{qn}]^T$$
$$\boldsymbol{x_d}' = \text{diag}[x_{di}'], \quad \boldsymbol{x_q} = \text{diag}[x_{qi}]$$

From equations (10.30) and (10.31) we can obtain

$$\boldsymbol{I_g} = \boldsymbol{C_d}[\boldsymbol{E_q}' - j(\boldsymbol{x_q} - \boldsymbol{x_d}')\boldsymbol{I_q} + \boldsymbol{C_E}V_E + \boldsymbol{C_B}V_B] \tag{10.32}$$

where

$$\boldsymbol{C_d} = (\boldsymbol{C}^{-1} + j\boldsymbol{x_d}')^{-1}, \boldsymbol{C_E} = \boldsymbol{C}^{-1}\boldsymbol{F_E}, \boldsymbol{C_B} = \boldsymbol{C}^{-1}\boldsymbol{F_B}$$

In $d_i - q_i$ coordinates,

$$I_{Gi} = I_{gi}e^{j\delta_i} = \sum_{k=1}^{n} C_{dik}[E_{qk}'e^{j(90^0+\delta_k-\delta_i)} + (x_{qk} - x_{dk}')e^{j(\delta_k-\delta_i)}I_{qk}$$
$$+ C_{Ek}V_E e^{j\delta_i} + C_{Bk}V_B e^{j\delta_i}] \tag{10.33}$$

By denoting

$$C_{dik} = C_{dik}e^{j\beta dik}, C_{Ek} = C_{Ek}e^{j\beta Ek}, C_{Bi} = C_{Bk}e^{j\beta Bk}$$

Equation (10.23) becomes

$$I_{di} = \sum_{k=1}^{n} C_{dik}[-E_{qk}'\sin\delta_{ikd} + (x_{qk} - x_{dk}')I_{qk}\cos\delta_{ikd} + C_{Ek}V_E\cos\delta_{Ek} + C_{Bk}V_B\cos\delta_{Bk}$$
$$I_{qi} = \sum_{k=1}^{n} C_{dik}[E_{qk}'\cos\delta_{ikd} + (x_{qk} - x_{dk}')I_{qk}\sin\delta_{ikd} + C_{Ek}V_E\sin\delta_{Ek} + C_{Bk}V_B\sin\delta_{Bk}$$
$$\tag{10.34}$$

where

$$\delta_{ikd} = \delta_k - \delta_i + \beta_{dik}, \delta_{Ek} = \delta_E + \delta_i + \beta_{dik} + \beta_{Ei}, \delta_{Bk} = \delta_B + \delta_i + \beta_{dik} + \beta_{Bi}$$

Linearizing equation (10.24) we can have

$$\Delta I_q = Y_q \Delta\delta + F_q \Delta E_q' + G_q \Delta v_{dc} + H_{Eq} \Delta m_E + H_{Bq} \Delta m_B + R_{Eq} \Delta\delta_E + R_{Bq} \Delta\delta_B$$
$$\Delta I_d = Y_d \Delta\delta + F_d \Delta E_q' + G_d \Delta v_{dc} + H_{Ed} \Delta m_E + H_{Bd} \Delta m_B + R_{Ed} \Delta\delta_E + R_{Bd} \Delta\delta_B \quad (10.35)$$

Substituting equation (10.35) into equation (10.18) we can obtain

$$\Delta T_e = K_1 \Delta\delta + K_2 \Delta E_q' + K_{pd} \Delta v_{dc} + K_{pe} \Delta m_E + K_{pde} \Delta\delta_E + K_{pb} \Delta m_B + K_{pdb} \Delta\delta_B$$
$$\Delta E_q = K_4 \Delta\delta + K_3 \Delta E_q' + K_{qd} \Delta v_{dc} + K_{qe} \Delta m_E + K_{qde} \Delta\delta_E + K_{qb} \Delta m_B + K_{qdb} \Delta\delta_B$$
$$\Delta V_t = K_5 \Delta\delta + K_6 \Delta E_q' + K_{vd} \Delta v_{dc} + K_{ve} \Delta m_E + K_{vde} \Delta\delta_E + K_{vb} \Delta m_B + K_{vdb} \Delta\delta_B \quad (10.36)$$

Substituting equation (10.36) into equation (10.18) we can obtain the linearized Phillips-Heffron model of the n-machine power system installed with the UPFC to be

$$\begin{bmatrix} \Delta\dot{\delta} \\ \Delta\dot{\omega} \\ \Delta\dot{E}_q' \\ \Delta\dot{E}_{fd} \end{bmatrix} = \begin{bmatrix} 0 & \omega_o \mathbf{I} & 0 & 0 \\ -\mathbf{M}^{-1}\mathbf{K}_1 & -\mathbf{M}^{-1}\mathbf{D} & -\mathbf{M}^{-1}\mathbf{K}_2 & 0 \\ -\mathbf{T}_{do}'^{-1}\mathbf{K}_4 & 0 & -\mathbf{T}_{do}'^{-1}\mathbf{K}_3 & \mathbf{T}_{do}'^{-1} \\ -\mathbf{T}_A^{-1}\mathbf{K}_A\mathbf{K}_5 & 0 & -\mathbf{T}_A^{-1}\mathbf{K}_A\mathbf{K}_6 & -\mathbf{T}_A^{-1} \end{bmatrix} \begin{bmatrix} \Delta\delta \\ \Delta\omega \\ \Delta\mathbf{E}_q' \\ \Delta\mathbf{E}_{fd} \end{bmatrix}$$
$$+ \begin{bmatrix} 0 & 0 & 0 & 0 & 0 \\ -\mathbf{M}^{-1}\mathbf{K}_{pd} & -\mathbf{M}^{-1}\mathbf{K}_{pe} & -\mathbf{M}^{-1}\mathbf{K}_{pde} & -\mathbf{M}^{-1}\mathbf{K}_{pb} & -\mathbf{M}^{-1}\mathbf{K}_{pdb} \\ -\mathbf{T}_{do}'^{-1}\mathbf{K}_{qd} & -\mathbf{T}_{do}'^{-1}\mathbf{K}_{qe} & -\mathbf{T}_{do}'^{-1}\mathbf{K}_{qde} & -\mathbf{T}_{do}'^{-1}\mathbf{K}_{qb} & -\mathbf{T}_{do}'^{-1}\mathbf{K}_{qdb} \\ -\mathbf{T}_A^{-1}\mathbf{K}_A\mathbf{K}_{vd} & -\mathbf{T}_A^{-1}\mathbf{K}_A\mathbf{K}_{ve} & -\mathbf{T}_A^{-1}\mathbf{K}_A\mathbf{K}_{vde} & -\mathbf{T}_A^{-1}\mathbf{K}_A\mathbf{K}_{vb} & -\mathbf{T}_A^{-1}\mathbf{K}_A\mathbf{K}_{vdb} \end{bmatrix} \begin{bmatrix} \Delta v_{dc} \\ \Delta m_E \\ \Delta\delta_E \\ \Delta m_B \\ \Delta\delta_B \end{bmatrix} \quad (10.37)$$

From equation (10.29) we have

$$I_{E2} = \frac{1}{x_\Sigma}[-jx_E V_1 + j(x_{1E} + x_E)V_2 - jx_{1E}V_E + j(x_{1E} + x_E)V_B]$$

$$I_E = I_{1E} - I_{E2} = \frac{1}{x_\Sigma}[-j(x_{E2} + x_B)V_1 - jx_{1E}V_2 + j(x_{1E} + x_{E2} + x_B)V_E - jx_{1E}V_B]$$

Thus

$$I_{E2d} = \frac{1}{x_\Sigma}[x_E V_{t1q} - (x_{1E} + x_E)V_{t2q} + x_{1E}\frac{m_E v_{dc}}{2}\sin\delta_E - (x_{1E} + x_E)\frac{m_B v_{dc}}{2}\sin\delta_B]$$

$$I_{E2q} = \frac{1}{x_\Sigma}[-x_E V_{t1d} + (x_{1E} + x_E)V_{t2d} - x_{1E}\frac{m_E v_{dc}}{2}\cos\delta_E + (x_{1E} + x_E)\frac{m_B v_{dc}}{2}\cos\delta_B]$$

$$I_{Ed} = \frac{1}{x_\Sigma}[(x_{E2} + x_B)V_{t1q} + x_{1E}V_{t2q} - (x_{1E} + x_{E2} + x_B)\frac{m_E v_{dc}}{2}\sin\delta_E + x_{1E}\frac{m_B v_{dc}}{2}\sin\delta_B]$$

$$I_{Eq} = \frac{1}{x_\Sigma}[-(x_{E2} + x_B)V_{t1d} - x_{1E}V_{t2d} + (x_{1E} + x_{E2} + x_B)\frac{m_E v_{dc}}{2}\cos\delta_E - x_{1E}\frac{m_B v_{dc}}{2}\cos\delta_B]$$

(10.38)

where $V_{tjq} = \Delta E_{qj}' - x_{dj}'\Delta I_{dj}, V_{tjd} = x_{qj}\Delta I_{qj}, j = 1,2$. By using equation (10.35) and equation (10.38), from equation (10.10) we can have

$$\Delta v_{dc} = \frac{1}{K_9 + s}(\boldsymbol{K_7 \Delta\delta} + \boldsymbol{K_8 \Delta E_q}' + K_{ce}\Delta m_E + K_{c\delta e}\Delta\delta_E + K_{cb}\Delta m_B + K_{c\delta b}\Delta\delta_B)$$

(10.39)

Therefore, the linearized dynamic model of equation (10.37) and equation (10.39) can also be shown by Figure 10.4, where

$$\boldsymbol{\Delta f_*} = [\Delta v_{dc} \quad \Delta u_k]$$

$$\boldsymbol{K_P} = \begin{bmatrix} \boldsymbol{M^{-1}K_{pd}} \\ \boldsymbol{M^{-1}K_{pu_k}} \end{bmatrix}, \qquad \boldsymbol{K_q} = \begin{bmatrix} \boldsymbol{T_{d0}'^{-1}K_{qd}} \\ \boldsymbol{T_{d0}'^{-1}K_{qu_k}} \end{bmatrix}, \qquad \boldsymbol{K_V} = \begin{bmatrix} \boldsymbol{T_A^{-1}K_A K_{vd}} \\ \boldsymbol{T_A^{-1}K_A K_{vu_k}} \end{bmatrix}$$

(10.40)

10.3 Analysis and design of FACTS-based stabilizers

For the analysis and design of FACTS-based stabilizers, eigenvalue (or oscillation modes) computation and assignment can be used. However, in this section, the application of a traditional and effective technique, damping torque analysis and phase compensation method, is demonstrated. This technique was proposed originally based on the traditional Phillips-Heffron model of single-machine infinite-bus power systems installed with Power System Stabilizers (PSS) [1–3]. Therefore, it will be applied for the augmented Phillips-Heffron model of power systems installed with FACTS-based stabilizers introduced in the section above.

- The demonstration is presented for SVC, TCSC, and TCPS-based stabilizers. Obviously, however, the technique can also be applied for the damping function of a UPFC;

- Phillips-Heffron model is not a limitation on the application of the technique introduced in this section. It can be used for a generalized model of power systems of equation (10.1), which will be demonstrated in the next section.

10.3.1 Analysis of damping torque contribution by FACTS-based stabilizers installed in single-machine infinite-bus power systems

The damping torque contributed by an additional damping controller to the electromechanical oscillation loop of the generator is

$$\Delta T_D = T_D \omega_0 \Delta\omega \tag{10.41}$$

where T_D is the damping torque coefficient. Assuming that the feedback signal of a FACTS-based stabilizer is y_f, and its transfer function $K_* C(s)$, where K_* is the gain of the stabilizer, then according to the principle of control signal decomposition in the $\Delta\delta - j\Delta\omega$ plane, the output of the stabilizer can be expressed in the frequency domain as

$$\Delta f_* = K_* C(j\omega_s)\Delta\phi = K_* K_S \Delta\delta + K_* K_D \omega_0 \Delta\omega \tag{10.42}$$

where ω_s is the angular frequency of the oscillation. From Figure 10.2 it can be seen that the damping torque contribution by the FACTS-based stabilizer can be considered to be in two parts. The first part directly applies to the electromechanical oscillation loop of the generator and its sensitivity is mainly measured by coefficient K_P (associated with the deviation in transferred power caused by the damping control of the FACTS devices as can be seen from equation (10.9b)), which is named the direct damping torque. The second part applies through the field channel of the generator and its sensitivity is related to the deviation of field voltage as shown by equation (10.9b), which is referred to as the indirect damping torque. Usually, the direct damping torque is much greater than the indirect one, which is attenuated by the two filters before it forms the damping torque as can be seen from Figure 10.2. Therefore, from equations (10.9a), (10.41) and (10.42) we can have

$$\Delta T_D \approx K_P K_* K_D \omega_0 \Delta\omega = \frac{\partial P_e}{\partial \Delta f_*} K_* K_D \omega_0 \Delta\omega \tag{10.43}$$

From equation (10.43) we can conclude immediately that

1. The damping function of the FACTS-based stabilizer is performed mainly through the changes of the power delivered along the transmission line;
2. The damping torque provided by the FACTS-based stabilizer is proportional to its gain. The strongest damping control will result in a 'bang-bang' control because of the physical limitation of the FACTS devices. This has been found to be true in the case of the SVC damping control. Here it is concluded analytically not only for SVC but also for TCSC and TCPS.

For simplicity of the expression, we denote P_e in equation (10.8a) as

$$P_e = P_{e1} - P_{e2} \tag{10.44}$$

where P_{e1} and P_{e2} are shown in Figure 10.6.

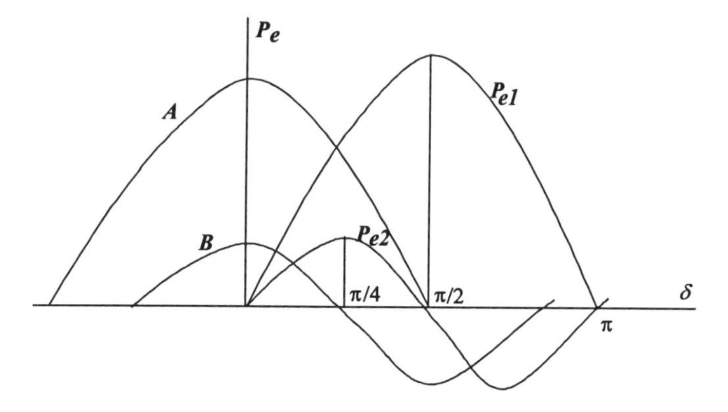

$$P_{e1} = \frac{V_b E_q' \sin\delta}{X_{d\Sigma}'}, P_{e2} = \frac{V_b^2 (X_q - X_d') \sin 2\delta}{2 X_{d\Sigma}' X_{q\Sigma}}$$

$$A = \frac{F_0 V_b E_q' \cos\delta}{X_{d\Sigma}'}, B = \frac{F_0 V_b^2 (X_q - X_d') \cos 2\delta}{2 X_{d\Sigma}' X_{q\Sigma}}$$

Figure 10.6 P–δ relationship

On the basis of equation (10.43) we can carry out some further detailed analysis as follows.

For SVC, from equations (10.4), (10.8) and (10.9) we can have

$$K_P = \frac{X_{sb}}{CX_{SVC}} \left(\frac{X_{ts} + X_d{}'}{X_{d\Sigma}{}'} P_e - \frac{X_{ts} + X_q}{X_{q\Sigma}} P_{e2} \right) \tag{10.45}$$

Since $P_e > P_{e2}$ as shown by Figure 10.6 and equation (10.44) and usually $X_q > X_d{}'$ which results in

$$\frac{X_{ts} + X_q}{X_{q\Sigma}} > \frac{X_{ts} + X_d{}'}{X_{d\Sigma}{}'}$$

Equations (10.43) and (10.45) indicate that there is a possibility of the existence of a dead point of the damping function of the SVC when $K_P = 0$ for certain parameters of the generator and a load condition. For a fixed-parameter SVC-based stabilizer, above the load condition associated with the dead point, the controller may provide the power system with positive damping and below the load condition with negative damping.

As far as the robustness of the SVC-based stabilizer to the variations of the system operating conditions is concerned, it has been found in many cases by numerical calculation and simulation that the stabilizer is more effective when the system load condition increases. This can also be concluded from equation (10.45), since the difference between P_e and P_{e2} increases when the system operates at a higher load condition as show by Figure 10.6, which results in a higher value of K_P. From equation (10.45) we also can have $\frac{\partial K_P}{\partial X_L} > 0$, which means that the SVC-based stabilizer will be more effective when the impedance of the transmission line increases.

For the TCSC, from equations (10.6), (10.8) and (10.9) we can obtain:

$$K_P = \frac{X_{CSC}}{X_{d\Sigma}{}'} P_e - \frac{X_{CSC}}{X_{q\Sigma}} P_{e2} > 0 \tag{10.46}$$

Equation (10.46) demonstrates that the TCSC-based stabilizer can always supply positive damping torque to the power system provided that the TCSC-based stabilizer is properly designed to make sure that $K_D > 0$, which means that the TCSC-based stabilizer should be designed in such a way that the output control signal can be guaranteed to have a positive component in phase with the rotor speed of the generator at the oscillation frequency on the $\Delta\delta - j\Delta\omega$ plane.

At a higher level of transferred power, the difference between P_e and P_{e2} is greater as shown by Figure 10.6, and equation (10.46) shows that more damping torque can be provided by the TCSC-based stabilizer. In other words, at a heavier load condition, the TCSC-based stabilizer is more effective. Furthermore,

$$\frac{\partial K_P}{\partial X_L} = -\left(\frac{P_e}{X_{d\Sigma}'^2} - \frac{P_{e2}}{X_{q\Sigma}^2}\right) < 0,$$

which indicates that the TCSC-based stabilizer will be less effective when the network condition of the power system changes to that with a longer transmission line provided that the same amount of load is demanded to be transferred after the change.

For the TCPS, from equations (10.8) and (10.9) we can have:

$$K_P = \frac{F_0 V_b E_q' \cos\delta}{X_{d\Sigma}'} - \frac{F_0 V_b^2 (X_q - X_d')\cos 2\delta}{2X_{d\Sigma}' X_{q\Sigma}} = A - B \qquad (10.47)$$

From Figure 10.6, we know that $K_P > 0$. Therefore, the same conclusion can be obtained about how the TCPS-based stabilizer should be designed to guaranteed that it can always provide positive damping as was the case for the TCSC-based stabilizer above. Furthermore, because

$$\frac{\partial K_P}{\partial X_L} = -\left(\frac{A}{X_{d\Sigma}'^2} - \frac{B}{X_{q\Sigma}^2}\right) < 0 \text{ (refer to Figure 10.6)} \qquad (10.48)$$

TCPS-based stabilizer will be less effective when the network conditions of the power system changes to that with a longer transmission line. From Figure 10.6 it can be seen that the difference between A and B does not change much with the load conditions of the power system. This means that the effectiveness of the TCPS-based stabilizer does not depend on the system load conditions. If $A >> B$, the same conclusion with that obtained for the TCSC-based stabilizer concerning the load conditions can be drawn.

10.3.2 Design of robust FACTS-based stabilizers installed in single-machine infinite-bus power systems by the phase compensation method

From the discussion above we know that the effectiveness of a FACTS-based stabilizer changes with the variations of power system operating conditions. To guarantee the robustness of the FACTS-based stabilizer over a set of known system operating conditions, Ω_0, we should choose the operating condition, $\mu_r \in \Omega_0$, at which the FACTS-based stabilizer is least effective so that once the stabilizer is designed at $\mu_r \in \Omega_0$ properly, it can work effectively over Ω_0. For example, from the discussion above we know that a TCSC-based stabilizer installed in a single-machine infinite-bus power system is less effective at lighter load condition with a weaker system connection to the infinite-bus bar. Therefore, it should be designed at known lightest load condition and weakest system connection so as to ensure its robustness. This strategy results in the following general method to design a robust FACTS-based stabilizer by the phase compensation.

10.3.2.1 Design procedure

If the transfer function of a FACTS-based stabilizer installed in a single-machine infinite-bus power system is denoted generally to be $F_{facts}(s) = K_* C(s)$

$$\Delta f_* = F_{facts}(s) y_f \tag{10.49}$$

and the relationship between the feedback signal of the FACTS-based stabilizer and the rotor speed of the generator is $y_f(s)$

$$y_{facts} = y_f(s)\omega_0 \Delta\omega \tag{10.50}$$

the damping torque provided by the FACTS-based stabilizer to the generator is

$$\Delta T_D = \mathrm{Re}[F_{ex}(j\omega_s) y_f(j\omega_s) F_{facts}(j\omega_s)]\omega_0 \Delta\omega \tag{10.51}$$

where $F_{ex}(s)$ is the transfer function of the forward path of the output control signal of the FACTS-based stabilizer to the electromechanical oscillation loop of the generator.

From equation (10.51) we know that with the variations of system operating conditions, the changes of the damping torque contributed by the FACTS-based stabilizer are mainly determined by the variations of the term $F_{ex}(j\omega_s)y_f(j\omega_s)$. At an operating condition $\mu_k \in \Omega_0$, if we denote

$$F_{ex}(j\omega_s)y_f(j\omega_s) = H_k\angle\varphi_k \tag{10.52}$$

$$F_{facts}(j\omega_s) = F\angle-\phi \tag{10.53}$$

from equation (10.51) we can obtain the damping torque provided by the FACTS-based stabilizer to the generator at $\mu_k \in \Omega_0$ to be

$$\Delta T_D = H_k F\cos(\varphi_k - \phi)\omega_0\Delta\omega \tag{10.54}$$

If we have

$$\varphi_0 = \max(\varphi_k), \mu_k \in \Omega_0 \tag{10.55}$$

in order to ensure the FACTS-based stabilizer providing the power system with positive synchronizing torque at all known operating conditions $\mu_k \in \Omega_0$, we should choose the compensation phase of the FACTS-based stabilizer, ϕ, to be

$$|\phi| \geq |\varphi_k| \qquad \mu_k \in \Omega_0 \tag{10.56}$$

According to equations (10.55) and (10.56) we may simply take

$$\phi = \varphi_0 \tag{10.57}$$

Therefore, from equations (10.54) and (10.57) we can obtain the damping torque contributed by the FACTS-based stabilizer to be

$$\Delta T_D = H_k F\cos(\varphi_k - \varphi_0)\omega_0\Delta\omega \tag{10.58}$$

From equation (10.58) we can see that before the FACTS-based stabilizer is designed, we can predict its effectiveness at $\mu_k \in \Omega_0$ by the term $H_k\cos(\varphi_k - \varphi_0)$. If we have

$$H_r\cos(\varphi_r - \varphi_0) = \min_k[H_k\cos(\varphi_k - \varphi_0)] \tag{10.59}$$

We know that the FACTS-based stabilizer will provide the system with least damping torque so that it will be least effective at the operating condition $\mu_r \in \Omega_0$. Therefore, we can choose $\mu_r \in \Omega_0$ to be the robust operating condition. Once the FACTS-based stabilizer is designed at $\mu_r \in \Omega_0$, equations (10.58) and (10.59) ensure that at any other operating condition, $\mu_k \in \Omega_0$, the FACTS-based stabilizer will provide the power system with more damping torque than it does at $\mu_r \in \Omega_0$ so that the effectiveness of the FACTS-based stabilizer is maintained. Thus the FACTS-based stabilizer will be robust to the variations of system operating conditions.

10.3.2.2 An example

An single-machine infinite-bus power system is to be installed with a TCR-FC type of SVC with stabilizing control loop, which is referred as SVC-based stabilizer, to damp low-frequency oscillation occurring in the power system. The parameters of the power system are

Generator:

$H = 4.0s, D = 0.0, X_d = 1.0, X_q = 0.6, X_{fd} = 1.03$

$X_{ad} = 0.85, X_{kd} = 0.95, X_{aq} = 0.45, X_{kq} = 0.7$

$R_A = 0.005, R_{fd} = 0.00065, R_{kd} = 0.0015, R_{kq} = 0.0014$ which leads

$X_d' = 0.3, T_{d0}' = 5.04s.$

Exciter (IEEE-ST1 type):

$K_a = 10.0, K_f = 0.02\ \ T_a = 0.01s, T_b = 10.0s, T_c = 1.0s, T_f = 1.0s.$

Transmission line (transformer included, one line): 0.0+j0.8

SVC voltage control: $T_{sv} = 0.15s$, $K_{sv} = 5.0p.u.$

It is known that the active power delivered along the transmission lines usually is $P_{e0} = 0.8$p.u. but may change in the range from $P_{e0} = 0.4$p.u. to $P_{e0} = 1.0$p.u. to meet the varying requirement of power supply. The normal configuration of the power system is of two parallel transmission lines. However, the system may operate with a single transmission line. Therefore, the set of the known operating conditions of the example power system is

$$\Omega_0 = \{(V_{t0}, V_{s0}, V_{b0}, P_{e0}, X_{line}) \\ \mid V_{t0} = V_{s0} = V_{b0} = 1.0\text{p.u.}, 0.4\text{p.u.} \le P_{e0} \le 1.0\text{p.u.}, X_{line} = X_L \, or \, X_{line} = 2X_L$$

Table10.1 presents the results of oscillation mode calculation and the value of $H_k \cos(\varphi_k - \varphi_0)$ at $\mu_k \in \Omega_0$. From Table10.1 we can see that there are two choices of the operating conditions at which the power system may be linearized for the design of the SVC-based stabilizer.

μ_8 : $P_{e0} = 1.0$p.u. (one). This is the operating condition at which the system oscillation mode is of poorest damping;
μ_1 : $P_{e0} = 0.4$p.u. (two). This is the robust operating condition selected by equation (10.59).

Table 10.1

Ω_0	operating condition*	oscillation mode	$H_k\angle\varphi_k$	$H_k\cos(\varphi_k - \varphi_0)$
μ_1	$P_{e0} = 0.4$p.u. (two)	$-0.0139 \pm j6.3586$	$1.3985\angle 102.89^0$	1.3806
μ_2	$P_{e0} = 0.6$p.u. (two)	$-0.0243 \pm j6.4387$	$2.3672\angle 104.32^0$	2.3456
μ_3	$P_{e0} = 0.8$p.u. (two)	$-0.0293 \pm j6.5023$	$3.5620\angle 105.63^0$	3.5396
μ_4	$P_{e0} = 1.0$p.u. (two)	$-0.0249 \pm j6.5241$	$4.8530\angle 106.54^0$	4.8305
μ_5	$P_{e0} = 0.4$p.u. (one)	$-0.0015 \pm j5.3070$	$1.9932\angle 110.53^0$	1.9925
μ_6	$P_{e0} = 0.6$p.u. (one)	$0.0074 \pm j5.2837$	$3.2514\angle 111.52^0$	3.2513
μ_7	$P_{e0} = 0.8$p.u. (one)	$0.0337 \pm j5.1943$	$4.5256\angle 112.06^0$	4.5256
μ_8	$P_{e0} = 1.0$p.u. (one)	$0.0825 \pm j4.9854$	$5.5425\angle 111.86^0$	5.5425

* (two) denotes the system configuration with two parallel transmission lines
(one) denotes the system configuration with single transmission line

In order to demonstrate that the robust operating condition (B) , $\mu_1 \in \Omega_0$, is a better selection for the design of the robust SVC-based stabilizer, the stabilizer is designed separately at μ_8 and μ_1 with a same setting target to increase the damping of the oscillation mode to around 0.1. The conventional transfer function of the stabilizer is adopted here for the SVC-based stabilizer

$$F_{facts}(s) = \frac{sT_w}{1+sT_w}\frac{K_{facts}}{1+sT_{ss}}\frac{(1+sT_2)(1+sT_4)}{(1+sT_1)(1+sT_3)}$$

With

$$K_{facts} = 26.31, T_1 = 0.5s, T_2 = 0.079s, T_3 = 0.9s, T_4 = 0.041s, T_{ss} = 0.01s, T_w = 10.0s$$

at μ_8, the stabilizer moves the oscillation mode to $-0.6003 \pm j5.2870$.

With

$$K_{facts} = 151.97, T_1 = 0.5s, T_2 = 0.031s, T_3 = 0.9s, T_4 = 0.053s, T_{ss} = 0.01s, T_w = 10.0s$$

at μ_1, the stabilizer moves the oscillation mode to $-0.6935 \pm j6.6338$.

Obviously, both of these designs result in effective SVC-based stabilizer at the operating condition selected. However, their robustness to the variations of system operating conditions is different, as shown by the results of eigenvalue calculation presented in Table 10.2.

Table 10.2 The results of oscillation mode calculation

Ω_0	selection (A)	selection (B)
μ_1	**$-0.1159 \pm j6.4297$**	$-0.6935 \pm j6.6338$
μ_2	**$-0.1940 \pm j6.5532$**	$-1.1418 \pm j6.8588$
μ_3	**$-0.2804 \pm j6.6667$**	$-1.6579 \pm j7.0715$
μ_4	**$-0.3677 \pm j6.7439$**	$-2.2105 \pm j7.2427$
μ_5	**$-0.2236 \pm j5.4068$**	$-1.3584 \pm j5.6210$
μ_6	**$-0.3568 \pm j5.4433$**	$-2.1352 \pm j5.7521$
μ_7	**$-0.4900 \pm j5.4208$**	$-2.8967 \pm j5.8150$
μ_8	$-0.6003 \pm j5.2870$	$-3.5425 \pm j5.7883$

From Table 10.2 we can see that the effectiveness of the SVC-based stabilizer designed at the operating conditions selected by (A) is not maintained at some operating conditions over Ω_0. However, the SVC-based stabilizer designed at the robust operating condition is robust to the variations of system operating conditions. These results justify the selection of the robust operating condition for the design of SVC-based stabilizer.

Figure 10.7 presents the results of a non-linear simulation. System oscillation is started by a three-phase to-earth short circuit at 1.0 second of simulation, which is cleared after 100ms. In Figure 10.7(a), the faulted line is switched back to service. In Figure 10.7(b) it is switched out of service. Obviously, the non-linear simulation confirms the results in Table 10.2 that the robust operating condition is a better selection which results in a robust SVC-based stabilizer.

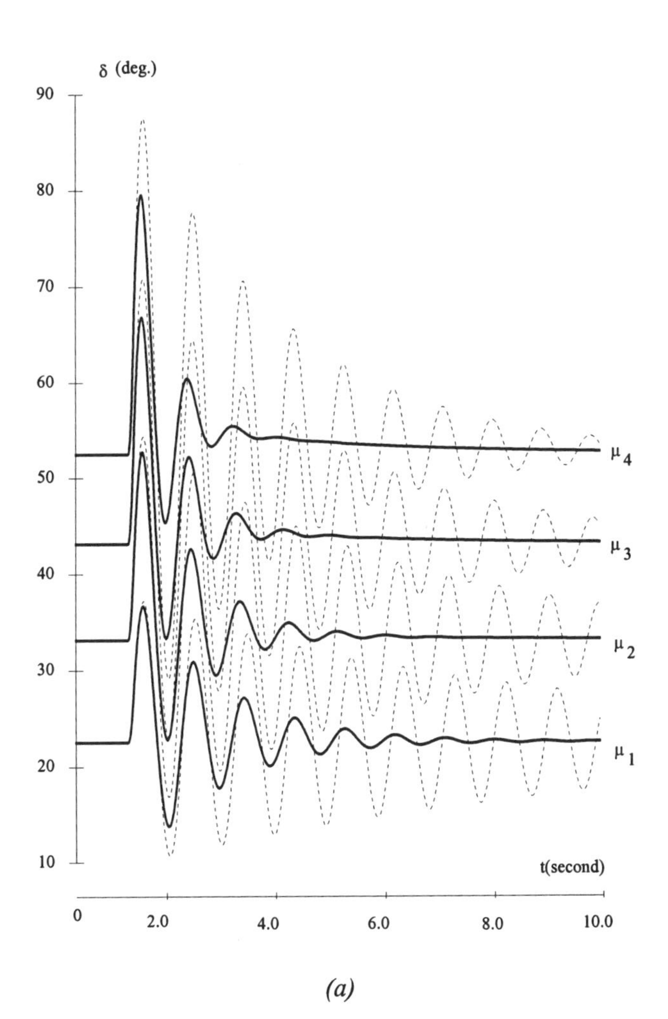

(a)

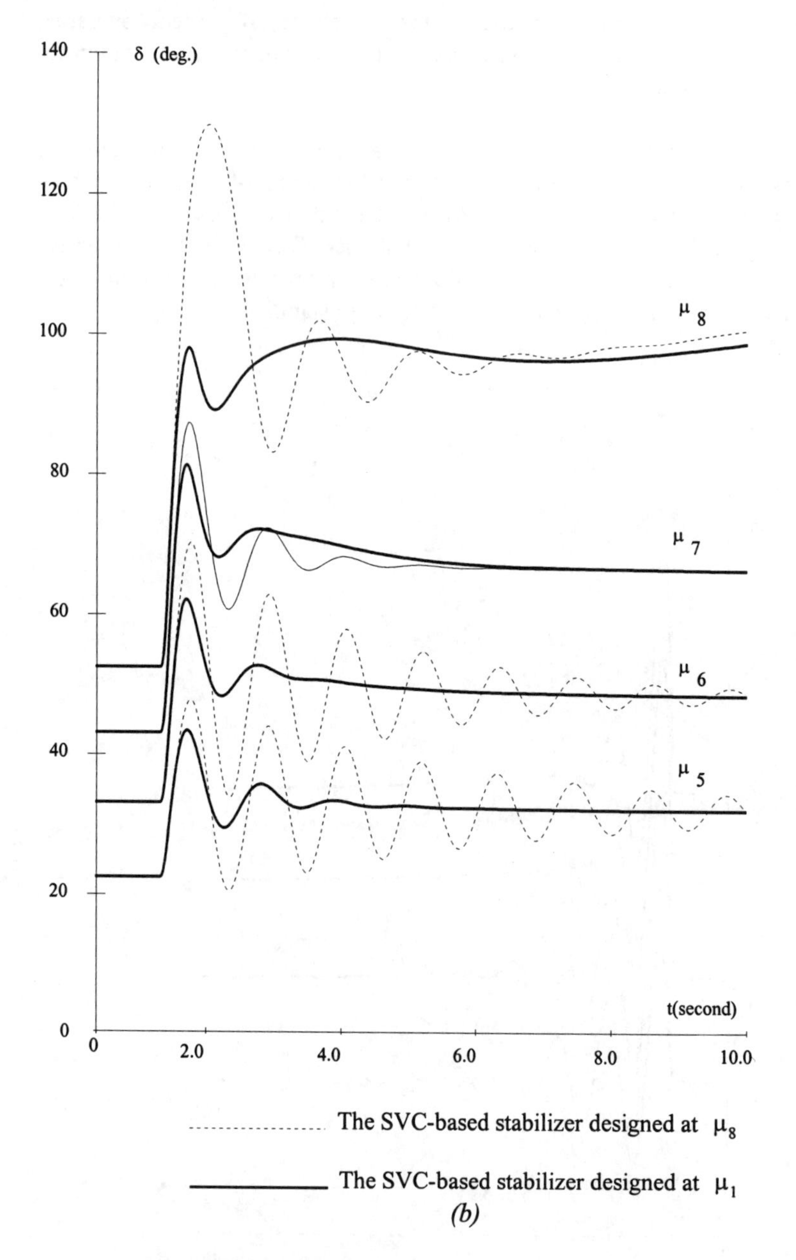

(b)

Figure 10.7 Non-linear simulation

10.3.3 Analysis of damping torque contribution by FACTS-based stabilizers installed in multi-machine power systems

From last two equations in equation (10.24) we can obtain

$$\begin{aligned}\Delta \mathbf{E_q}' = &-[(\mathbf{K_3}+s\mathbf{T_D})+\mathbf{EX}(s)\mathbf{K_6}]^{-1}[\mathbf{K_4}+\mathbf{EX}(s)\mathbf{K_5}]\Delta\delta \\ &-[((\mathbf{K_3}+s\mathbf{T_D})+\mathbf{EX}(s)\mathbf{K_6}]^{-1}[\mathbf{k_q}+\mathbf{EX}(s)\mathbf{k_V}]\Delta F \ = \mathbf{E}(s)\Delta\delta+\mathbf{f}(s)\Delta F\end{aligned} \tag{10.60}$$

where $\boldsymbol{k_q} = \boldsymbol{K_q I_n}$, $\boldsymbol{k_v} = \boldsymbol{K_V I_n}$. If we denote

$$\boldsymbol{M^{-1}K_1} = \begin{bmatrix} \boldsymbol{K_{1Jul}} & \boldsymbol{k_{1Jup}} & \boldsymbol{K_{1Jur}} \\ \boldsymbol{k_{1Jlt}} & \dfrac{K_{1jj}}{M_j} & \boldsymbol{k_{1Jrt}} \\ \boldsymbol{K_{1Jll}} & \boldsymbol{k_{1Jlo}} & \boldsymbol{K_{1Jlr}} \end{bmatrix}, \boldsymbol{M^{-1}K_2} = \begin{bmatrix} \boldsymbol{K_{2Jup}} \\ \boldsymbol{k_{2J}} \\ \boldsymbol{K_{2Jlo}} \end{bmatrix}, \boldsymbol{k_p} = \begin{bmatrix} \boldsymbol{k_{pJup}} \\ k_{pj} \\ \boldsymbol{k_{pJlo}} \end{bmatrix}$$

$$\boldsymbol{E}(s) = \left[\boldsymbol{E_{Jlt}}(s) \quad \boldsymbol{e_J}(s) \quad \boldsymbol{E_{Jrt}}(s)\right]$$

$$\boldsymbol{E}(s) = \begin{bmatrix} \boldsymbol{E_{Jup}}(s) \\ \boldsymbol{e_J}(s) \\ \boldsymbol{E_{Jlo}}(s) \end{bmatrix}, \boldsymbol{\Delta\delta} = \begin{bmatrix} \boldsymbol{\Delta\delta_{Jup}} \\ \Delta\delta_j \\ \boldsymbol{\Delta\delta_{Jlo}} \end{bmatrix}, \boldsymbol{\Delta\omega} = \begin{bmatrix} \boldsymbol{\Delta\omega_{Jup}} \\ \Delta\omega_j \\ \boldsymbol{\Delta\omega_{Jlo}} \end{bmatrix}, \boldsymbol{M_J}^{-1}\boldsymbol{D_J} = diag(\frac{D_i}{M_i}),$$

$i = 1,2,\cdots n, i \neq j$

and choose a pair $\Delta\delta_j, \Delta\omega_j$, system model of equation (10.24) can be expressed as

$$\begin{aligned} s\Delta\omega_j &= -\frac{K_{1jj}}{M_j}\Delta\delta_j - \frac{D_j}{M_j}\Delta\omega_j - \left[\boldsymbol{k_{1Jlt}} \quad \boldsymbol{k_{1Jrt}}\right]\begin{bmatrix} \boldsymbol{\Delta\delta_{Jup}} \\ \boldsymbol{\Delta\delta_{Jlo}} \end{bmatrix} - \boldsymbol{k_{2J}\Delta E_q}' - k_{pj}\Delta F \\ s\begin{bmatrix} \boldsymbol{\Delta\omega_{Jup}} \\ \boldsymbol{\Delta\omega_{Jlo}} \end{bmatrix} &= -\begin{bmatrix} \boldsymbol{k_{1Jup}} \\ \boldsymbol{k_{1Jlo}} \end{bmatrix}\Delta\delta_j - \begin{bmatrix} \boldsymbol{K_{1Jul}} & \boldsymbol{K_{1Jur}} \\ \boldsymbol{K_{1Jll}} & \boldsymbol{K_{1Jlr}} \end{bmatrix}\begin{bmatrix} \boldsymbol{\Delta\delta_{Jup}} \\ \boldsymbol{\Delta\delta_{Jlo}} \end{bmatrix} - \begin{bmatrix} \boldsymbol{K_{2Jup}} \\ \boldsymbol{K_{2Jlo}} \end{bmatrix}\boldsymbol{\Delta E_q}' - \begin{bmatrix} \boldsymbol{k_{pJup}} \\ \boldsymbol{k_{pJlo}} \end{bmatrix}\Delta F \\ \boldsymbol{\Delta E_q}' &= \left[\boldsymbol{E_{Jlt}}(s) \quad \boldsymbol{E_{Jrt}}(s)\right]\begin{bmatrix} \boldsymbol{\Delta\delta_{Jup}} \\ \boldsymbol{\Delta\delta_{Jlo}} \end{bmatrix} + \boldsymbol{e_J}(s)\Delta\delta_j + \boldsymbol{f}(s)\Delta F \end{aligned} \tag{10.61}$$

From the last two equations in equation (10.61) we can have

$$\{\frac{s^2}{\omega_0}I + \begin{bmatrix} K_{1Jul} & K_{1Jur} \\ K_{1Jll} & K_{1Jlr} \end{bmatrix} + \begin{bmatrix} K_{2Jup} \\ K_{2Jlo} \end{bmatrix} [E_{Jlt}(s) \quad E_{Jrt}(s)]\} \begin{bmatrix} \Delta\delta_{Jup} \\ \Delta\delta_{Jlo} \end{bmatrix} =$$

$$-\{\begin{bmatrix} k_{1Jup} \\ k_{1Jlo} \end{bmatrix} + \begin{bmatrix} K_{2Jup} \\ K_{2Jlo} \end{bmatrix} e_J(s)\}\Delta\delta_j - \{\begin{bmatrix} k_{pJup} \\ k_{pJlo} \end{bmatrix} + \begin{bmatrix} K_{2Jup} \\ K_{2Jlo} \end{bmatrix} f(s)\}\Delta F$$

Thus we can obtain

$$\begin{bmatrix} \Delta\delta_{Jup} \\ \Delta\delta_{Jlo} \end{bmatrix} = h_{1J}(s)\Delta\delta_j + h_{2J}(s)\Delta F \tag{10.62}$$

Substituting equation (10.62) into the last equation in equation (10.61) and (10.24) and then into the first equation in equation (10.61) we can obtain

$$\begin{aligned} \Delta E_q' &= h_{3J}(s)\Delta\delta_j + h_{4J}(s)\Delta F \\ \Delta E_{FD} &= h_{5J}(s)\Delta\delta_j + h_{6J}(s)\Delta F \\ \Delta F &= h_{J7}(s)\Delta\delta_j \end{aligned} \tag{10.63}$$

So if the feedback signal of the FACTS-based stabilizer is

$$y_f = C_1\Delta\delta_j + C_2 \begin{bmatrix} \Delta\delta_{Jup} \\ \Delta\delta_{Jlo} \end{bmatrix} + C_3\Delta E_q' + C_4\Delta E_{FD} \tag{10.64}$$

then substituting equations (10.62) and (10.63) into equation(10.64) we can have

$$y_f = \kappa_j(s)\Delta\delta_j = \frac{\omega_0\kappa_j(s)}{s}\Delta\omega_j \tag{10.65}$$

Equation (10.65) indicates that for any feedback signal of a FACTS-based stabilizer installed in an n-machine power system, it can be expressed to be

$$y_f = \gamma_j(s)\omega_0\Delta\omega_j \tag{10.66}$$

From Figure 10.4 we can obtain the forward path of the output control signal of the FACTS-based stabilizer to the electromechanical oscillation loop to be

$$\mathbf{F}(s) = k_p - K_2[K_3 + sT_D + EX(s)K_6]^{-1}[k_q + EX(s)k_V] \tag{10.67}$$

where $k_p = K_p I_n$

Since $\Delta F = F_{facts}(s)y_f$, from equations (10.66) and (10.67) we can obtain the damping torque provided by the FACTS-based stabilizer to the electromechanical oscillation loop of the *j*th generator in the power system to be

$$\Delta T_{Dij} = \mathrm{Re}[F_j(\lambda_i)\gamma_j(\lambda_i)F_{facts}(\lambda_i)]\omega_0\Delta\omega_j, j = 1,2,\cdots n \tag{10.68}$$

where $F_j(\lambda_i)$ is the *j*th element of $\boldsymbol{F}(\lambda_i)$ in equation (10.67) and λ_i is the oscillation mode of interest (or the complex oscillation frequency).

Equation (10.68) indicates that in the multi-machine power system, the damping torque is contributed by the FACTS-based stabilizer to every machine through a single forward channel

$$F_j(\lambda_i)\gamma_j(\lambda_i) = H_j\angle\varphi_j \tag{10.69}$$

So there are total n channels through which the FACTS-based stabilizer provides all machines with the damping torque.

However, in the multi-machine power system, the sources of the oscillations associated with different oscillation modes may be different. If a FACTS-based stabilizer provides a machine with a certain amount of damping torque which results in little improvement of the damping of an oscillation mode, we can believe that to the oscillation mode, the machine is not the 'source'. Based on this understanding, to the oscillation mode, $\lambda_i = -\xi_i \pm j\omega_i$, the damping of which is mainly characterised by ξ_i, we can define the sensitivity of ξ_i to an addition of the damping torque on the jth machine, $D_{Gij}\Delta\omega_j$, as S_{ij}

$$S_{ij} = \frac{\partial\xi_i}{\partial D_{Gij}} \tag{10.70}$$

So, the total improvement of the damping of the oscillation mode, λ_i, due to the addition of a damping torque on all machines is

$$\Delta\xi_i = \sum_{j=1}^{n} \frac{\partial\xi_i}{\partial D_{Gij}}\Delta D_{Gij} = \sum_{j=1}^{n} S_{ij}\Delta D_{Gij} \tag{10.71}$$

From equation (10.68) we can have

$$\Delta D_{Gij} = \mathrm{Re}[F_j(\lambda_i)\gamma_j(\lambda_i)F_{facts}(\lambda_i)]$$

Thus we have

$$\Delta\xi_i = \sum_{j=1}^{n} S_{ij} \operatorname{Re}[H_j \angle \varphi_j \Delta F_{facts}(\lambda_i)] \tag{10.72}$$

If there are M FACTS-based stabilizers installed in the n-machine power system, at the complex oscillation frequency λ_i

$$\Delta D_{Gij} = \sum_{k=1}^{M} \operatorname{Re}[F_{kj}(\lambda_i)\gamma_{kj}(\lambda_i)\Delta F_{factsk}(\lambda_i)] \tag{10.73}$$

so the total damping contribution to the oscillation mode of interest is

$$\Delta\xi_i = \sum_{j=1}^{n} S_{ij} \sum_{k=1}^{M} \operatorname{Re}[H_{kj} \angle \varphi_{kj} \Delta F_{factsk}(\lambda_i)] \tag{10.74}$$

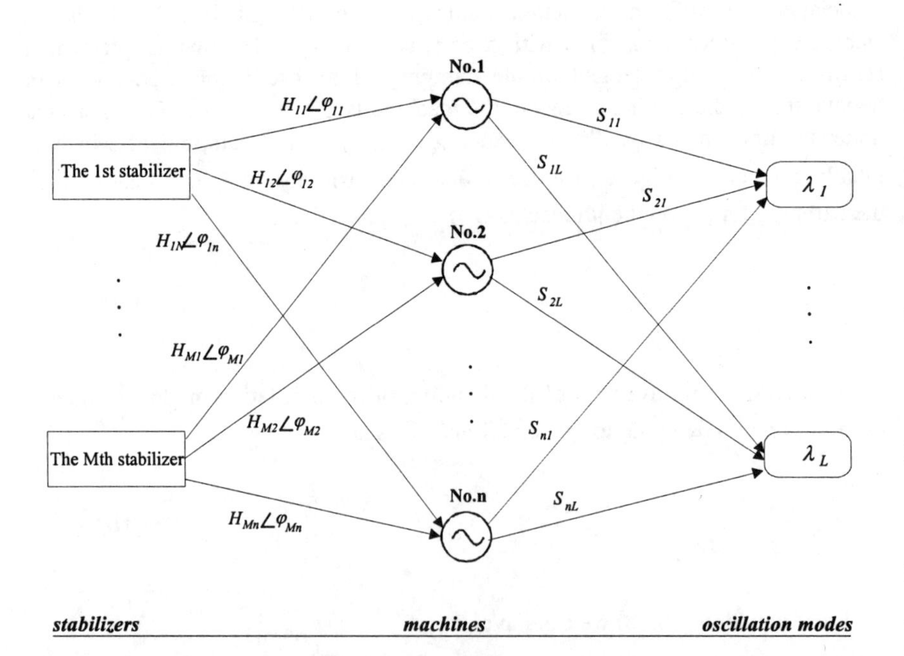

Figure 10.8 The multi-channel model of FACTS-based stabilizers providing damping to an oscillation mode in the multi-machine power system

In equations (10.73) and (10.74), the subscript k is added onto all variables to denote the kth FACTS-based stabilizer in the power system.

According to equation (10.74), a model describing the pattern that M FACTS-based stabilizers in the n-machine power system provide the oscillation mode with damping can be shown by Figure 10.8.

From Figure 10.8 it can be seen that a FACTS-based stabilizer in the n-machine power system provides damping to an oscillation mode through two groups of n channels. The first group of n channels are the forward paths through which the FACTS-based stabilizer supplies damping torque to every machine in the power system. The second group of n channels are the connections between the damping of the oscillation mode and the addition of damping torque on machines, through which the damping torque is converted into the damping of the oscillation mode. Therefore, this model presents a full and clear picture of how a FACTS-based stabilizer distributes and contributes damping to oscillation modes in the multi-machine power system.

10.3.4 Design of robust FACTS-based stabilizers installed in multi-machine power systems

Assume that the set of all known operating conditions of the n-machine power system is Ω_0. For the kth stabilizer and the ith oscillation mode, at all operating conditions $\mu_{ik}(m) \in \Omega_0$, we can calculate the weights attached to each channel in the multi-channel model of Figure 10.8, $w_{jik} = S_{ji}H_{kj}, j = 1,2,\cdots n$, and then obtain a total weight

$$W_{ik}(m) = \sum_{j=1}^{n} w_{jik} = \sum_{j=1}^{n} S_{ji}H_{kj}, \mu_{ik}(m) \in \Omega_0 \tag{10.75}$$

W_{ik} measures the damping provided by the kth FACTS-based stabilizer to the ith oscillation mode in the n-machine power system. So it presents an index to estimate the effectiveness of the kth FACTS-based stabilizer to suppress the oscillation associated with the ith oscillation mode in the power system. If we have

$$W_{ik}(m_t) = \min_{m}(W_k), \mu_{ik}(m) \in \Omega_0 \tag{10.76}$$

We know that the kth FACTS-based stabilizer is least effective in damping the oscillation at the operating condition $\mu_{ik}(m_t) \in \Omega_0$. So if we choose $\mu_{ik}(m_t) \in \Omega_0$ as the operating condition to design the kth FACTS-based stabilizer, an effective design of the kth FACTS-based stabilizer at $\mu_{ik}(m_t) \in \Omega_0$ can ensure its effectiveness at all operating conditions $\mu_{ik}(m) \in \Omega_0$ due to equation (10.76). Therefore, by selecting a robust operating condition according to equation (10.76), the robustness of the design of a single FACTS-based stabilizer to the variations of power system operating conditions is achieved.

However, in the case of the design of multiple FACTS-based stabilizers or PSSs, obviously, equation (10.76) may result in a different selection of the robust operating condition for different stabilizers and oscillation modes. So, it may be difficult or impossible to select a common operating condition at which all stabilizers can be designed simultaneously in co-ordination. On the other hand, if they are designed in co-ordination at a common operating condition simultaneously, their robustness may not be ensured due to the difference in the changing pattern of their effectiveness to the variations of system operating conditions. To solve this problem, the following method is suggested for the design of multiple stabilizers in two stages.

10.3.4.1 Sequential setting of the phase of the stabilizers

For simplicity of expression, we use the conventional form of the transfer function of stabilizers

$$F_{factsk}(s) = K_k \frac{sT_{wk}}{1+sT_{wk}} \frac{1}{1+sT_k} \frac{(1+sT_{k2})(1+sT_{k4})}{(1+sT_{k1})(1+sT_{k3})} = K_k T_k(s) \qquad (10.77)$$

It is assumed that the kth stabilizer is designed mainly to improve the damping of the ith oscillation mode in the power system, $\lambda_i = -\xi_i \pm j\omega_i$. Without loss of generality, we assume that at the operating condition of the kth stabilizer selected by use of equation (10.76), $S_{i1} \geq S_{i2} \geq \cdots \geq S_{in}$. Then a compensation angle ϕ_k is chosen to satisfy

$$90^0 > \phi_k - \varphi_{kj} > 0^0, \qquad j = 1,2,\cdots n \qquad (10.78)$$

If such a compensation angle can not be obtained, the selection of the compensation angle ϕ_k can be made among the first (n-1) machines to satisfy

$$90^0 > \phi_k - \varphi_{kj} > 0^0, \qquad j = 1,2,\cdots n-1 \qquad (10.79)$$

This procedure can be continued until such a compensation angle is obtained. Then $T_k(s)$ is set to have a phase angle $-\phi_k$, that is, $T_k(\lambda_i) = T_k \angle -\phi_k$. So from equation (10.74), the damping contribution by the kth stabilizer to λ_i can be expressed as

$$\Delta\xi_i(k) = \sum_{j=1}^{n} S_{ij} H_{kj} T_k \cos\beta_{kj} \Delta K_k, \quad j = 1,2,\cdots n \tag{10.80}$$

where $\beta_{kj} = -\phi_k + \varphi_{kj}$. The procedure above for selecting the compensation angle ensures that $0 > \beta_{kj} > -90^0$ is tenable for the machines which are more sensitive to λ_i so that they are provided with a positive synchronizing torque. Equation (10.80) shows the damping contribution to the ith oscillation mode by the kth stabilizer.

10.3.4.2. Simultaneous gain tuning

From equation (10.73) and equation (10.80) we have

$$\Delta\xi_i = \sum_{k=1}^{M} \Delta\xi_i(k) = \sum_{k=1}^{M} \sum_{j=1}^{N} S_{ij} H_{kj} T_k \cos\beta_{kj} \Delta K_k \tag{10.81}$$

Equation (10.81) demonstrates the total damping improvement by M stabilizers. It also indicates that after every $T_k(s)$ is set, the damping improvement of the oscillation mode is determined by tuning the stabilizers' gains, $\Delta K_k = K_k$.

Assume that the target damping of the oscillation modes by the co-ordinated design of M stabilizers is $-\xi_i{}^*, i = 1,2,\cdots L$. An objective function can be formed to be

$$J(\boldsymbol{K}) = \sum_{i=1}^{L} Q_i[\xi_i(\boldsymbol{K}) - \xi_i{}^*]^2 \tag{10.82}$$

where **K** is the gain vector of the stabilizers $\boldsymbol{K} = [K_1 \quad K_2 \quad \cdots \quad K_M]^T$. To find the solution of the objective function, the steepest descent algorithm can be used

$$\boldsymbol{K}(n+1) = \boldsymbol{K}(n) - st \times \nabla J[\boldsymbol{K}(n)] \tag{10.83}$$

where st is the single-dimension optimal searching length and $\nabla J[\boldsymbol{K}(n)]$ is the gradient of $J(\boldsymbol{K})$ with respect to $\boldsymbol{K}$ as

$$\nabla J[\boldsymbol{K}(n)] = \sum_{i=1}^{L} 2Q_i[\xi_i(\boldsymbol{K}) - \xi_i{}^*]\frac{\partial \xi_i(\boldsymbol{K})}{\partial \boldsymbol{K}}$$

$$\frac{\partial \xi_i(\boldsymbol{K})}{\partial \boldsymbol{K}} = \begin{bmatrix} \frac{\partial \xi_i(\boldsymbol{K})}{\partial K_1} & \frac{\partial \xi_i(\boldsymbol{K})}{\partial K_2} & \cdots & \frac{\partial \xi_i(\boldsymbol{K})}{\partial K_M} \end{bmatrix}^T \tag{10.84}$$

By using equations (10.82) and (10.83), the gains of M stabilizers are tuned simultaneously which will assign L oscillation modes accurately to positions on the complex plane with target damping.

This simultaneous tuning of all gains of stabilizers is conducted at the operating condition selected by use of equation (10.76) for each stabilizer. At the operating condition selected for the first stabilizer, the initial values of the gains are simply taken to be zero. From the operating condition selected for the second stabilizer on, the gain values set by the previous tuning are taken to be the initial values to start the current tuning. Also in the tuning process, the following constraint is checked

$$|\xi_i(\boldsymbol{K})| \geq |\xi_i{}^*|, i = 1,2,\cdots L \tag{10.85}$$

If equation (10.85) is satisfied for any $\lambda_1, 1 = 1,2,\cdots L$, which is supposed to be damped by the kth stabilizer, we set $Q_1 = 0$ and $\frac{\partial \xi_1(\boldsymbol{K})}{\partial K_k} = 0$ in equation (10.83). This arrangement frees the eigenvalue drift towards the 'good' direction, the left half of the complex plane over the target value. Therefore, a unique final solution of gain tuning can be obtained.

10.3.4.3 An example

An example of a three-machine power system is shown by Figure 10.9, the parameters of which are

Generator: $H_1 = 30.17s$, $H_2 = 13.5s$, $H_3 = 30.17s$, $D_1 = D_2 = D_3 = 0.0$,

$T_{d01}' = 7.5s$, $T_{d02}' = 4.7s$, $T_{d03}' = 7.5s$, $X_{d1} = 0.19$, $X_{d2} = 0.41$, $X_{d3} = 0.19$,

$X_{q1} = 0.163$, $X_{q2} = 0.33$, $X_{q3} = 0.163$,

$X_{d1}' = 0.0765$, $X_{d2}' = 0.173$, $X_{d3}' = 0.0765$

Exciter: $Ex_1(s) = Ex_3(s) = \dfrac{K_{Ai}}{1+sT_{Ai}}$, $T_{A1} = T_{A3} = 0.05s$,

$K_{A1} = K_{A2} = K_{A3} = 100$

Transmission lines: $Z_{12} = 0.04 + j1.2$, $x_{T1} = x_{T2} = j0.03$
The load condition: $L2 = 0.5 + j0.2$, $L3 = 1.0 + j0.6$.

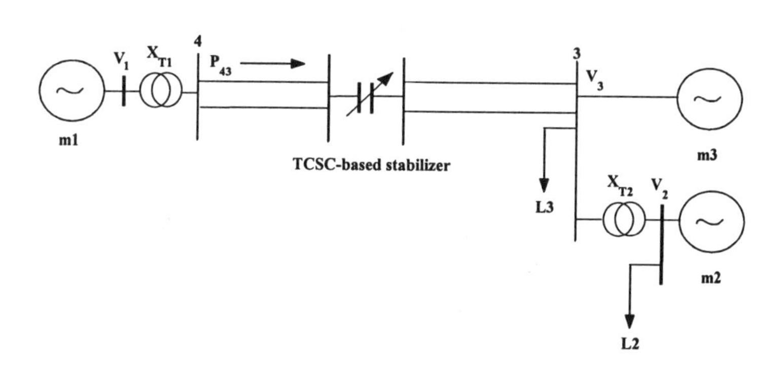

Figure 10.9 The example power system

The active power transferred along the main transmission line from bus 4 to 3 varies from $P_{43} = 0.1p.u.$ to $P_{43} = 0.9p.u.$ The example power system has one local mode and one inter-area mode poorly damped as shown in the third column of Table 10.3. It is decided that a PSS is to be installed on machine 3 to damp the local oscillation mode and a TCSC-based stabilizer is to be installed on line 3-4 to improve the damping of the inter-area. The locally available active power delivered along the transmission line is taken as the feedback signal of the stabilizers.

From Table 10.3 it can be seen that in this case of two-mode oscillations, the damping of two oscillation modes changes with system operating conditions in a different pattern. There is no common operating condition at which two modes are of worst damping. Therefore, even the simple procedure to select a common operating condition according to the damping of the oscillation modes can not be used. Here, for demonstration, a compromised operating condition for both oscillation modes, μ_3, is selected to be the common operating condition. At μ_3 two stabilizers are designed in co-ordination by use of eigenvalue assignment, which results in

Table 10.3

Ω_0	P_{43}	Without any stabilizer	With stabilizers designed at common operating conditic μ_3
μ_1	$P_{43} = 0.1$p.u.	$\lambda_{local} = -0.3925 \pm j8.6483$ $\lambda_{area} = -0.0564 \pm j4.3977$	$\lambda_{local} = -1.2107 \pm j8.6064$ $\lambda_{area} = -0.1766 \pm j4.4071$
μ_2	$P_{43} = 0.3$p.u.	$\lambda_{local} = -0.3981 \pm j8.5666$ $\lambda_{area} = -0.0578 \pm j4.4925$	$\lambda_{local} = -1.0483 \pm j8.5468$ $\lambda_{area} = -0.3246 \pm j4.5516$
μ_3	$P_{43} = 0.5$p.u.	$\lambda_{local} = -0.4000 \pm j8.5686$ $\lambda_{area} = -0.0543 \pm j4.4460$	$\lambda_{local} = -0.8574 \pm j8.5740$ $\lambda_{area} = -0.8574 \pm j8.5740$
μ_4	$P_{43} = 0.7$p.u.	$\lambda_{local} = -0.4002 \pm j8.6028$ $\lambda_{area} = -0.0462 \pm j4.3787$	$\lambda_{local} = -0.6895 \pm j8.6278$ $\lambda_{area} = -0.5593 \pm j4.5751$
μ_5	$P_{43} = 0.9$p.u.	$\lambda_{local} = -0.3987 \pm j8.6755$ $\lambda_{area} = -0.0319 \pm j4.2681$	$\lambda_{local} = -0.5172 \pm j8.7161$ $\lambda_{area} = -0.6530 \pm j4.5553$

Table 10.4

Ω_0	$W_{local-PSS}(m)$	$W_{area-TCSC}(m)$	With robust stabilizers
μ_1	4.8521	0.2997	$\lambda_{local} = -3.4821 \pm j7.2360$ $\lambda_{area} = -0.4514 \pm j4.5155$
μ_2	3.9417	0.9527	$\lambda_{local} = -2.9983 \pm j7.6606$ $\lambda_{area} = -0.7824 \pm j4.8455$
μ_3	2.8903	1.7007	$\lambda_{local} = -2.3118 \pm j8.1779$ $\lambda_{area} = -1.0067 \pm j5.0143$
μ_4	2.0081	2.5512	$\lambda_{local} = -1.6438 \pm j8.5517$ $\lambda_{area} = -1.1440 \pm j5.1281$
μ_5	1.2308	3.7461	$\lambda_{local} = -1.2474 \pm j8.8460$ $\lambda_{area} = -1.2474 \pm j5.2117$

PSS:

$K_{PSS2} = 1.8 p.u.,\ T_1 = 0.2s,\ T_3 = 00.79s,\ T_2 = 0.001s,\ T_4 = 0.79,\ T_{PSS} = 0.01s,\ T_w = 10.0s$

TCSC:

$K_{PSS2} = 1.8 p.u.,\ T_1 = 0.2s,\ T_3 = 00.79s,\ T_2 = 0.001s,\ T_4 = 0.79s,\ T_{PSS} = 0.01s,\ T_w = 10.0s$

They move two modes to:

$\lambda_{local} = -0.8574 \pm j8.5740$, $\lambda_{area} = -0.4571 \pm j4.5710$,

both of which are of good damping around 0.1. However, the results of eigenvalue calculation presented in the last column of Table 10.3 show that this design at μ_3 does not provide robust stabilizers to the variations of power system operating conditions.

Then the method proposed in the paper is used for the co-ordinated design of robust stabilizers.

1. From the results of calculating $\mathrm{W_{local-PSS}(m)}$ and $\mathrm{W_{area-TCSC}(m)}$ in Table 10.4, it can be seen that at μ_5 PSS is least effective to damp the local mode and at μ_1 the TCSC-based stabilizer is least effective to damp the inter-area mode. This estimation is confirmed by the results in the last column of Table 10.3, which demonstrate that the lower the level of power delivery on the main transmission line P_{43}, the less effective the TCSC-based stabilizer is and the higher P_{43}, the less effective the PSS. The changing pattern of the effectiveness of the stabilizers is totally different. Therefore, μ_5 is chosen to be the robust operating condition of the PSS and μ_1 the robust operating condition of the TCSC-based stabilizer.

2. At μ_5, the parameters of the PSS except its gain are set to be

$K_{PSS2} = 1.8\text{p.u.},\ T_1 = 0.2s,\ T_3 = 00.79s,\ T_2 = 0.001.,\ T_4 = 0.79s,\ T_{PSS} = 0.01s,\ T_w = 10.0s$

and at μ_1, those of the TCSC-based stabilizer except its gain are set to be

$K_{PSS2} = 1.8\text{p.u.},\ T_1 = 0.2s.,\ T_3 = 00.79s.,\ T_2 = 0.001s.,\ T_4 = 0.79,\ T_{PSS} = 0.01s.,\ T_w = 10.0s.$

3. At μ_5, the gains of the PSS and the TCSC-based stabilizer are tuned jointly by the algorithm of equations (10.82–10.85). The target damping for two oscillation modes is 0.1. The solution is $K_{PSS} = 1.28\text{p.u.}$, $K_{TCSC} = 2.26\text{p.u.}$, which moves two oscillation modes to

$\lambda_{local} = -0.8810 \pm j8.814$, $\lambda_{area} = -0.4410 \pm j4.4140$.

4. At μ_1, the gains of the PSS and the TCSC-based stabilizer are tuned jointly again by the algorithm of equations (10.82–10.85). With the same target damping of 0.1. The initial values of the gains are taken to be $K_{PSS} = 1.28 p.u.$, $K_{TCSC} = 2.26 p.u.$ The final solution is $K_{PSS} = 1.28 p.u., K_{TCSC} = 12.0 p.u.$

The last column in Table 10.4 presents the results of the robustness of the stabilizers to the variations of system operating conditions. Obviously, both stabilizers are robust.

Figure 10.10 is the results of non-linear simulation. The oscillations are triggered by a three-phase short circuit occurring at bus 3 in the example power system at 0.1 second of the simulation and is cleared after 120ms. They confirm all the results obtained above based on the eigenvalue calculation.

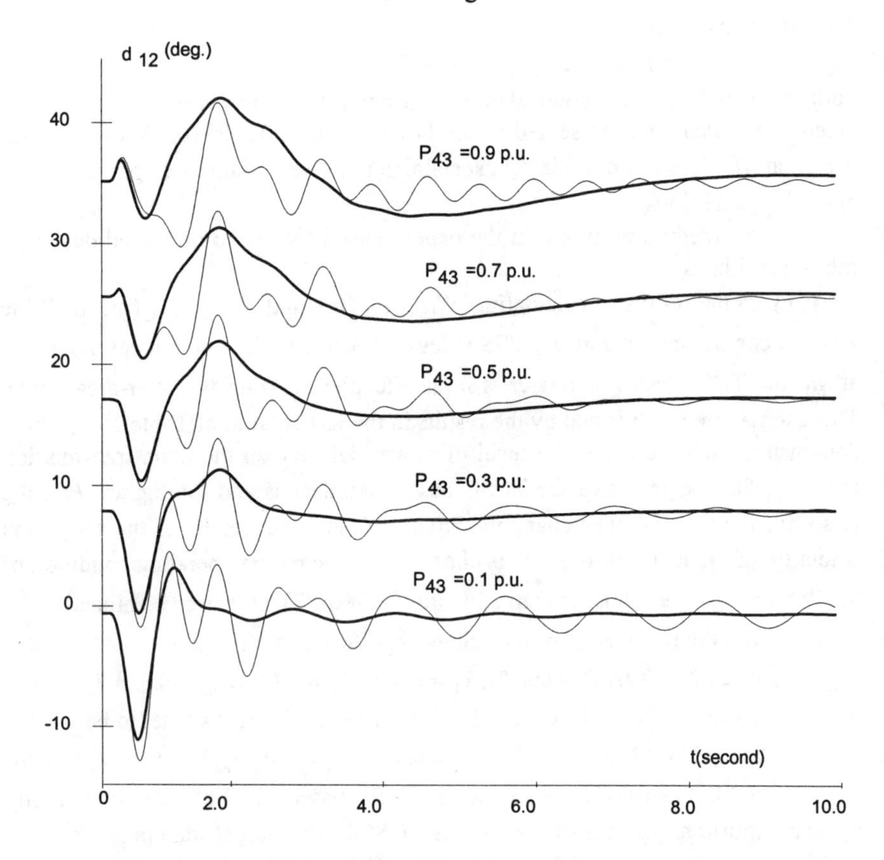

Figure 10.10 Non-linear simulation

10.4 Selection of installing locations and feedback signals of FACTS-based stabilizers

Before the parameters of a FACTS-based stabilizer installed in a multi-machine power system are set, one of the first stages in the design of the stabilizer is the selection of its location and feedback signal. At this stage, the detailed structure and parameters of the stabilizer are usually unknown, so that the closed-loop system equations cannot be formed. Therefore, the initial design to select the installing locations and feedback signals of the FACTS-based stabilizer has to be performed with information provided by the open-loop system equations.

Among the most popular techniques, the modal control analysis and the damping torque analysis methods have been applied both for FACTS-based stabilizers and PSS. In the literature, various indices may be found, based on the model control analysis and the damping torque analysis methods or a combination of both methods. The modal control analysis method is from the modal control theory of linear time-invariant systems, whereas the damping torque analysis method is based on a physical understanding of the electromechanical oscillations of power systems. They were thought to be two different techniques and have been used for the selection of installing locations and feedback signals of FACTS-based stabilizers. So far, however, the selection has been concentrated on ensuring the effectiveness of the installing locations and feedback signals of FACTS-based stabilizers. The criterion of the selection has been the maximum capability of FACTS-based stabilizers to damp power system oscillations.

However, for a good design of stabilizers, besides the maximum effectiveness of the stabilizers, the robustness of stabilizers to the variations of power system operation conditions is an equally important factor to be considered. This means that at the stage of selecting the installing locations and feedback signals of the stabilizers, not only the effectiveness of the stabilizers at a typical operating condition, where the stabilizers are designed, but also their robustness over all the range of power system operating conditions should be examined.

Therefore, in this section, the following two subjects about the selection of installing locations and feedback signals of FACTS-based stabilizers are discussed:

- The connection between the modal control analysis and damping torque analysis;
- Selection of robust installing locations and feedback signals of FACTS-based stabilizers.

The general linearized model of power systems installed with FACTS-based stabilizers of equation (10.1) is used throughout the discussion.

10.4.1 The connection between the modal control analysis and the damping torque analysis method

10.4.1.1 Damping torque analysis

To derive the damping torque contribution to a generator, for example, the jth generator, from a FACTS-based stabilizer (the *k*th stabilizer) installed in an n-machine power system, by choosing the pair of state variables on the jth generator in the power system, '$\Delta\delta_j, \Delta\omega_j$', we can rearrange equation (10.2.1) into the following form

$$\begin{bmatrix} \Delta\dot{\delta}_j \\ \Delta\dot{\omega}_j \\ \dot{z} \end{bmatrix} = \begin{bmatrix} 0 & \omega_0 & \boldsymbol{0} \\ -k_j & -d_j & \boldsymbol{A}^T{}_{J23} \\ \boldsymbol{A}_{J31} & \boldsymbol{A}_{J32} & \boldsymbol{A}_{J33} \end{bmatrix} \begin{bmatrix} \Delta\delta_j \\ \Delta\omega_j \\ z \end{bmatrix} + \begin{bmatrix} 0 \\ -B_{J2} \\ \boldsymbol{B}_{J3} \end{bmatrix} \Delta u_k$$

$$y_k = [C_{J1}, C_{J2}, \boldsymbol{C}_{J3}{}^T] \begin{bmatrix} \Delta\delta_j \\ \Delta\omega_{ji} \\ z \end{bmatrix} \tag{10.86}$$

which can be shown by Figure 10.11, where

$$K_j(s) = k_j + \frac{s}{\omega_0} d_j + \omega_0 \boldsymbol{A}_{J23}{}^T (s\boldsymbol{I} - \boldsymbol{A}_{J33})^{-1} (\boldsymbol{A}_{J31} + \frac{s}{\omega_0} \boldsymbol{A}_{J32})$$

$$K_{oj}(s) = (\frac{\omega_0}{s} C_{J1} + C_{J2}) + \boldsymbol{C}_{J3}{}^T (s\boldsymbol{I} - \boldsymbol{A}_{J33})^{-1} (\frac{\omega_0}{s} \boldsymbol{A}_{J31} + \boldsymbol{A}_{J32}) \tag{10.87}$$

$$K_{cj}(s) = \boldsymbol{A}_{J23}{}^T (s\boldsymbol{I} - \boldsymbol{A}_{J33})^{-1} \boldsymbol{B}_{J3} + B_{J2}$$

$$K_{ILj}(s) = \boldsymbol{C}_{J3}{}^T (s\boldsymbol{I} - \boldsymbol{A}_{J33})^{-1} \boldsymbol{B}_{J3}$$

From Figure 10.11 we can obtain the electric torque provided by the FACTS-based stabilizer to the electromechanical oscillation loop of the jth generator to be

$$\Delta T_j = \frac{K_{cj}(\lambda_i) K_{oj}(\lambda_i) F_{factsk}(\lambda_i)}{1 - K_{ILj}(\lambda_i) F_{factsk}(\lambda_i)} \Delta\omega_j = D_j(\lambda_i) \omega_0 \Delta\omega_j \tag{10.88}$$

The real part and imaginary part of the electric torque is the damping and synchronizing torque contributed by the damping controller. Under the conditions of the open-loop system with $F_{factsk}(\lambda_i) = 0$ when the installing location and feedback signal of the FACTS-based stabilizer are selected, the damping torque provided by the FACTS-based stabilizer to the *j*th generator is measured by the term $|K_{cj}(\lambda_i) K_{oj}(\lambda_i)|$.

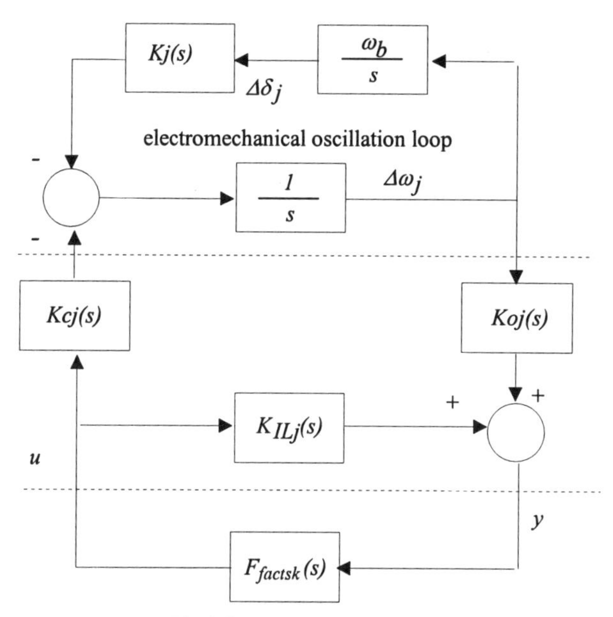

Figure 10.11 The multi-machine power system installed with a FACTS-based stabilizer

10.4.1.2 The modal control analysis

For a linear system of equation (10.1) is written as,

$$\begin{bmatrix} \Delta\dot{\delta}_j \\ \Delta\dot{\omega}_j \\ \dot{z} \end{bmatrix} = \boldsymbol{A}\begin{bmatrix} \Delta\delta_j \\ \Delta\omega_j \\ z \end{bmatrix} + \boldsymbol{B}\Delta u_k, \qquad y_k = \boldsymbol{C}^{\boldsymbol{T}}\begin{bmatrix} \Delta\delta_j \\ \Delta\omega_{ji} \\ z \end{bmatrix} \tag{10.89}$$

the controllability and observability index of a FACTS-based stabilizer associated with the oscillation mode of interest, $\lambda_i = -\sigma_i + j\omega_i$, is

$$b_i = \boldsymbol{W}_i^{\boldsymbol{T}}\boldsymbol{B}, \quad c_i = \boldsymbol{C}^{\boldsymbol{T}}\boldsymbol{V}_i \tag{10.90}$$

where $\boldsymbol{V}_i$ and $\boldsymbol{W}_i^{\boldsymbol{T}}$ are the right and left eigenvectors of the state matrix **A** with respect to the eigenvalue $\lambda_i = -\sigma_i + j\omega_i$, i.e.

$$\begin{aligned} &\boldsymbol{A}\boldsymbol{V}_i = \lambda_i\boldsymbol{V}_i \\ &\boldsymbol{W}_i^{\boldsymbol{T}}\boldsymbol{A} = \lambda_i\boldsymbol{W}_i^{\boldsymbol{T}} \\ &\boldsymbol{W}_i^{\boldsymbol{T}}\boldsymbol{V}_j = 1, \textit{if } i = j, \textit{0 otherwise} \end{aligned} \tag{10.91}$$

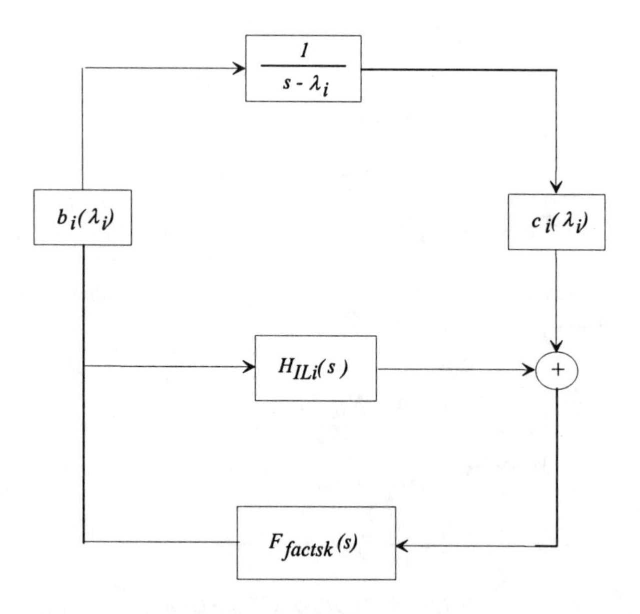

Figure 10.12 The full modal decomposition

The residue is the product of the controllability and observability index

$$R_i = b_i(\lambda_i)c_i(\lambda_i) \tag{10.92}$$

The full form of the modal control analysis is shown by Figure 10.12, from which we have

$$\Delta\lambda_i = b_i(\lambda_i)\frac{F_{factsk}(\lambda_i)}{1 - H_{ILi}(\lambda_i)F_{factsk}(\lambda_i)}c_i(\lambda_i) \tag{10.93}$$

Under the condition of the open-loop system with $F_{factsk}(\lambda_i) = 0$, the residue can be used to estimate the effectiveness of the FACTS-based stabilizer to be installed so that its installing location and feedback signal can be selected.

10.4.1.3 The connection

If we denote the left and right eigenvector in equation (10.91) to be

$$\boldsymbol{V}_i = \begin{bmatrix} v_{i1} \\ v_{i2} \\ \boldsymbol{V}_{i3} \end{bmatrix}, \boldsymbol{W}_i^T = \begin{bmatrix} w_{i1} & w_{i2} & \boldsymbol{W}_{i3}^T \end{bmatrix}$$

From equations (10.86) and (10.91) we have

$$\begin{bmatrix} w_{i1} & w_{i2} & \boldsymbol{W}_{i3}^T \end{bmatrix} \begin{bmatrix} 0 & \omega_0 & \boldsymbol{0} \\ -k_j & -d_j & -\boldsymbol{A}_{J23} \\ \boldsymbol{A}_{J31} & \boldsymbol{A}_{J32} & \boldsymbol{A}_{J33} \end{bmatrix} = \lambda_i \begin{bmatrix} w_{i1} & w_{i2} & \boldsymbol{W}_{i3}^T \end{bmatrix} \tag{10.94}$$

from which we can obtain

$$-\boldsymbol{A}_{J23} w_{i2} + \boldsymbol{W}_{i3}^T \boldsymbol{A}_{J33} = \lambda_i \boldsymbol{W}_{i3}^T \tag{10.95}$$

Solving equation (10.95) results in

$$\boldsymbol{W}_{i3}^T = -\boldsymbol{A}_{J23}(\lambda_i \boldsymbol{I} - \boldsymbol{A}_{J33})^{-1} w_{i2} \tag{10.96}$$

From equations (10.86), (10.90) and (10.96) we have

$$b_i(\lambda_i) = \begin{bmatrix} w_{i1} & w_{i2} & \boldsymbol{W}_{i3}^T \end{bmatrix} \begin{bmatrix} 0 \\ -B_{J2} \\ \boldsymbol{B}_{J3} \end{bmatrix}$$

$$= [-B_{J2} - \boldsymbol{A}_{J23}(\lambda_i \boldsymbol{I} - \boldsymbol{A}_{J33})^{-1} \boldsymbol{B}_{J3}] w_{i2} = -K_{cj}(\lambda_i) w_{i2} \tag{10.97}$$

Similarly, from equations (10.86) and (10.91) we have

$$\begin{bmatrix} 0 & \omega_0 & \boldsymbol{0} \\ -k_j & -d_j & -\boldsymbol{A}_{J23} \\ \boldsymbol{A}_{J31} & \boldsymbol{A}_{J32} & \boldsymbol{A}_{J33} \end{bmatrix} \begin{bmatrix} v_{i1} \\ v_{i2} \\ \boldsymbol{V}_{i3} \end{bmatrix} = \lambda_i \begin{bmatrix} v_{i1} \\ v_{i2} \\ \boldsymbol{V}_{i3} \end{bmatrix} \tag{10.98}$$

from which we can obtain

$$\begin{aligned} &\omega_0 v_{i2} = \lambda_i v_{i1} \\ &\boldsymbol{A}_{J31} v_{i1} + \boldsymbol{A}_{J32} v_{i2} + \boldsymbol{A}_{J33} \boldsymbol{V}_{i3} = \lambda_i \boldsymbol{V}_{i3} \end{aligned} \tag{10.99}$$

Solving equation (10.99) results in

$$\begin{aligned} &v_{i1} = \frac{\omega_0}{\lambda_i} v_{i2} \\ &\boldsymbol{V}_{i3} = (\lambda_i \boldsymbol{I} - \boldsymbol{A}_{J33})^{-1}(\boldsymbol{A}_{J31} \frac{\omega_b}{\lambda_i} + \boldsymbol{A}_{J32}) v_{i2} \end{aligned} \tag{10.100}$$

From equations (10.86), (10.90) and (10.100) we have

$$c_i(\lambda_i) = [C_{J1}, C_{J2}, \boldsymbol{C}_{J3}] \begin{bmatrix} v_{i1} \\ v_{i2} \\ \boldsymbol{V}_{i3} \end{bmatrix}$$

$$\begin{aligned} &= [(\frac{\omega_0}{\lambda_i} C_{J1} + C_{J2}) + \boldsymbol{C}_{J3}(\lambda_i \boldsymbol{I} - \boldsymbol{A}_{J33})^{-1}(\boldsymbol{A}_{J31} \frac{\omega_0}{\lambda_i} + \boldsymbol{A}_{J32})] v_{i2} \\ &= K_{oj}(\lambda_i) v_{i2} \end{aligned} \tag{10.101}$$

From equations (10.97) and (10.101) we can establish the connection among the electric torque and the residue to be

$$R_i = -K_{cj}(\lambda_i)K_{oj}(\lambda_i)v_{i2}w_{i2} = -K_{cj}(\lambda_i)K_{oj}(\lambda_i)P_i \tag{10.102}$$

where $P_i = v_{i2}w_{i2}$ is the participation of the jth generator associated with the ith oscillation mode.

10.4.2 Selection of robust installing locations and feedback signals of FACTS-based stabilizers

If we denote the set of the candidate installing locations and feedback signals of a FACTS-based stabilizer in a multi-machine power system is $\Phi(\varphi)$ and that of power system operating conditions $\Omega(\mu)$, the effectiveness of the stabilizer is the function of φ and μ, $C(\varphi, \mu)$. If only the maximum capability of the stabilizer to damp oscillations is considered, the criterion of selecting the installing locations and feedback signals of the stabilizer is

$$\underset{\varphi}{Max}[C(\varphi,\mu_0)], \quad \mu_0 \in \Omega(\mu) \tag{10.103}$$

However, the following two criteria are useful in the selection of both locations and feedback signals of a FACTS-based stabilizer, to ensure both the effectiveness and the robustness of the stabilizer

$$\underset{\varphi}{Max}[\underset{\mu}{\min} C(\varphi,\mu)] \tag{10.104}$$

$$\underset{\varphi}{Min}[\underset{\mu}{\max} C(\varphi,\mu) - \underset{\mu}{\min} C(\varphi,\mu)] \tag{10.105}$$

1. The objective of choosing an effective installing location or feedback signal for the stabilizer is to reduce the control cost. Therefore, by applying the criterion of equation (10.104) so that the most effective installing location or feedback signal is selected at the operating condition where the stabilizer is least effective, it is guaranteed that the minimum control cost is achieved.
2. A good design of the stabilizer requires that it provides steady damping over all the range of power system operating conditions. If the damping contribution of the stabilizer increases greatly, on one hand, with the variations of power system operating conditions, the control could be over-strong at some operating conditions, which would pose much unwanted influence on other modes in the power system. On the other hand, a sharp drop in the damping contribution from the stabilizer with the changes of power system operating conditions results in poor robustness of the stabilizer. Therefore, the criterion of equation (10.105) requires that the damping contribution by the stabilizer changes as little as possible with the variations of power system operating conditions. However, this criterion should be applied jointly with

that of equation (10.104), since failing to meet the requirement of the effectiveness obviously is not a proper selection.

Obviously, to apply the criteria of equations (10.104) and (10.105), the effectiveness of the FACTS stabilizer has to be examined over all the range of power system operating conditions. The modal control analysis and the damping torque analysis methods introduced above can be used for this application. However, since these two methods require the eigensolution of system state matrix, it would result in a heavy computational burden.

Therefore, an eigensolution free method is introduced as follows which is based on the assumption that at least one of the generators in the power system, which are sensitive to the oscillation mode of interest, is known. Without loss of generality, it can be assumed to be the *j*th generator in the power system and the oscillation mode is λ_i.

In the selection of installing locations and feedback signals of the FACTS-based stabilizer to be installed in the multi-machine power system, the residue, $R_i = b_i(\lambda_i)c_i(\lambda_i)$, are calculated for comparison among various candidate locations and feedback signals. For example, if there are two candidate installing locations or feedback signals, A and B, and

$$|b_{iA}(\lambda_i)|\,|c_{iA}(\lambda_i)| > |b_{iB}(\lambda_i)||c_{iB}(\lambda_i)| \tag{10.106}$$

then A is considered to be better than B as the installing location or the feedback signal of the FACTS-based stabilizer. Therefore, it is the ratio, $\dfrac{|b_{iA}(\lambda_i)|}{|b_{iB}(\lambda_i)|}\dfrac{|c_{iA}(\lambda_i)|}{|c_{iB}(\lambda_i)|}$, not the values of the residue, that determines the selection.

On the other hand, in system state equation of equation (10.86) or equation (10.89), if the installing location or the feedback signal of the FACTS-based stabilizer is different, the control and the output vectors, $\boldsymbol{B}$ and $\boldsymbol{C}^T$, are not the same but the open-loop state matrix $\boldsymbol{A}$ is unchanged provided that the operating point of the system remains the same. That is,

$$v_{i2A} = v_{i2B}, \qquad w_{i2A} = w_{i2B} \tag{10.107}$$

Therefore, from equations (10.102) and (10.107), we can obtain

$$\frac{|b_{iA}(\lambda_i)|}{|b_{iB}(\lambda_i)|}\frac{|c_{iA}(\lambda_i)|}{|c_{iB}(\lambda_i)|} = \frac{|K_{biA}(\lambda_i)|}{|K_{biB}(\lambda_i)|}\frac{|K_{ciA}(\lambda_i)||w_{i2A}|}{|K_{ciB}(\lambda_i)||w_{i2B}|}\frac{|v_{i2A}|}{|v_{i2B}|} = \frac{|K_{biA}(\lambda_i)|}{|K_{biB}(\lambda_i)|}\frac{|K_{ciA}(\lambda_i)|}{|K_{ciB}(\lambda_i)|} \tag{10.108}$$

Equation (10.108) shows that $|K_{bi}(\lambda_i)|\,|K_{ci}(\lambda_i)|$ can replace $|b_i(\lambda_i)||c_i(\lambda_i)|$ to be the index to measure the effectiveness of the FACTS-based stabilizer so as to select the installing locations and feedback signals. Since in most cases, the oscillation mode of interest is lightly damped, i.e., $\lambda_i \approx j\omega_i$, so

$$|K_{bi}(\lambda_i)|\,|K_{ci}(\lambda_i)| \approx |K_{bi}(j\omega_i)|\,|K_{ci}(j\omega_i)| \tag{10.109}$$

The index can be calculated without knowing the eigensolution of the open-loop state matrix $\boldsymbol{A}$. Therefore, it is an eigensolution free index. The objective to choose the pair of $\Delta\delta_j$, $\Delta\omega_j$ on the jth generator, which is sensitive to the oscillation mode of interest, is to avoid $|v_{i2}| \approx 0, |w_{i2}| \approx 0$ so that the replacement of the ratio of $|b_i(\lambda_i)||c_i(\lambda_i)|$ by that of $|K_{bi}(\lambda_i)|\ |K_{ci}(\lambda_i)|$ in equation (10.108) is correct.

To select the robust installing locations and feedback signals of the FACTS-based stabilizer by use of $|K_{bi}(j\omega_i)|\ |K_{ci}(j\omega_i)|$, according to equation (10.104) and (10.105), the criteria are

$$\underset{\varphi}{\text{Max}}\{\underset{\mu}{\text{Min}}[|K_{bi}(j\omega_i)||K_{ci}(j\omega_i)|], \text{ for all} \mu \in \Omega(\mu)\}, \text{ for all} \varphi \in \Phi(\varphi) \quad (10.110)$$

$$\underset{\varphi}{\text{Min}}\{\underset{\mu}{\text{Max}}[|K_{bi}(j\omega_i)||K_{ci}(j\omega_i)|] - \underset{\mu}{\text{Min}}[|K_{bi}(j\omega_i)||K_{ci}(j\omega_i)|], \text{ for all} \mu \in \Omega(\mu)\}, \text{ for all} \varphi \in \Phi(\varphi) \quad (10.111)$$

Obviously, applying the criteria to select the robust installing locations and feedback signals in the set of system operating conditions, $\Omega(\mu)$, over the candidate set, $\Phi(\varphi)$, the eigensolution free method greatly reduces the computational cost.

10.4.3 An example

An example three-machine power system is shown by Figure 10.13. Its parameters are

Generator: $H_1 = 30.17s$, $H_2 = 13.5s$, $H_3 = 30.17s$, $D_1 = D_2 = D_3 = 0.0$,

$T_{d01}' = 7.5s$, $T_{d02}' = 4.7s$, $T_{d03}' = 7.5s$, $X_{d1} = 0.19$, $X_{d2} = 0.41$, $X_{d3} = 0.19$,

$X_{q1} = 0.163$, $X_{q2} = 0.33$, $X_{q3} = 0.163$,

$X_{d1}' = 0.0765$, $X_{d2}' = 0.173$, $X_{d3}' = 0.0765$

Exciter: $Ex_1(s) = \dfrac{K_{A1}}{1 + sT_{A1}}$, $Ex_2(s) = \dfrac{K_{A2}}{1 + sT_{A2}}$, $Ex_3(s) = \dfrac{K_{A3}}{1 + sT_{A3}}$,

$T_{A1} = T_{A2} = T_{A3} = 0.05s$, $K_{A1} = K_{A2} = K_{A3} = 100$

Transmission lines: $Z_{12} = Z_{32} = 0.0 + j0.2$, $X_{T1} = X_{T3} = j0.03$

A low-frequency oscillation of about 0.65 Hz has been observed in the power system. It is decided that a TCSC (Thyristor-Controlled Series Compensator)-based stabilizer is to be installed in the power system to damp the low-frequency oscillation. The candidate installing locations are lines between node 1 and 2 and that between node 3 and 2. The locally available active line power or the magnitude of the bus voltage of TCSC could be the feedback signal of the stabilizer.

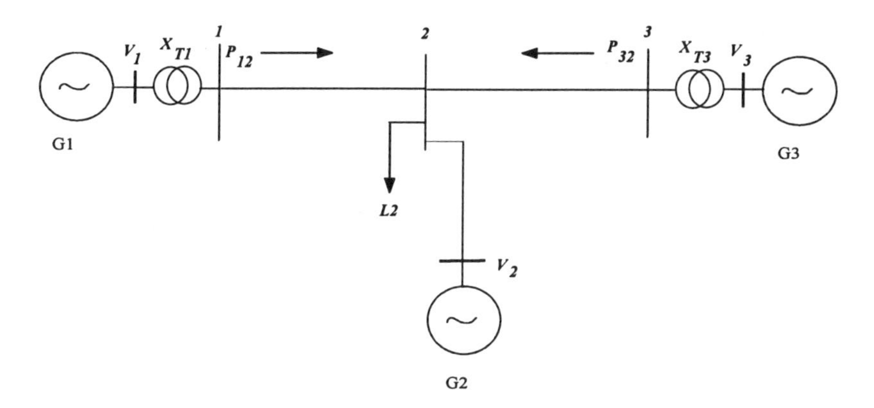

Figure 10.13 An example three-machine power system

Table 10.5

The candidate set $\Phi(\varphi)$	φ_1	φ_2	φ_3	φ_4
Installing location	Line 1-2	Line 3-2	Line 1-2	Line 3-2
Feedback signal	P_{12}	P_{32}	Bus voltage	Bus voltage
$\lvert K_{bi}(j\omega_i)\rvert$	18.11	14.21	18.11	14.21
$\lvert K_{ci}(j\omega_i)\rvert$	0.92	1.01	0.13	0.09
$\lvert K_{bi}(j\omega_i)\rvert\,\lvert K_{ci}(j\omega_i)\rvert$	**16.69**	**14.38**	**2.31**	**1.28**

10.4.3.1 Selection by using the conventional criterion of equation (10.103)

The typical operating condition of the example power system is μ_0:

$\overline{V_1} = 1.05\angle 28.7^0, \overline{V_3} = 1.05\angle 21.1^0, \overline{V_2} = 1.0\angle 0^0, P_{12} = 0.8\text{p.u.}, P_{32} = 0.6\text{p.u.}$

It is known that G_3 is sensitive to system oscillation. So by choosing the pair of variables, $\Delta\delta_3$, $\Delta\omega_3$, and taking $\lambda_i \approx j\omega_i = j2\pi \times 0.65 = j4.1$, the computational results of the eigensolution free method are presented in Table 10.5. From Table 10.5 it can be seen that the best installing location and feedback signal for the TCSC-based stabilizer are line 1-2 and P_{12}.

To confirm the prediction made by the eigensolution free method of modal control analysis, the TCSC-based stabilizer is installed on line 1-2 with P_{12} as the feedback signal and on line 3-2 with P_{32} separately. The conventional stabilizer structure is adopted here to be

$$G_{TCSC}(s) = K\frac{sT_w}{1+sT_w}\frac{1}{1+sT}\frac{(1+sT_2)(1+sT_4)}{(1+sT_1)(1+sT_3)}$$

When the TCSC-based stabilizer is installed on line 1-2 and takes P_{12} as the feedback signal,

$K = 0.90, T_1 = 0.7s,\ T_2 = 0.37s,\ T_3 = 0.9s, T_4 = 0.34s,\ T = 0.025s,\ T_w = 3.0s$

the oscillation mode is moved from $\lambda = -0.13 \pm j4.14$ to $\lambda = -0.52 \pm j3.88$.

When the TCSC-based stabilizer is installed on line 3-2 and takes P_{32} as the feedback signal,

$K = 2.50, T_1 = 0.7s,\ T_2 = 0.42.,\ T_3 = 0.9s,\ T_4 = 0.45s,\ T = 0.025s,\ T_w = 3.0s$

the oscillation mode is moved to $\lambda = -0.52 \pm j3.74$.

Figure 10.14 is the result of non-linear simulation. The oscillation is triggered by a three-phase short circuit occurring at bus 2 in the example power system at 0.1 second of the simulation and is cleared after 120ms. The results above demonstrated that if the TCSC-based stabilizer is installed on line 3-2 and takes P_{32} as the feedback signal, higher control cost (higher gain value of the stabilizer) is needed to damp the oscillation. Therefore, at this operating condition, μ_0, line 1-2 and P_{12} is the best installing location and feedback signal of the TCSC-based stabilizer.

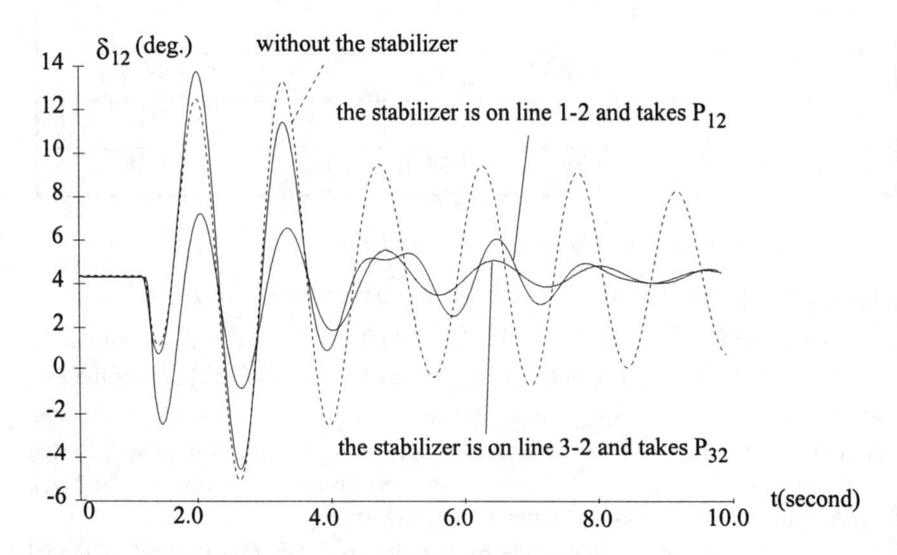

Figure 10.14 Non-linear simulation

10.4.3.2 The selection of robust installing location and feedback signal by using the new criterion of equations (10.104) and (10.105)

The selection of installing location and feedback signal above is carried out at the typical operating condition of the power system, μ_0, according to the criterion of equation (10.103). However, in the example power system of Figure 10.13, the power supply from two end generators to the load centre at bus 2 is known to vary with the changes of load requirement. The variations of power supply are in the range

$0.2\text{p.u.} \leq P_{12} \leq 1.2\text{p.u.}; 0.6\text{p.u.} \leq P_{32} \leq 0.8\text{p.u.}$

Therefore, the set of known system operating conditions is

$\Omega(\mu) = \{\mu(P_{12}, P_{32}) : 0.2\text{p.u.} \leq P_{12} \leq 1.2\text{p.u.}; 0.6\text{p.u.} \leq P_{32} \leq 0.8\text{p.u.}\}$

The selection carried out at the typical operating condition $\mu_0 \in \Omega(\mu)$ cannot ensure that the installing location line 1-2 and feedback signal P_{12} is the best at every operating condition of the power system $\mu \in \Omega(\mu)$. Therefore, the robustness of the candidate installing locations and feedback signals to the variations of system operating conditions should be examined so that a robust installing location and feedback signal is chosen. Figure 10.15 and 10.16 show the results of the examination by use of the eigensolution free method. From Figure 10.15 and 10.16 it can be seen that

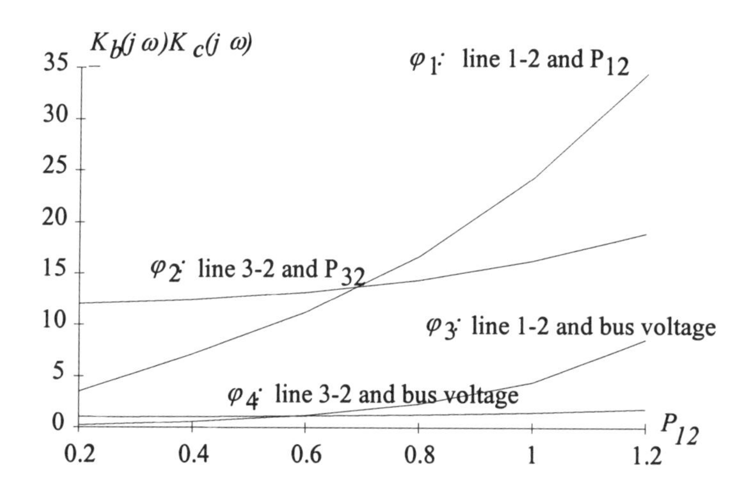

Figure 10.15 The computational results with variations of P_{12} at P_{32} =0.6p.u. by the eigensolution free method

$$\underset{\mu}{Min}[|K_{bi}(j\omega_i)||K_{ci}(j\omega_i)|] at \varphi_2 \in \Phi(\varphi) >$$
$$\underset{\mu}{Min}[|K_{bi}(j\omega_i)||K_{ci}(j\omega_i)|] at \varphi_1 \in \Phi(\varphi) >$$
$$\underset{\mu}{Min}[|K_{bi}(j\omega_i)||K_{ci}(j\omega_i)|] at \varphi_4 \in \Phi(\varphi) >$$
$$\underset{\mu}{Min}[|K_{bi}(j\omega_i)||K_{ci}(j\omega_i)|] at \varphi_3 \in \Phi(\varphi) \tag{10.112}$$

$$\underset{\mu}{Max}[|K_{bi}(j\omega_i)||K_{ci}(j\omega_i)|] - \underset{\mu}{Min}[|K_{bi}(j\omega_i)||K_{ci}(j\omega_i)|] at\, \varphi_1 \in \Phi(\varphi) >$$
$$\underset{\mu}{Max}[|K_{bi}(j\omega_i)||K_{ci}(j\omega_i)|] - \underset{\mu}{Min}[|K_{bi}(j\omega_i)||K_{ci}(j\omega_i)|] at\, \varphi_3 \in \Phi(\varphi) >$$
$$\underset{\mu}{Max}[|K_{bi}(j\omega_i)||K_{ci}(j\omega_i)|] - \underset{\mu}{Min}[|K_{bi}(j\omega_i)||K_{ci}(j\omega_i)|] at\, \varphi_2 \in \Phi(\varphi) >$$
$$\underset{\mu}{Max}[|K_{bi}(j\omega_i)||K_{ci}(j\omega_i)|] - \underset{\mu}{Min}[|K_{bi}(j\omega_i)||K_{ci}(j\omega_i)|] at\, \varphi_4 \in \Phi(\varphi) \tag{10.113}$$

Therefore, according to the criterion of equation (10.110), the best installing location and feedback signal is φ_2: line 3-2 and P_{32}. Although φ_2 does not provide the best result based on criterion (10.111), it does provide adequate insensitivity and is judged overall to be the best.

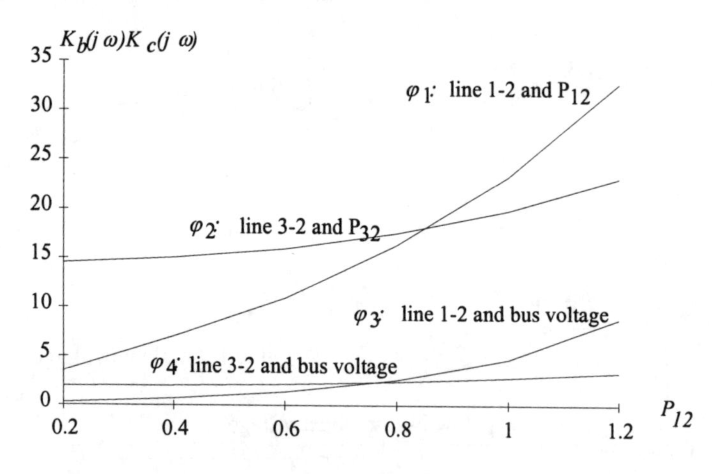

Figure 10.16 The computational results with variations of P_{12} at P_{32} =0.8 p.u. by the eigensolution free method

To confirm the conclusion that line 3-2 and P_{32} is the robust installing location and feedback signal for the TCSC-based stabilizer, Table 10.6 presents the results of the changes of the damping of the oscillation mode by eigenvalue calculation over $\mu \in \Omega(\mu)$ when the TCSC-based stabilizer is designed at $\mu_0 \in \Omega(\mu)$. From Table 10.6 it can be seen that when the power transferred along line 1-2 is below 0.6p.u., the damping provided by the stabilizer to the oscillation mode is not enough (less than 0.1) if it is installed on line 1-2 and takes P_{12} as the feedback signal. The changing range of the damping of the oscillation mode is (0.2471-0.0381)=0.209 with the variations of system operating conditions. However, when the stabilizer is installed at line 3-2 and takes P_{32} as the feedback signal, it is robust to the variations of system operating conditions. The changing range of the damping of the oscillation is only (0.2054-0.1143)=0.0911.

Table 10.6

P_{12} (p.u.)	The stabilizer installed at line 1-2 and takes P_{12} as the feedback signal		The stabilizer installed at line 3-2 and takes P_{32} as the feedback signal	
	P_{32}=0.6p.u.	P_{32}=0.8p.u.	P_{32}=0.6p.u.	P_{32}=0.8p.u.
0.2	**0.0428**	**0.0381**	0.1575	0.1982
0.4	**0.0678**	**0.0630**	0.1553	0.2010
0.6	**0.0970**	**0.0918**	0.1485	0.2047
0.8	0.1340	0.1314	0.1379	0.2054
1.0	0.1811	0.1784	0.1284	0.1872

Figure 10.17 gives non-linear simulation at P_{32}=0.6p.u., P_{12}=0.2p.u., which demonstrates that when the TCSC-based stabilizer is installed at line 1-2 and takes P_{12} as the feedback signal, its effectiveness is lost and if it is installed at line 3-2 and P_{32} is taken as the feedback signal, system oscillation is still damped effectively.

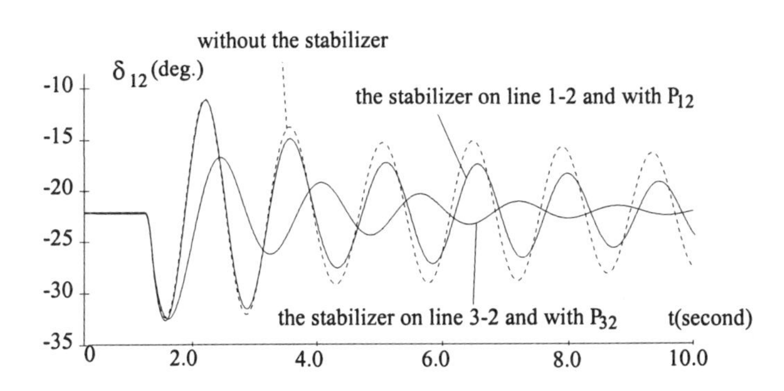

Figure 10.17 Non-linear simulation

10.5 Summary

This chapter introduces oscillation stability analysis and control of power systems installed with FACTS-based stabilizers. The discussions cover the subjects of modelling, analysis, and design of FACTS-based stabilizers installed in single-machine infinite-bus and multi-machine power systems. Examples are given when it is appropriate to demonstrate the analytical studies.

For a good design of FACTS-based stabilizers in a power system, three factors must be considered: the effectiveness, the robustness and the interactions among stabilizers. In this chapter, some aspects of these factors are explored. However, a great effort needs to be directed to investigate the following problems in future:

- Design of a single FACTS-based stabilizer in the power system, which is effective, robust, and imposes non-negative interactions on other stabilizers in the system.
- Co-ordinated design of a group of FACTS-based stabilizers, which are effective, robust, and impose non-negative interactions on each other and other stabilizers in the power system excluded in the co-ordination.

The reason is that when a FACTS-based stabilizer or a group of FACTS-based stabilizers are to be put into a power system, often other stabilizers may not be adjustable or it may not be desirable to readjust them.

10.6 References

1 Heffron, W.G., and Phillips, R.A., "Effect of a modern amplidyne voltage regulator on under excited operation of large turbine generator", *AIEE Trans* 71, 1952.

2 De Mello, F.P. and Concordia, C., "Concept of synchronous machine stability as affected by excitation control", *IEEE Trans. on Power Apparatus and Systems* 2, pp.316-329, 1969.

3 Larsen, E.V. and Swann, D.A., "Applying Power System Stabilizer, Part I-III", *IEEE Trans. on Power Apparatus and Systems*, 6, pp.3017-3046, 1981.

4 Hingorani, N.G., "High power electronics and flexible AC transmission system", *IEEE Power Eng. Rev.*, pp.3-5, July, 1988.

5 Gyugyi, L., "Unified power-flow control; concept for flexible AC transmission systems", *IEE Proc.-Gener. Transm. Distrib.* 4, pp.323-331, 1992.

6 Larsen, E.V., Sanchez-Gasca, J.J. and Chow, J.H., "Concept for design of FACTS controllers to damp power swings", *IEEE Trans. on Power Systems*, 2, pp.948-956, 1995.

7 Noroozian, M. and Andersson, G., "Damping of power system oscillations by use of controllable components", *IEEE Trans. on Power Delivery*, 4, pp.2046-2054,1994.

8 Hammad, A.E., "Analysis of power system stability enhancement by static var. compensator", *IEEE Trans. on Power Systems*, 4, pp.222-227, 1986.

9 Zhou, E.Z., "Application of static var. compensators to increases power system damping", *IEEE Trans. on Power Systems*, 2, pp.655-661, 1993.

10 Nyati, S., Wegner, C.A., Delmerico, R.W., Baker, D.H., Piwko, R.J. and Edris A., "Effectiveness of thyristor controlled series capacitor in enhancing power system dynamics: an analog simulator study", *IEEE Trans. on Power Delivery*, 2, pp.1018-1027, 1994.

11 Pourbeik, P. and Gibbard, M.J., "Damping and synchronising torque induced on generators by FACTS stabilizers in multi-machine power systems", *IEEE Trans. On PWRS*, 4, pp.1920-1925, 1996.

12 Gyugyi, L., Rietman,T.R., Edris, A., Schauder, C.D., Torgerson, D.R. and Williams, S.L., "The unified power flow controller: a new approach to power transmission control", *IEEE Trans. on Power Systems*, 2, pp.1085-1097, 1995.

13 Nabavi-Niaki, A. and Iravani, M.R., "Steady-state and dynamic models of unified power flow controller (UPFC) for power system studies", *IEEE Trans. on Power Systems*, 4, pp.1937-1943, 1996.

14 Arabi, A. and Kundur, P., "A versatile FACTS device model for power flow and stability simulation", *IEEE Trans. on Power Systems*, 4, pp.1944-1950, 1996.

15 Papic, I., Zunko, P., Povh, D. and Weinhold, M., "Basic control of unified power flow controller", *IEEE Trans. on Power Systems*, 4, pp.1734-1739, 1997.

16 Mihalic, R., Zunko, P. and Povh, D., "Improvement of transient stability using unified power flow controller", *IEEE Trans. on Power Delivery*, 1, pp.485-492, 1996.

17 Yu, Y.N., *Electric power system dynamics*, Academic Press, 1983.

18 Wang, H.F., "Selection of robust installing locations and feedback signals of FACTS-based stabilizers in multi-machine power systems", *IEEE Trans. on Power Systems*, 2, pp.569-574, 1999.

19 Wang, H.F., Swift, F.J., and Li, M., "A unified model for the analysis of FACTS devices in damping power system oscillations Part II: multi-machine power systems", *IEEE Trans. on Power Delivery,* 4, pp.1355-1362, 1998.

20 Wang, H.F. and Swift, F.J., "An unified model for the analysis of FACTS devices in damping power system oscillations Part I: single-machine infinite-bus power systems", *IEEE Trans. on Power Delivery,* 2, pp.941-946, 1997.

21 Wang, H.F., "Partial modal decomposition and the modal control analysis", Power Engineering Letter on IEEE PE Review, 4, pp.61-63, 1999.

22 Wang, H.F., "Selection of operating conditions for the co-ordinated setting of robust fixed-parameter stabilizers", *IEE Proc.-Gener. Transm. Distrib.,* 2, pp.111-116, 1998.

23 Swift, F.J. and Wang, H.F., "Application of the controllable series compensator in damping power system oscillations", *IEE Proc.-Gener. Transm. Distrib.,* 4, pp.359-364, 1996.

24 Wang, H.F. and Swift, F.J., "The capability of the static var compensator in damping power system oscillations", *IEE Proc.-Gener. Transm. Distrib.,* 4, pp.353-358, 1996.

25 Wang, H.F., Swift, F.J. and Li, M., "The indexes for selecting the best locations of PSS or FACTS-based stabilizers in multi-machine power systems: a comparison study", *IEE Proc.-Gener. Transm. Distrib.*, 2, pp.155-159, 1997.

26 Wang, H.F. and Swift, F.J., " A comparison on the modal controllability of PSS and FACTS-based stabilizer in multi-machine power systems", *IEE Proc.-Gener. Transm. Distrib.*, 6, pp.575-581, 1996.

27 Wang, H.F., Swift, F.J. and Li, M., "Selection of installing locations and feedback signals of FACTS based stabilizers in multi-machine power systems by reduced-order modal analysis", *IEE Proc.-Gener. Transm. Distrib.*, 3, pp.263-270, 1997.

28 Wang, H. F., "Damping funtion of unified power flow controller", *IEE Proc.-Gener. Transm. Distrib.*, 1, pp.81-88, 1999.

29 Wang, H. F., "Applications of modelling UPFC into multimachine power systems", *IEE Proc.-Gener. Transm. Distrib.*, 3, pp.306-312, 1999.

Chapter 11

Transient stability control

R. Mihalič, D. Povh, and P. Žunko

11.1 Introduction

The *stability* of an electric power system (EPS) may be defined as an EPS's ability to remain in synchronous operation under normal operating conditions as well as after being subjected to a disturbance. After a disturbance a stable EPS will regain a pre-fault or a new acceptable state of equilibrium. Stability problems may be manifested in many different ways depending on the EPS configuration, mode of operation and the nature of the disturbance. As such, stability problems may considerably influence EPS operation and control and have to be considered during the planning period. The loss of EPS stability (instability) may have dramatic consequences resulting in blackout of parts or of the whole system (who has not heard about the "famous" New York blackout).

According to the nature of the phenomena, following a disturbance in an EPS, stability problems are often divided into various categories (e.g. steady-state stability, oscillatory stability, transient stability, voltage stability, etc.), although they are more or less connected to each other. According to the "nature of the stability problem" some phenomena predominate and in any given situation the response of only a limited amount of equipment or devices may be significant. Therefore specific methods are normally applied when studying specific stability categories.

Transient stability (referred to also as *1st swing stability* – although multi-machine systems may become transiently instable in the second or even later swing) may be defined as the ability of an EPS to remain in synchronism after being subjected to a major system disturbance such as a short circuit on transmission facilities, loss of generating unit or large load. The reaction of an EPS after such a disturbance is characterised by large deviations in: machine rotor relative angular positions, power flows, system voltage profiles etc. The criteria for a system to be transiently stable is angular separation between the machines. If they remain within certain bounds, an EPS keeps synchronism and is consequently stable. It should be noted that the rotor angular speed or velocity can not be a criteria for stability (instability) thus angular speed or velocity of the system's rotating masses may change more or less simultaneously (system frequency deviations). In this case the system may, in fact, be stable, although such operating conditions may not be tolerated.

As already reported in numerous references, FACTS devices (also referred to as FACTS controllers) are, first of all, effective tools for dynamic power flow control in an EPS. As is known, and is the subject of discussion in the following sections, power flow is closely related to a system's transient stability problems. As a result of these considerations FACTS devices may be an effective tool to mitigate transient stability problems in EPS.

In Chapter 11 our intention is to show how with various FACTS controllers the transient stability margin of an EPS may be enhanced. Because, as already mentioned, transient stability is concerned with a large disturbance and, consequently, large deviations of electric quantities, it is influenced by the non-linear characteristics of an EPS. The equations describing phenomena can not be linearized. Therefore in our theoretical explanation a few assumptions will be made which will enable us to explain basic phenomena applying the equal-area criterion. The detailed description of methods for estimation of EPS transient stability is beyond the scope of this work.

11.2 Basic theoretical considerations

The aim of this chapter is to provide an explanation of how and why mechanical movement of the generator rotor is influenced by electromagnetic effects and how it is related to the problem of transient stability. The majority of the explanation will be restricted to a simplified machine model and to an equal area transient stability criterion. The theory relating to the equal area criterion is old and has been described in numerous works. Nevertheles this approach is still useful because of the following reasons: first it is easy to understand; second the transient stability problem of two generators (which can be representations of coherent groups of machines swinging simultaneously) can be transformed to the "one-machine infinite-bus" system [1], and third, there are methods existing which enable the transformation of the transient stability problem of a multimachine system to the "one-machine infinite-bus" system with time-varying parameters (applying the so-called generalized equal-area criterion [2, 3, 4, 5]).

11.2.1 Generator behaviour under transient conditions

For the purpose of studying electromechanical phenomena the generator can be represented by a driven rotating mass (equivalent to all turbines, shafts, and generator rotors) which is braked by an electromagnetic field. In steady state operation the mechanical power delivered to the rotating mass equals the electric power produced by the rotor electromagnetic field. In this equilibrium point the mechanical turbine torque τ_m is equal to the electric torque τ_e + mechanical damping synchronous speed torque τ_d (rotational losses) and no relative rotor motion appears. As soon as mechanical and electric torque are no more in equilibrium the rotating masses are accelerated or decelerated following Newton's law

$$J\frac{d\omega_r}{dt} + D\Delta\omega_r = \tau_m - \tau_d - \tau_e \qquad (11.1)$$

where

$$\omega_r = \omega_0 + \Delta\omega_r \qquad (11.2)$$

J represents the total moment of inertia of rotating masses (kg m^2), ω_r is the rotor angular velocity (rad/s), ω_0 is synchronous speed (rad/s), $\Delta\omega_r$ is the rotor angular speed deviation (rad/s), D is the damping-torque coefficient (Nms) and τ_m, τ_e and τ_d (Nm) are the torques as already explained. The mechanical damping torque τ_d is small and can be neglected for all practical purposes [6]. The main source of damping in equation 11.1 ($D\Delta\omega_r$) is a generator damping winding. In synchronous operation there is no damping thus $\Delta\omega_r$ equals 0. In transient conditions which are interesting for phenomena related to the transient stability the generator air gap flux penetrates the damper winding and induces voltage (emf) whenever $\omega_r \neq \omega_0$. As a consequence of this voltage, current flows in the damper winding which further causes a torque opposite to the change of rotor's relative angle (according to Lenz's Law). This torque can for small speed deviations be assumed to be proportional to $\Delta\omega_r$ and can be referred to as *asynchronous torque* (the phenomenon is similar to that in a running asynchronous machine). For convenience and clarity of explanation in our further considerations let this damping be neglected. In this way our calculations may be considered to be "on the safe side".

Considering these assumptions, equation 11.2 and the fact that ω_0 is a constant, equation 11.1 can be written:

$$J\frac{d\Delta\omega_r}{dt} = \tau_m - \tau_e\,. \qquad (11.3)$$

Let δ_r be defined as a rotor angle with respect to the synchronous rotating reference axis. Then:

$$\Delta\omega_r = \frac{d\delta_r}{dt} \qquad (11.4)$$

and according to equation 11.3

$$J\frac{d^2\delta_r}{dt^2} = \tau_m - \tau_e \tag{11.5}$$

Multiplying equation 11.5 by the synchronous velocity ω_0 and taking into consideration that power is a product between torque and angular velocity equation 11.5 can be rewritten as follows:

$$J\omega_0\frac{d^2\delta_r}{dt^2} = \frac{\omega_0}{\omega_r}P_m - \frac{\omega_0}{\omega_r}P_e\ , \tag{11.6}$$

where P_m is shaft power provided to the generator and P_e is the electrical air-gap power. In all practical cases it can be assumed that the rotor speed of a synchronous machine is so close to the synchronous speed that

$$\frac{\omega_0}{\omega_r} \approx 1\ . \tag{11.7}$$

Considering also that the product $J\omega_0$ equals the angular momentum M_r (kg m^2/s) finally the basic equation is obtained that describes rotor dynamics – the so-called *swing equation*

$$M_r\frac{d^2\delta_r}{dt^2} = P_m - P_e\ . \tag{11.8}$$

Often rotor angular momentum M_r is expressed either with:

- *normalized inertia constant* H(s), defined as a stored kinetic energy of rotating masses in mega Joules at synchronous speed, normalized with the machine rating S_N

$$H = \frac{1}{2}\frac{J\omega_0{}^2}{S_N} \Rightarrow M_r = \frac{2HS_N}{\omega_0} \quad \text{or with the} \tag{11.9}$$

- *mechanical time constant* T_a (s), defined as the time in which a generator rotating mass would reach the synchronous speed if the nominal mechanical torque (S_N/ω_0) was suddenly applied to the turbine shaft of the generator at rest

$$T_a = 2H \Rightarrow M_r = \frac{T_a S_N}{\omega_0}\ . \tag{11.10}$$

The changes in the mechanical power P_m are dependent upon the turbine power (frequency) controller. The time constants of mechanical power control are high compared to the rotor initial-swing time interval, therefore during the transients, characteristic for transient stability, P_m can in our theoretical considerations be assumed constant (pre-disturbance steady-state value).

The remaining term of the swing equation that still has to be discussed is the air-gap electrical power P_e. If generator resistances are neglected (and in our considerations they are) then P_e also represents the generator electrical power delivered into electric network.

As is well known from the theory of synchronous machines the steady state electric generator power P_G can under the described assumptions be expressed with the following equation:

$$P_G = P_e = \frac{|\underline{E}||\underline{U}_G|}{X_d}\sin\delta_G + \frac{|\underline{U}_G|^2}{2}\left(\frac{1}{X_q} - \frac{1}{X_d}\right)\sin 2\delta_G \qquad (11.11)$$

In equation 11.11 X_d and X_q represent direct- and quadrature synchronous reactances respectively, $\underline{U}_G$ is the generator terminal voltage, $\underline{E}$ is the generator air gap emf and δ_G is the phase shift between the phasors $\underline{U}_G$ and $\underline{E}$.

Equation 11.11, however, can not be used for transient conditions. During these conditions the armature flux is forced into high reluctance paths outside the field winding. Therefore the reactances $X_d'$and X_q' (transient machine reactances in the direct and quadrature axes respectively) associated with the flux path in transient conditions differ essentially from those of the corresponding steady state (synchronous) reactances [6, 7]. If the rotor flux linkages in both axes are assumed to remain constant during transients, a generator can in such conditions be represented by constant transient emf $\underline{E}'$ acting behind X_d' and X_q'. In this case the transient power equation 11.12 has for the salient pole generators [6] the same form as the steady state power equation 11.11, however synchronous quantities have to be substituted by transient values ($\underline{E}$ by $\underline{E}'$, X_d by X_d'). $\underline{U}_G$ should be considered a "post-disturbance" generator terminal voltage and δ_G the angle betwen the phasors $\underline{E}'$ and $\underline{U}_G$. Thus the armature flux associated with the quadrature axis current component is by the salient pole generators not linked to the rotor field winding in the transient power equation, and X_q remains unchanged. We obtain:

$$P_G = P_e = \frac{|\underline{E}'||\underline{U}_G|}{X_d'}\sin\delta_G + \frac{|\underline{U}_G|^2}{2}\left(\frac{1}{X_q} - \frac{1}{X_d'}\right)\sin 2\delta_G \qquad (11.12)$$

It should be noted that the transient power equation for round-rotor generators differs from that of 11.12. The transient emf $\underline{E}'$ can then be calculated from the pre-fault conditions from equation 11.13 as follows:

$$|\underline{E}'| = \frac{X_d'}{X_d}|\underline{E}| + \frac{X_d - X_d'}{X_d}|\underline{U}_G|\cos\delta_{G0} \qquad (11.13)$$

where $\underline{U}_{G0}$ is the pre-fault generator terminal voltage and δ_0 the angle between $\underline{E}$ and $\underline{U}_{G0}$. Assuming $X_d' \approx X_q'$ ($\approx X_q$ for salient pole generators) the second term in equation 11.12 disappears and the transient power equation simplifies to:

$$P_G = P_e = \frac{|\underline{E}'||\underline{U}_G|}{X_d'} \sin\delta_G \ . \qquad (11.14)$$

This equation is for the assumptions adopted valid also for round-rotor generators. The simplification introduced does not significantly affect the qualitative considerations regarding transient stability for the two following reasons. Firstly the neglected terms in the transient power equation are relatively small compared to the remaining term (equation 11.14). Secondly they (or a part of them in the case of a round-rotor generator) have double frequency which means that the error introduced by the simplifications is of different sign in the region $0° < \delta_G' < 90°$ to that in the region $90° < \delta_G' < 180°$. In this way during the angular swings ranging through these regions (they are the most interesting when considering transient stability limits) the error has a tendency to average out to zero.

11.2.2 Equal area criterion

Now let us assume that a generator is connected to the infinite bus via a transformer and a transmission line, as shown in Figure 11.1a. Let the infinite bus voltage $\underline{U}_2$ be the reference phasor. The angle difference between the generator transient voltage $\underline{E}'$ and the reference phasor is referred to as δ and the sum of system reactances $X_d' + X_T + X_L$ is referred to as X. The transmission characteristic is presented in Figure 11.1b.

Let us examine the behaviour of the system if the mechanical power is stepwise changed from P_{m0} to P_{m1}. In this case the system operating point follows the transient transmission characteristic from the old equilibrium point (point 0 on the characteristic) to the new equilibrium point (point 1). However the rotor cannot be pushed to a new point instantaneously and in its way from 0 to 1 the mechanical power accelerates the rotating masses because there is an excess of mechanical power over electrical power. This excess is manifested in the enlarged kinetic energy of the rotating masses. Therefore the rotor moves forward until its excess of kinetic energy is not transformed into potential energy at point 2. From point 1 to 2 the generator is decelerating because there is an excess of electric power over mechanical power P_{m1}. In 2 the relative rotor movement direction is turned around and the rotor swings back. Without any damping the rotor would swing around the point 1 between the points 0 and 2. It is obvious that the point 3 of a transmission characteristic is a point of no return. By passing over point 3 the electrical power becomes higher than the mechanical power and the rotating masses are accelerated. The system is unstable.

(a)

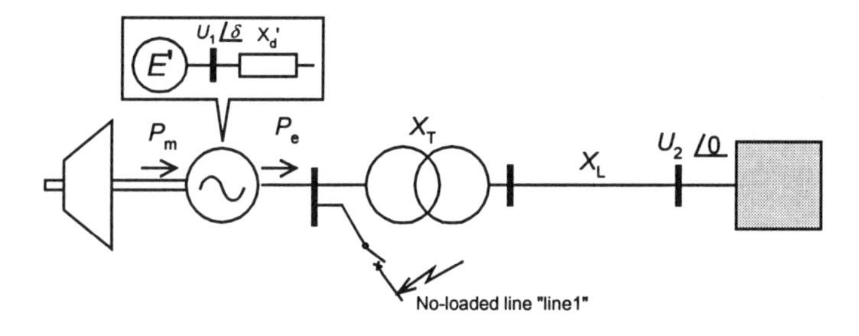

(b)

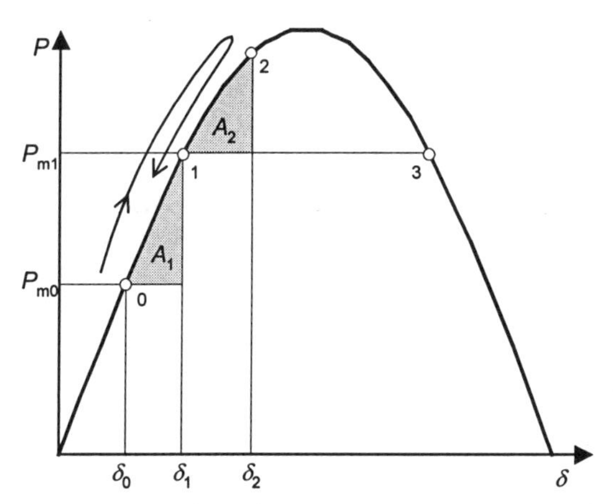

Figure 11.1 Generator-infinite bus system (a) Network configuration (b) Related transmission characteristic

Now, let the swing equation be multiplied by $\mathrm{d}\delta/\mathrm{d}t$. We obtain:

$$M_\mathrm{r} \frac{\mathrm{d}\delta}{\mathrm{d}t}\frac{\mathrm{d}^2\delta}{\mathrm{d}t^2} = \left(P_\mathrm{m} - P_\mathrm{e}\right)\frac{\mathrm{d}\delta}{\mathrm{d}t} \quad \text{or} \tag{11.15}$$

$$M_\mathrm{r} \frac{1}{2}\frac{\mathrm{d}}{\mathrm{d}t}\left(\frac{\mathrm{d}\delta}{\mathrm{d}t}\right)^2 = \left(P_\mathrm{m} - P_\mathrm{e}\right)\frac{\mathrm{d}\delta}{\mathrm{d}t} \ . \tag{11.16}$$

Integrating gives:

$$\left(\frac{\mathrm{d}\delta}{\mathrm{d}t}\right)^2 = \int \frac{2}{M_r}(P_\mathrm{m} - P_\mathrm{e})\mathrm{d}\delta \ . \tag{11.17}$$

In steady state operation the speed deviation dδ/dt is zero. At the point of rotor movement direction turn around the speed deviation dδ/dt is also zero. If the rotor continues moving (passing point 3) the system is not stable. Therefore the stability criteria may, according to Figure 11.1, be written as:

$$\int_{\delta_0}^{\delta_2} \frac{2}{M_r}(P_\mathrm{m} - P_\mathrm{e})\mathrm{d}\delta = 0 = \int_{\delta_0}^{\delta_2}(P_\mathrm{m} - P_\mathrm{e})\mathrm{d}\delta \ . \tag{11.18}$$

It means that the area under the function P_m - P_e must be 0 to maintain the system stable. In other words the accelerating area A_1 must be equal to the decelerating area A_2 i.e.:

$$\int_{\delta_0}^{\delta_1}(P_\mathrm{m} - P_\mathrm{e})\mathrm{d}\delta = \int_{\delta_1}^{\delta_2}(P_\mathrm{m} - P_\mathrm{e})\mathrm{d}\delta \ . \tag{11.19}$$

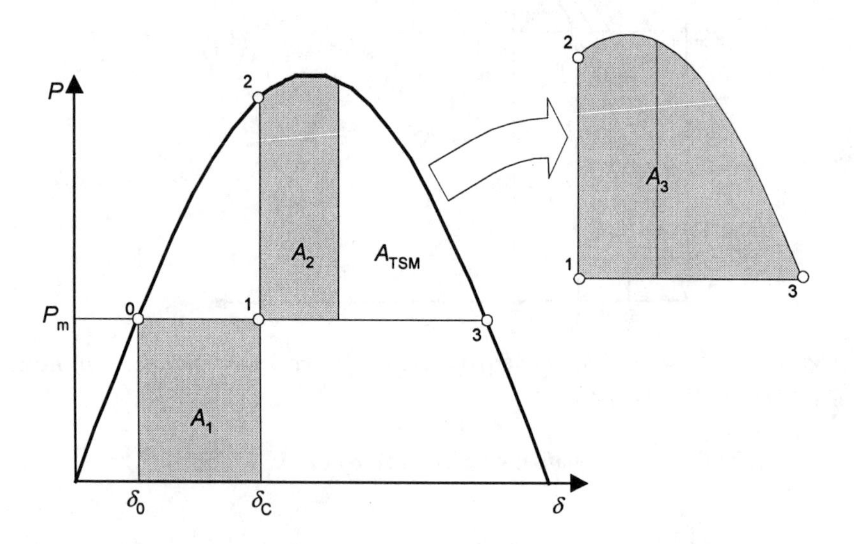

Figure 11.2 Illustration of transient stability

Now let us assume a three phase fault occurs on line 1 near the generator bus (c.f. Figure 11.1a) at time t_0 ($\delta = \delta_0$) and is cleared after a certain time t_C ($\delta = \delta_\mathrm{C}$). During this fault let the generator electrical power be zero. Transmission characteristics for such operation are presented in Figure 11.2. It is obvious that according to previous considerations the system is stable as long as the area A_1 is smaller than the area bounded by points 1-2-3 (area A_3). The difference between the areas A_3 and A_2 can be denoted as the transient stability margin A_{TSM}. As soon as this stability margin becomes negative, the system is unstable.

11.3 Analysis of power systems installed with FACTS devices

As presented in Section 11.2.2. the criteria for the system to remain stable is "area A_1 < area A_3". Let us assume that the stability margin is to be enhanced applying FACTS devices. The main way to achieve this, is to modify the system transmission characteristic so as to enlarge the area A_3 and consequently the transient stability margin A_{TSM}. The aim of this section is to explain how with FACTS devices the system transmission characteristics can be modified. Such proper modification plays a key role in transient stability margin enhancement. In order to be able to determine the "optimal" controllable parameters of FACTS devices in the transient stability enhancement sense, static models are developed, which describe the interdependence between the power transmission characteristic, the simplified system parameters, and the controllable parameters of the FACTS devices. This further serves as a basis for the determination of the power swing damping strategy.

11.3.1 System model and basic transmission characteristics

Let the simplest model of a generating unit, supplying a stiff grid via transmission facility (Figure 11.1a) serve as a basis for considerations regarding impact of FACTS devices on transmission characteristics. The electrical scheme is presented in Figure 11.3a.

Such a model may be a representation of the two idealized cases, i.e.:

- Generator is equipped with an idealised voltage regulator which holds the voltage of the generator bus constant. In this case X represents the sum of X_T and X_L (c.f. Figure 11.1); $\underline{U}_1$ ($\underline{U}_1 = U_1\ e^{j\delta}$) represents a generator bus voltage.
- Generator voltage regulator is very slow and does not react during the 1st swing. In this case $\underline{U}_1$ represents the generator transient emf $\underline{E}'$, while X is the sum of X_d', X_T and X_L.

As is well known for the presented system, the transmitted active power is a function of the transmission angle δ as follows:

$$P_1 = P_2 = P = \frac{U_1 U_2}{X}\sin(\delta)\ , \tag{11.20}$$

and this is further referred to as the basic transmission characteristic. As in the presented test system losses are neglected, the transmitted active power does not change along the transmission corridor. The transmission characteristics are going to be presented in P.U. system, the basis being the maximum value of the basic transmission characteristic P_{MAX}, as presented in Figure 11.3b.

(a)

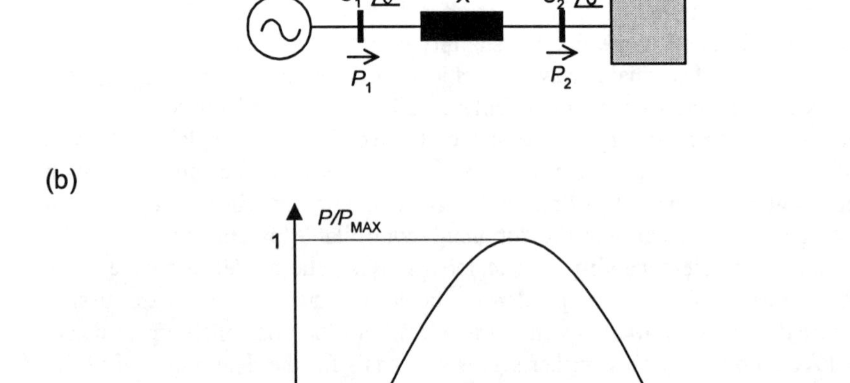

Figure 11.3 Model of the transmission system (a) Network scheme (b) Basic transmission characteristic

11.3.2 Power transmission control using controllable series compensation (CSC)

From the system point of view CSC can be represented as a controllable capacitance connected in series in the line. The model of the network with CSC included is presented in Figure 11.4a. The CSC controllable parameter may be assumed to be its reactance X_{CSC}. If CSC is operating in capacitive mode X_{CSC} is negative, while in inductive operating mode it is positive. The transmission characteristic is determined with:

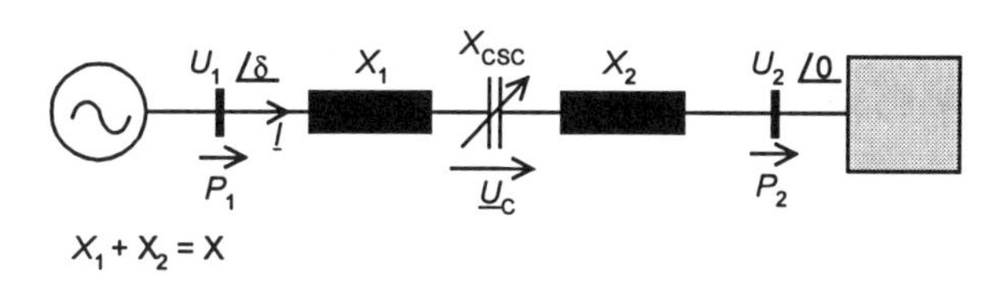

(b)

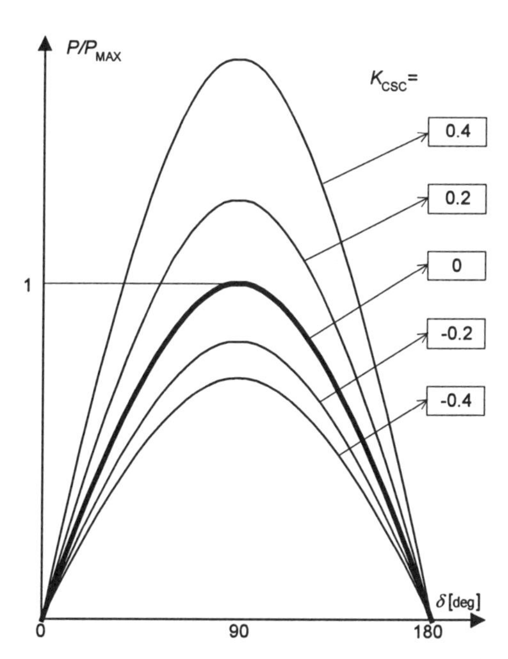

Figure 11.4 Model of the transmission system with CSC (a) Network scheme (b) Transmission characteristics

$$P_1 = P_2 = P = \frac{U_1 U_2}{X + X_{CSC}} \sin(\delta) = \frac{U_1 U_2}{X(1 - K_{CSC})} \sin(\delta) \qquad (11.21)$$

K_{CSC} being the so called series compensation degree ($K_{CSC} = -X_{CSC} / X$). Location of a CSC does not affect a system transmission characteristic therefore the ratio between X_1 and X_2 (c.f. Figure 11.4a) is arbitrary, their sum being equal to X. The transmission characteristics for various compensation degrees K_{CSC} are presented in Figure 11.4b.

11.3.3 Power transmission control using static series synchronous compensator (SSSC)

As is known SSSC may be assumed a series connected reactive voltage source [8, 9, 10]. In the major part of its operating area SSSC injected voltage $\underline{U}_T$ is independent of the throughput current. Therefore the SSSC controllable parameter may be assumed to be an injected voltage magnitude U_T. If the device is assumed without losses, the phasor $\underline{U}_T$ is perpendicular to the SSSC throughput current. The network scheme is presented in Figure 11.5a. According to this figure the following equation may be written:

$$\underline{I} = \frac{1}{j(X_1 + X_2)}\left((\underline{U}_1 - \underline{U}_2) - U_T \frac{(\underline{U}_1 - \underline{U}_2)}{|\underline{U}_1 - \underline{U}_2|}\right) = \frac{j(\underline{U}_2 - \underline{U}_1)}{(X_1 + X_2)}\left(1 - \frac{U_T}{|\underline{U}_1 - \underline{U}_2|}\right) \tag{11.22}$$

In equation (11.22) the term "$(\underline{U}_1$ - $\underline{U}_2)$" represents the phasor difference between $\underline{U}_1$ and $\underline{U}_2$. Without SSSC this would be the voltage drop on reactance X. The injected voltage phasor $\underline{U}_T$ has the same direction because it is a reactive voltage source. With the term $(\underline{U}_1$ - $\underline{U}_2)/|\underline{U}_1$ - $\underline{U}_2|$ this dirrection is determined. Multiplication with the injected voltage magnitude U_T mathematically describes the phasor $\underline{U}_T$. Now, the difference between $(\underline{U}_1$ - $\underline{U}_2)$ and $\underline{U}_T$ is the sum of voltage drops on reactances X_1 and X_2 in the SSSC presence.

The transmission characteristic can be obtained from the following equation:

$$P_1 = P_2 = P = \mathrm{Re}(\underline{U}_1 I^*) = \mathrm{Re}(\underline{U}_2 I^*) = U_2\,\mathrm{Re}(I) \tag{11.23}$$

(Note: $\underline{U}_2$ has been chosen as a reference phasor, therefore: $\underline{U}_2 = U_2\, e^{j0} = U_2$, $\underline{U}_1 = U_1\,(\cos(\delta) + j\sin(\delta))$.)

After a litle algebra, taking into consideration that $|\underline{U}_1 - \underline{U}_2| = \sqrt{U_1^{\,2} + U_2^{\,2} - 2U_1U_2\cos(\delta)}$ and $X_1 + X_2 = X$, it is not to hard to calculate the final result:

$$P_1 = P_2 = P = \frac{U_1U_2\sin(\delta)}{X}\left(1 - \frac{U_T}{\sqrt{U_1^{\,2} + U_2^{\,2} - 2U_1U_2\cos(\delta}}\right) \tag{11.24}$$

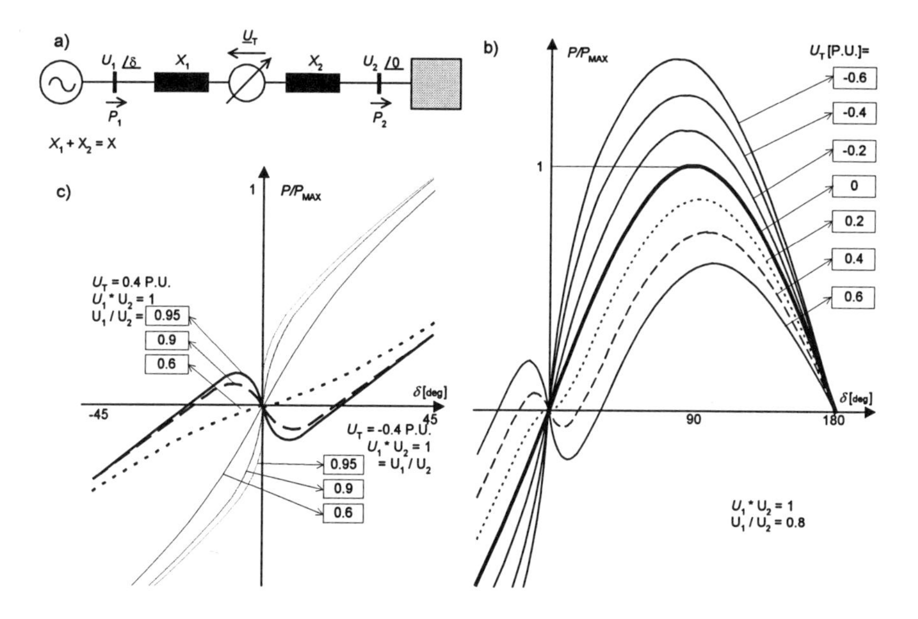

Figure 11.5 Model of the transmission system with SSSC (a) Network scheme (b) Transmission characteristics (c) Impact of the system terminal voltage ratio on transmission characteristics

The group of transmission characteristics is shown in Figure 11.5b. From equation (11.24) it is obvious that the SSSC location in the present theoretical case has no impact on power transmission characteristics, on the other hand the change of the ratio between the system terminal voltage magnitudes impacts transmission characteristics (especially in the area of low transmission angles), although their product remains constant. Let us assume: $U_1 / U_2 = \mathrm{U}_{\mathrm{ratio}}$ and $U_1 U_2 = 1$, then $U_1 = (\mathrm{U}_{\mathrm{ratio}})^{1/2}$, $U_2 = 1/(\mathrm{U}_{\mathrm{ratio}})^{1/2}$. Changing of the transmission characteristic with $\mathrm{U}_{\mathrm{ratio}}$ variation is illustrated in Figure 11.5c.

11.3.4 Power transmission control using static var compensator (SVC)

There are two possible explanations of the influence of an SVC on the real power flow in the system. The first one is the so-called constant voltage principle, the second one is the representation of the SVC by a parallel connected controllable susceptance B_{SVC} corresponding to the instantaneous operating point of the SVC [11]. To explain both principles consider the system presented in Figure 11.6a.

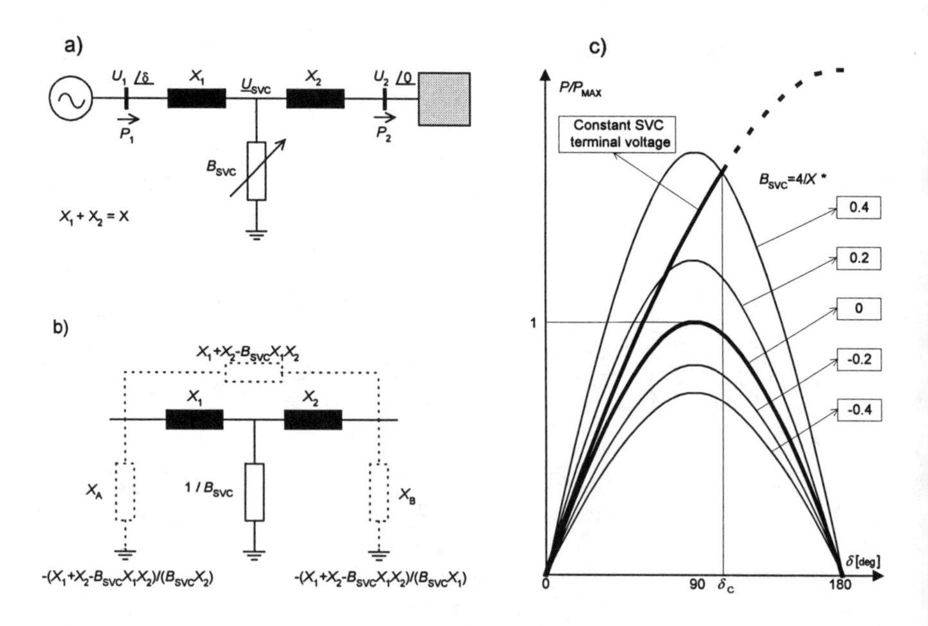

Figure 11.6 Model of the transmission system with SVC (a) Network scheme (b) Network impedance scheme (c) Transmission characteristics

11.3.4.1 Constant voltage principle

The main SVC task in electric power systems is voltage control. Without additional signals (besides the voltage magnitude signal), and supplementary control loops, SVC may be assumed to keep the terminal voltage magnitude U_{SVC} at a predefined constant level. In this case the transmission line is divided into two sections and the transmission system may be assumed to be electrically shortened. The power transfer is dictated by the electrically longer of the two sections. Let us assume the system terminal voltage magnitudes U_1 and U_2 are equal to 1 p.u. and that an SVC is located in the electrical middle of the system (both X_1 and X_2 equal to X/2 – c.f. Figure 11.6a) and keeps terminal voltage U_{SVC} at 1 p.u. In this case the system is electrically shortened by a factor of 2 and the transmission characteristic is described by the equation (11.25).

$$P_1 = P_2 = P = \frac{U_1 U_2 \sin(\delta/2)}{X/2} \tag{11.25}$$

The transmission characteristic for this case is presented in Figure 11.6c.

11.3.4.2 Controllable parallel susceptance

If the SVC terminal voltage is not kept at a constant level (and, as it will be shown, for transient stability margin purposes should not), the SVC may be represented as a parallel susceptance B_{SVC} (losses in the SVC being neglected) corresponding to the SVC operating point. The impedance scheme of such a system is shown in Figure 11.6b. This impedance scheme can be transformed by "Y - D" transformation (dotted-line elements in Figure 11.6b). The parallel reactances X_A and X_B in the present case do not play any role thus U_1 and U_2 are assumed constant. The effect on the transmission line is in this case the same as if series compensation with the impedance $X_{CSC} = - X_1 * X_2 * B_{SVC}$ were used. From this equation it is evident that B_{SVC} is in direct proportion to the reduction of the series system reactance X_R ($X_R = X_1 * X_2 * B_{SVC}$). The transmission characteristic is, applying the same logic as in the case of CSC, described by the equation (11.26).

$$P_1 = P_2 = P = \frac{U_1 U_2}{X - X_1 X_2 B_{SVC}} \sin(\delta) \tag{11.26}$$

The question still to be answered is, where in the transmission corridor should the SVC be placed in order to achieve its maximum impact on transmission characteristics. Let the ratio between X_1 and X_2 be referred to as X_{RATIO}. Then applying the equations:

$$X_1 + X_2 = X; \frac{X_1}{X_2} = X_{RATIO} \Rightarrow X_1 = \frac{X X_{RATIO}}{1 + X_{RATIO}}; X_2 = \frac{X}{1 + X_{RATIO}} \tag{11.27}$$

the system series reactance reduction X_R can, applying a little algebra, be expressed with X and X_{RATIO} as follows:

$$X_R = X_1 X_2 B_{SVC} = \frac{B_{SVC} X^2 X_{RATIO}}{(1 + X_{RATIO})^2} \tag{11.28}$$

In order to maximize SVC impact on transmitted power, X_{RATIO} should be chosen so as to maximize this expression. As is well known, such “optimal” X_{RATIO} can be calculated from the following derivative equalled to 0.

$$\frac{\partial X_R}{\partial X_{RATIO}} = \frac{B_{SVC} X^2 (1 - X_{RATIO})}{(1 + X_{RATIO})^3} = 0 \Rightarrow (1 - X_{RATIO}) = 0 \tag{11.29}$$

According to (11.29) X_{RATIO} should be 1 i.e. X_1 and X_2 should both be equal to $X / 2$. In other words, SVC should be positioned in the electrical middle of the system. Transmission characteristics for various SVC susceptances B_{SVC}, with SVC in the middle of the system, are presented in Figure 11.6c.

For economic reasons in practice SVCs are not rated as high as to be able to keep voltage at the desired level over the whole range of transmission angles (δ from 0 to 180°). At a certain angle δ_C SVC reaches its limit and the constant voltage principle can not be satisfied. From this angle on SVC behaves as a parallel connected capacitor. If e.g., according to Figure 11.6c, maximal SVC susceptance B_{SVC} equals $4/X$*0.4, then from δ_C on the transmission characteristics follows the characteristic "$B_{SVC} = 4/X * 0.4$".

As is shown from figure 11.6c, stability margins can be enhanced by applying both of the principles, i.e. constant voltage principle as well as susceptance principle. However, more effective is the last one ("jump" to the SVC capacitive limit) thus also area between the curves "constant SVC terminal voltage" and "B_{SVC}=.... " can be used too.

11.3.5 Power transmission control using static synchronous compensator (STATCOM)

From the system point of view STATCOM may be treated as a parallel connected current source because in the major part of its operating area its current is independent of the terminal voltage magnitude $U_{STATCOM}$ [11, 12, 13]. The STATCOM controllable parameter may therefore be assumed its current magnitude I_Q. When losses are neglected, $\underline{I}_Q$ represents the STATCOM reactive current phasor which is perpendicular to the terminal voltage phasor $\underline{U}_{STATCOM}$.

The network scheme is presented in Figure 11.7a. According to Kirchoffs laws the following three equations can be written:

$$\underline{U}_{STATCOM} = \underline{U}_1 - j\underline{I}_1 X_1 \tag{11.30}$$

$$\underline{I}_2 = \frac{\underline{U}_{STATCOM} - \underline{U}_2}{jX_2} \tag{11.31}$$

$$\underline{I}_2 = \underline{I}_1 - \underline{I}_Q \tag{11.32}$$

By equating the right-hand terms of (11.31) and (11.32) and by considering (11.30), an equation is defined, from which current $\underline{I}_1$ can be easily calculated as presented in (11.33).

$$\underline{I}_1 = \frac{\underline{U}_1 - \underline{U}_2}{j(X_1 + X_2)} + \underline{I}_Q \frac{X_2}{(X_1 + X_2)} \tag{11.33}$$

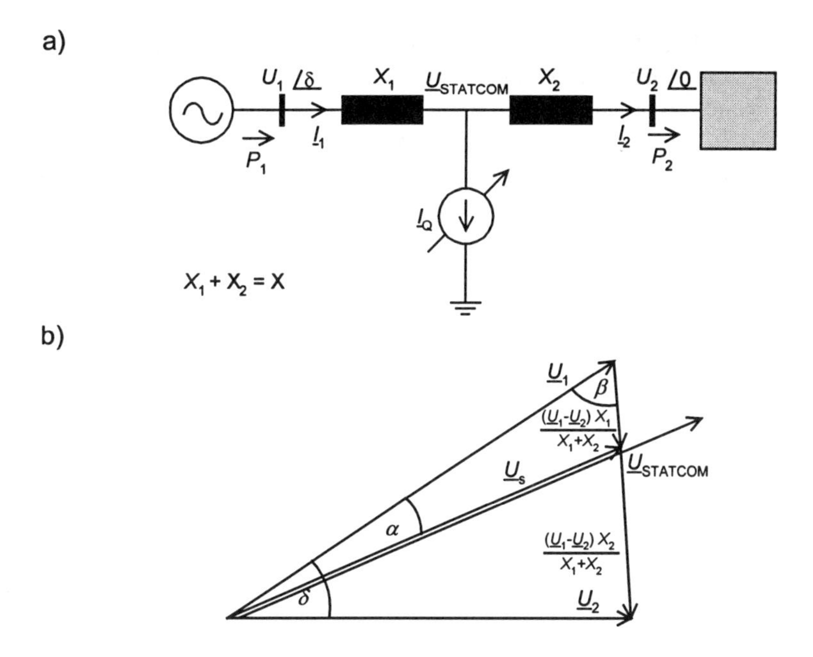

Figure 11.7 Model of the transmission system with statcom (a) Network scheme. (b) Corresponding phasor diagram

Now $\underline{U}_{\text{STATCOM}}$ can be defined in accordance with (11.30)

$$\underline{U}_{\text{STATCOM}} = \underline{U}_1 - \frac{(\underline{U}_1 - \underline{U}_2)X_1}{(X_1 + X_2)} - \text{j}\underline{I}_{\text{Q}} \frac{X_1 X_2}{(X_1 + X_2)} = \underline{U}_{\text{S}} - \text{j}\underline{I}_{\text{Q}} \frac{X_1 X_2}{(X_1 + X_2)} \tag{11.34}$$

Taking into consideration that:

$$\underline{I}_{\text{Q}} = \text{j}I_{\text{Q}} \frac{\underline{U}_{\text{S}}}{U_{\text{S}}} \tag{11.35}$$

(11.34) can be rewritten as follows:

$$\underline{U}_{\text{STATCOM}} = \underline{U}_{\text{S}} + I_{\text{Q}} \frac{\underline{U}_{\text{S}}}{U_{\text{S}}} \frac{X_1 X_2}{(X_1 + X_2)} = \underline{U}_{\text{S}} \left(1 + \frac{I_{\text{Q}}}{U_{\text{S}}} \frac{X_1 X_2}{(X_1 + X_2)}\right) \tag{11.36}$$

As shown in (11.34) a new variable $\underline{U}_S$ has been introduced. It is in fact the STATCOM terminal voltage if STATCOM is out of operation (is not connected to the system, i.e. $\underline{I}_Q = 0$). Equation (11.35) describes the fact that $\underline{I}_Q$ is shifted by 90° with regard to $\underline{U}_S$ ("$\underline{U}_S$ / U_S" being a unit phasor in $\underline{U}_S$ direction). This can be proved in the following way. Let us suppose that STATCOM is not represented by a reactive current source but with its operating point reactance $X_{STATCOM}$. Then in (11.34) $\underline{I}_Q$ should be replaced by the term $\underline{U}_{STATCOM}$ / (j $X_{STATCOM}$). Now (11.34) can be transformed into the form $\underline{U}_{STATCOM}\ K = \underline{U}_S$. Thus K is a scalar quantity, $\underline{U}_{STATCOM}$ and $\underline{U}_S$ have the same phase and $\underline{I}_Q$ is perpendicular to both, the terminal voltage phasor $\underline{U}_{STATCOM}$ as well as $\underline{U}_S$.

The corresponding voltage phasor diagram is presented in Figure 11.7b. According to this diagram, applying sine law, the following two equation can be written:

$$\frac{\sin(\beta)}{U_2} = \frac{\sin(\delta)}{\left|\underline{U}_1 - \underline{U}_2\right|}; \frac{\sin(\alpha)}{\left|\underline{U}_1 - \underline{U}_2\right| \dfrac{X_1}{(X_1 + X_2)}} = \frac{\sin(\beta)}{U_S} \tag{11.37}$$

from which the sine of α can be calculated:

$$\sin(\alpha) = \frac{U_2 \sin(\delta) X_1}{U_S (X_1 + X_2)} \tag{11.38}$$

The transmitted power can (taking (11.20) and (11.38) into consideration) be calculated from:

$$P_1 = P_2 = P = \frac{U_{STATCOM} U_1}{X_1} \sin(\alpha) = \frac{U_1 U_2 \sin(\delta)}{(X_1 + X_2)} \frac{U_{STATCOM}}{U_S} \tag{11.39}$$

Considering (11.36), this equation can be rewritten as:

$$P = \frac{U_1 U_2 \sin(\delta)}{(X_1 + X_2)} \frac{\left|\underline{U}_S \left(1 + \dfrac{I_Q}{U_S} \dfrac{X_1 X_2}{(X_1 + X_2)}\right)\right|}{U_S} = \frac{U_1 U_2 \sin(\delta)}{(X_1 + X_2)} \left(1 + \frac{I_Q}{U_S} \frac{X_1 X_2}{(X_1 + X_2)}\right) \tag{11.40}$$

Thus (note: $\underline{U}_2$ is the reference phasor therefore $\underline{U}_2 = U_2$, $\underline{U}_1 = U_1\, e^{j\delta}$)

$$U_S = |\underline{U}_S| = \left|\frac{\underline{U}_1 X_2 + \underline{U}_2 X_1}{(X_1 + X_2)}\right| = \frac{\sqrt{U_1^{\,2} X_2^{\,2} + U_2^{\,2} X_1^{\,2} + 2U_1 U_2 X_1 X_2 \cos(\delta)}}{(X_1 + X_2)} \tag{11.41}$$

finally the transmission characteristic is described by the following equation:

$$P_1 = P_2 = P = \frac{U_1 U_2 \sin(\delta)}{(X_1 + X_2)}\left(1 + \frac{I_Q X_1 X_2}{\sqrt{U_1^{\,2} X_2^{\,2} + U_2^{\,2} X_1^{\,2} + 2U_1 U_2 X_1 X_2 \cos(\delta)}}\right) \tag{11.42}$$

From this equation it is evident that the rise in transmission characteristic is in direct proportion to the STATCOM current. The question regarding the location can be answered by applying the same procedure as in Section 11.3.4. X_1 and X_2 should in (11.42) be replaced in accordance with (11.26). Then the derivative of this expression is set to 0 and finally X_{RATIO} is calculated. The expressions are quite comprehensive and detailed explanation is out of scope of this work. It should only be noted that X_{RATIO} is dependent on system terminal voltages U_1 and U_2 as well as on transmission angle. In case the terminal voltages in question are equal to each other, X_{RATIO} equals to 1 and thus in this case the most efficient STATCOM location is the electrical middle of the system.

The group of transmission characteristics for STATCOM being located in the middle of the system is shown in Figure 11.8a. It can be noted that there appears to be some kind of dualism with SSSC. The form of transmission characteristics is similar as in SSSC case, however STATCOM characteristics are a "mirror" picture of those in SSSC case. Also here the ratio between the system terminal voltage magnitudes affects the transmission characteristics (especially in the area of transmission angles around 180°), although their product remains constant. Let the same assumptions be applied as in Section 11.3.3 i.e. $U_1/U_2 = \mathrm{U}_{ratio}$, $U_1 U_2 = 1$, $U_1 = (\mathrm{U}_{ratio})^{1/2}$, $U_2 = 1/(\mathrm{U}_{ratio})^{1/2}$. Changes of the transmission characteristic with U_{ratio} variation is illustrated in Figure 11.8b.

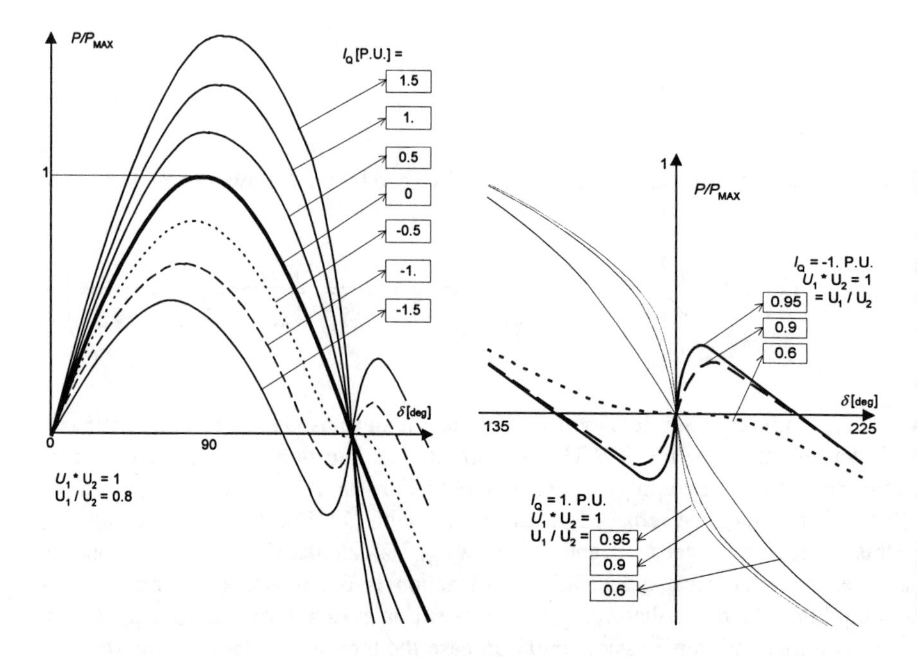

Figure 11.8 Model of the transmission system with STATCOM (a) Transmission characteristics (b) Impact of the system terminal voltage ratio on transmission characteristics

11.3.6 Power transmission control using phase shifting transformer (PST)

The term Phase Shifting Transformers (PST) denotes devices which, from the system point of view, have the ability to introduce a phase shift between terminal voltage phasors more or less independently of throughput current. If losses in the devices (active and reactive) are neglected then PSTs do not produce nor consume active and reactive power (PST power is balanced). They can, among others, be modelled as a combination of a series injected voltage source $\underline{U}_T$ and a parallel connected current source $\underline{I}_T$ [11, 14, 15, 16]. The $\underline{I}_T$ magnitude and phase as well as $\underline{U}_T$ phase are determined the system parameters and the type of PST, which any model should take account of. In terms of their basic structure various PST types may be realised. For practical application two types are most interesting i.e. the so called Phase Angle Regulator (PAR) and Quadrature Boosting Transformer (QBT). Regarding their impact on power flow PAR and QBT may differ considerably and will therefore be discussed separately. The model of the system with PST included is presented in Figure 11.9a.

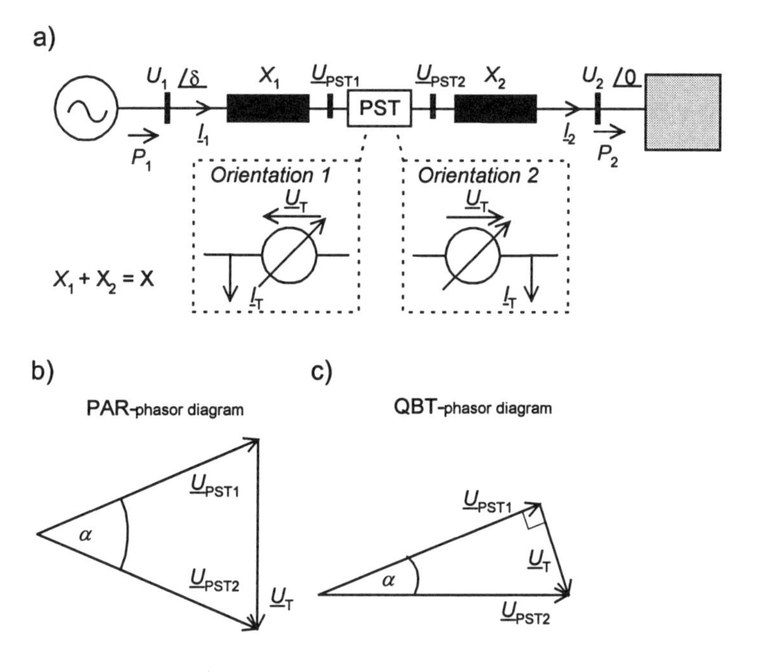

Figure 11.9 Model of the transmission system with PST (a) Network scheme (b) PAR phasor diagram (c) QBT phasor diagram

11.3.6.1 Phase angle regulator (PAR)

The PAR is a type of PST, which, due to its structure, has the ability to separate the phase of its terminal voltage phasors without changing their magnitude (do not forget: losses are not taken into consideration). The PAR phasor diagram is presented in Figure 11.9b. The controllable PAR parameter may be assumed to be the terminal voltage phasor separation i.e. the angle α. According to basic relations the following equations can be written:

$$\underline{U}_{\mathrm{PST1}} = \underline{U}_1 - \underline{I}_1 \mathrm{j} X_1 , \tag{11.43}$$

$$\underline{U}_{\mathrm{PST2}} = \underline{U}_{\mathrm{PST1}} \mathrm{e}^{\mathrm{j}\alpha} , \tag{11.44}$$

$$\underline{I}_2 = \underline{I}_1 \mathrm{e}^{\mathrm{j}\alpha} = \frac{\underline{U}_{\mathrm{PST2}} - \underline{U}_2}{\mathrm{j} X_2} . \tag{11.45}$$

As already mentioned PST powers are balanced therefore PAR input and output powers are equal i.e. $\underline{U}_{PST1}\underline{I}_1{}^* = \underline{U}_{PST2}\underline{I}_2{}^*$. From this relation it is easy to prove that the phasors $\underline{I}_1$ and $\underline{I}_2$ are shifted by the same angle α as the PAR terminal voltages, their magnitudes being equal – equation (11.45). From (11.45), considering both previous equations, it is not hard to calculate $\underline{I}_1$ and consequently $\underline{I}_2$ as follows:

$$\underline{I}_1 = \frac{-\mathrm{j}}{(X_1 + X_2)}\left(\underline{U}_1 - \underline{U}_2 \mathrm{e}^{-\mathrm{j}\alpha}\right) \Rightarrow \underline{I}_2 = \frac{-\mathrm{j}}{(X_1 + X_2)}\left(\underline{U}_1 \mathrm{e}^{\mathrm{j}\alpha} - \underline{U}_2\right) \quad (11.46)$$

Considering again $\underline{U}_2$ as a reference phasor ($\underline{U}_2 = U_2$), the transmitted power can be determined as follows:

$$P_1 = P_2 = P = \mathrm{Re}\left(\underline{U}_2 \underline{I}_2{}^*\right) = U_2 \mathrm{Re}\left(\underline{I}_2{}^*\right) = \frac{U_1 U_2}{(X_1 + X_2)} \sin(\delta + \alpha) \quad (11.47)$$

Evidently PAR shifts the "basic" transmission characteristic (α = 0) for transmission angle α in the "δ direction" without changing its form.

11.3.6.2 Quadrature boosting transformer (QBT)

Like PAR, QBT also separates its terminal voltage phasors but the injected voltage $\underline{U}_T$ phase is fixed with regard to the input voltage phasor $\underline{U}_{PST1}$ i.e. $\underline{U}_T$ is perpendicular to $\underline{U}_{PST1}$. The QBT phasor diagram is presented in Figure 11.9c. As in the PAR case, the controllable PAR parameter here may be assumed to be the terminal voltage phasor separation i.e. the angle α. The test system is described with the following equations:

$$\underline{U}_{PST1} = \underline{U}_1 - \underline{I}_1 \mathrm{j} X_1 \quad (11.48)$$

$$\underline{U}_{PST2} = \underline{U}_{PST1} \mathrm{e}^{\mathrm{j}\alpha} \frac{1}{\cos(\alpha)} \quad (11.49)$$

$$\underline{I}_2 = \underline{I}_1 \mathrm{e}^{\mathrm{j}\alpha} \cos(\alpha) = \frac{\underline{U}_{PST2} - \underline{U}_2}{\mathrm{j} X_2} \quad (11.50)$$

Equation (11.50) is, as in PAR case, derived from balanced PST power conditions i.e. $\underline{U}_{PST1}\underline{I}_1{}^* = \underline{U}_{PST2}\underline{I}_2{}^*$. Now $\underline{I}_1$ and $\underline{I}_2$ can be calculated.

$$
\begin{aligned}
\underline{I}_1 &= \frac{-\mathrm{j}}{\left(\dfrac{X_1}{\cos(\alpha)} + X_2\cos(\alpha)\right)}\left(\frac{\underline{U}_1}{\cos(\alpha)} - \underline{U}_2\mathrm{e}^{-\mathrm{j}\alpha}\right) \Rightarrow \\
\underline{I}_2 &= \frac{-\mathrm{j}}{\left(\dfrac{X_1}{\cos(\alpha)} + X_2\cos(\alpha)\right)}\left(\underline{U}_1\mathrm{e}^{\mathrm{j}\alpha} - \underline{U}_2\cos(\alpha)\right)
\end{aligned}
\tag{11.51}
$$

In the same way as in PAR case transmitted power can be calculated

$$
P_1 = P_2 = P = \mathrm{Re}\left(\underline{U}_2\underline{I}_2^*\right) = U_2\mathrm{Re}\left(\underline{I}_2^*\right) = \frac{U_1U_2}{\left(\dfrac{X_1}{\cos(\alpha)} + X_2\cos(\alpha)\right)}\sin(\delta+\alpha)
\tag{11.52}
$$

From the equations, describing transmission characteristics of the system model with the two PST types included ((11.47) for PAR and (11.52) for QBT), the following can be observed:

- PAR location has under these assumptions no impact on transmission characteristics,
- QBT location influences transmission characteristics essentially,
- PAR is a "symmetrical" device,
- QBT is a "nonsymmetrical" device.

A device is denoted as "symmetrical" when transmission characteristics do not depend on orientation (how the PST terminals are connected to the system – Figure 11.9a) and as "nonsymmetrical" in the opposite case. The equations (11.47) and (11.52) describing PAR and QBT transmission characteristics respectively have been derived taking "orientation 1" into consideration. If they are valid for "orientation 2" too, then the system in Figure 11.9a should be reversed, i.e. U_1 should be replaced with U_2, X_1 with X_2, δ with $-\delta$, and α with $-\alpha$. If those replacements are done, then in the case of a "symmetrical" device the transmission characteristics remain unchanged (power flows in the opposite direction to the model, and therefore it takes a negative sign). Of course for a "nonsymmetrical" device this is not the case.

A set of the test system transmission characteristics with PAR included is, for various α, shown in Figure 11.10a (only positive region). In Figure 11.10b transmission characteristics are presented for QBT positioned in the middle of the system (in this particular case orientation does not play any role). Figures 11.10c and 11.10d represent transmission characteristics (only positive region) when QBT is positioned at the grid terminals, for both orientations.

(a)

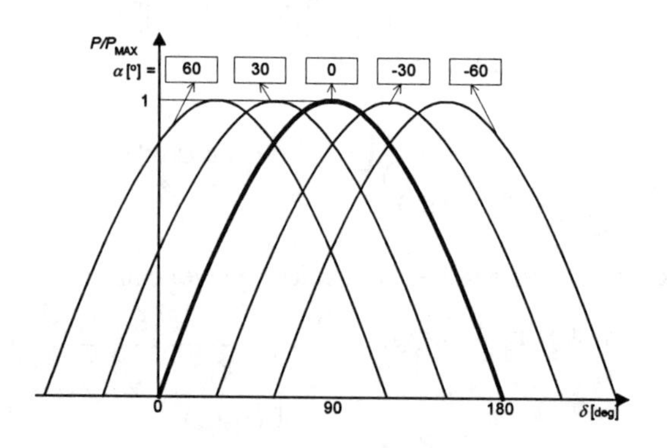
P/P_MAX
α [°] =
60
30
0
-30
-60
1
0
90
180
δ [deg]

(b)

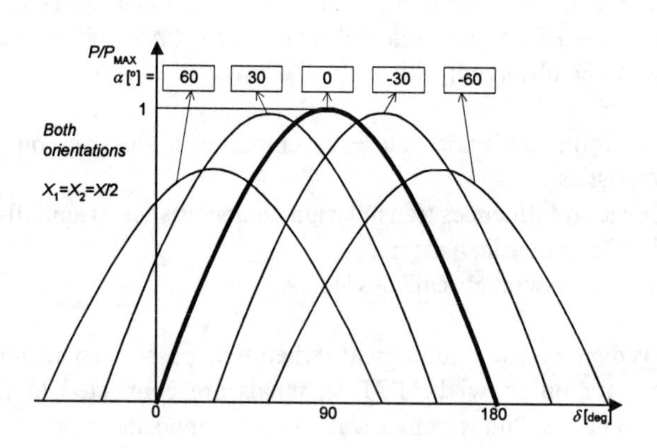
P/P_MAX
α [°] =
60
30
0
-30
-60
1
Both orientations
X1=X2=X/2
0
90
180
δ [deg]

(c)

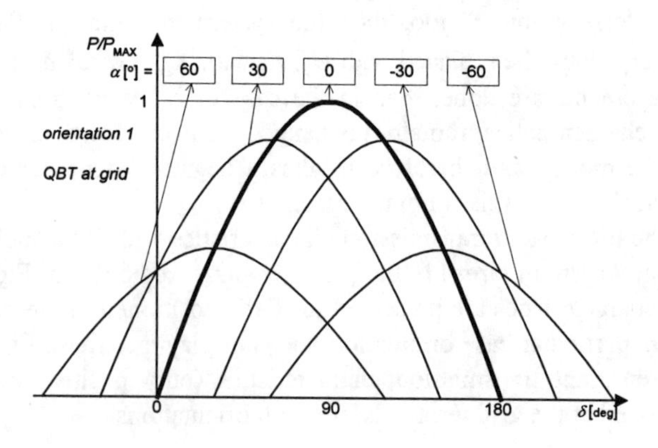
P/P_MAX
α [°] =
60
30
0
-30
-60
1
orientation 1
QBT at grid
0
90
180
δ [deg]

(d)

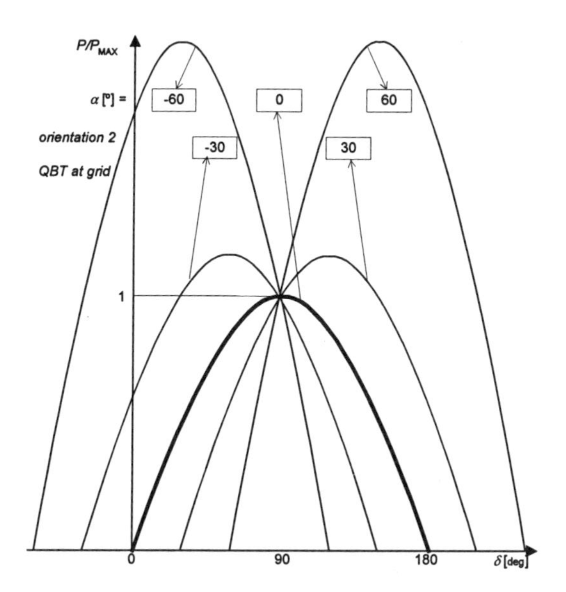

Figure 11.10 PAR and QBT transmission characteristics (a) Transmission characteristics – PAR (b) Transmission characteristics – QBT in the electrical middle of the system (c) Transmission characteristics – QBTat grid, orientation 1 (d) Transmission characteristics – QBTat grid, orientation 2

11.3.7 Power transmission control using unified power flow controller (UPFC)

UPFC is considered a universal tool for power flow control because it has an ability to simultaneously and independently control all three system parameters which affect power flow, i.e. transmission angle, terminal voltage and system reactance [17]. According to its impact on the system it might be modelled as a combination of a series voltage source, an active and a reactive current source [18]. The scheme is presented in Figure 11.11a. According to their structure UPFCs resemble PSTs, however, when active and reactive losses are neglected, their apparent power is not balanced. The injected voltage phasor $\underline{U}_T$ can have any phase with regard to its throughput current ($\underline{I}_2$ or $\underline{I}_1$ – depending on orientation – c.f. Figure 11.11a). The active power inserted into the system via $\underline{U}_T$ is balanced by the current source $\underline{I}_T$. Here $\underline{I}_Q$ represents a reactive current source and is independent of $\underline{U}_T$. Its presence might be considered as identical to a STATCOM (not taking dimensioning of the UPFC into consideration), the impact of which on transmission characteristics has already been discussed. The UPFC controllable parameters are thus: $\underline{U}_T$ magnitude U_T, $\underline{U}_T$ phase φ_T and current $\underline{I}_Q$ magnitude I_Q. The system presented in Figure 11.11a can be described with the following set of equations:

$$\underline{U}_{UPFC1} = \underline{U}_1 - \underline{I}_1 jX_1 \tag{11.53}$$

$$\underline{U}_{UPFC2} = \underline{U}_{UPFC1} + \underline{U}_T \tag{11.54}$$

$$\underline{I}_2 = \frac{\underline{U}_{UPFC2} - \underline{U}_2}{jX_2} = \underline{I}_1 - \underline{I}_Q - \underline{I}_T \tag{11.55}$$

where (do not forget: injected active power is balanced by $\underline{I}_T$ i.e. $\mathrm{Re}(\underline{U}_T\ \underline{I}_2^*) = \underline{U}_{UPFC1}\ \underline{I}_T^*$):

$$\underline{I}_T = \frac{\mathrm{Re}\left(\underline{U}_T \underline{I}_2^*\right)}{\underline{U}_{UPFC1}^*} \tag{11.56}$$

By splitting all complex quantities into their real and imaginary parts and by replacing $\underline{I}_Q$ with a parallel susceptance, from (11.55) it is possible to determine currents $\underline{I}_1$ and $\underline{I}_2$ as functions of system parameters and UPFC controllable parameters [18]. Then it is easy to calculate the transmitted power. However, the derivation, as well as the result (described in [19]) are quite long and are out of scope of this work. On the other hand, for a special case when UPFC is positioned at the system terminal, it is quite easy to determine the transmission characteristics. According to Figure 11.11a the reactance X_1 is taken out and X_2 is replaced with X. For this case the phasor diagram presented in Figure 11.11b is valid. It is obvious that in this case transmitted power can be expressed as:

$$P_1 = P_2 = P = \frac{U_{UPFC2} U_2}{X} \sin(\delta + \alpha) \tag{11.57}$$

where with a little trigonometry $\underline{U}_{UPFC2}$ and α can be obtained

$$U_{UPFC2} = \sqrt{U_1^2 + U_T^2 + 2U_1U_T\cos(\varphi_T - \delta)};\ \alpha = \arctan\left(\frac{U_T\sin(\varphi_T - \delta)}{U_1 + U_T\cos(\varphi_T - \delta)}\right) \tag{11.58}$$

In the sense of effective transient stability enhancement, the UPFC controllable parameters should be determined so as to achieve maximum impact on transmission characteristics. In general the "optimal" UPFC parameters can be determined from the following system of equations:

$$\frac{\partial P}{\partial \varphi_T} = 0; \frac{\partial P}{\partial I_Q} = 0; \frac{\partial P}{\partial U_T} = 0 \tag{11.59}$$

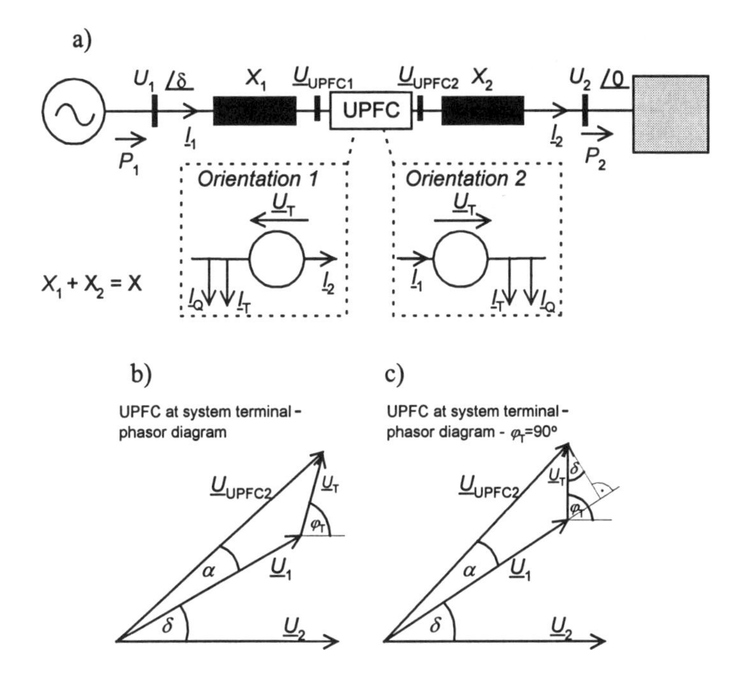

Figure 11.11 Model of the transmission system with UPFC (a) Network scheme. (b) Phasor diagram – UPFC at the generator terminals (c) Phasor diagram – UPFC at the generator terminals, "optimal" φ_T

Calculations have shown that the system does not have a solution and that maximum impact on transmission characteristics is achieved if U_T and I_Q are set to their maximum values determined by the device rating [19]. If the UPFC is located at the system terminals, optimal φ_T can be calculated from the first equation of the set (11.59). In our simple case optimal φ_T can be analytically calculated and equals ± 90°. For this case the phasor diagram presented in Figure 11.11c can be drawn. According to this diagram it can be concluded that:

$$\sin(\alpha) = \pm\frac{U_T\cos(\delta)}{U_{UPFC2}}; \qquad \cos(\alpha) = \pm\frac{U_1 + U_T\sin(\delta)}{U_{UPFC2}} \tag{11.60}$$

Now from (11.57) it is not to hard to calculate the "optimal" transmission characteristic

$$P_1 = P_2 = P = \frac{U_1U_2}{X}\sin(\delta) \pm \frac{U_2U_T}{X} \tag{11.61}$$

A positive sign in (11.60) and (11.61) is valid for maximum transmitted power and a negative sign for minimum transmitted power. As shown, the "power transfer benefit" with UPFC included is constant, i.e. it is possible to shift transmission characteristics "up and down". The family of "optimal" transmission characteristics (only positive region) is presented in Figure 11.12. A negative sign of U_T denotes minimum transmitted power. It is mathematically the same if $\underline{U}_T$ is rotated 180° or the sign of U_T is changed.

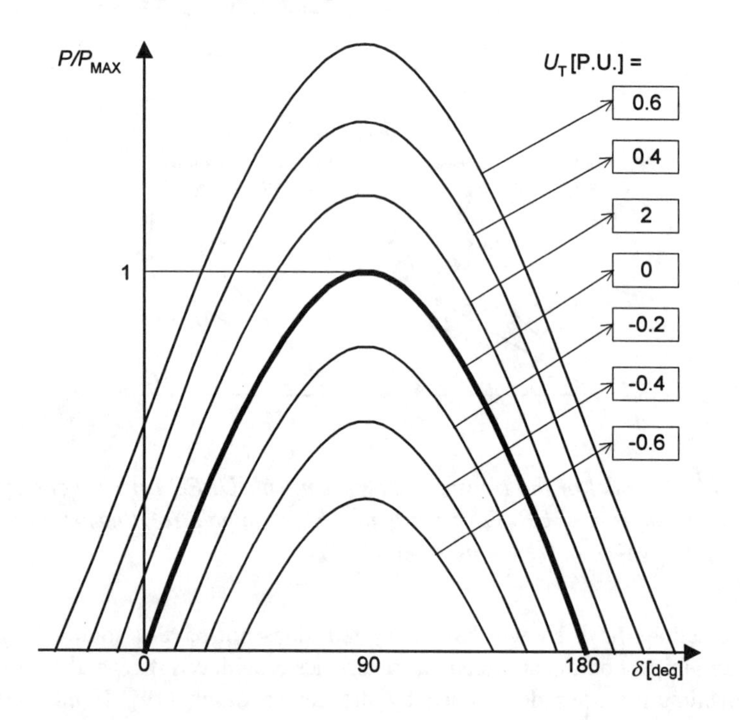

Figure 11.12 "Optimal" UPFC transmission characteristics

From numerical calculations with UPFC "inside" the system (not at the system terminal) the following conclusions could be drawn (but have not been analytically proved yet).

- If I_Q is set to 0, then the "optimal" transmission characteristics are, regardless of their position and orientation, the same as if UPFC were positioned at the system terminal as long as the system terminal voltages U_1 and U_2 are equal (characteristics in Figure 11.12 are valid). In this case "optimal" φ_T depends on UPFC location and transmission angle δ.

- If terminal voltages are not equal, "optimal" transmission characteristics depend on UPFC location. "Optimal" φ_T depends on location and transmission angle. Calculations have shown that UPFC impact on transmitted power is larger if it is positioned near the node with the lower voltage.

If UPFC is not positioned at the system terminals I_Q impacts transmission characteristics as though a STATCOM were present. Its effect can be considered as additional to that of UPFC with I_Q=0. From the previous section it is known that, in the case where the system terminal voltages are equal, optimal STATCOM location is in the middle of the system. Keeping in mind that, under these circumstances, location of a UPFC with I_Q=0 does not play any role, optimal UPFC location is also the middle of the system.

11.4 Control of FACTS devices for transient stability improvement

As already noted, the main method of extending the transient stability margin by the insertion of FACTS devices, is to modify the system transmission characteristic so as to extend the rotor-decelerating-area (area A_3 in Figure 11.2). In order to make the most of applied FACTS devices, they should be controlled so as to assure maximisation of the rotor-decelerating-area at a given device rating i.e. at a given region of possible controllable parameters of a device. In this way the necessary FACTS device rating to assure transient stability of a system will be minimised. The considerations described in Section 11.3 will serve as a basis for determination of the control strategy of FACTS devices during the transient period. Although the FACTS devices considered in Chapter 11 differ essentially from each other in structure, in the electric parameters they influence, as well as in their impact on system power transfer the main features of control actions during transient periods will be the same and are described in the following Section 11.4.1. Specific control features of various devices are described in the subsequent chapters.

11.4.1 General consideration of FACTS devices control strategy

The main points of the control strategy of FACTS devices, that may be fulfilled during the transient period, are listed below. For a hypothetical example the possible control actions are described in order to illustrate the goals in question.

1. **System should maintain stability after a major disturbance.** This is actually the main goal to be achieved and all other goals considered in

Chapter 11 are of a subordinate importance. In terms of the equal-area criterion, transmission characteristics should during the 1st swing, after a fault is cleared, be "raised" as high as possible with a FACTS device of a given rating (i.e. given area of possible controllable parameters). In Figure 11.13 a typical general case is demonstrated. Let us, for the reader's convenience, assume the FACTS device has only three possible controllable parameter values i.e. "PAR1", "PAR2" and "0" (FACTS device in neutral position, i.e. no influence on power transfer). The system transmission characteristics for those parameters are presented in the figure. The system operating point in the instance of a fault is the point "0". Under assumptions from Section 11.2.2, in the instance of a fault the operating point "jumps" to "1" and during the fault "travels" toward "2" where the fault is cleared. At this moment the controllable parameter should be "PAR1" in order for the operating point to "jump" to "3". Now the operating point should be kept on the "highest" of the transmission characteristics. In sense of this, at the point "4" the controllable parameter should change from "PAR1" to "0" at "5" from "0" to "PAR2" and at "6" from "PAR2" to "0". Let "7" be the 1st swing limit. "8" represents a theoretical stability limit. If the operating point crosses this point the system is not stable.

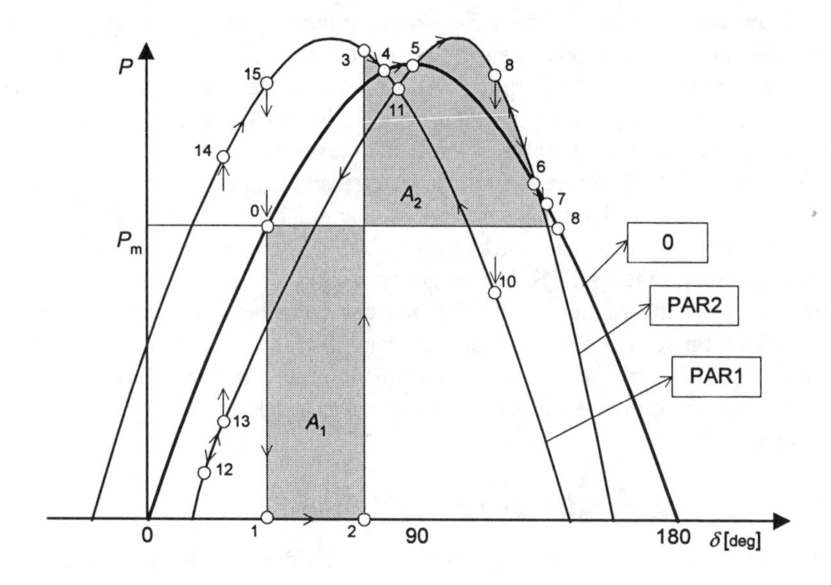

Figure 11.13 Demonstration of the general FACTS devices control strategy

2. **The system should not persevere near the maximum of the 1st swing.** The phenomena of the generator rotor persevering near the 1st swing could be considered acceptable in a two-machine system. On a more practical system, containing many machines, the interaction of machines swinging with different periods could well cause instability. Now, the question is, how to control the FACTS device after reaching the 1st swing limit "7". The worst action would be a "jump" in the "PAR1" characteristic. This would cause the rotor to be further accelerated and consequently synchronism would be lost. If the operating point should move from the area of large transmission angles quickly, the decelerating power should be at its maximum at the beginning of the rotor back-swing. Therefore the FACTS controllable parameter should remain "0" and changes to "PAR2" at "6".
3. **The rotor back swing should be small.** In this way the subsequent swings can be effectively damped, and the whole transient phenomena in general is smoother. Consequently the components in the system are subject to less stress. In terms of the equal-area criterion the decelerating area i.e. the area between P_m and the operating point trajectory should be small. However, this requirement contradicts the previous one, therefore a certain compromise has to be found. Let us say: on its "way back", as it reaches point "9", it jumps to "PAR1" characteristic (point 10). Then it remains with this characteristic until point "11" is reached. From there it may follow the "PAR2" characteristic until the back swing limit is reached at point "12".
4. **Subsequent oscillations should be effectively damped.** In a system that is transiently stable then, after a disturbance, generator rotors oscillate around their equilibrium points. These oscillations are more or less damped and might in a system with low damping persist for a long times, which is undesirable. Therefore FACTS devices, which are used for transient stability reasons, may be applied also for damping of the subsequent swings (rotor swings after the 1st swing). Ratings for FACTS devices which would assure systems to be transiently stable would normally be quite high compared to those ones which are used to assure oscillatory stability of the systems. Therefore damping of subsequent swings should normally not be a problem although an "optimal" damping control is not applied. Although the problem of oscillation damping is beyond the scope of this section, for readers convenience a brief insight is provided. In the case presented (Figure 11.13), it is dealt with by "bang-bang" control as we have only three possible states ("PAR1", "PAR2" and "0"). Let us assume the point "12" is positioned "behind" "0". Then, in order to achieve effective damping, during the 2nd swing forward movement, at the proper moment, e.g. at "13", a jump to "14" is made. In the theoretical case "13" should be chosen so that the 2nd swing accelerating energy (area between "12" to "13" and P_m) is balanced by the decelerating energy accumulated in the rotating masses between "14" and

"15". At "15" relative rotor movement is stopped and, before it swings back, the jump from "15" to equilibrium point "0" should be made. In this way the system is completely damped. If "12" were positioned "in front of" "0" then from "12" the operating point should jump to "PAR1". That would cause the operating point to decelerate again. At the right moment a jump to "PAR2" should be performed. Now the rotor has been decelerated and relative movement would be 0 at the equilibrium transmission angle. At this moment a jump to "0" would be performed. In practice, however, it is not possible to determine the right jumping moments to achieve complete damping therefore the subsequent swings have to be damped more or less by the system natural damping. If continuous control is possible, then any operating points between the "PAR1" and "PAR2" characteristics can be reached. The vertical distance between the basic transmission characteristic and the operating point can then be a continuous function of the damping signal (e.g. frequency difference, power or transmission angle gradient, generator speed deviation, etc.). In this way effective damping is possible also as the swings become small.

To extend the transient stability margin of a two-machine system only 1 is relevant. It also defines the ratings of FACTS devices that would assure transient stability. Actions described in 2 to 4 help to stabilise the system after the 1st swing. They may be of crucial importance in multi-machine systems or in systems with negative damping. Their realisation may differ considerably depending on specific needs and circumstances.

In our further considerations the continuous control of FACTS devices will be assumed.

11.4.2 CSC, SSSC, SVC, STATCOM and UPFC control strategy

The devices in question can be discussed together because the basic strategy to fulfil the requirements of Section 11.4.1 is the same for all of them. The reason for this is, that for all devices, change of their controllable parameter causes change of power transfer in the direction, which is, in the region of interest for transient stability, always the same. In Figure 11.14 the basic and the two "extreme" transmission characteristics ("MAX" for maximum and "MIN" for minimum power transmission) are presented. From Section 11.3 it can be concluded that for all the devices considered, the extreme characteristics are achieved if the absolute values of their controllable parameters are set to their maximums, defined by the device ratings. In the case of a UPFC, the "optimal" angle φ_T is assumed (see Section 11.3.7). The settings of parameters according to Section 11.3 are summarised in Table 11.1.

Table 11.1 Setting of controllable parameters of FACTS devices to achieve "extreme" transmission characteristics

Device:	Controllable parameter	Parameter setting when maximal transmission is required	Parameter setting when minimal transmission is required
CSC	K_{CSC}	to maximum	to minimum
SSSC	U_T	to minimum	to maximum
SVC	B_{SVC}	to maximum	to minimum
STATCOM	I_Q	to maximum	to minimum
UPFC	U_T	to maximum	to minimum
	I_Q	to maximum	to minimum

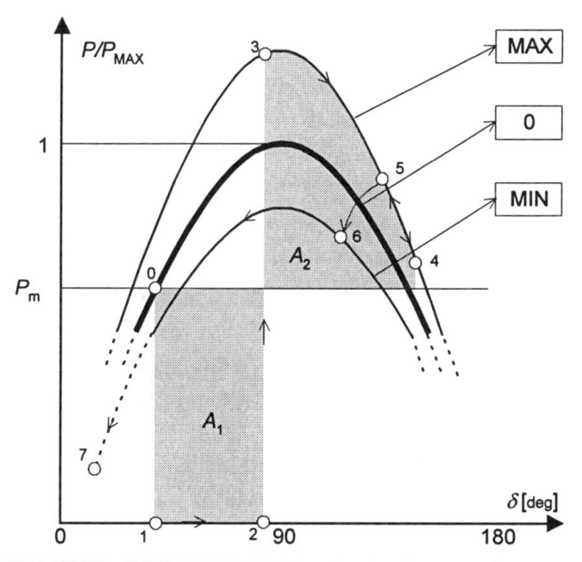

Figure 11.14 CSC, SSSC, SVC, STATCOM and UPFC control strategy demonstration

In accordance with the first requirement of the general control strategy (Section 11.4.1) between points "3" and the 1st swing limit "4", the "MAX" characteristic is chosen. In order to fulfil the second goal of the general control strategy the operating point should persist with this characteristic until e.g. "5" is reached. The third goal is fulfilled if from "5" the operating point goes over to "MIN" characteristic ("6") and remains there until the back swing limit "7" is reached. It should be noted that care has to be taken if the "MIN" characteristic lies well below the P_m level. In this case "7" may be positioned at relatively high δ angle. Bringing the system into equilibrium point might take longer and might be more difficult than necessary. In this case choosing the trajectory along P_m level would not be bad idea. From "7" on, the oscillation damping strategy takes over.

If "7" is positioned in the region $\delta < 0$ then in the vicinity of $\delta = 0$ the operating point should go from one to other extreme characteristic in all cases except in the UPFC case. The reason for that is the "intersection" of characteristics (see Section 11.3) [20].

11.4.3 PAR control strategy

As discussed in Section 11.3.6.1 with PAR it is possible to "shift" the transmission characteristic along the "δ" axis. The outline of a possible control strategy is discussed with the help of Figure 11.15(a). At the instant of fault occurrence the operating point "jumps" from the pre-fault point "0" to "1" and moves towards "2" during the fault. In order to maximize decelerating area at the moment of the fault clearance the controllable parameter α should be set so as the maximize the transmission characteristic at the transmission angle of the point "2" $\delta(2)$. In this case the equation $\delta(2) + \alpha = 90°$ has to be fulfilled and thus at "3" α should be set to $90° - \delta(2)$. Between the points "3" and "4" the parameter α is changed simultaneously with δ i.e. $\delta + \alpha = 90°$ has to be fulfilled at each moment. Let us assume α_{min} represents the device limit and "5" the 1st swing limit. Then "6" represents the point of "no return". In this way the first goal of the general control strategy (Section 11.4.1) is satisfied. In order to satisfy the second requirement α might remain constant (α_{min}) until "7" is reached. Then α is changed simultaneously with δ until e.g. point "8" (in this case at the same location as "0" that is at $\alpha = 0$) is reached. Here α is blocked and from the back swing limit "9" on oscillation damping strategy takes over.

It should be emphasised that the proposed strategy represents only one of the possibilities and might be useful in the case of relatively high pre-fault power transfer levels (P_m) and high phase shift limits α_{min}. In a variant of the proposed strategy the power level between "7" and "8" might be chosen below P_m in order to minimise back swing. If the pre-fault level as well as α_{min} are relatively low then the strategy illustrated in Figure 11.15(b) might be applied. The operating point moves from "4" to "5" and back to "7" (between "4" and "7" $\alpha = \alpha_{min}$), then to "8" (α controlled so that: $d\alpha/dt > d\delta/dt$), from "8" to "9" ($\alpha = \alpha_{max}$) and from "9" to the back swing limit "10" ($\alpha = \alpha_{min}$) where oscillation damping strategy takes over. Some other considerations, regarding 1st swing stability enhancement using PAR can e.g. be found in [21, 22, 23, 24].

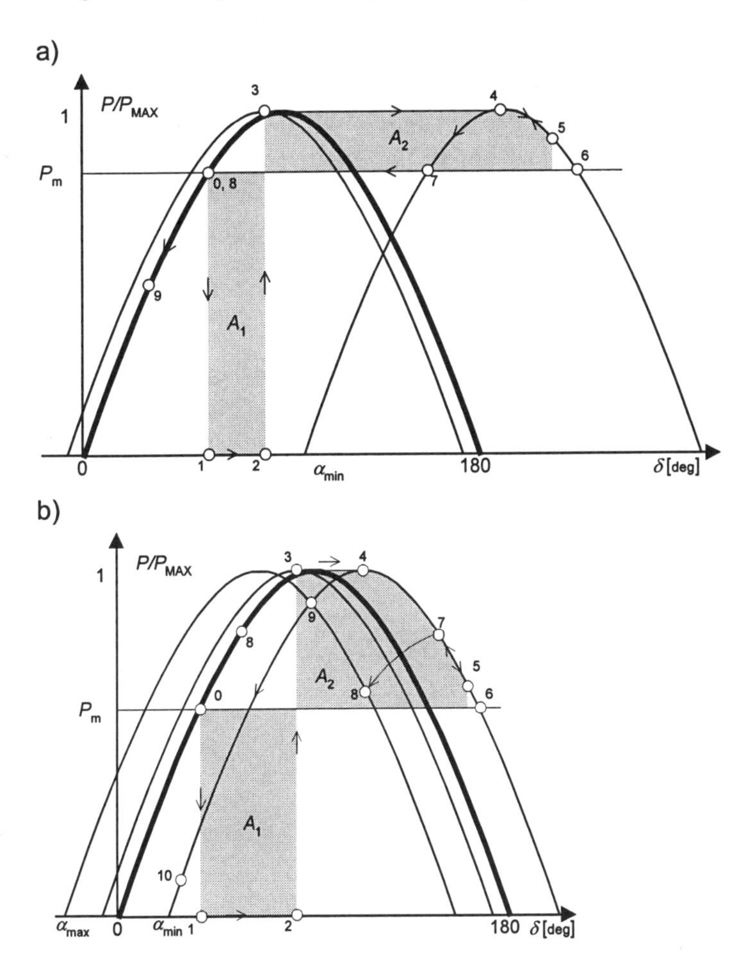

Figure 11.15 PAR control strategy demonstration (a) One of the possibilities (b) Another posibility

11.4.4 QBT control strategy

According to calculations from Section 11.3.6.2 it is obvious that the most appropriate QBT location is at the voltage source location, a parallel branch (modelled as the current source $\underline{I}_{PAR}$) being connected to the source terminal (the case presented in Figure 11.10(d)). In this case the situation is very similar to that presented in Section 11.4.1. At the moment of fault clearing α should be set to its limit ($\alpha = \alpha_{min}$) in order to reach “3”. At “4” α is switched from α_{min} to α_{max} (α changes its sign) and remains unchanged until, during the back swing, e.g. “7” is reached (“5” being the 1st swing limit and “6” the stability margin limit). Between “7” and “8” α is continuously changed from α_{max} to α_{min}. Between “8” and “9” it is constant α_{min} and switches to α_{max} at “9” (actually the same point as “4”). “10” represents the back-swing limit where oscillation damping strategy takes over. A few more details regarding QBT control strategy can e.g. be found in [16, 25, 26].

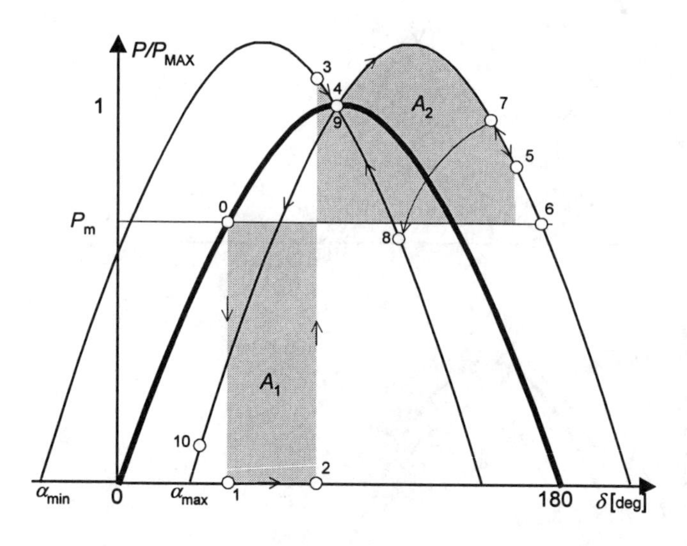

Figure 11.16 QBT control strategy demonstration

11.5 Transient stability analysis and dynamic models of FACTS devices

The main question to be answered, when studying transient stability, is: does a system maintain stability under certain system structure, operating point and contingency? As a mean of introducing basic concepts the one-machine infinite-bus system has been considered in previous chapters. With some simplifications the graphical approach (comparing the accelerating and decelerating areas) could be used for analysis. In contrast to such a simplified system, practical power systems are characterised by complex network structures. The previously adopted simplifications can lead to uncertain results. Some of the factors present in real power systems and neglected in our basic theoretical considerations are:

- **Flux decrement** Until now, transient voltage E' has been considered constant during transients, which may be a good approximation for short fault clearing times. However, for long clearing times the flux decrement and consequently decay of E' can have considerable effect. In such circumstances assuming E' to be a constant value may lead to an optimistic assessment of the transient stability margin.
- **Automatic Voltage Regulator (AVR)** During a fault the terminal voltages of generators drop and that causes AVRs to react by raising the excitation current. Consequently, with a certain time delay E' increases too. Fast acting AVRs and exciters can therefore decrease the accelerating area (A_1) as well as enhance the possible decelerating area (A_3) [6].
- **Changes in mechanical power** In our considerations the turbine mechanical power has been assumed constant (P_m). During longer transients this assumption might not be correct any more and mechanical power changes may considerably affect transient stability. This situation is met especially when in the system (steam) turbines with fast response are present (fast valving).
- **Load characteristics** Loads are in real systems voltage dependent. During transients, in a power system voltages can vary considerably which results in load variations.
- **FACTS devices characteristics** In previous sections idealized FACTS devices are presented. In real devices, limitations may be present, which are dependent on a device and the system operating point. Although FACTS devices are normally "fast-acting" devices, they nevertheless have time constants and "bang-bang" type switching between various operating points is not possible. From a modelling point of view, firstly, this means that the model limitations are not always constant but may in certain operating areas be dependent on the operating point, and, secondly, that the operating point can not be moved inside the operating area without a time delay. For these reasons the FACTS impact on transmission characteristic might in dynamic conditions differ from the idealized one.

The methods for transient stability analysis may in general be divided into two groups, i.e.:

- direct methods and
- indirect methods.

It is beyond the scope of this book to go into the details of computation methods, however a brief explanation is given.

Direct methods determine transient stability without explicitly solving system differential equations. They are based on the so-called energy function. The energy-based methods may be considered a special case of the more general Lyapunov's method, energy function being one of the possible Lyapunov functions. For description of the main principles of the transient energy function approach the rolling ball analogy can be used (Figure 11.17) [27, 28].

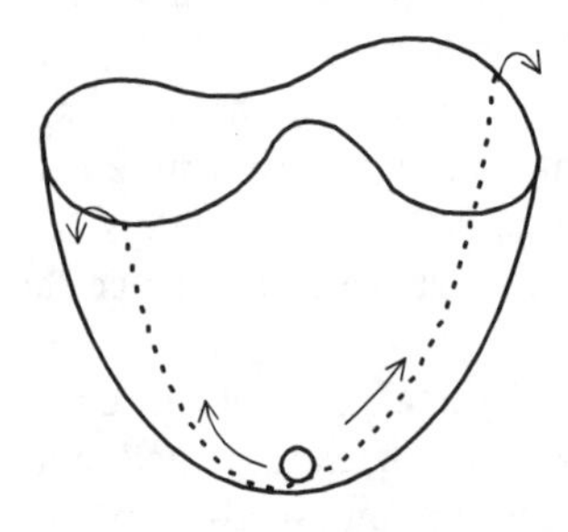

Figure 11.17 A rolling ball analogy presentation

A bowl is an analogy of the system potential energy surface. Let us consider a pre-fault operating point (stable equilibrium) is present with the ball at the bottom of the bowl. During the fault, the system gains kinetic and potential energy and ball rolls up the bowl. As the fault is cleared the ball is no more "pushed" upwards, but it still has some speed (kinetic energy). If this energy is higher than the potential difference between the current ball position and the bowl rim, the ball will "jump" out of the bowl and the system is not stable. The rim represents the maximum potential energy-absorbing capability of the post-disturbance system.

Indirect methods require solving the system differential equations. Of course such systems of (non-linear) differential equations cannot be solved analytically. Their solution is performed by applying step-by-step integration procedures. The most practical approach is time-domain simulation by applying (existing) computer programs for digital simulation of electric power systems.

In recent years direct methods have been accomplished significantly. Nevertheless at present they remain of minor importance due to the modelling limitations and the unreliability of computational techniques. In the vicinity of stability limits of the multi-machine systems direct methods are vulnerable to numerical problems resulting in unrealistic results [6, 28]. Direct methods might be the favourite tool for on-line computations because they should be faster than indirect methods and they give an answer about the degree of stability (how far from the bowl rim the ball turns around) [2].

In contrast to direct methods, by using the time-domain simulation, systems can be modelled as accurately as needed (including all controllers, rotating masses, nonlinearities, protection actions etc.) [29] and is today mostly applied. However the answer is more or less of a "yes/no" type and therefore the stability margin has to be determined by subsequent simulation runs.

The best course in development of tools for transient stability analysis might be a hybrid approach in which the transient energy calculation is incorporated into a time-domain simulation [30, 31].

In our numerical examples time-domain simulation methods have been applied.

11.5.1 Dynamic models

Since the time-domain simulation approach is predominating in studying transient stability enhancement using FACTS devices, the models presented here are suitable for application in digital time-domain simulation tools. There are two possible ways to carry out time-domain studies [32].

The first is referred to as the "momentary mode". When working in this mode, electric power systems including FACTS devices have to be modelled more or less in detail. The results should represent the time functions of physical quantities "as they really are" with all fast transients etc. Voltages and currents are e.g. in steady state conditions represented as sine functions. When applying this mode, the integration step has to be short and modelling is more or less comprehensive. Simulation procedures are time consuming. As an electric power system dynamic in a transient stability problem is relatively slow, such an approach is not the most suitable one. Exceptions may be certain special situations that have to be studied in detail.

The second approach is the so called "stability mode". In this mode an electric power system is modelled as single-phase, and electric quantities are represented with their effective values (phasors). Since sine quantities are not dealt with, the integration step may be longer, and modelling is simpler and the simulation procedure is faster than in the momentary mode. Assuming proper modelling, the results, for electromechanical dynamics (transient stability), should be very close to those achieved in the momentary mode. For these reasons this mode is chosen as a basis for our further considerations.

11.5.1.1 Elements of a control system

As already presented in Section 11.3 FACTS devices are for our purposes presented as controllable impedances, voltage sources and/or current sources [15]. The basic strategies for transient stability enhancement (Section 11.4.1) are the same for all the devices considered. Therefore certain parts of their control may have similar or identical tasks. The main structure of the possible control model for FACTS devices is presented in Figure 11.18.

It is assumed that local and remote electrical system quantities are available (that is nowadays, due to modern telecommunication facilities, technically no problem). The "SYSTEM OPERATING POINT MONITORING" module serves to determine the operating point location i.e., the controller determines in which part of a transmission characteristic it is located. For example: if a fault has been detected (voltage drop) and the voltage rises, it is an indication that fault is cleared; if the transmission angle time derivative changes its sign for the 1st time (and a fault has happened) it means that the 1st swing limit has been reached; if the sign in question is changed for the 2nd time it means that the back swing limit is reached etc. Of course such operating point monitoring determination is only one of several possibilities and has been chosen for ease of explanation without taking technical and/or economic factors into consideration.

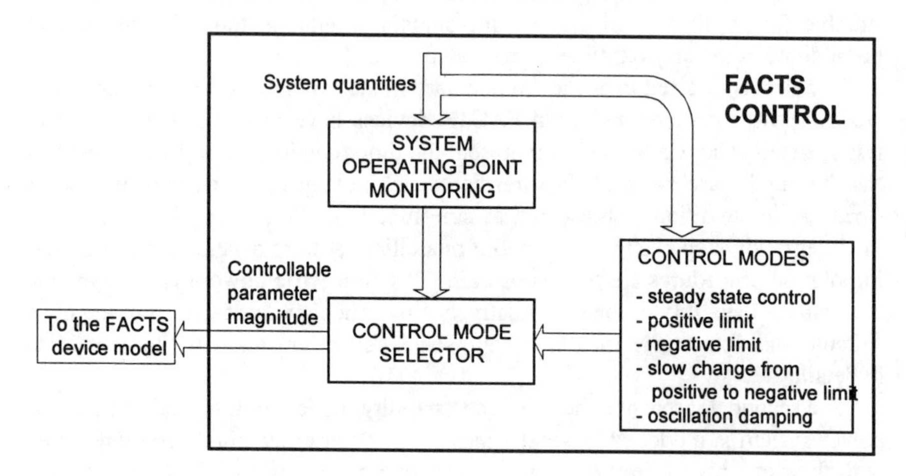

Figure 11.18 General structure of the FACTS devices control

The "CONTROL MODES" module contains signals for various operating modes. Normally a FACTS device has to fulfil a task in steady state conditions therefore one of the control modes is steady state control (e.g. voltage control of an SVC or STATCOM). This mode is not of interest in our considerations therefore in the pre-fault conditions FACTS devices are assumed to be in neutral position (out of operation). Positive and negative limit denote operation with the controllable parameter at the device limit. Slow change from positive to negative limit is used to realise the mode between points "5" and "6" (Figure 11.4) and between "7" and "8" (Figures 11.15b and 11.16) respectively. The controllable parameter change can be a function of time or of the transmission angle δ.

Oscillation damping strategy can be used as a controllable parameter being in proportion to the damping signal. Although theoretically the bang-bang control is the most effective damping mode, in practice proportional control may be more appropriate [33]. The reason for that is, that optimal switching times for bang-bang control are not known and with such a control mode it is not possible to damp oscillations completely. In our numerical studies proportional control has been applied. The damping signal may derive from frequency difference, power gradient, rotor speed deviation, transmission angle difference deviation etc. In this module also the device's controllable parameter limitations may be realised. They are in general time and operating point dependent [34] (e.g. transient rating of the device may be essentially higher than the steady-state rating).

The "CONTROL MODE SELECTOR" chooses the control mode that has to be applied according to the location of the operating point and the control strategy. It also introduces "smooth" transitions from one mode to another by introducing time constants (e.g. switching between α_{max} and α_{min} at the point "4" – Figure 11.16 does not happen instantaneously but some time is involved).

The output of the "FACTS CONTROL" is a controllable parameter which is introduced into the network representation of the FACTS device model.

11.5.1.2 Controlable series Capacitor (CSC) model

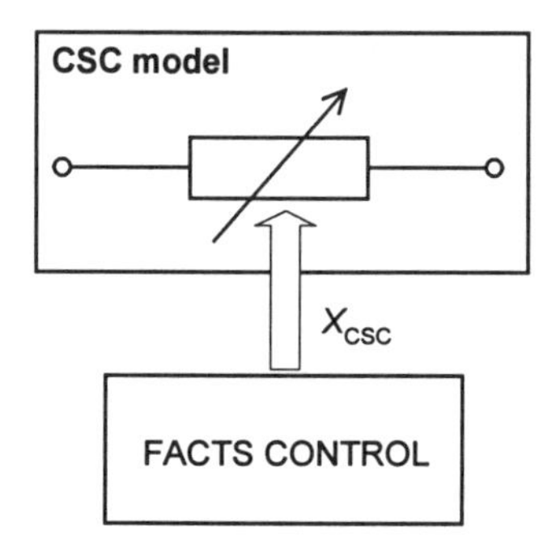

Figure 11.19 CSC model

CSC can in the stability mode of the time domain simulation be represented by a series connected controllable impedance. Its magnitude is determined by the "FACTS CONTROL" block. "Positive limit" and "negative limit" represent the maximum and minimum reactance. Schematically the model is presented in Figure 11.19.

11.5.1.3 SSSC model

An SSSC can be represented as a reactive voltage source connected in series with the transmission line. The controllable parameter is the injected voltage magnitude U_T. “Positive limit” and “negative limit” represent operating modes with opposite U_T signs (phases), U_T magnitude being at its maximum. An idealized SSSC does not consume active power, and the injected voltage phasor $\underline{U}_T$ has to be perpendicular to the SSSC throughput current phasor $\underline{I}_{SSSC}$. Therefore the “SSSC COORDINATOR” monitors the current phasor $\underline{I}_{SSSC}$ and adjusts the corresponding phase of the phasor $\underline{U}_T$. The output may be $\underline{U}_T$ in magnitude and phase or $\underline{U}_T$ real and imaginary components (determined with regard to the system reference phasor). The scheme is presented in Figure 11.20.

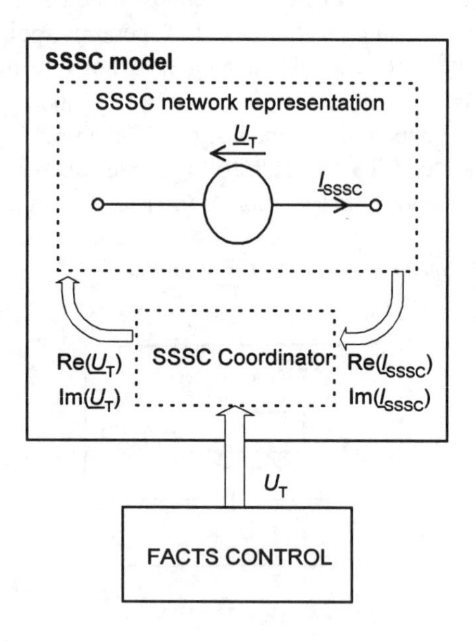

Figure 11.20 SSSC model

11.5.1.4 SVC model

According to its operating characteristics an SVC in stability mode of the time domain simulation can be represented by a parallel connected susceptance. Its magnitude is determined by the “FACTS CONTROL” block. “Positive limit” and “negative limit” represent the maximum and minimum susceptance. Schematically the model is presented in Figure 11.21.

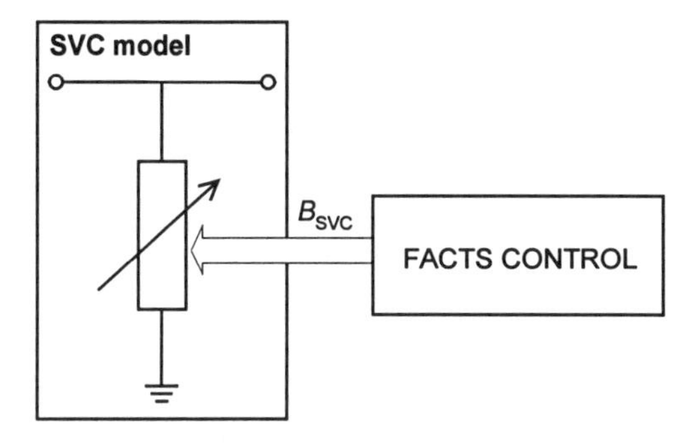

Figure 11.21 SVC model

11.5.1.5 STATCOM model

A STATCOM may be represented as a reactive current source connected in parallel with the transmission line. The controllable parameter is the injected current magnitude I_Q. Both, "positive limit" and "negative limit" represent operation with maximum $\underline{I}_Q$ magnitude, but with opposite signs (phases). As in the SSSC case, the "STATCOM COORDINATOR" has been introduced which monitors the terminal voltage phasor $\underline{U}_{STATCOM}$ and adjusts the phase of the phasor $\underline{I}_Q$ so as to be perpendicular to $\underline{U}_{STATCOM}$. In this way, STATCOM is considered to be without losses. The coordinator network input may be terminal voltage phasor data while outputs are $\underline{I}_Q$ magnitude and phase or $\underline{I}_Q$ real and imaginary components. The scheme is presented in Figure 11.22.

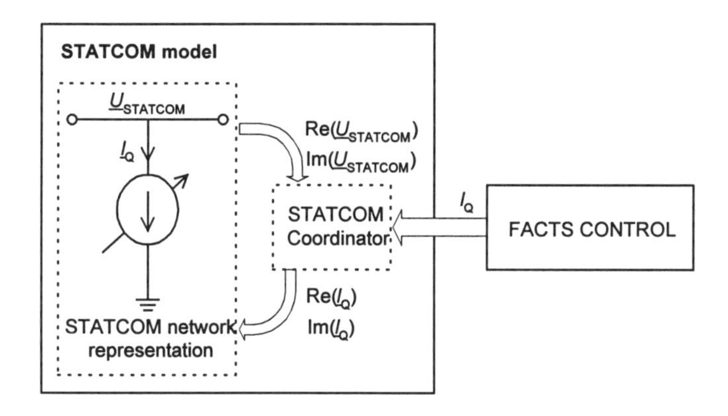

Figure 11.22 STATCOM model

11.5.1.6 PAR and QBT model

A possible representation of PAR and QBT (phase shifting transformers in general) may be a combination of a series connected voltage source $\underline{U}_T$, representing a boosting transformer, and current source $\underline{I}_T$, representing PAR and QBT parallel transformers. The scheme is presented in Figure 11.23. The controllable parameter is the voltage phasor shift α, therefore the "positive limit" and "negative limit" represent the operating modes with the maximum and minimum α respectively (taking its sign into consideration, of course). In the present structure the "PAR, QBT COORDINATOR" has to fulfil two roles. First it has to keep the injected voltage phasor $\underline{U}_T$ in proper phase with regard to the PAR or QBT input voltage phasor $\underline{U}_A$ (perpendicular in the QBT case and at 90° - α / 2 in PAR case). Second, it has to determine the phase and the magnitude of the current source $\underline{I}_T$. With $\underline{I}_T$, the power ($\underline{S}_T$) is taken out from the system that is injected into it via voltage source $\underline{U}_T$ (do not forget: phase shifting transformers do not produce or consume active and/or reactive power – neglecting losses, of course). In this way $\underline{S}_{IN}$ and $\underline{S}_{OUT}$ are equal. The coordinator network inputs are the input voltage and power (or current) data, while the outputs are $\underline{U}_T$ and $\underline{I}_T$ data.

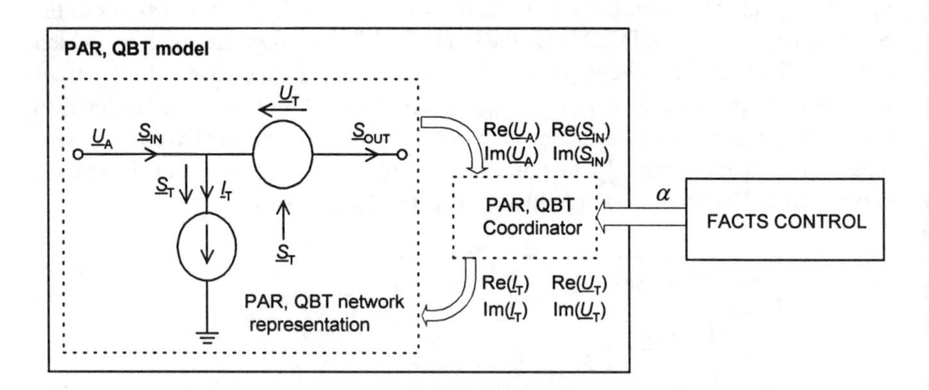

Figure 11.23 PAR and QBT model

Another possibility (not discussed here) to model controllable phase shifting transformers is to represent them as controllable transformers with complex turn ratios [35] (the magnitude and the modus of the complex turn ratio being dependent on transformer type – e.g. PAR or QBT – and phase shift) or in the form of the controlled power injections [36, 37].

11.5.1.7 UPFC model

A UPFC may, on account of its basic structure, be considered a kind of phase shifting transformer (PST). Therefore it may also be modelled in a similar manner. The difference from the PST model is that the UPFC parallel branch is modelled with two current sources. One of them ($\underline{I}_T$) is an active current source and balances *active power* which is injected into the system via a series branch (P_T) while the other one ($\underline{I}_Q$) is a representation of the parallel branch reactive current control (c.f. Section 11.3.7). The "UPFC COORDINATOR" has to fulfil two roles. First, it has to determine the phase and the magnitude of the current source $\underline{I}_T$ and second it determines the phase of the reactive current source $\underline{I}_Q$ in an identical way to the case of a STATCOM. As in the PST case, the coordinator network inputs may be the input voltage and power (or current) data, while the outputs are $\underline{I}_T$ and $\underline{I}_Q$ data. Data regarding injected voltage $\underline{U}_T$ (U_T and φ_T), given from the "FACTS CONTROL" module, are eventually transformed and transmitted to the model.

The control modes "positive limit" and "negative limit" (c.f. Figure 11.18) in general represent operating modes in which, at given device ratings, transmitted power should be maximised or minimised respectively. In these two modes the $\underline{U}_T$ and $\underline{I}_Q$ magnitudes (U_T and I_Q) should be set to their maximal values. For $\underline{I}_Q$ the phase (sign of I_Q) determines if the power flow should be maximised or minimised. $\underline{U}_T$ phase determination, in order to achieve extreme transmitted powers (maximum or minimum), is not so simple. As already discussed in Section 11.3.7, the "optimal" angle φ_T has to be determined. A possible way for longitudinal transmission systems, to determine "optimal" φ_T, is described below.

It has been proven [18, 19, 38] that it is possible to express the terminal powers of the so called "basic model" (consisting of two Π sections connected to voltage sources with a UPFC in between) as the analytic functions of the "basic model" network parameters and UPFC parameters. This offers the possibility of searching for optimal φ_T if the following procedure, illustrated in Figure 11.24, is carried out in the UPFC controller "CONTROL MODES". First, the simulated system is transformed into the form of the "basic model" (by calculating equivalent impedances) thus determining its parameters. Because the simulated system consists in part of non-linear elements, this procedure must be carried out on-line (for each integration step of the simulation), thus equivalent impedances change with the time (e.g. the equivalent impedances of the loads change with the terminal voltages, which further change with time). On the basis of the mathematical model, the optimal φ_T is determined (from the mathematical model the controller calculates at which φ_T the basic model terminal powers would be maximal or minimal, applying any mathematical procedure for determination of function extremes). Such "optimal" φ_T represents one of the three outputs of "FACTS CONTROL" module when extreme power flows are required. The procedure presented is executed inside of each integration step of the digital simulation.

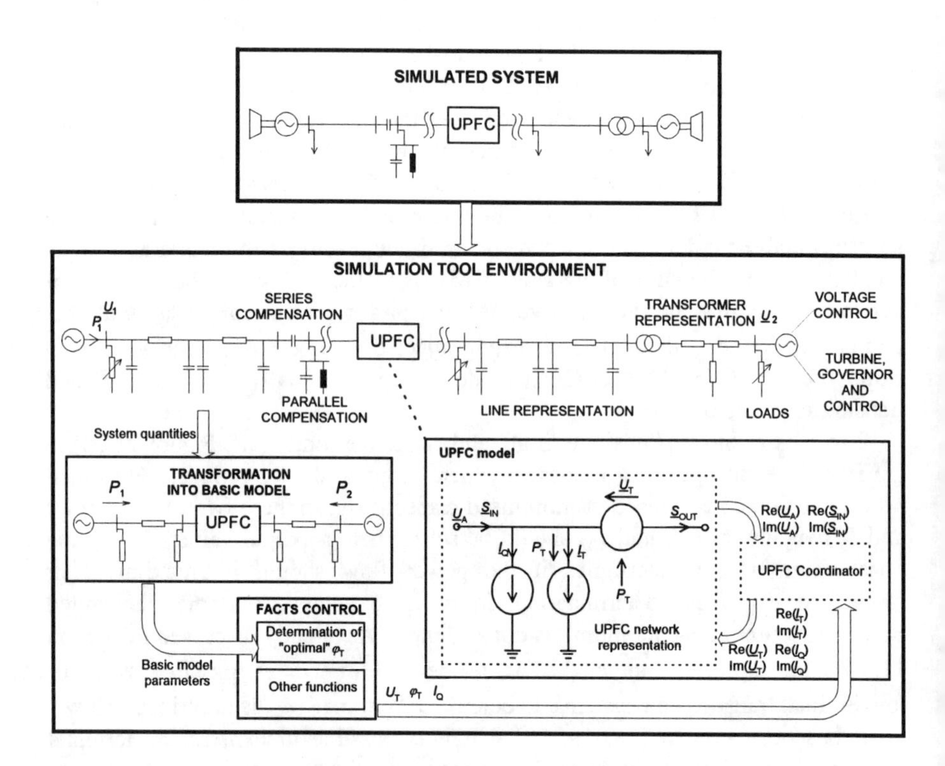

Figure 11.24 UPFC model and basic outlines of digital simulation modelling of a longitudinal system with UPFC included

The UPFC network representation (two current sources and one voltage source) is only one of the possibilities. Another possibility is to represent it by controlled power injections [39].

11.5.1.8 Some further considerations

In our models, active and reactive losses of the FACTS devices have not been taken into consideration. They may be modelled in the form of additional impedances (reactances for reactive and resistances for active power losses) which may be constant or dependent on operating conditions (e.g. a PST's reactance depends on its operating point [40]).

In the majority of the proposed models "coordinators" have been applied which, on the basis of system quantities, adjust the phase of the model voltage and/or current sources. This action can not, however, be carried out in the actual integration step because system quantities (e.g. throughput current phase) have to be known first and only then the controllable parameter phase can be determined (it is applied one integration step later). Because of this integration step delay, some error is present. Numerical calculations have shown that, normally, this can be neglected. If, however, the error in question should be reduced, then the integration step could be reduced (resulting in longer digital simulation times) and/or the system parameters of the following integration step could be predicted by e.g. linear interpolation and used in controllable parameter phase determination. In this way error is drastically reduced [14].

It should be noted that, due to the integration step delay in question, in some operating areas, some of the proposed models may become numerically unstable, if in the "coordinator" the phase of the controllable parameter is determined in an improper manner. A typical example is the determination of the SSSC injected voltage ($\underline{U}_T$) phase. If in "SSSC COORDINATOR" it is determined simply by adding 90° to the SSSC throughput current $\underline{I}_{SSSC}$ phase, then numerical instability appears when the operating point is within the area "negative power at positive δ" (Figure 11.5 - e.g. U_T=0.6, δ between 0 and ca. 35°). The reason for that is, the $\underline{U}_T$ phase change causes the $\underline{I}_{SSSC}$ phase to change in the next integration step, then a new $\underline{U}_T$ phase is calculated and the procedure repeats. In the area in question, the $\underline{I}_{SSSC}$ phase is obviously very sensitive to the $\underline{U}_T$ phase changes and "runs away" faster than (with one integration step delay) the $\underline{U}_T$ phase can follow. A solution to avoid such numeric instability may be to apply the "two-axis theory" [9, 10].

11.6 Numerical studies

11.6.1 Test system and system behaviour without power flow control

For demonstration of transient stability enhancement, by application of FACTS devices, a longitudinal transmission test system, has been chosen. It consists of a generator, which can be a representation of an electrically concentrated subsystem, a transmission system, consisting of two sections, and of a stiff system. A scheme and the characteristic data are presented in Figure 11.25. The transmitted pre-fault real power (P_{GEN}) is equal to 1350 MW (90% of the generator rated power).

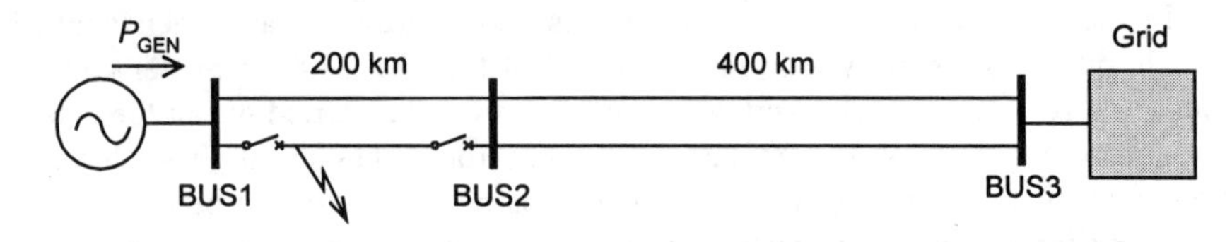

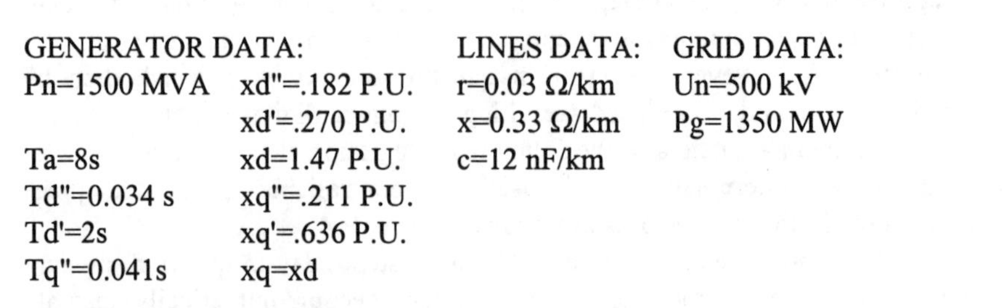

GENERATOR DATA:		LINES DATA:	GRID DATA:
Pn=1500 MVA	xd"=.182 P.U.	r=0.03 Ω/km	Un=500 kV
	xd'=.270 P.U.	x=0.33 Ω/km	Pg=1350 MW
Ta=8s	xd=1.47 P.U.	c=12 nF/km	
Td"=0.034 s	xq"=.211 P.U.		
Td'=2s	xq'=.636 P.U.		
Tq"=0.041s	xq=xd		

Figure 11.25 The test system

Both the generator and the excitation control (a detailed model of a real excitation controller) are modeled in detail. The turbine and governor are modeled in a simplified manner using constant turbine power P_m. The lines are modeled as series-connected Π sections (1 section per 100 km).

The three possible locations for FACTS devices to be installed are: the generator terminals ("BUS1"), a location between the two transmission line sections ("BUS2") and a location near the grid ("BUS3"). The disturbance in the system is the three phase fault near the generator terminals and disconnection of the line.

System behaviour is going to be presented in the form of oscillograms and dynamic transmission characteristics. In the oscillograms the characteristic system quantities are presented in the form of time functions (the bottom axis represents the time with marks in seconds). The following quantities are considered the most representative: Generator speed deviation (in p.u. of the nominal speed), Transmission angle δ (voltage angle between the generator emf and grid – in degrees), the Generator active power (in MW) and the FACTS device characteristic. The dynamic transmission characteristic represents interdependence between the generator active power and the transmission angle in question. With "double arrow" the direction of the operating point movement is denoted. They are plotted at time intervals of 0.8 s.

The maximum fault duration time at which the system maintains synchronism, without FACTS devices included, is about 75 ms (oscillograms and transmission characteristic are presented in Figures 1.26a and b respectively). As presented in Figure 1.26c, fault duration of 76 ms causes loss of synchronism in 1st swing. As shown in 11.26a and b, the system is also at the limit of oscillatory stability.

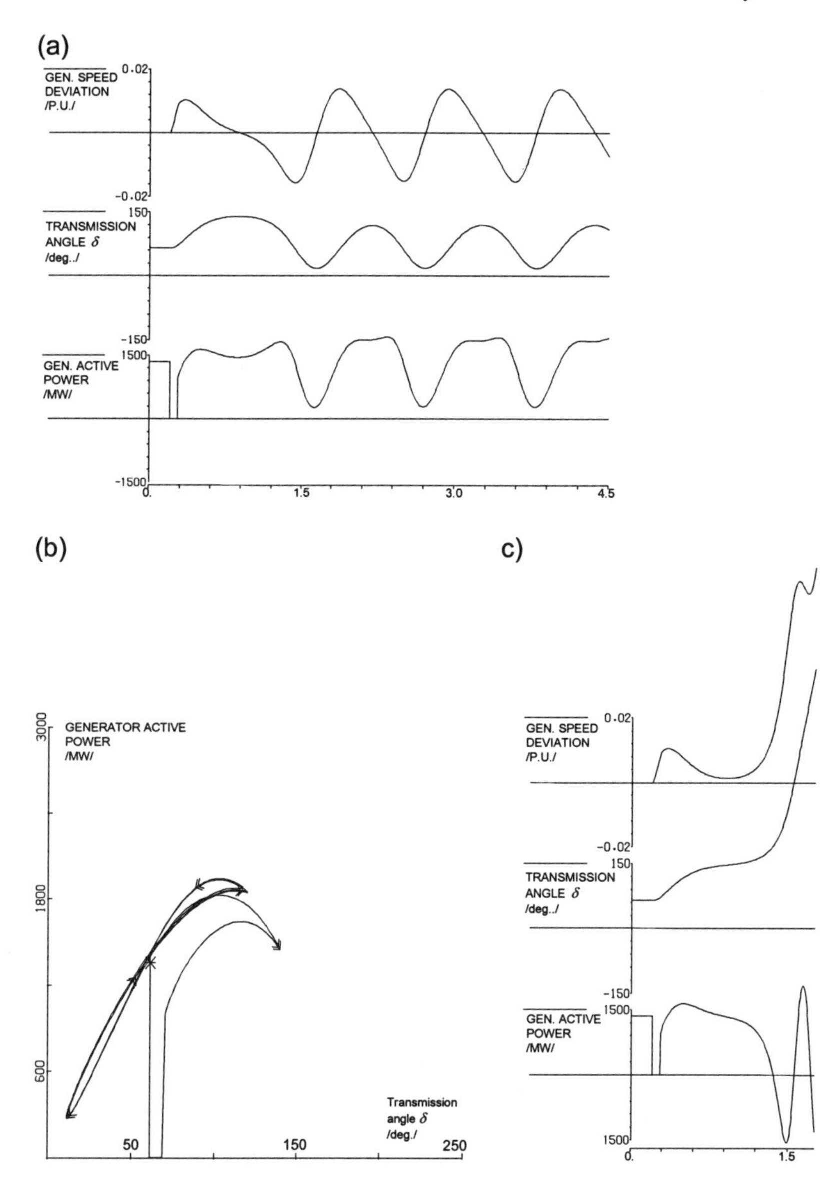

Figure 11.26 Results of digital simulation – no FACTS devices present (a) Oscillograms – 75 ms fault (b) $P_{GEN}(\delta)$ characteristic – 75 ms fault (c) Oscillograms – 76 ms fault

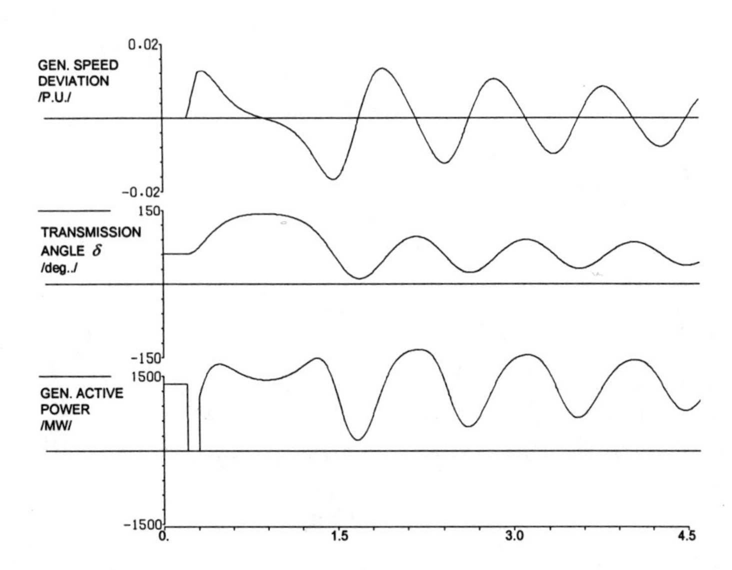

(b)

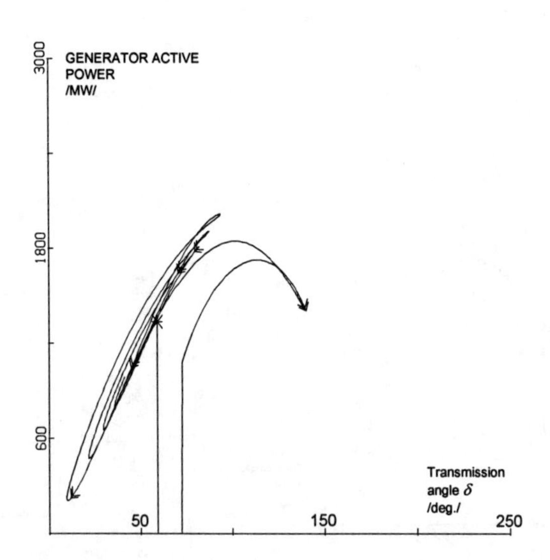

Figure 11.27 Results of digital simulation – CSC, 15% compensation (a) Oscillograms – 100 ms fault (b) $P_{GEN}(\delta)$ characteristic – 100 ms fault

11.6.2 Maintaining system stability using FACTS devices

The aim of the computations was to find the minimum size of FACTS devices, which, for the test system, assure system stability after 100, 150 and 200 ms faults, and to demonstrate the proposed control strategies. Not all, but the most significant results are presented. The FACTS device ratings are summarised at the end of this section.

11.6.2.1 System behaviour using CSC

In calculations it is assumed that all the lines are equipped with CSC. For demonstration reasons the CSCs are set to their maximum values during the fault and do not change any more (it could be represented by switched series compensation). The characteristic oscillograms and the transmission characteristic are for a 100 ms fault (limit case) presented in Figure 1.27. As shown, the system damping is poor although the electrical line distance is shortened due to CSC insertion.

11.6.2.2 System behaviour using SSSC

The most suitable location for an SSSC installation is “BUS2”, although at both other locations SSSC efficiency is not much lower. The characteristic SSSC quantity is injected voltage U_T. The oscillograms and the transmission characteristic are presented in Figure 11.28 for a 100 ms fault. In the presented results the characteristic points, according to the control strategy definition (Section 11.4.2, Figure 11.14), are marked. As shown, the oscillations are damped almost immediately.

11.6.2.3 System behaviour using SVC and STATCOM

Both SVC and STATCOM are best located at “BUS2”. Their efficiency is lower at the generator bus, and at “BUS3” they are useless. The characteristic quantity is in the SVC case its admittance, and in the STATCOM case the injected current I_Q. The oscillograms and the transmission characteristic are for a 100 ms fault with SVC included, presented in Figure 11.29. Here also the characteristic points, according to the control strategy definition, are marked. The results if a STATCOM is included are very similar to those presented in Figure 11.29.

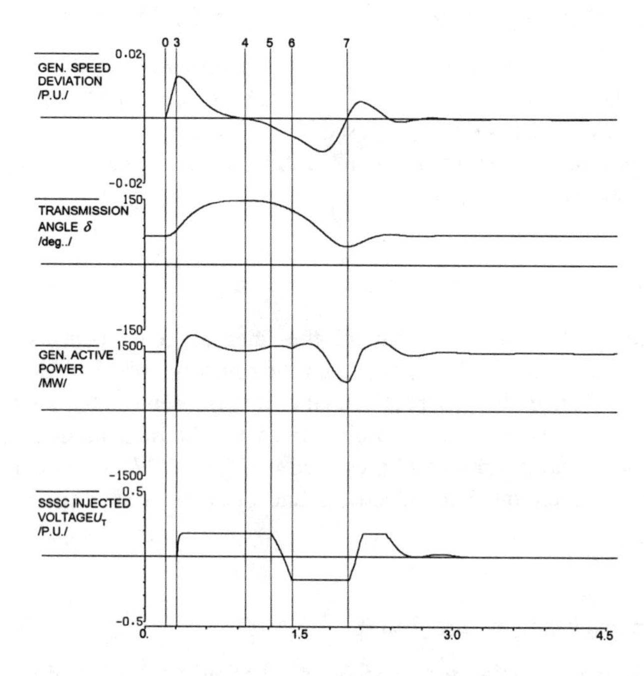

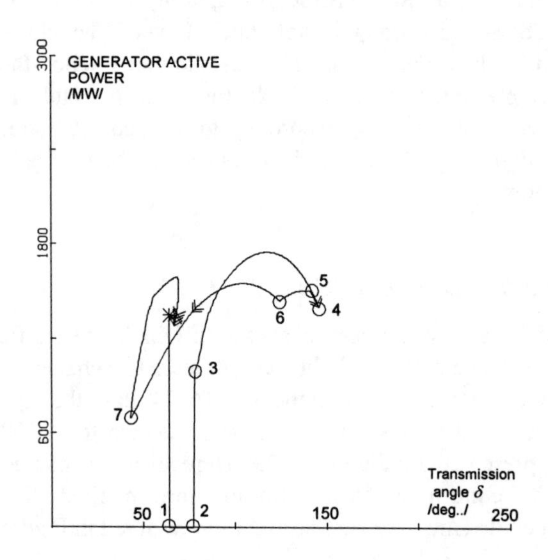

Figure 11.28 Results of digital simulation – SSSC, $U_T = 0.177$ p.u. (a) Oscillograms – 100 ms fault (b) $P_{GEN}(\delta)$ characteristic –100 ms fault

(a)

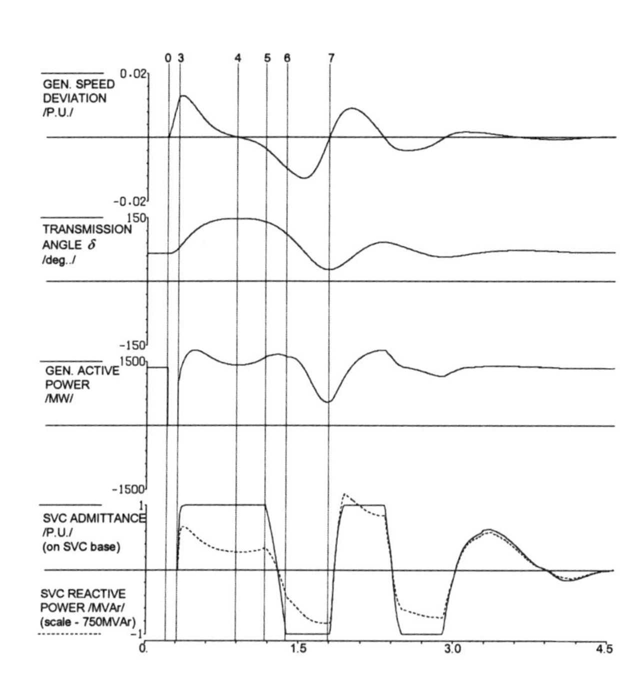

(b)

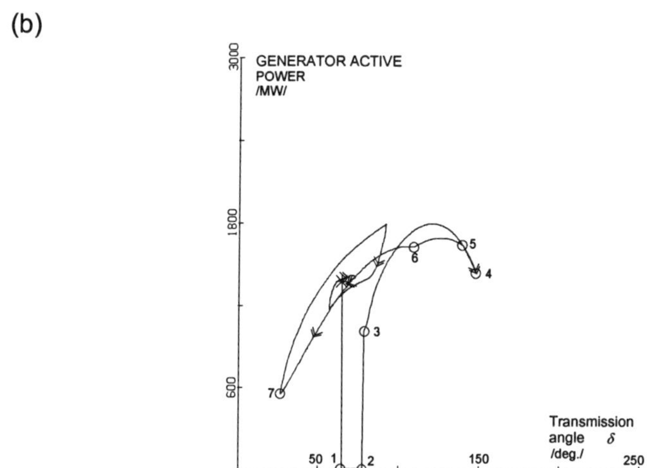

Figure 11.29 Results of digital simulation – SVC, Rating 585 MVAr (a) Oscillograms – 100 ms fault (b) $P_{GEN}(\delta)$ characteristic – 100 ms fault

(a)

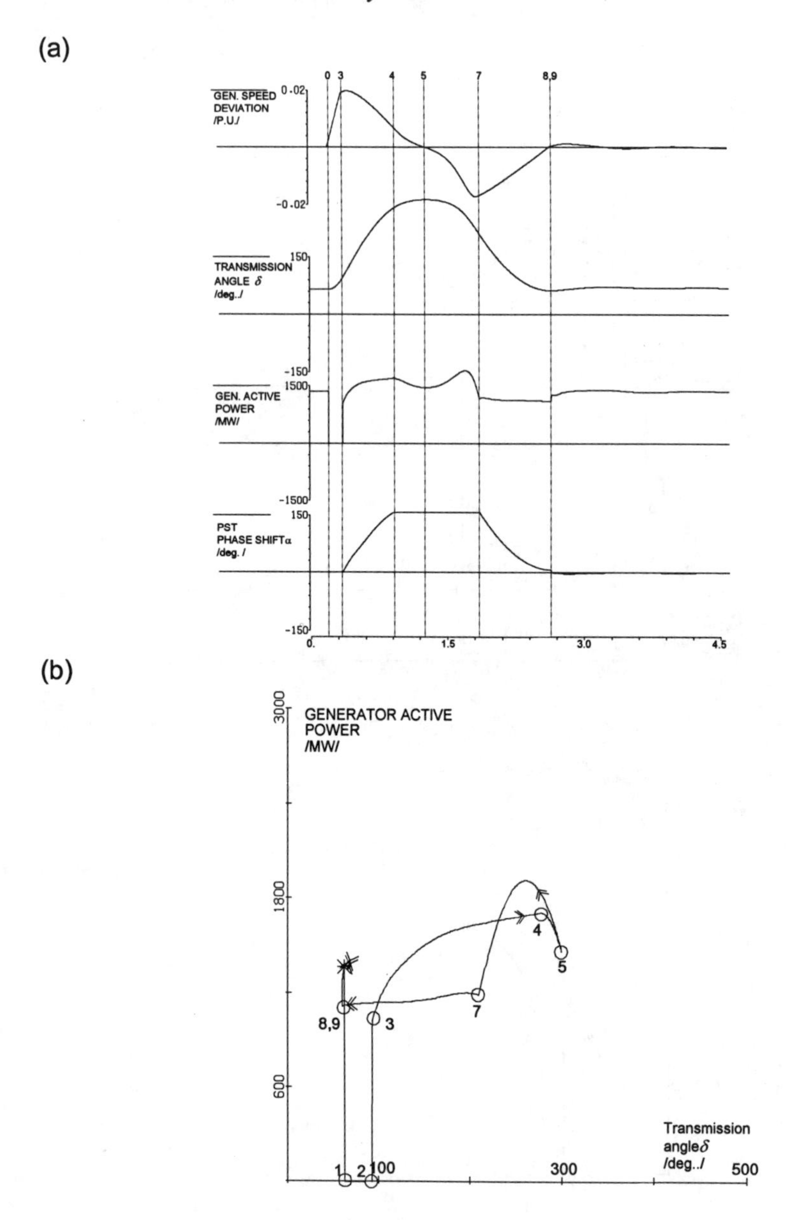

Figure 11.30 Results of digital simulation – PAR, α_{min}=157° (a) Oscillograms – 150 ms fault (b) $P_{GEN}(\delta)$ characteristic –150 ms fault

11.6.2.4 System behaviour using PAR

In the case of a PAR the impact of the location is of minor importance. Its characteristic parameter is the phase shift α. In the example results are presented in Figure 11.30, the PAR is located at "BUS3". The first of the control strategies presented in Section 11.4.3 (Figure 11.15a) is applied. This strategy assumes that between "3" and "4" the PAR keeps transmission characteristic at its maximum. In steady state conditions, presented in Section 11.4.3, means "$\alpha = 90° - \delta(2)$". In dynamic conditions it is not so easy to achieve the maximum in question. In the calculations presented besides the α signal, determined by the equation mentioned, an additional signal has been introduced into the α determination module. This signal is based on measurement of generator active power. The time derivative of the power is introduced into a PI controller. In this way the operating point is "pushed" towards the maximum power (as the power derivative differs from 0 this signal changes and "pushes" the operating point towards larger power). The signal in question had to be added with a certain delay after a fault clearance. The proposed strategy is only one of the possibilities and brought quite good results. Theoretical considerations regarding maximum transmittable power during dynamic conditions, using PAR, might be an interesting question for academic research work. As shown in Figure 11.30, the chosen power between "7" and "8" is below the turbine power. In this way the back-swing and oscillations have been practically eliminated.

11.6.2.5 System behaviour using QBT

The demonstration oscillograms and the transmission characteristic (100 ms fault) for the QBT installation are presented in Figure 11.31. As assumed, according to theoretical considerations in Section 11.3, the favourite location is at "BUS3". In oscillograms, the series branch injected voltage U_T has been chosen as a characteristic quantity. Because QBT is connected to the stiff system, the phase shift equals to Arctan(U_T /Grid voltage). As shown in Figure 11.31, at points "4" and "9" the change of the QBT controllable parameter is not momentary but some time constant is introduced (see Section 11.5.1.1).

11.6.2.6 System behaviour using UPFC

Finaly, one of the results, when applying UPFC as a means for transient stability enhancement, is presented in Figure 11.32. Calculations have shown that the favourite UPFC location is "BUS1". For the controllable parameters U_T and I_Q the same (already known) strategy has been applied. During the simulation, the controllable parameter φ_T has been calculated "on-line" so as to achieve extreme transmitted power at given U_T magnitude. The change of U_T sign in the oscillogram means, the desired extreme power in the φ_T calculation changes from maximum to minimum or vice versa.

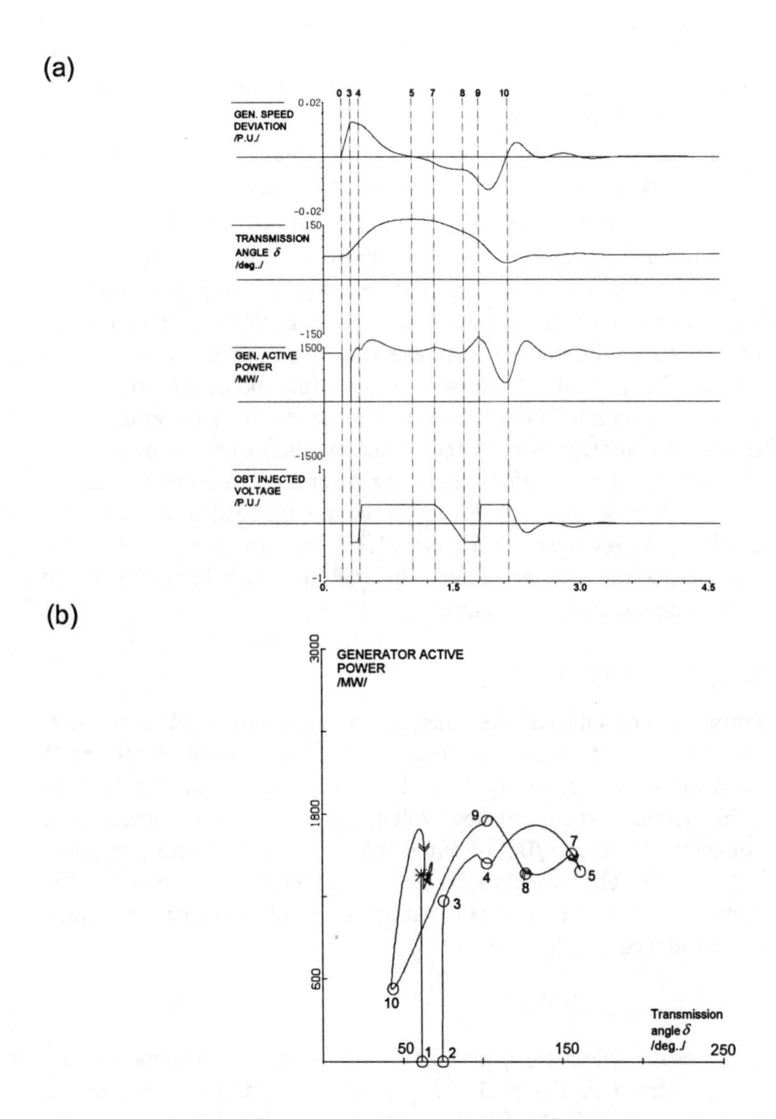

Figure 11.31 Results of digital simulation – QBT, U_T = 0.348 p.u. (a) Oscillograms – 100 ms fault (b) $P_{GEN}(\delta)$ characteristic – 100 ms fault

(a)

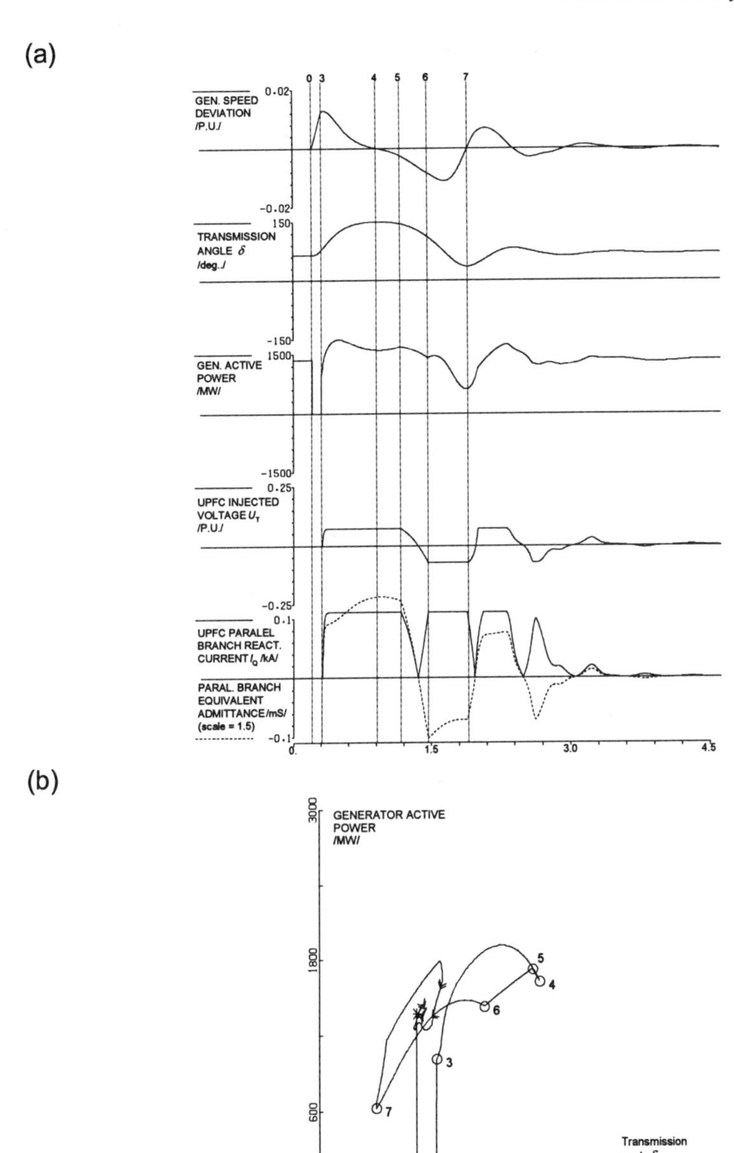

(b)

Figure 11.32 Results of digital simulation – UPFC, U_T = 0.072 p.u. (a) Oscillograms – 100 ms fault (b) $P_{GEN}(\delta)$ characteristic – 100 ms fault

11.6.3 Ratings of FACTS devices maintaining the system stability

The (approximate) minimal FACTS device ratings to maintain synchronism in the presented test system, are presented in Table 11.2. As shown, the MVA ratings of the devices are also presented. In determining the ratings in question, a few assumptions have been made. As a basic throughput current the generator rated current is considered (1.732 kA). In the CSC case, the devices are rated assuming the whole current I_{BASE} as their rated current because of a line tripping, following a fault. Ratings of other series connected devices (SSSC, PAR, QBT, UPFC) are calculated as a product of injected voltage U_T and the rated current I_{BASE}. In the PAR, QBT and UPFC cases it should be noted that the calculated rating represents only the series branch rating. PAR and QBT parallel branch ratings are (neglecting losses) equal to their series branch ratings respectively (do not forget: if losses are neglected PSTs do not produce or consume power). The UPFC parallel branch reactive current is a controllable parameter. In the calculations presented it has been assumed the parallel branch maximal reactive power equals the series branch MVA rating. Thus maximal parallel branch active power P_T (cf. Figure 11.24) also equals series branch MVA rating, the parallel branch rating equals and the $\sqrt{2}$ times the series branch MVA rating.

Table 11.2 Comparison of FACTS device ratings to maintain synchronism

	CSC		SSSC		SVC	STATCOM		PAR			QBT	UPFC	
fault duration /ms/	% comp.	rated power /MVAr/	U_T /p.u./	rated power /MVA/	rated power /MVAr/	I_Q rated /kA/	rated power /MVAr/	phase shift /°/	rated power /MVA/	U_T /p.u./	rated power /MVA/	U_T /p.u./	rated power /MVA/
100	15	535	0.177	265	585	0.538	470	39	1000	0.348	522	0.072	108
150	47	1670	0.751	1126	1860	2.080	1875	157	2940	0.850	1287	0.310	465
200	78	2780	2.025	3040	3420	5.660	5100	322	3000	1.256	1884	1.040	1560

The maximal PAR rating equal to 3000 MVA. This rating is met at shift α = 180°. Under such circumstances the PAR injected voltage is maximal. If α exceeds 180°, the injected voltage becomes smaller. Theoretically, with a 3000 MVA PAR, stability can be assured for any fault duration. Such PAR may (theoretically) be used as a means for connection of two asynchronous operating systems. The frequency difference of the two systems in question compensates PAR by rotating the terminal voltage phasors.

As anticipated, the most effective device is the UPFC. It is shown that stability can be maintained also in the case of a 150 ms or 200 ms fault. However, very high FACTS device ratings are needed.

It should be noted that the present study is for demonstration purposes and it should give the reader a feeling about FACTS devices possibilities on the field of power system transient stability enhancement. The results can not be generalised.

11.7 Summary

In this chapter first the reader has been introduced to the basic theory of electromechanical transients in power systems. The relative generator rotor motion has been described by the swing equation. The problem of transient stability has been explained on the basis of the well known but, especially from the tutorial point of view, still valuable equal-area criteria. According to this criteria, transient stability of a power system is maintained if the accelerating area (which is in proportion to the additional kinetic energy stored in rotating masses during the fault) equals the decelerating area (which is in proportion to the excess of generator electric power over the mechanical power supplied by turbine) during the 1st rotor swing following the fault clearance.

In order to be able to determine the control strategy for FACTS devices of when they are used for transient stability enhancement, their impact on power flow has been studied. To be more concrete, the impact of the FACTS devices controllable parameters on power transmission characteristics has been studied. As a basis, the simplified model of a transmission corridor, consisting of two voltage sources connected via a reactance (representing transmission facilities) has been studied. The transmitted power has been expressed in the form of analytical functions of the system model parameters and FACTS device parameters. In this way it was possible to determine the FACTS device parameters as well as the optimum location of the FACTS device installation in order to achieve their maximal effect on power transmission change. Calculations showed that (for such a simplified system) the optimum location of CSC, SSSC and PAR does not play any role. The optimal SVC location is in the electrical middle of the transmission corridor, while QBT should with the "parallel branch side" be connected to the voltage source. Optimal STATCOM or UPFC location depends on system parameters. If extreme power transfers are to be achieved, then in CSC, SSSC, SVC, and STATCOM case, their controllable parameters should be set to their maximums (or minimums) determined by device ratings. In the UPFC case two of the three controllable parameters (series branch injected voltage magnitude and parallel branch reactive current) should be set to their maximums while the "optimal" phase of the series branch injected voltage depends on the system operating point. The situation is similar in the PAR and QBT cases, where the "optimal" controllable parameter is a function of system operating point location.

The idea of a FACTS device control strategy for maintaining system transient stability has been chosen, which should fulfil the following roles:

- the system should maintain stability after a major disturbance,
- the system should not persevere near the maximum of the 1st swing,
- the rotor back swing should be small,
- subsequent oscillations should be effectively damped.

Although only the first task is essential for maintaining transient stability of the longitudinal system, the other three might be important in meshed systems or they may lower stress on power system components.

In general, the methods for transient stability analysis can be divided into direct and indirect methods. At the present stage of knowledge, the indirect methods (time-domain simulation approach) have advantages compared to the direct ones, when studying transient stability applying FACTS concepts. The possible models of FACTS devices for time-domain simulation purposes have been presented. In general, they may consist of controllable admittances or reactances, controllable voltage sources, controllable current sources or combinations of these. These elements should be controlled so as to simulate real FACTs device behaviour during transients. As electromechanical swings are relatively slow, the proposed models are suitable for calculations in "stability modes" of electric power systems simulation programs.

Finally, a numerical study has been carried out, using a power systems digital simulation tool. The longitudinal system, consisting of a power generation facility supplying a stiff system via a transmission corridor, has been taken as an example. The elements of this system have been modelled with only a few simplifications. The goal of the study was to determine minimal ratings of FACTS devices in order to assure transient stability when the fault occurs and consequently a part of the transmission facility is disconnected. From the typical oscillograms and dynamic transmission characteristics presented it can be concluded that the proposed control strategies of FACTS devices are effective in the presented case, although they have been developed on the basis of static models.

From results, summarized in the table, it can be concluded that the most effective device is the UPFC. This has been anticipated, since it combines properties of controllable parallel and series compensation as well as properties of controllable phase shifting transformers. However, a comparison has been carried out on the basis of device ratings which does not necessarily match with the economic comparison. In real systems there are many factors influencing transient stability therefore the presented study should be considered as a demonstration the example results of which should not be generalized.

11.8 References

1. Kimbark, E.W., *Power system stability*, John Wiley & Sons Ltd, New York, 1948, 1950, 1956.
2. Xue, Y., Wehenkel, L., Euxibie, E., Heilbronn, B., and Lesigne, J.F., "Extended equal area criterion revisited", *IEEE Transactions on Power Systems*, 7, (3), pp. 1012-1022, 1992.
3. Dong, Y. and Pota, H.R., "Transient stability margin prediction using equal-area criterion", *IEE PROCEEDINGS-C,* 140, (2), pp. 96-104, 1993.
4. Pai, M.A., *Power system stability – analysis by the direct method of Lyapunov*, North - Holland Publishing Company, New York, 1981.
5. Rahim, I. A., "Generalized equal-area criterion: a method for on-line transient stability analysis" *IEEE International Conference on Systems, Man and Cybernetics*, Los Angeles, USA, pp. 684-688, 1990.
6. Machowski, J., Bialek, J.W., and Bumby, J.R., *Power system dynamic and stability,* John Wiley & Sons Ltd, Chichester, 1997.
7. Elgerd, O.I., *Electric energy systems theory,* McGRAW-HILL, New York, 1971.
8. Gyugyi, L., Schauder, C.D., and Sen, K.K., "Static synchronous series compensator: a solid-state approach to the series compensation of transmission lines", *IEEE Trans. on Power Delivery*, 12, (1), pp. 406-417, 1997.
9. Mihalič, R., and Papič, I., "Static synchronous series compensator – a mean for dynamic power flow control in electric power systems", *Electric Power System Research*, 45, pp. 65-72, 1998.
10. Mihalič, R., and Papič, I., "Power flow control using static synchronous series compensator", *32nd Universities Power Engineering Conference UPEC'97*, Manchester, UK, pp. 174-177, 1997.
11. Povh, D. et al., "Load flow control in high voltage power systems using FACTS controllers", *CIGRE brochure*, Jan. 1996.
12. Gyugyi, L., Hingorani, N.G., Nannery, P.R., and Tai, N., "Advanced static var compensator using gate turn-off thyristors for utility applications", *CIGRE 1990 Session*, 26th August – 1st September, Paris, France, paper 23-203.
13. Larsen, E., Miller, N., Nilsson, S., and Lindgreen S., "Benefits of GTO - based compensation systems for electric utility applications", *IEEE Trans on Power Delivery*, 7, (4), pp.2056-2063, 1992.
14. Mihalič, R., "Determination of parameters for phase shifting transformers and UPFC to increase power system transmission capability", Ph.D. Thesis, Ljubljana, Slovenia, 1993, [in Slovenian language].
15. Povh, D. and Mihalič, R., "Simulation of power electronic equipment", *EPRI Workshop*, October 7-9 1994, Baltimore, USA.

16 Mihalič, R., and Žunko, P., "Phase shifting transformer with fixed phase between terminal voltage and voltage boost – tool for transient stability margin enhancement", *IEE proceedings Generation, Transmission and Distribution*, 142, (3), pp. 257-262, 1995.
17 Gyugyi, L., Schauder, C.D., Williams, S.L., Rietman, T.R., Torgerson, D.R., and Edris, A., "The unified power flow controller: a new approach to power transmission control", *IEEE/PES 1994 Summer Meeting*, July 24-28 1994, San Francisco, USA, paper 474-7 PWRD.
18 Mihalič, R., Žunko, P., and Povh, D., "Improvment of transient stability using unified power flow controller", *IEEE Trans. on Power Delivery*, 11, (1), pp. 485-491, 1996
19 Mihalič, R., and Žunko, P., "Streckenmodell zur einstellung eines univerasalen lastflußreglers", *Archiv für Elektrotechnik (Electrical Engineering)*, 78, (2), pp. 133-140, 1995.
20 Mihalič, R., Povh, D., Žunko, P., and Papič, I., "Improvement of transient stability by insertion of FACTS devices", *Athens Power Tech* - APT'93, IEEE - NTUA joint international power conference, September 1993, Athens, Greece, pp. 521-525.
21 Iravani, M.R., Dandeno, P.L., Nguyen, K.H., Zhu, D., and Maratukulam, D., "Application of static phase shifters in power systems", *IEEE Transactions on Power Delivery*, 9, (3), pp. 1600-1608, 1994.
22 Baker, R., Gürth, G. Egli W., and Eglin, P., "Control algorithm for a static phase shifting transformer to enhance transient and dynamic stability of large power systems", *IEEE Trans. on Power Apparatus and Systems*, PAS-101, (9), September 1982.
23 Edris, E., "Enhancement of First-swing stability using a high-speed phase shifter", *IEEE Trans*, PWRS-6, (3), pp. 1113-1118, 1991.
24 O'Kelly, D. and Musgrave, G., "Improvement of power-system transient stability by phase-shift insertion", *PROC. IEE*, 120, (2), 1973.
25 Arnold, C.P., Duke, R.M., and Arrillaga, J., "Transient stability improvement using thyristor controlled quadrature voltage injection", *IEEE Transactions on Power Apparatus and Systems*, PAS-100, (3), pp. 1382-1388, 1981.
26 Fang, Y.J. and McDonald, D.C., "Dynamic and transient stability enhancement by use of fast controlled quadrature boosters in power systems", *12th Power System Computation Conference*, Dresden, Germany, August 19-23, 1996, pp. 1083-1089.
27 Fuad, A.A. and Vittal, V., *Power system transient stability analysis using the transient energy function method*, Prentice-Hall, 1992.
28 Kundur, P., *Power system stability and control*, McGraw-Hill, New York, 1994.
29 Martinez-Velasco, J.A. (Eds), "Computer analysis of electric power system transients" IEEE Inc., New York, 1997.

30 Kundur, P., "A practical view of the applicability of the direct methods", *IEEE Trans*, PAS-193, (9), pp. 1634-1635, 1984.
31 Maria, G., Tang, C., and Kim, J., "Hybrid transient stability analysis", *IEEE Trans*, PWRS-5, (2), pp. 384-393, 1990.
32 Povh, D., "Application of FACTS systems", *EPSOM'98*, Zurich, Switzerland, September 1998.
33 Bayer, W. and Sihombing, P., "Aspects of damping of power oscillations by power modulation", *7th Conference on Electric Power Supply Industry*, Brisbane, Australia, October 15-22, 1988.
34 Thérond, G. et al., "Modeling of power electronics equipment (FACTS) in load flow and stability programs", CIGRE TF 38-01-08 Report 1998, [to be published as CIGRE brochure].
35 Xing, K. and Kusic, G., "Application of thyristor-controlled phase shifters to minimize real power losses and augment stability of power systems", *IEEE Trans. on Energy Conversion*, 3, (4), pp. 792-798, 1988.
36 Srinivasan, N., Prakasa, K.S., Indulkar, C.S., and Venkata, S.S., "On-line computation of phase shifter distribution factors and lineload alleviation", *IEEE Trans. on Power Apparatus and Systems*, PAS-104, (7), pp. 1656-1662, 1985.
37 Youssef, R.D., "Phase shifting transformers in load flow and short-circuit analysis: modelling and control", *IEE PROCEEDINGS - C*, 140, (4), pp. 331 - 336, 1993.
38 Povh, D., Mihalič, R., and Papič, I., "FACTS equipment for load flow control in high voltage systems", *Proceedings of CIGRE Symposium Power Electronics in Electric Power Systems*, Tokyo, Japan, May 22-24, 1995, paper 310-06.
39 Noroozian, M., Ängquist, L., Ghandhari, M., and Andersson G., "Use of UPFC for optimal power flow control", *IEEE Transactions on Power Delivery*, 12, (4), pp. 1629-1634, 1997.
40 Nelson, R.J., "Transmission power flow control: electronic vs. electromagnetic alternatives for steady state operation", *IEEE/PES 1994 Winter Meeting*, January 30 – February 3, 1994, New York, USA, paper 067-9 PWRD.

Chapter 12

Protection for EHV transmission lines with FACTS devices

Q.Y. Xuan, Y.H. Song, and A.T. Johns

12.1 Introduction

In recent years, because of energy, environmental, and regulatory concerns, the growth of electric power transmission facilities has been restricted. The result can be a transmission bottleneck, under utilisation and/or uncontrolled use of facilities. These difficulties have hastened the development of flexible ac transmission systems (FACTS) techniques to obtain increased utilisation and control of existing transmission systems by using power electronic devices. However, the employment of FACTS devices has a profound impact on the operation of other equipment in the system, such as protection. Adamik et al. [1] identified the protection requirement for flexible ac transmission systems and pointed out its importance. For example, controllable series compensation (CSC) as one of the main FACTS techniques has the ability to direct and control power flow by changing the firing angle. However, this causes problems for conventional distance protection schemes [2, 3] because the application of the CSC technique often causes rapid changes in the apparent impedance measured. Figure 12.1 illustrates the impedance characteristics of the CSC circuit. It is clear that with the changing of firing angle, the reactance of the CSC varies from Inductive to Capacitive. Furthermore, the complex variation of the line impedance is accentuated as the Capacitor protection device operates nonlinearly under fault conditions. Figure 12.2 gives a typical nonlinear characteristic of a ZnO resistor in a Capacitor protection device. Under fault conditions, the voltage across the Capacitor is kept below the protected level. Also, the parameters of the bypass filter used to damp Subsynchronous Resonance (SSR) and harmonics have an effect on the apparent impedance measured. In this respect, some adaptive protection schemes [3, 4] have been proposed for such systems.

The majority of power system protection techniques are involved in defining the system state through identifying the pattern of the associated voltage and current waveforms. This means that the development of adaptive protection can be essentially treated as a problem of pattern classification/recognition. However, due to the many possible causes of faults and the nonlinear operation of some power system devices, conventional pattern recognition methods may not be satisfactory in applications involving complex transmission lines. Successful applications of neural networks in the area of power engineering have demonstrated that they can be employed as an alternative method for solving certain long-standing problems where conventional techniques have experienced difficulties. In the context of protection, neural networks have been applied to (i) the recognition of high impedance faults in distribution feeders; (ii) fault identification in AC/DC transmission systems; (iii) accurate fault location in transmission lines; (iv) transformer protection; (v) adaptive autoreclosure; (vi) fault classification and directional discrimination for protection for EHV transmission lines and (v) protection for series compensated transmission lines. More encouragingly, some of the techniques have been hardware prototyped and installed on actual systems to evaluate their performance under service conditions.

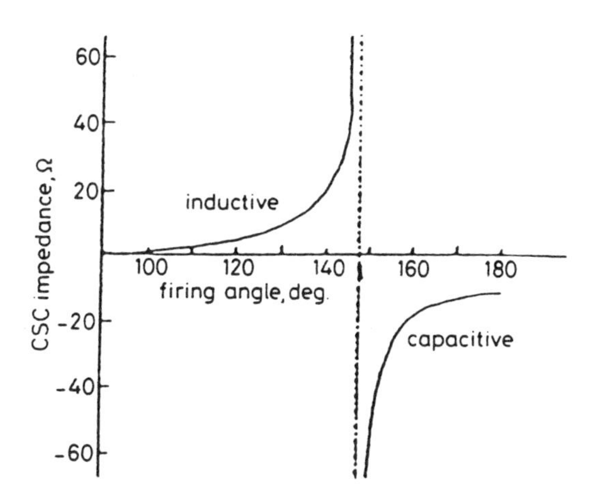

Figure 12.1 Impedance characteristics of the controllable part of CSC system

Very little work has been done on the design and development of protection for CSC systems and this chapter therefore describes a novel protection scheme which employs an artificial neural network method that extracts features in a certain frequency range under fault conditions. This is different from conventional schemes which are based on deriving implicit mathematical equations by complex filtering techniques. In this chapter, the results of a digital simulation for a CSC transmission system are presented, and the feature extraction and topology of suitable ANNs is discussed. The chapter concludes with presentation of the overall performance of the protection scheme.

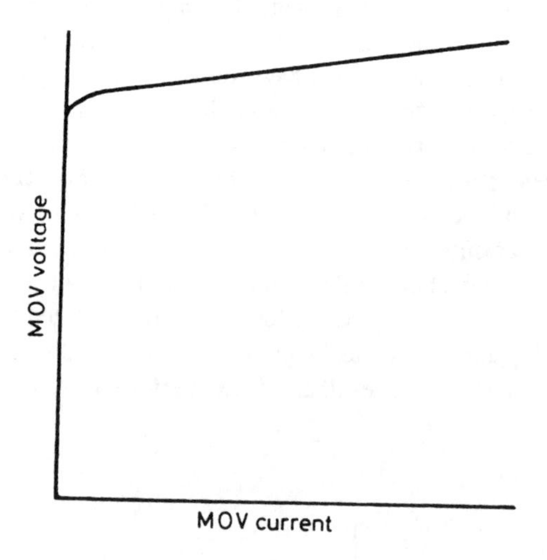

Figure 12.2 Voltage and current characteristics of a ZnO varistor in the capacitor protection device of CSC system

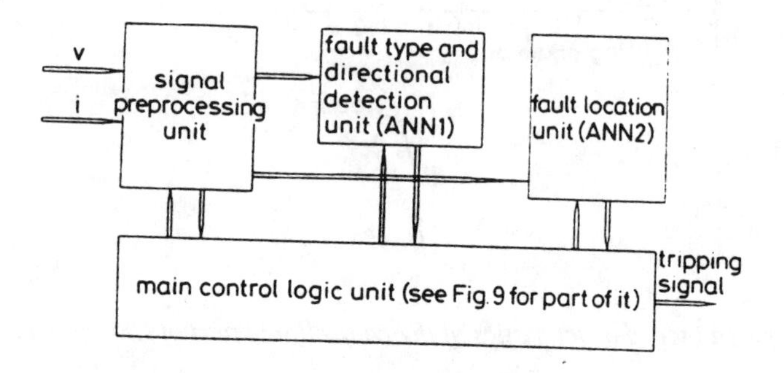

Figure 12.3 Block diagram of the proposed protection scheme

12.2 Artificial neural network based protection scheme

Figure 12.3 illustrates the functional parts of a digital protective relay. The first part is concerned with fault detection and classification and the second part is for fault location. From the point view of artificial neural network (ANN) convergence, the fault type is an issue of "type determination", but the fault location needs to be determined with an emphasis on high accuracy. Hence, the two variables (fault type and fault location) cannot be employed as outputs in one multi-layer network. The reason is that the same ANN which is used to solve distinctly different problems cannot converge. Therefore, the protection scheme is composed of two artificial neural networks (ANN1, ANN2) and a final logical comparison to produce a trip signal. This means that the fault types will be detected first by ANN1 and the second network (ANN2) is used to detect the fault location.

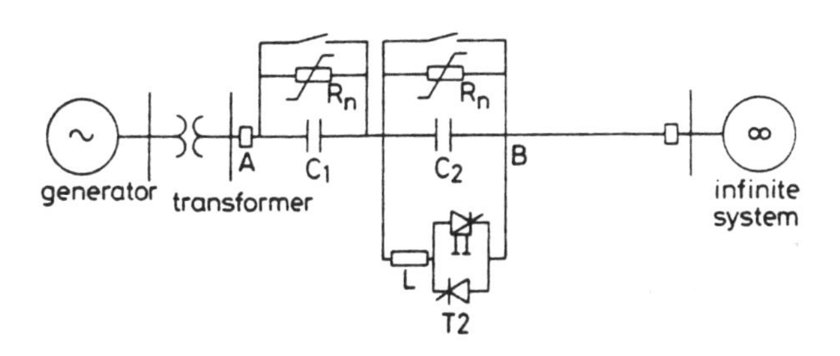

Figure 12.4 The system studied

12.3 Generation of training and testing data

12.3.1 Digital simulation of faulted systems

Owing to a lack of field data, digital simulation using the Electromagnetic Transient Program (EMTP) is used to generate the sample data required to set up the training/testing data for the neural networks. The 500 kV 60 Hz power system, which is illustrated in Figure 12.4, is used for the study of controllable series compensated transmission systems. It is a modified version of the widely used IEEE benchmark system for series compensation studies [5], which consists of a generator, a transformer, circuit breakers, capacitors and their protection components, a 380 km transmission line, and an assumed infinite source. The system parameters are: X_d=1.79 p.u. X'_d=0.169 p.u. X''_d=0.135 p.u. X_q=1.71 p.u. X'_q=0.228 p.u. X''_q=0.200 p.u. T_{do}=4.3 s T''_{do}=0.032 s T_{q0}=0.85 s T''_{q0}=0.05s R_T=0.01 p.u. X_T=0.14 p.u. X_L=0.50 p.u. and R_L=0.02 p.u. There are two capacitors: one is a fixed capacitor C_1 which provides 26% compensation and the other is a controllable capacitor C_2 which provides variable compensation starting from 10%. All the equipment components, including coupling capacitor voltage transformer (CVT) and current transformer (CT) models, are modelled using the Electromagnetic Transient

Program (EMTP). The transducers that have an effect on the primary waveforms are necessarily considered before training the ANN, because the CVT has a low-pass filtering characteristic. A detailed description of the digital simulation can be found in reference [6, 7]. Figures 12.5 and 12.6 show the results of a simulation of an a-phase-to-b-phase-to-ground fault occurring at the middle of the line when the firing angle is 165^0. Three phase voltages at busbar end S1 and three phase line currents are shown in Figures 12.5 and 12.6 respectively. It is obvious that when the fault occurs, the faulted phase voltages suddenly collapse and the faulted phase currents increase.

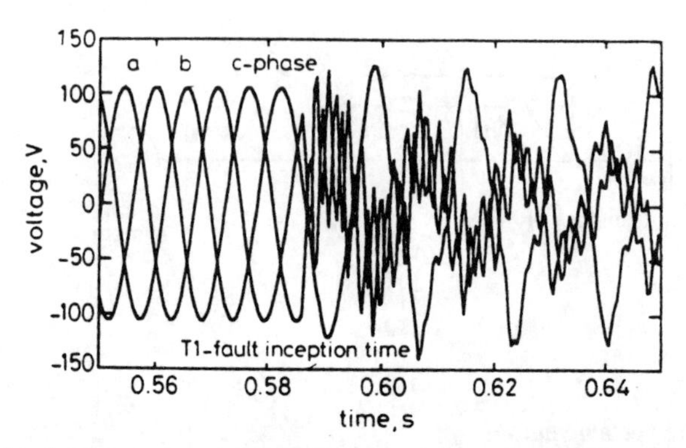

Figure 12.5 a-b-phase-to-ground fault occurring at the middle of the line showing three-phase voltages at sending end busbar

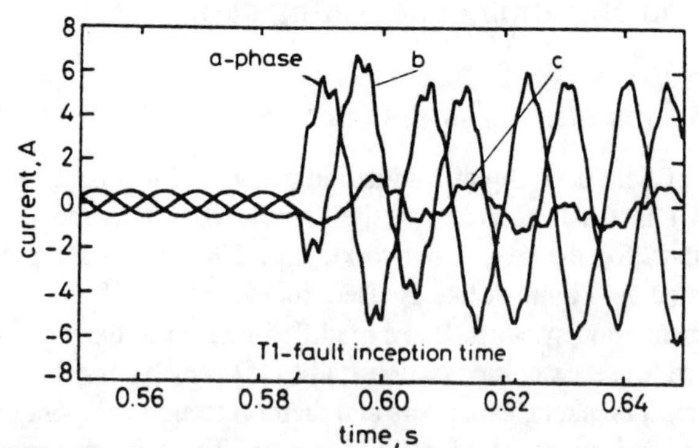

Figure 12.6 a-b-phase-to-ground fault occurring at the middle of the line showing three-phase line currents

12.3.2 Input selection of the neural networks

The application of a pattern classification requires a selection of features that contain the information needed to distinguish between classes and which permit efficient computation to limit the amount of training data and the size of the network required. For reasons of simplicity and availability of measuring elements of conventional protection, only the line voltages on the bus side of the CSC, the line currents, and firing angle are employed as input signals. The firing angle (or any equivalent) is used to enable the controlled compensation degree to be approximated.

In general, when a fault occurs on a transmission line, DC offset, fundamental frequency and non-fundamental frequency components are produced. It has been shown that the components change as the fault position and fault type etc. vary. Figure 12.7 expresses the spectra of a-phase voltages (two cycle window after fault inception) for different fault types. It is important to note that different faults produce different non-power Hz frequency components. Figure 12.8 shows the spectra of a-phase voltages for an *a*-phase-to-*b*-phase fault but at different fault locations (60%, 80% and 100% of the line). It is evident that the non-fundamental frequency changes as the fault position varies. Furthermore, the spectral variation of the signal is affected by the fault inception angle.

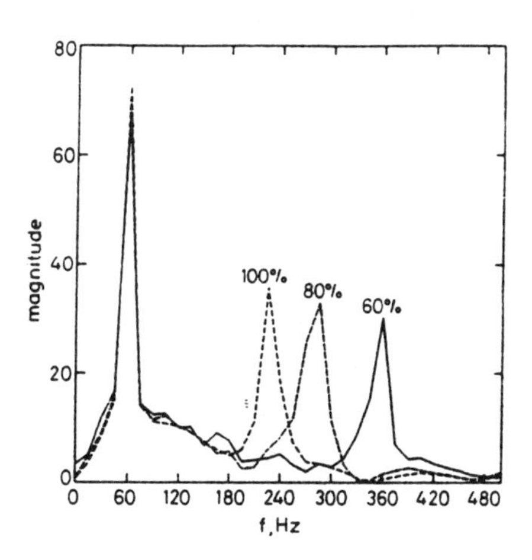

Figure 12.7 Spectra for faulted voltages under different types of fault

Since an ANN has multi-input parallel processing ability, the data window length is also a major factor which needs to be considered. The pattern space is essentially the domain which is defined by the discretization of measured data observing the specific system and its dimensions must necessarily reflect that fact. However, the sampling rate is dependent on the feature space. Theoretically, a long data window can produce more selective frequency response characteristics, but, in general, a fast trip signal is needed, and it follows that a short data window which reduces the size of the ANN is required. The appropriate data window length must meet both of these requirements and an extensive series of studies has revealed that a data window of 4 samples at 960 samples/sec is necessary and sufficient because the dominant non-fundamental frequency components generated by faults are below 500Hz [8]. This sampling rate is compatible with these sampling rates commonly used in digital relays.

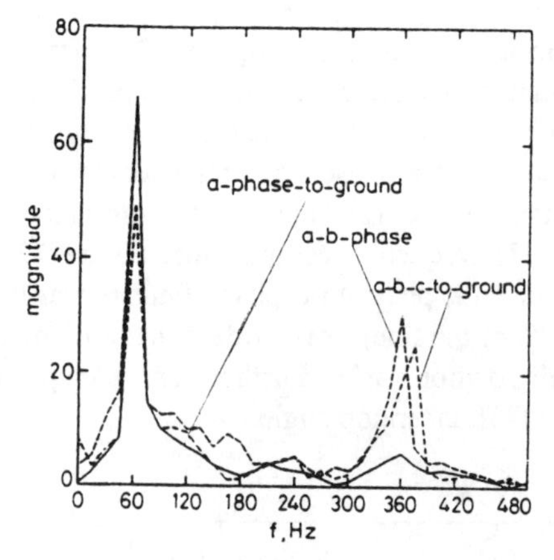

Figure 12.8 Spectra for faulted voltages under different fault locations

Therefore, the three phase voltages and line currents sampled at 16 samples can be used as the input signals to ANN1. The moving data window contains 4 samples and at each moving takes one new sample in. For simplicity, the firing angle α is used as an input to indicate the degree of series compensation. In practice, the controlled impedance, which can be easily obtained from the output of the CSC controller, can be used as an equivalent signal [1].

The training and test patterns are generated by sampling the simulation data from the EMTP. In order to cover the typical scenario of interest, various conditions are simulated, which include the effect of variations in: (i) source parameters, (ii) fault location, (iii) fault inception angle, (iv) prefault loading, (v) fault resistance and (vi) fault types. There are over 5000 patterns generated by the EMTP, some of which have been used for training and some for testing and performance evaluation.

12.4 Artificial neural network 1 (ANN1) for fault type and directional detection

12.4.1 Network structure and training

A total of 25 signals comprising of 24 sampled voltages and currents plus the firing angle have been employed as inputs to the neural networks. The outputs of ANN1 consist of A, B, C (*a*, *b*, *c* three phase operation states), G (connected ground state) and D (fault direction). Any of the A, B and C approaching to 1 expresses a fault in that phase. If G approaches 1, it indicates that the fault is connected to ground. A value of D approaching 0 means that the fault occurs in the reverse direction, and ANN2 is then blocked. Conversely, if output D approaches 1 the fault is taken as occurring forward of the measuring point. Thus for example, the data A B C G D = 1 0 0 1 0 represents an *a*-phase-to-ground fault in the reverse direction, whereas 0 1 1 0 1 represents a *b*-phase-to-*c* phase fault clear of ground in the forward direction.

Since the ANN1 is designed for detection of fault type and direction, the network which is trained should test the fault data under many fault cases. Examples of 2000 fault cases were used to train ANN1. These comprised the 10 basic fault types that appeared at a number of forward and reverse fault locations for a large number of system operating conditions including variations in source capacity, fault resistance, fault inception angle and CSC firing angles.

As discussed above, there are 25 inputs and five outputs. The selection of the number of layers and the number of neurons required in each layer of ANNs are open issues and must consequently, be determined by experimentation involving training and testing different network configurations. The process is terminated when a suitable network with a satisfactory performance is established. In this study, one hidden layer with 12, 14, and 16 hidden neurons was first considered. It was found that a network with 14 hidden neurons, had an acceptable performance which converged in a shortest time when a Hyperbolic tangent transfer function, which has a better convergence performance than the commonly used sigmoidal function, is used. The sampled voltages and currents were scaled to have a maximum value of +1 and a minimum value of -1. The learning factor, which controls the rate of convergence and stability, was first chosen to be 0.5 and was gradually reduced to 0.01. The optimum momentum factor, which is added to speed up the training and avoid local minima, was found to be 0.4 and the training process was repeated until the root mean square (RMS) error between the actual output and the desired output reached an acceptable value of 0.01.

12.4.2 Test results

Validation data for fault cases which are different from those used for training was used to test the effectiveness of the ANN1. All the test results show that the ANN1 designed is suitable for the determination of fault types and their direction. Table 12.1 summarizes the misclassification rate of the 3000 test cases in terms of four types of fault when using one moving window output and using two of three consecutive window outputs respectively. It can be seen that the error rates vary with the type of fault when one moving window output was used and there is only a very small misclassification rate when two of three consecutive window outputs were used.

Table 12.1 Misclassification rate for ANN1

	Phase-to-ground faults	Phase-to-phase-to-ground faults	Phase-to-phase faults	Three phase faults	Average
Misclassification rate (using one window output)	2.77%	3.12%	6.79%	1.85%	3.63%
Misclassification rate (using three consecutive window outputs)	0.0%	0.0%	0.5%	0.1%	0.15%

12.5 Artificial neural network 2 (ANN2) for fault location

12.5.1 Network structure and training

As will be seen in section 12.6, the outputs of ANN1 activate ANN2, which in turn is trained by data for known fault conditions. The input signals to ANN2 are selected to be sampled voltages, currents, and firing angle. The outputs of the ANN2 are composed of M and L which confirms the fault state and indicates the fault position respectively. Output M approaching 1 indicates that there is a fault on the line and fault location will be simultaneously given by L which expresses the percentage of the line length at which it occurs.

With the inputs to ANN2 consisting of only the sampled phase voltages and currents and the firing angle, the network performance was found not to reach the required accuracy. Extensive investigation revealed that by expanding the input space from 3 dimensions (v, i, α) to 5 dimensions (v, v^2, i, i^2, α), greatly enhanced the convergence speed and accuracy of ANN2. By experimental training, ANN2 with 16 hidden neurons was found to posses the best performance. The final network designed for fault location has 49 inputs and 16 hidden neurons and 2 outputs.

12.5.2 Test results

A large number of tests were performed, all of which are documented in detail in ref [2]. The analyses of the actual ANN2 output L for 3000 test data indicate that maximum error is 8.4%, minimum error is 0.21% and the overall average error is 5.01% when using the average value L of two consecutive window outputs.

12.6 Overall performance evaluation

The two neural networks ANN1 and ANN2 were extensively tested separately and were then embedded into the structure shown in Figure 12.3 to evaluate the overall performance and determine the necessary decision logic. After an extensive series of investigations, the decision logic shown in Fig. 12.9 was developed.

In order to ensure that the fault type and directional detection by ANN1 is fast and correct, three consecutive outputs from ANN1 were found to be needed and sufficient. The decision can be safely made if the outputs of at least two four-sample windows exceed the specified threshold of 0.9. Thus the number of samples required to determine fault type and direction is 6.

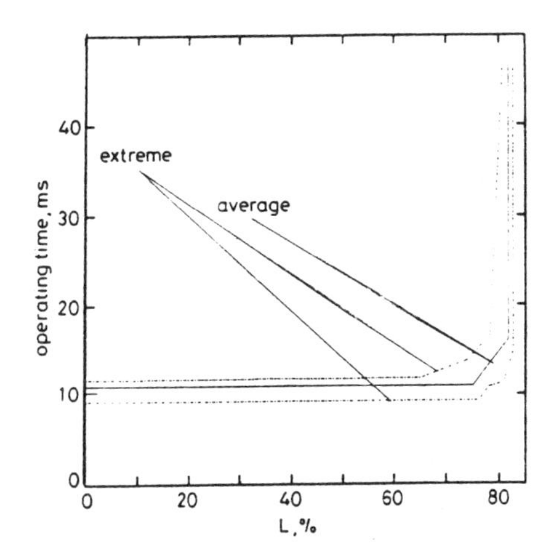

Figure 12.9 Decision logic for fault classification and location (zone-1)

Since the responsibility for the protection devices of a power system is defined in terms of zones of protection, the protection devices in the first zone must be designed to operate only for faults occurring in this zone. The setting value of the protection device operation in zone-1 is commonly 80% of the line length and in order to meet the reliability requirement of the relay, studies show that the average of two consecutive outputs is needed when the output L is less than 75% and the average of two additional consecutive outputs is required when the output is bigger than 80%. Thus a minimum of 2 windows and a maximum of 4 windows, as shown in Fig. 12.9, is required for fault location.

The overall number of samples required for the scheme is seen to be 8 or 10. With the sampling rate of 16 samples per cycle, 10 samples correspond to 10.42

milliseconds. The actual trip time is of course the sum of the time required to complete all the signal processing and computation which is approximately 10 to 12 milliseconds. As a guide to the typical performance evaluation, Figure 12.10 summarised the response for an "a"-phase-to-ground on the test system of Figure 12.4 under a number of conditions. It can be seen that the minimum operating time for zone-1 extends to typically 75% of the line length and it then takes a longer time from 75% to 80% so as to maintain the accuracy and security. An extensive evaluation in ref [2] indicates that the proposed technique works a well for a wide range of system and fault conditions. The results are very encouraging in both speed and security.

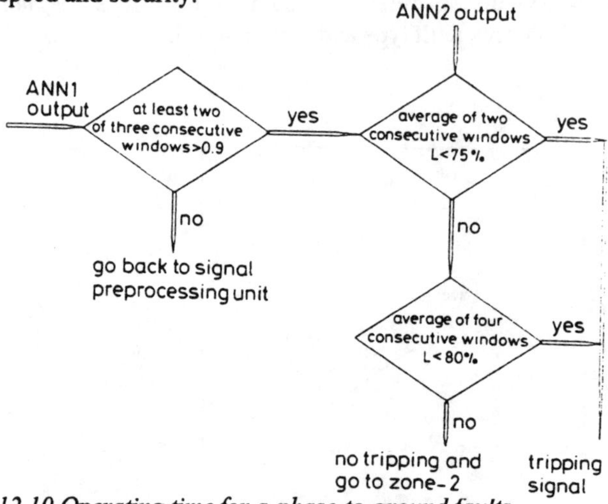

Figure 12.10 Operating time for a-phase-to-ground faults

12.7 Conclusions

A novel protection scheme for CSC EHV transmission systems has been presented in this chapter. In particular, the feature extraction, sampling rate, data window length and training and testing of the ANNs are developed and described in detail. The main idea of the protection scheme is to employ ANNs to capture and analyze the salient features of the faulted voltages and currents. This is different from conventional schemes which are based on deriving implicit mathematical equations involving complex filtering techniques. The trained neural networks have been fully tested and the overall performance of the proposed protection scheme has been extensively evaluated to determine the effects of various factors, such as different fault types, positions, resistances, inception angles, firing angles, source capacities, and load angles. All the test results demonstrate that the proposed protection scheme is, in particular, suitable for application to CSC transmission systems. However, it is now necessary to fully evaluate the proposed technique, using field test data from practical systems.

Research has also been directed to the development of protection for systems with other embedded FACTS devices such as UPFC [10].

12.8 References

1 Adamik, M. and Patterson, R., "Protection requirements for flexible ac transmission systems", CIGRE Paper #34-206, Plenary Session, Paris, 1992.

2 Xuan, Q.Y., "Adaptive protection and control schemes for controllable series compensated EHV transmission systems using neural networks", PhD Thesis, Liverpool John Moores University/ Bath University, 1995.

3 Johns, A.T., Xuan, Q.Y. and Song, Y.H., "Adaptive distance protection scheme for controllable series compensated ehv transmission systems", *Proc IPEC*, pp.621-626, 1993.

4 Girgis, A.A., Sallam, A.A. and Karim El-Din, A., "An adaptive protection scheme for advanced series compensated transmission lines", *IEEE Trans Power Delivery*, 13, (2), pp.414-420, 1998.

5 IEEE SSR Working Group, "First benchmark model for computer simulation of subsynchronous resonance", *IEEE -PAS*, 96, (5), pp.1565-1572, 1977.

6 Johns, A.T., and Xuan, Q.Y., "Digital study of thyristor controlled series capacitor compensated EHV transmission system", *Proc IEEE TENCON*, Beijing, Vol.5, pp.386-392, 1993.

7 Song, Y.H., Xuan, Q.Y. and Johns, A.T., "Comparison studies of five neural network based fault classifiers for complex transmission lines", *International Journal of Electric Power Systems Research*, 43, (2), pp.125-132, 1997.

8 Song, Y.H., Xuan, Q.Y. and Johns, A.T., "Protection scheme for EHV transmission systems with thyristor controlled series compensation using radial basis function neural networks", *International Journal of Electric Machines and Power Systems*, 25, (5), pp.553-565 , 1997.

9 Song, Y.H., Johns, A.T. and Xuan, Q.Y., "Artificial neural network based protection scheme for controllable series compensated EHV transmission lines", *Proc IEE - Generation, Transmission And Distribution*, Pt.C, pp.535-540, 1996.

10 Song, Y.H., Johns, A.T., Xuan, Q.Y. and Liu, J.Y., "Genetic algorithm based neural networks applied to fault classification for EHV transmission lines with UPFCs", *Proc IEE 6th International Conference on Developments in Power System Protection*, Nottingham, pp.278-281, 1997.

Chapter 13

FACTS development and applications

Yasuji Sekine and Toshiyuki Hayashi

13.1 Introduction

Japan's electric power system was established in parallel with industrial and economic development, thereby resulting in highly reliable, high quality systems. This power system incorporates nine individual power systems which are interconnected by 500kV AC trunk lines, 50/60Hz frequency converter (FC) stations and ±250kV HVDC link. The interconnection capacity between the three eastern power systems which operate at 50Hz and the six central/western power systems which operate at 60Hz will be increased to 1.2GW when the Higashi-Shimizu 300MW-125kV FC comes on line in 2001. With this, the Minami-Fukumitsu 300MW-125kV BTB began interconnection between the Chubu Electric Power System and the Hokuriku Power System in 1998, and the Kii-Channel HVDC system with a 50km long submarine cable will interconnect the Kansai Power System and the Shikoku Power System with 1400MW-±250 kV in 2000 in the first stage, and 2800MW-±250kV in the second stage.

These interconnected power systems are mainly composed of radial systems and partially looped systems. The stable operation of these interconnected systems is supported by the precise coordination of protection systems, power system stabilization systems and interconnection splitting systems. Moreover, effective analysis of power system dynamics with several types of hypothetical faults on principal power system operating conditions is indispensable for power system coordination.

The major power system stabilization schemes such as the quick response generator exciter with power system stabilizer (PSS), the stabilizing damping resister (SDR) and the system stabilizing controller (SSC) have been developed and put into practical use. Also, in accordance with the development of semi-conductor devices, measures to enhance the power system performance using power electronics technologies such as the static var compensator (SVC) are now expected to play important roles in power system stabilization.

Fundamentally, the HVDC systems also provide a stabilizing function to mitigate interactions between interconnected power systems by disallowing the fault conditions of one system to adversely affect another.

The need to improve the power system performance of interconnected systems becomes more pressing in the future because the deregulation of the power utility industry tends to cause more difficult technical problems. Uncertainty of power flow with open marketing of generations and consumers is apt to cause unstable operation through a lack of effective analysis of power system dynamics or ancillary facilities of the power systems.

The application of power electronics technologies such as FACTS and the HVDC system for power system performance enhancement will resolve the issues to secure a stable operation of open access power systems.

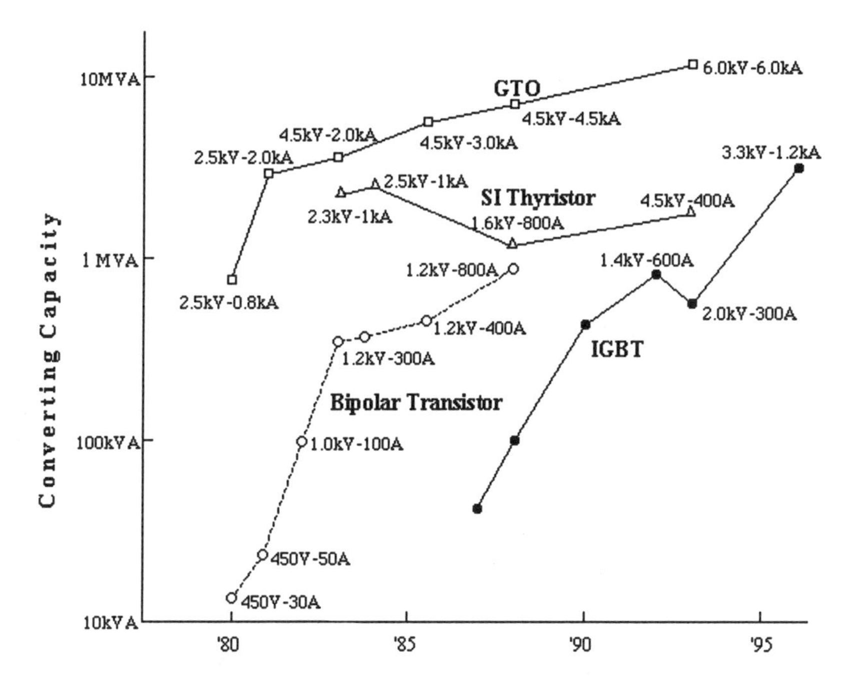

Figure 13.1 Development of self-extinguished devices

13.2 Development status of semi-conductor devices

The remarkable development of self-extinguished semi-conductor devices is shown in Figure 13.1. The development of the gate turn-off thyristor (GTO), the insulated gate bipolar transistor (IGBT), and the static induction (SI) thyristor, have promoted the application of the self-commutated (SC) converter to the motor drive systems in railway vehicles and iron mills. Using AC motors in place of DC motors makes it possible to save energy and reduce machine maintenance work.

The rating of GTO with n-p-n-p structure has reached the level of 6kV peak forward blocking voltage and 6kA controllable current by applying ion injunction to unify impurity distribution and anode-shorted structure to make it easier to extinguish the anode current. To reduce the on-state voltage drop, 'n+' buffer layer is introduced to 'n' base of the GTO to mitigate the electric field. While the turn-off time of a conventional GTO can only be controlled by carrier life time and segment width, a new type GTO, named the 'gate commutation turn-off thyristor' (GCT), with a turn-off time one tenth that of a conventional unit, is developed by ABB and Mitsubishi Electric Co.

Table 13.1 Recent characteristics of GTO, IGBT and SI Thyristor

	GTO		IGBT		SI Thyristor	
	Conventional	GCT	Conventional	IEGT	Normally-on	-off
Peak forward blocking voltage (kV)	6.0	4.5	3.3	4.5	4.5	4.0
Controllable current (kA)	6.0 (Cs=6μF)	3.0 (0μF)	1.2	1.5	0.4	1.0
On-state voltage drop (V)	4.0	3.5	4.4~4.8	(5.0)	4.6	(3.5)
Turn-on time (μs)	6.0	3.0	3.0	(1.0)	1.2	(1.1)
Turn-off time (μs)	30.0	2.5	4.0	(2.0)	2.0	2.0
Wafer size	6 inchϕ	3 inchϕ		12.7×12.7 mm^2	32mm	62 mm
Major application	Railway vehicle Iron mills SC-SVC	Sources for railway	Railway vehicle Active filter	(Experimental use)	(Experimental use)	

Table 13.2 SC converters for power system application

	Power quality		New energy conversion			SVC	
	AF	UPS	PV[1)]	BES[2)]	FC[3)]	Flicker	Imbalance
Capacity (kVA)	50~1000	several 10 ~1400	10~750	several 10 ~1000	~200, 11,800	13,000 ~27,000	16~40,000
Voltage (V)	200/400, 6600(AC)	200/400(AC)	100~430 (DC)	1000(DC)	100~240, 2400(DC)	3000(DC)	2000(DC)
SC Converter	IGBT module/ parallel	IGBT module/ parallel	GTO	IGBT/ GTO Quadruple	MOSFET/ IGBT,GTO 1S-6P,1S-1P	GTO 2 sextuple 1S-1P	GTO 3 sextuple 3S-1P
Control	PWM (~6kHz)	PWM (low input harmonics)	PWM (constant dc voltage)		PWM (1~2kHz)	PWM (390,540Hz)	PWM (180~500Hz)
Remarks	Harmonics; 80~% com-pensation	Over load cap; 120~150%			Efficiency; 91~94,96~%	Snubber regeneration	Negative seq. compensation

1) Photo Voltaic generation now in operation 2) Battery Energy Storage of experimental use
3) Fuel Cell generation

As an IGBT is a compounded device of bipolar transistor with p-n-p structure and MOSFET with n-p-n, it provides high-speed carrier discharge characteristics. However, its disadvantage is that it is difficult to increase the peak forward blocking voltage without increasing the on-state voltage drop and thus decreasing the high-speed carrier discharge characteristics. The injection enhanced gate transistor (IEGT) developed by the Toshiba Co. allows increasing the peak forward blocking voltage with a low on-state voltage drop and high speed carrier discharge by applying the electron injunction enhancement to the trench type gate.

The SI thyristor has the same structure as a GTO but different gate characteristics. It is operated normally in off-state by applying a positive voltage to the gate, and in on-state by removing the gate voltage. Even though the turn-off and turn-on times are respectively one tenth that of a GTO, and 4.5kV peak forward blocking voltage devices are developed, only small controllable current devices have so far been developed because of the narrow segment width. The characteristics of the SI thyristor have been improved by compounding it with MOSFET and adding a 'p-' layer so that it is operated normally in the on-state. In addition, controllable current is increased by applying electron irradiation and by optimization of impurity distribution.

Table 13.1 shows the recent typical characteristic data of the GTO, IGBT and SI thyristor.

13.3 Development of high performance SC converter

13.3.1 Application status of SC converter

The SC converter has been utilized to improve power quality in such systems as uninterrupted power sources (UPS), active filters (AF), and power converters for new energy resources. Also it is employed in SVC to compensate voltage flicker and voltage unbalance in the small capacity and low voltage distribution system. To improve power system stability, we need larger capacity and higher voltage SC converters with enhanced reliability and reduced operation loss.

SC converters utilized in the power system, as shown in Table 13.2, are quite different in their types and ratings. Small rating IGBT is mainly applied to small capacity high speed switching SC converters used for power quality improvement. On the other hand, large capacity higher voltage GTO is used for the SC converter applied to battery energy storage (BES), fuel cell (FC) energy conversion, and the SVC to compensate system voltage disturbances at distribution substations. To reduce the harmonics flowing into an interconnected power system, double, triple or sextuple connection of single or a three-phase converter with PWM control is adopted.

SVCs with SC converters as shown in Table 13.3 for power system stabilization, were developed by the Kansai Electric Power Co. (KEPCO) at Inuyama substation and by Tokyo Electric Power Co. (TEPCO) at the Shin-Shinano Substation. The series connection of the GTO in converter arms and the multiphase shifting connection or multiple cascading connection of converters are adopted to increase converter voltage and to reduce the harmonic voltage of outputs.

13.3.2 High performance SC converter

To utilize the SC converter in trunk power systems particularly for stability improvement of interconnected power system and for voltage stability mitigation of power wheeling system, it is necessary to develop high performance SC converters with larger capacities and higher voltages together with the necessary higher reliability and efficiency. In the R&D joint project for interconnection reinforcement with MITI subsidy, CRIEPI and ten electric power companies started in 1992 the development of a high performance SC converter.

The designing technologies of high performance SC converter have been developed by establishing 'elemental models', of which features are characterized by:

(1) Multiple series connection of GTOs for higher DC voltages.

(2) Gate power supply from anode-cathode voltage to reduce the number of converter elements.

(3) Regenerating the energy of snubber circuits and anode reactors to reduce the switching loss of converters.

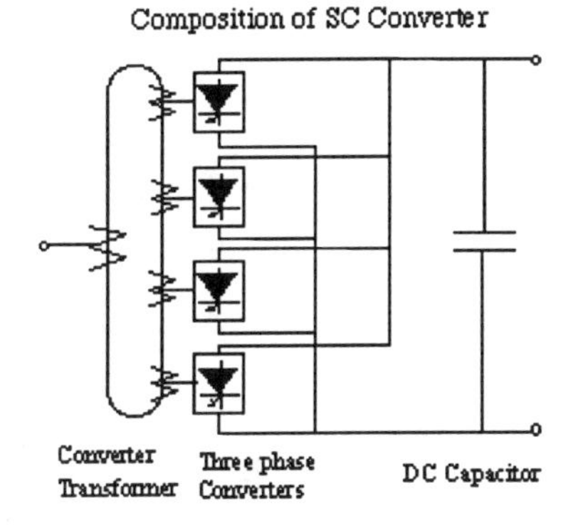

Rating & Specification of SC Converter

	Rating, Specification
· Capacity	Active Power; 37.5MW
	Reactive Power; 37.5MVar
· Type	Air Insulated, Pure Water Cooling
· Composition	Quadruple Cascade Connection
· DC ratinng	10.6kV, 3.6kA
· Devices	GTO(6kV-6kA, 500Hz)
· Arm Comp.	4 series -1 pallarel /module

Figure 13.2 High performance SC converter for verification test

Table 13.3 Development of SC-SVC

	Inuyama S.S. (KEPCO)	Shin-Shinano S.S.(TEPCO)
Circuit diagram	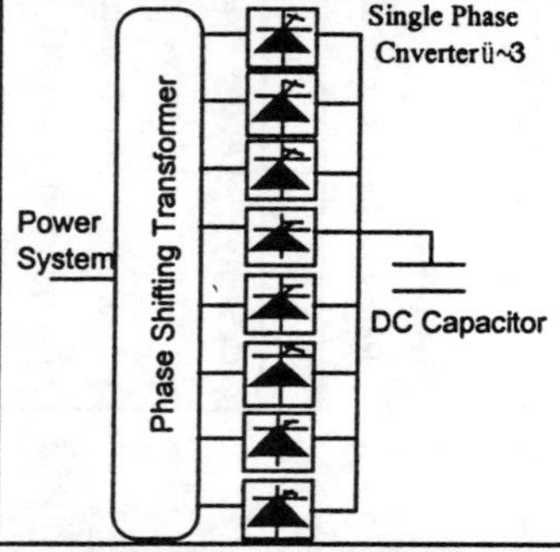 Power System; Phase Shifting Transformer; Single Phase Cnverterü~3; DC Capacitor	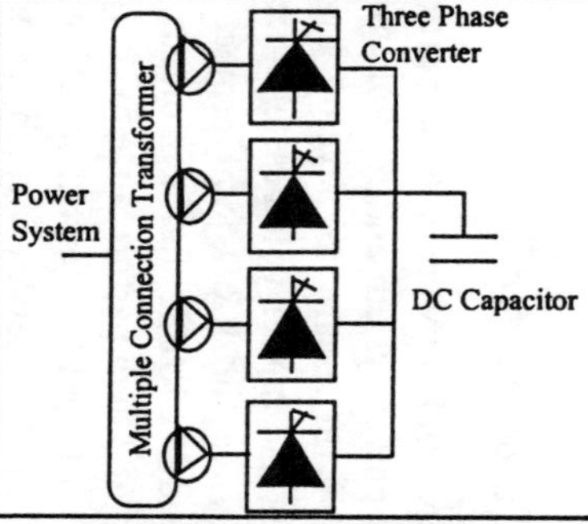 Power System; Multiple Connection Transformer; Three Phase Converter; DC Capacitor
Rating	• 10MVA×8, 4.15kVdc • Single phase converter 3, 8 conv. phase shifting connection, Pure water cooling • GTO(4.5kV-3kA), 3S-1P/Arm	• 12.5MVA×4, 16.8kVdc • Three phase converter×1, Quadruple cascade connection, Pure water cooling GTO(6kV-2.5kA), 8S-1P/Arm
Purpose	• Reactive power control • Power system stabilization (utilized in actual system)	• Series connection of GTO • Transformer flux bias suppression control

Table 13.4 Designing technologies for high performance SC converter

	Multiple Series Connection	Gate energy supply	Regeneration of snubber & anode reactor energy
Circuit diagram	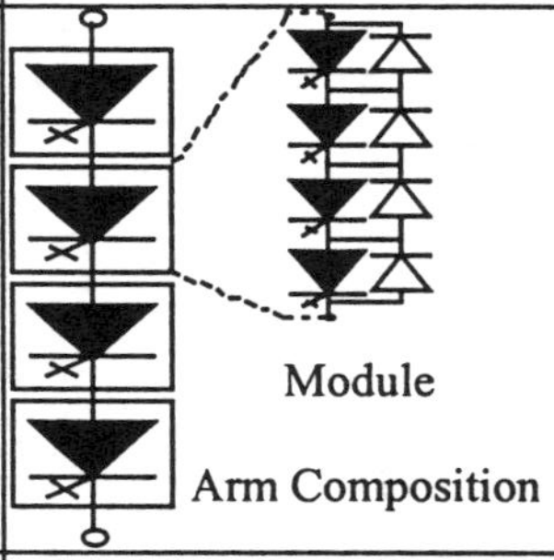 Module Arm Composition	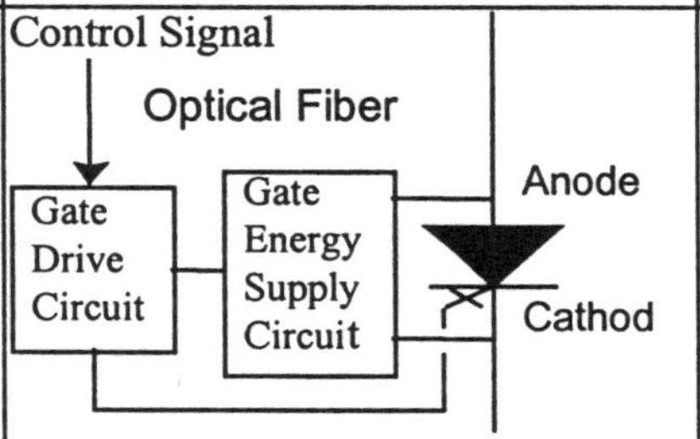 Control Signal Optical Fiber Gate Drive Circuit Gate Energy Supply Circuit Anode Cathod	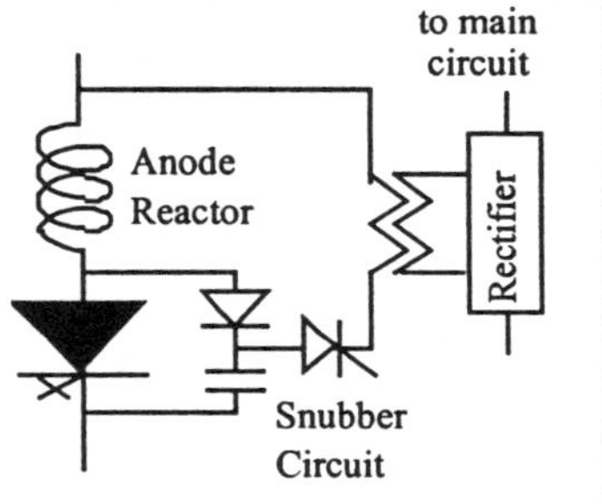 to main circuit Anode Reactor Rectifier Snubber Circuit
Major results	Voltage unbalanced ratio is suppressed to 10% by adjusting turn-off timing with 0.1μ second	Gate energy supplied from anode-cathode voltage is available in 100ms after main DC voltage drops	Snubber & anode reactor energy is regenerated to increase 3% of converter efficiency

Verification tests of these technologies applied to 37.5MW SC converters as shown in Figure 13.2 are being implemented in the Shin-Shinano Substation. They will be continued until 2000 followed by application of the results to the construction of a 300MW SC converter.

To establish multiple series connection technologies of GTOs, the turn-off timing adjusting circuit to suppress the unbalanced voltage of series elements was developed. The test results by 'elemental model' composed of 16 elements of 4.5kV-3000A GTOs showed that the imbalance ratio of each element can be suppressed to less than 10% of the turn-off peak voltages in the variation of turn-off timing less than 0.1 micro-second. Calculations with the assumption of the dispersion of element characteristics showed that up to 30 elements can be connected in series.

As the light direct triggered GTO has not yet been developed, it is necessary that the turn-on and turn-off energies are supplied through an insulated transformer from the controller base. If the energies are supplied from the anode-cathode voltage of the GTO itself and only turn-on and turn-off signals are transmitted from the controller, the GTO module would become more simplified and reliable. The 'elemental model' composed of 16 elements whose gate energies were supplied from anode-cathode voltage was developed to verify that adequate energy can be supplied until 0.1 second after the shut down of the DC main voltage.

The energy loss of the SC converter is larger than that of the line commutated (LC) converter because the switching loss caused by turn-on and turn-off according to PWM control is added to the loss caused by forward voltage drop which is almost the same as the LC converter. To reduce this switching loss which depends on the frequency of the PWM and the value of the snubber capacity and the anode reactor, a regeneration circuit is integrated into the main converter circuit. The 'elemental model' tests showed that almost 75% of switching loss can be recovered to the main DC circuit, representing an average of 3% increase in converter efficiency.

The results from development of 'elemental models' are summarized in Table 13.4.

13.3.3 Verification test of SC converter in actual field

Based on the technological results developed by 'elemental models' shown in Table 13.4, verification tests of SC converters are now being carried out at the Shin-Shinano Substation aiming at establishing the utilization technologies of the SC converter. More precisely, the aims of these tests are threefold; to verify the effectiveness of design technologies by exposing to actual system disturbances, to certify the reliability of facilities through long term operation, and to confirm the control schemes including the control system for the multi-terminal SC-HVDC system.

The verification test system in the field is composed of three terminal SC converters which can be tested individually as SC-SVC, combined with two terminals as SC-BTB or SC-HVDC system, together with three terminals as a multi-terminal SC-HVDC system.

The test items listed in Table 13.5 are composed of fundamental tests and operation tests. Fundamental tests are composed of the start and stop of individual and/or combined terminals, change of active power, reactive power or voltage preferences, and switching on and off of shunt capacitors, reactors and transformer. Operation tests are to certify the stable operation and the reliability of SC converters under AC line faults, or various operating conditions experienced in those intervals. Many analytical studies were carried out using the power system simulator and a digital simulation program with related power system models.

Table 13.5 Field verification tests for SC converter

	Operation mode	Test items
Fundamental tests	* SVC * BTB * Multi-terminal	* Charging DC capacitor * Start-Stop * Changing references of P,Q,V (including power reverse) * On-off nearby AC equipment
Operational tests	* BTB * Multi-terminal	* Start-Stop * Changing references of P,Q,V (including power reverse) * On-off nearby AC equipment * Cutting off a terminal

13.4 Application of power electronics equipment for power system performance enhancement

Power electronics technologies such as SVC with LC converters have been utilized to compensate system voltage fluctuation at low voltage level, to improve voltage stability in power wheeling systems, and to stabilize interconnected power systems at higher voltage levels. Power electronics technologies have also been applied to various types of generator exciters to improve the stability, as well as to variable speed machines (VSM) to control power system frequency.

13.4.1 Improvement of voltage stability by SVC

Voltage stability in trunk power system wheeling to large capacity consumer areas is seen as liable to cause wide area black-outs due to disturbances such as transmission line trip or high speed consumption power change. This voltage stability can be illustrated using the so-called '*P-V* nose curve' at a substation feeding to consumers.

If the *P-V* characteristics as shown in Figure 13.3 are sensitive to line faults or change of shunt elements, the operating point of terminal voltage moves from pre-fault point 'A' to post-fault point 'D' through points 'B', 'C' during the fault. The terminal voltage at point 'D' are going to decline following curve c. Applying SVC composed with a thyristor controlled reactor (TCR) and constant capacitor to the substation, the voltage moves to point 'D'' recovering to point 'A' following curve c'.

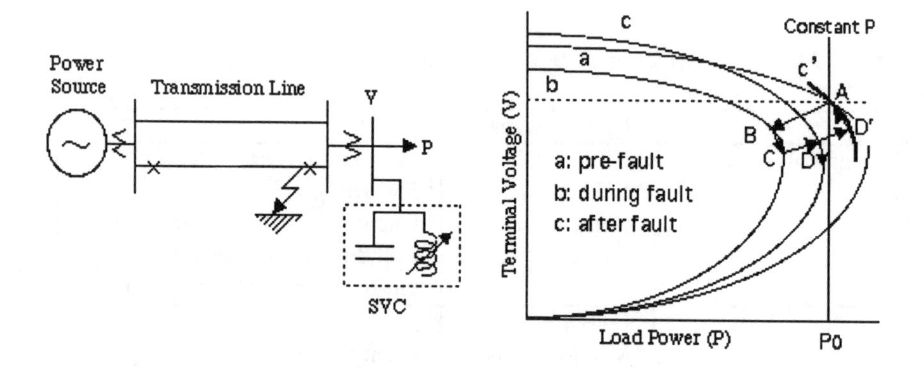

Figure 13.3 P-V characteristics and voltage stability

Since the voltage stability indices, which can be defined as marginal load power at each node to the spearhead of the nose curve shown in Figure 13.4, have different values at each node, it is extremely laborious to calculate indices of all nodes using the conventional power flow calculation method. CRIEPI has developed a new and convenient calculation method for the index of each node. Assuming that the load power at node 'n' becomes 'h' times as large as the current load S_n, the voltage change $V_n = X \times V_{no}$ can be obtained by the following equation;

$$hS_n = V_n \times conj\{(V_{no} - V_n) \times Y_n\} + S_{no}$$
$$= (X \times V_{no}) \times conj\{(1-X) V_{no} \times Y_n\} + S_{no}$$

where S_{no} and V_{no} are the load power and voltage at each node, respectively.

Y_n is the equivalent admittance at node 'n' including shunt elements which can be defined by power flow calculation of the original system.

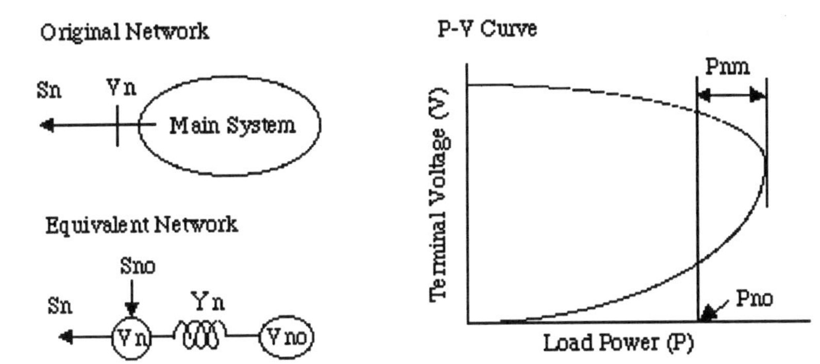

Figure 13.4 Calculation of voltage stability indices

The voltage stability index at node 'n' defined as $P_{nm}=(hP_{no}-P_{no})/P_{no}=(h-1)/h$ is the value of 'h' wherein the above described equation has multiple solutions as the load power increases and reaches the spearhead of the nose curve.

Several sets of SVC were introduced to improve the voltage stability of the power wheeling system after a large-scale outage of TEPCO's western area in 1987. To clarify the effect of SVC to voltage stability, verification studies using the 'AC/DC Power System Simulator' at CRIEPI representing the power system model with the SVC were carried out with the aid of the voltage stability index as described.

13.4.2 Power system stabilization by SVC

As the interconnected power systems may fall into instability due to interactions between constituent power systems, precise analysis considering the condition of generators, transmission lines and power flows is required to increase the transmission capability. The tripping of transmission lines and generators may particularly cause drastic changes in the power system conditions and accelerate poor damping oscillation due to the interactions of constituent power systems as experienced in many power systems in the world.

To suppress this type of poor damping oscillation, an optimal selection method for PSS constants of each generator exciter was developed based on eigenvalue analysis. This method was applied to study the effect of SVC on power system stability. The dominant eigenvalues and eigenvectors are analyzed using an interconnected system model which represents the western/central region of Japan as shown in Figure 13.5.

The low frequency oscillation of about 0.3 Hz caused by the interaction between the western/central power systems can be suppressed by PSS constants optimization using the following procedures;

(1) Evaluation function; $f = S \times \exp(k_n \times l_n)$, where '$l_n$' is eigenvalue and '$k_n$' is coefficient.
(2) Sensitivity analysis; $df/d\alpha = S \times k_n \times (dl_n/d\alpha) \times \exp(k_n \times l_n)$, where '$\alpha$' is PSS constants parameter.
(3) Maximum gradient method is applied to minimize 'f'.

The comparison of damping coefficients with and without optimization is shown in Figure 13.6.

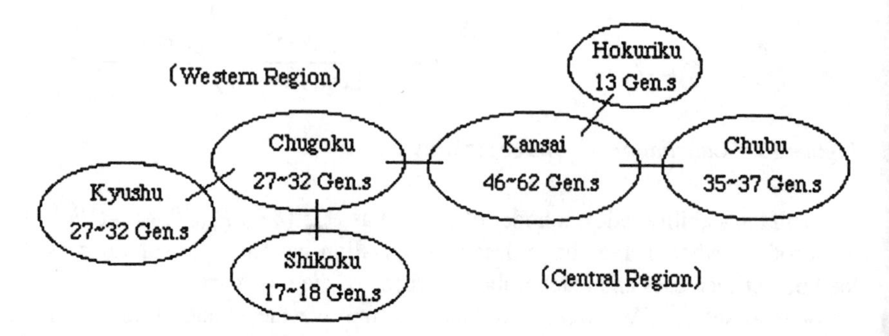

Figure 13.5 Interconnected power system model for analysis

Block diagram of PSS

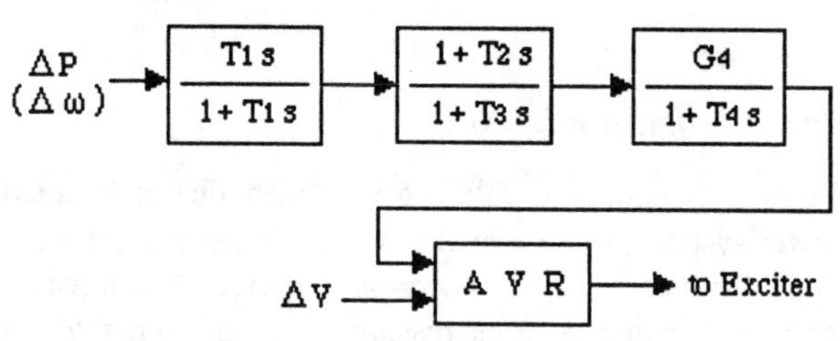

Damping coefficient and frequency

	Damping coef. (1/s)	Frequency (Hz)
without optimized PSS	0.033	0.315
with optimized PSS	-0.141	0.314

Figure 13.6 Damping coefficients with and without PSS optimization

The damping control of SVCs as applied to the interconnected power system model also can be optimized by the same procedures. The results given in Figure 13.7 show that the SVC installed at the sending end is effective for improving oscillation damping. The SVC installed at the receiving end is also effective for optimized damping control.

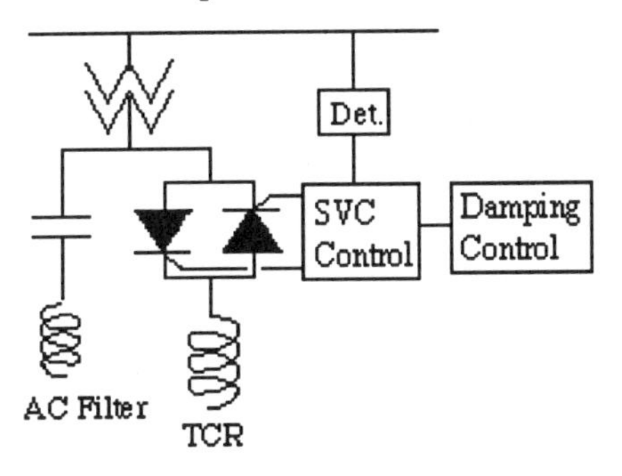

Damping Coefficient of Interconnected System

Installation Site	without SVC	with SVC	SVC with Damping Cntrl.
Western End	0.07	-0.04	-0.19
between Western and Central	//	0.1	0.1
Central End	//	0.08	-0.02

Figure 13.7 Damping coefficients with and without SVC

13.4.3 Power system frequency control by VSM

The exciter of synchronous machines has been improved to have a fast response time and to ease maintenance work by using thyristor converters. Replacement of a DC machine exciter by a thyristor exciter can stabilize generator operation with the PSS. In addition, use of a brushless exciter with an AC machine whose output voltage is rectified by a thyristor converter for supplying the field current makes it possible to ease the machine maintenance work.

In the variable speed machine, the conventional DC field winding is replaced by the three phase winding which induces a rotating flux facing the induced rotating flux of armature windings. It can control the output electrical power $P(\omega_s)$

independent of the input mechanical torque 'T' by using the converter which supplies the field current.

$P(\omega_s)= \omega_r \times T+P(\omega_c)$

where '$P(\omega_c)$' is the output power of converter, 'ω_s', 'ω_c' are the rotating speeds of fluxes induced by stator and converter, and 'ω_r', is the rotating speed of rotor.

As the converter of VSM is required to be able to change the power, the relation $\omega_s=\omega_r+\omega_c$ is kept in the region from $\omega_c=0$ at synchronous speed to $\omega_c=\omega_s-\omega_r$ at asynchronous operation.

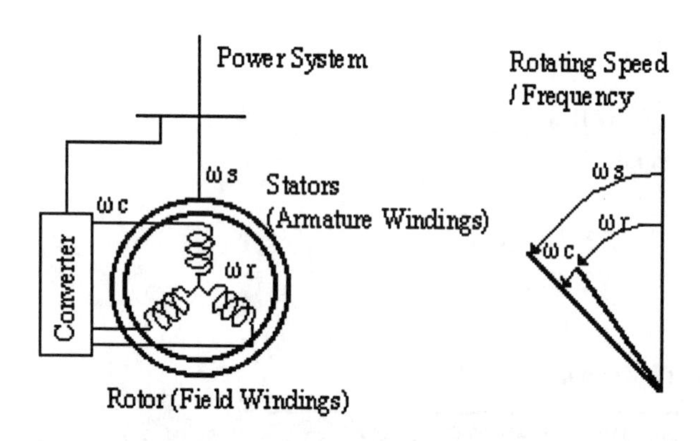

Figure 13.8 Schematic diagram of VSM

It is necessary for the pumping generator to be able to change output power with constant rotor or shaft speed in order to control the power system frequency. Moreover flywheel power storage, the rotating speeds of which is higher than the system frequency in the charged condition and lower in the discharged condition can be operated at rotating speeds different from the system frequency. Application examples of the VSM to pumping generator and flywheel power storage in Japan are shown in Table 13.6.

Cycloconverters are usually adopted for the converter of VSM as a standard means to supply low frequency power from the power system. In recent years, the SC converter as shown in Figure 13.9 has been applied to a pumping generator under the following design restrictions.

(1) The rectifier is able to take power reversing operation mode corresponding to the operation of the main SC converter supplying the field current whereas it reverses the power when 'ω_s' is lower than 'ω_r'.

(2) In case the SC converter is adopted to the rectifier, its current rating should be large enough not to trip the SC converter at the AC system faults.

(3) It is necessary that the main SC converter can continuously yield output power at zero frequency, or DC power.

(4) The over-voltage in case of power rejection due to the control and protection of the exciter should be suppressed.

Table 13.6 Application of VSM to pumping station and flywheel storage

	Ohkawachi pumping station (KEPCO)	Shiobara pumping station (TEPCO)	Okukiyotu II pumping station (EPDC)	Nakagusuku-Bay substation (Okinawa)
Main generator	395MVA 330~390rpm	360MVA 345~390rpm	345MVA 407~450rpm	26.5MVA 510~690rpm
Converter	72MVA-5.2kV Cycloconverter (4 bridges /phase) -5~+5Hz	51.1MVA-5.9kV Cycloconverter (4 bridges /phase) 0.25~+4Hz	31.5MVA-3.04kV SC Converter (6 parallel 3Φbridges) −2.5~+2.5Hz	25MVA-1.93kV Cycloconverter (Non-circulating) 0.25~9Hz
Control	3Φ independent control Power predictive control		50Hz PWM control Snubber energy regeneration	

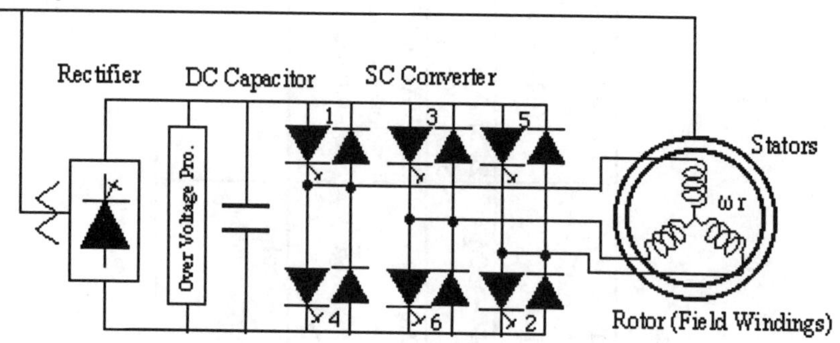

Figure 13.9 SC converter for VSM

13.5 Development of FACTS control schemes with power system model

The enhancement measures for power system performance applying power electronics technologies, so-called FACTS, are promising means for reinforcing the transmission capability. The influence of FACTS technologies on the future power system in Japan and the benefits they have for utilizing the existing power system to the maximum under the new circumstances of electric power deregulation and electric power costs reduction should be studied. Subsidized by MITI, research to establish FACTS control schemes and analytical methods required for the application of FACTS technologies to future power systems in Japan has been carried out jointly by CRIEPI and ten electric power companies.

13.5.1 Selection of power system model

Power systems in Japan are basically composed of radial interconnected power systems in the western and central regions and a looped power system in the eastern region. These power systems are modified as shown in Figure 13.10. They have the following features.

(1) The power flow of the radial system model is basically from system 1 to systems 4 and 6. As the increase of this power flow is liable to cause low

frequency oscillation, an increased transmission capability to 3GW from system 1 to systems 4 and 6 will make it possible to postpone the construction of new transmission lines.
(2) As the generator capacity of the looped system model is difficult to allocate in system 6, it is necessary to reinforce the power system to move 3GW of generation from system 6 to system 2.

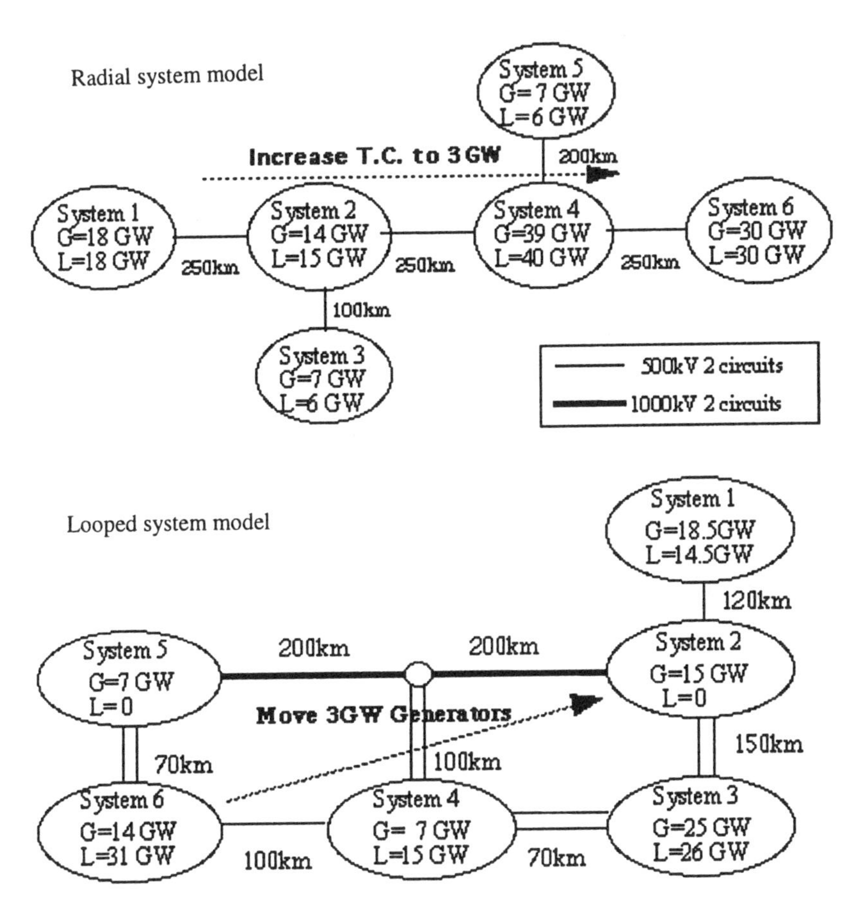

Figure 13.10 Interconnected power system models for evaluation

To make clear the multi-functional effects of FACTS technologies such as generator stability and voltage stability improvement, two model systems are set up. One is called the 'long distance transmission system from remote generator' model and the other is the 'power supply system to large consumers' model.

Table 13.7 Development and research status of FACTS equipment

	Thyristor controlled series capacitor (TCSC)	Thyristor controlled phase shifter (UPFC)	Self-commutated SVC
Circuit diagram	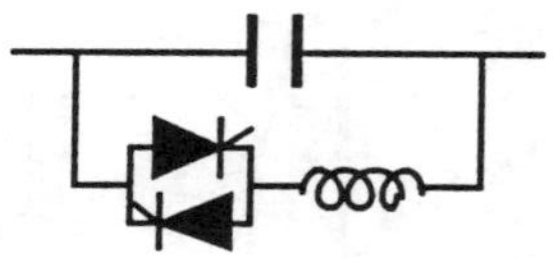	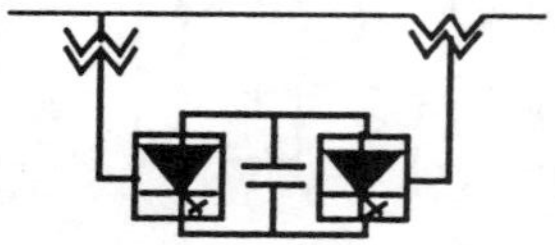	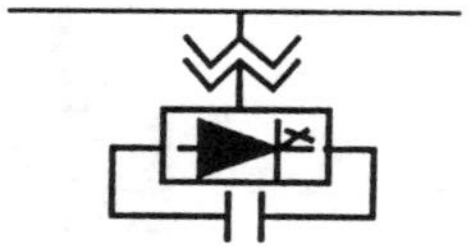
Development status	• BPA; Actual field test, Slatt S.S. (500kV,202 MVA) • WAPA; Actual field test, Kayenta S.S. (275kV,330MVA) • R.D. joint project for interconnection reinforcement	• WAPA; Analysis of actual system • AEP; Actual field test, Inez S.S. (138kV,160 MVA) • EDF; Proto-type model (500kV,7MVA • R.D. joint project for interconnection reinforcement	• KEPCO; Actual field test, Inuyama S.S. (154kV,80MVA) • TEPCO; actual field test, Shin-Shinano S.S. (275kV,50MVA) • TVA; Actual field test, Sullivan S.S. (275kV, 500MVA) • R.D. joint project for inteconnection reinforcement

Table 13.8 Evaluation results of power system reinforcement measures

<table>
<tr><th colspan="2"></th><th colspan="3">Radial system model</th><th colspan="3">Looped system model</th></tr>
<tr><th colspan="2">Reinforcement measures</th><th>Increase T.C. to 3GW (Capacity)</th><th>Multi. func.</th><th>Total</th><th>Move 3GW gen. (Capacity)</th><th>Multi. func.</th><th>Total</th></tr>
<tr><td rowspan="2">Conventional measures</td><td>• Quick response exciter with PSS</td><td>| (18,000 MW)*</td><td>|</td><td>|</td><td>(3,000 MW)*</td><td>|</td><td>|</td></tr>
<tr><td>• SVC</td><td>(2,400 MVA)</td><td></td><td></td><td>(1,800 MVA)</td><td></td><td></td></tr>
<tr><td rowspan="4">Power electronics measures</td><td>• SC-SVC</td><td>(1,500 MVA)</td><td></td><td></td><td>(1,200 MVA)</td><td></td><td></td></tr>
<tr><td>• TCSC</td><td>(600 MVA)</td><td></td><td></td><td>(1,000 MVA)</td><td></td><td></td></tr>
<tr><td>• UPFC</td><td>| (1,100 MVA)</td><td></td><td>|</td><td>(650 MVA)</td><td></td><td></td></tr>
<tr><td>• VSM with F.W.</td><td>(1,200 MVA)</td><td>|</td><td>|</td><td>(900 MVA)</td><td>|</td><td>|</td></tr>
</table>

* Generator capacity; valuable; | less valuable
Evaluation of increase T. C. and Move Gen. are based on cost evaluation of equipment capacity

13.5.2 Evaluation of transmission capability reinforcement

The results of R & D work on FACTS technologies have been applied to an actual system after verification tests in an actual power system as shown in Table 13.7.

To clarify the effects of FACTS equipment for increased transmission capability in the above described power system models, digital simulation for the faults of transmission lines in the power system models was carried out to determine the required equipment capacity. The 'Power System Dynamics Analysis Program (Y Method)' developed by CRIEPI was used for this purpose, including both FACTS equipment and conventional power system reinforcement measures.

The evaluation of FACTS equipment in each of the two power system models as described in Section 13.5.1 has been made in terms of required capacity and cost of each set of equipment as shown in Table 13.8. SC-SVC and TCSC are valuable for the increase of transmission capability in the radial power system model, in comparison with the more expensive UPFC. Moreover, SC-SVC , TCSC and UPFC are almost equally effective for the looped power system model.

Furthermore, each measure is almost equally valuable for multi-functional evaluation on the stabilization of the 'long distance transmission system from remote generator', or the voltage stability mitigation in the 'power supply system to large consumers',

On the whole, SC-SVC, TCSC, and UPFC are valuable for the reinforcement of transmission capability in both power system models except UPFC in the radial power system model.

13.5.3 Verification test using APSA (Advanced Power System Analyser)

As the control characteristics of FACTS equipment affect the transmission capability such as power system stability and voltage stability, it is necessary to study the subjects not only from the standpoints of application to the future power system in Japan, but also for the development of future FACTS equipment.

Verification tests were carried out using 3-27VA 50Vac miniature models developed for the analysis by 'APSA' at the Kansai Electric Power Co. The test results are shown in Table 13.9. together with power system models and FACTS equipment models.

(1) The prospect for utilization of FACTS equipment is shown by the operational characteristics confirmation test of miniature models on 'APSA', such as the stable operation of SC-SVC, the impedance controllability of TCSC, and the fast restart function of UPFC in an AC line fault.

Table 13.9 Verification test results using 'APSA'

	SC-SVC	TCSC	UPFC
Long distance transmission system from remote generators	*Stable operation after AC faults *Improvement of T.S. *Improvement of D.S.	*Impedance control after AC faults *Improvement of T.S. *Improvement of D.S. *SSR mitigation	*Fast restart after AC faults *Improvement of T.S. *Improvement of D.S. by damping & braking control
Power supply system to large consumers	*AC voltage stability after 1cct trip or load increase	[same as left]	
Radial power system	*Required cap.: 1500MVA *Damping of LFO	*Required cap.;600MA *Effect of damping & braking control	
Looped power system	*Required cap.: 900~1200MVA *Damping of LFO	*Required cap.: 1000MVA *Effect of damping & braking control	*Required cap.: 650MVA *Effect of damping & braking control
Subjects for application	*Minimum loss operation *Active filter function *Rational design by over current suppress	[same as left] *SSR mitigation [same as left]	*Optimum operation of series & parallel conv. *Short circuit detection [same as left]
Subjects for development	*High performance control with less pulse numbers *High performance converter	*Stable APPS system *Impedance detection *Elements (Arrester, Bypass Switch)	*High performance control with less pulse numbers *High performance converter

T.S.; Transient Stability, D.S.; Dynamic Stability, LFO; Low Frequency Oscillation

(2) Improvement of stability in the 'long distance transmission system from remote generator' model by each equipment, suppression of oscillation in the interconnected power system by damping control, and the mitigation of voltage stability in the 'power supply system to large consumers' model are clarified.
(3) The study using 'APSA' gives almost the same results as the 'Y Method' as regards the capacities required for increasing the transmission capability in the radial and looped system.
The subjects for application to future power systems and further development of FACTS equipment are also clarified as shown in Table 13.9.

13.6 Digital simulation program for FACTS analysis

To analyze the effect of FACTS equipment, a digital simulation program with SC-SVC, TCSC and UPFC based on the 'Y Method' was developed. In this program, the quantities such as P, Q, V are expressed by root mean square values. In addition, the dynamic characteristics of generators, HVDC systems, and loads are calculated by numeric integration with control and protection means. The symmetrical coordinate method was applied in the network equation.

With the dynamic characteristics of SC-SVC and UPFC based on the dynamics of the SC converter, the model for SC converter was developed taking into account the control and protection means. The dynamics of TSCS were the same as the conventional SVC except that the connection in the former is in series to line while the latter is between line to ground. Therefore it is very easy to derive the model of TCSC from that of SVC.

13.6.1 Modelling of SC converter

The SC converter typically configured with quadruple three phase bridge converters with nine pulse PWM control is shown in Figure 13.11. The fundamental component of output AC voltage 'V_c' can be obtained by multiplying DC voltage 'V_{dc}' by constant coefficient 'k' and modulation ratio 'm'. The fundamental AC current 'I' can be calculated taking into account the difference of output AC voltage 'V_{ac}' and terminal AC voltage 'V_n' together with tranformer reactance.

The *d-q* axis currents can be independently controlled, (named 'independent vector control') with the following d-q axis relation of AC side values;

$$V_{cd}=V_{nd}+\omega L\times I_q+L\times dI_d/dt$$
$$V_{cq}=V_{nq}-\omega L\times I_d+L\times dI_q/dt$$

where $V_c=V_{cd}+jV_{cq}$, $V_n=V_{nd}+jV_{nq}$, and $I=I_d+jI_q$

The terms '$L\times dI_d/dt$' and '$L\times dI_q/dt$' are so small as compared with fundamental components that these can be ignored. The current controllers of SC converter are simplified as shown in Figure 13.12.

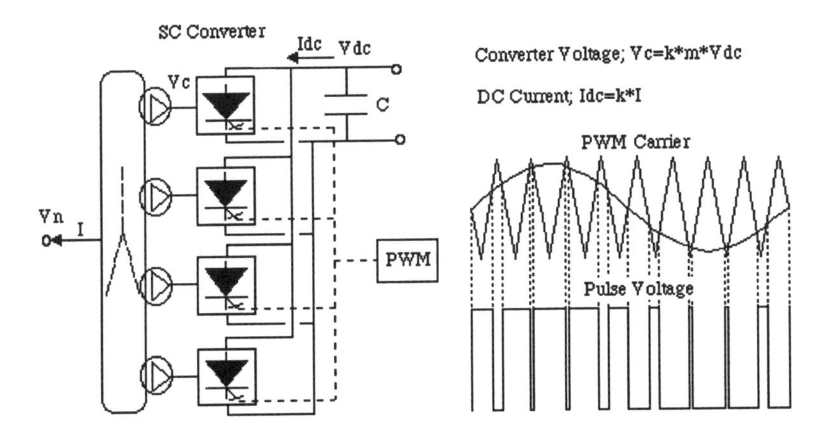

Figure 13.11 Schematic diagram of SC converter

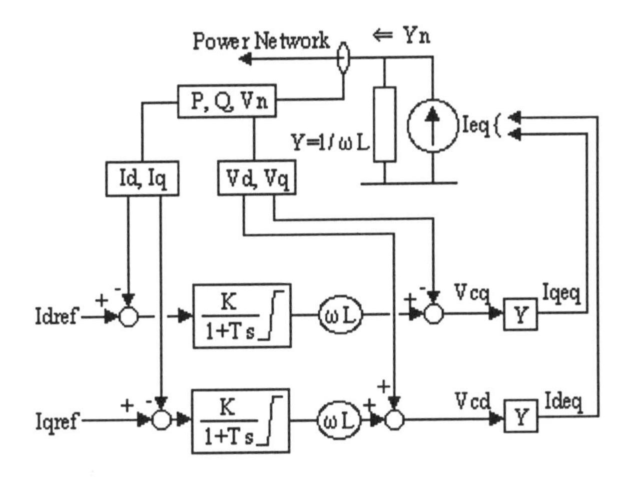

Figure 13.12 Block diagram of SC converter model

To connect with the symmetrical coordinate equation of the network, the converter voltages 'V_{cd}' and 'V_{cq}' are transformed to the current sources 'I_{deq}' and 'I_{qeq}' injected to the network. The iteration at node 'n' is proceeded to adjust the node injection currents to current sources by the following equations;

Network equation; $[I_n]=[Y_n]*[V_n]$

Node iteration; $V_n(i+1)=V_n(i)+\varepsilon\times\{I_{eq}-I_n(i)\}$ => $I_n(i+1)=Y_n\times V_n(i+1)$

where 'i' is the iteration number, and 'ε' is a convergence acceleration constant.

The dynamics of the SC converter regulators whose outputs correspond to 'I_{dref}' and 'I_{qref}' in Figure 13.12 are calculated by time step numerical integration method together with the dynamics of DC capacitance by solving $I_{dc}=-C\times dV_{dc}/dt$ with $I_{dc}=k\times I$.

13.6.2 Modelling of FACTS equipment

The main regulator AVR of SC-SVC in Figure 13.13, controls 'V_c' to keep the terminal voltage 'V_n' constant by adjusting 'I_{qref}'. The supplemental damping control applied to the reference signal of AVR increases the damping of oscillation in power system. The detailed regulator model can be simulated by the 'Y Method', in which current limiter and over-current protection are precisely simulated.

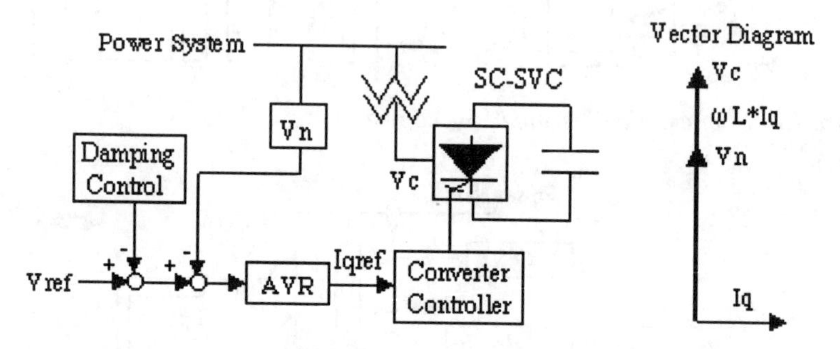

Figure 13.13 Schematic diagram of SC-SVC

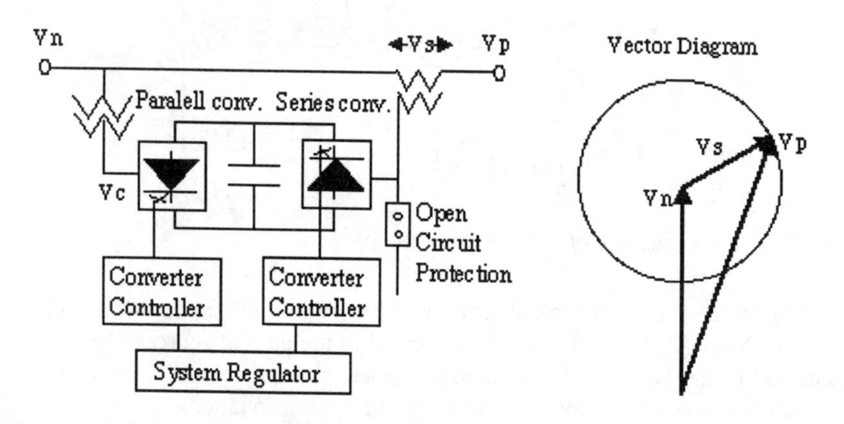

Figure 13.14 Schematic diagram of UPFC

The parallel converter controller in UPFC as shown in Figure 13.14, is basically the same as that of SC-SVC. The controller of a series converter makes it possible for it to function as a phase shifter to control the power flow thus improving power system stability. Supplemental controls are also added to each regulator to damp or brake the power system oscillation. The series converter has to be provided with a special protection scheme to prevent open circuiting of the series transformer in abnormal converter operation.

As the TCSC simply shown in Figure 13.15 is obtained by connecting SVC in series to the transmission line, the thyristor controlled reactor TCR is represented by a current source to combine the network equations as follows;

Network equation; $[I_n]=[Y_n]\times[V_n]$, $[I_m]=[Y_m]\times[V_m]$

Node iteration; $V_n(i+1)=V_n(i)+\varepsilon\times\{I_{eq}+I_n(i)\}$ $\Rightarrow I_n(i+1)=Y_n\times V_n(i+1)$

$V_m(i+1)=V_m(i)+\varepsilon\times\{I_{eq}-I_m(i)\}$ $\Rightarrow I_m(i+1)=Y_m\times V_m(i+1)$

where 'i' is iteration number, and 'ε' is convergence acceleration constant.

The equivalent current source 'I_{eq}' can be calculated using the TCR control angle 'β' as defined by the dynamics of the controller as well as terminal voltages 'V_n' and 'V_m', that is $I_{eq}=f(V_n, V_m, \beta)$

The simulation model for TCSC is also shown in Figure 13.15

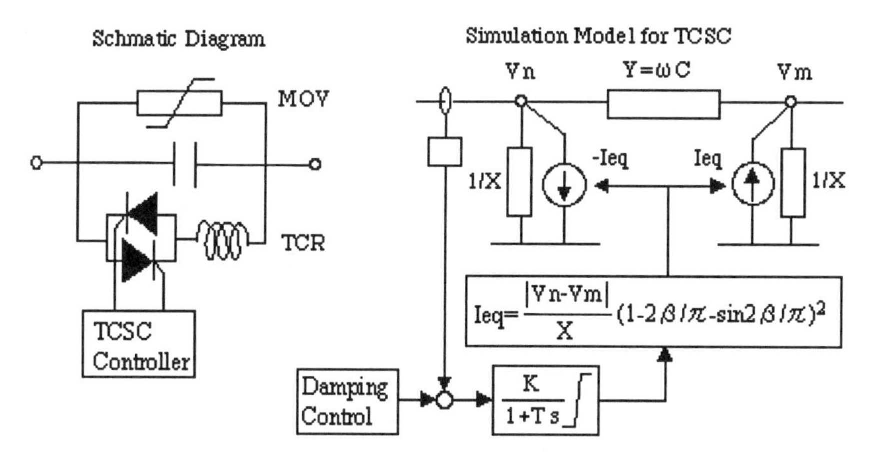

Figure 13.15 Schematic diagram and simulation model of TCSC

13.7 Conclusion

This chapter reviews the status of power electronics technologies including the development of semiconductor devices for power application, the development of high performance SC converters, and research on FACTS equipment for power system enhancement. Remarkable progress in this field will be useful not only for improving the transmission capability of trunk power systems but also to serve for upgrading the distribution systems through the improvement of power quality.

Amidst the business environments such as the deregulation of electric power utilities, global environmental restrictions and energy security for electric power resources, electric power utilities in Japan are facing many social and technical problems. The utilization of power electronics technologies is indispensable for reducing the cost of electric power, for enhancement of transmission capability to adapt to free power wheeling, and for preventing wide area black-out due to inadequate coordination of power system planning and operation. It is also indispensable for the improvement of power quality for meeting higher demand as imposed by the expanded use of upgraded information equipment.

The authors expect that the new power electronics devices such as the GCT and IEGT will be used for more advanced SC converters, and that the results from R&D joint projects for interconnection reinforcement in Japan will be useful for developing better power systems in the future.

13.8 References

1 Gruing, H., *et.al.*, “High-Power Hard-Driven GTO Module for 4.5kV/3kA Snubberless Operation”, *Power Conversion-May 1996, Proceedings,* p.169-183, 1996.

2 Nakagawa, T., *et.al.*, “A New High Power Low Loss GTO”, *ISPSD '95*, p.84-88, 1995.

3 Takahashi, Y., *et.al.*, “2.5kV-100A Power Pack IGBT”, *ISPSD '96*, p.299,1996.

4 Nishizawa, J., *et.al.*, “A Low-Loss High-Speed Switching Device: The 2500-V 300-A Static Induction Thyristor”, *IEEE. Trans. on ED*, ED-33, (4), p.507-515.

5 Ballad, J. P., *et.al.*, “Power Electronic Devices and Their Impact for Power Transmission”, *IEE Conference ‘AC and DC Power Transmission’, Conference Publication,* No.423, p.245-250, 1996.

6 Bose, B. K., “Power Electronics – A Technology Review”, *Proceedings of the IEEE*, 80, (8), 1992.

7 Sekine, Y., *et.al.*, “Application of Power Electronics Technologies to Future Interconnected Power System in Japan”, *CIGRE Symposium Tokyo 1995*, No.210-03, 1995.

8 Kinoshita, N., *et.al.*, “Development of High Performance Self-Commutated Power Converter”, *CIGRE Symposium Tokyo 1995*, No.510-05, 1995.

9 Nagao, T., *et.al.*, “Development of Static and Simulation Program for Voltage Stability Studies of Bulk Power System”, IEEE, 1996.

10 Suzuki, H., *et.al.*, "Development and Testing of Prototype Models for a High Performance 300MW Self-Commutated AC/DC Converter", *IEEE, 96 SM 448-1 PWRD*, 1996.
11 Sakamoto, K., *et.al.*, "Development of Control System for High Performance Self- Commutated AC/DC Converter", *IEEE, PE-790-PWRD-0-04*, 1997.
12 Mori, S., *et.al.*, "Development of a Large Static Var Generation using Self-Commutated Inverters for Improving Power System Stability", *IEEE Trans. on Power System*, 8, (1), 1993.
13 Uzuka, T., *et.al.*, "A Static Voltage Fluctuation for Electric Railway using Self-commutated Inverters", *IPEC -Yokohama*, No.185-7, 1995.
14 Iizuka, A., *et.al.*, "Self-Commutated Static Var Generator at Shintakatuka Substation", *IPEC -Yokohama*, pp.609-614, 1995.
15 Schauder, C., *et.al.*, "Operation of +-100Var TVA Statcon", *IEEE Trans. on Power Delivery*, 12, (4), 1997.
16 Rahman, M., *et.al.*, "UPFC Application on the AEP System; Planning Considerations", *IEEE Trans. on Power Delivery*, 12, (4), 1997.
17 Lombard, X., "Control of Unified Power Flow Controller: Comparison of Methods on the Basis of a Numerical Method", *IEEE, 96-SM 511-6 PWRS*, May 1997.

Chapter 14

Application of power electronics to the distribution system

N. Jenkins

14.1 Introduction

The concept of Flexible AC Transmission Systems (FACTS), as the name implies, was originally developed for transmission networks but similar ideas are now starting to be applied in distribution systems. However, when considering the use of high power electronics on the distribution network it is important to recognise another strand of technology development which comes from Uninterruptable Power Supplies (UPS), large variable speed machine drives and Active Filters. Thus, the new high power electronic systems now being applied to distribution systems owe something to the ideas described in earlier chapters of this book but also use concepts and techniques developed for power electronic systems with lower voltage and current ratings. In both cases, voltage source converters are used as the main component of the systems but the implementation and control of the converters differs considerably [1].

Transmission level FACTS equipment presently uses GTO thyristors as the switching elements due to the high voltage and current ratings of the devices which are available. These are generally switched only rather slowly in order to reduce losses; either at fundamental frequency or using a limited number of notches to give selective harmonic elimination. Generation of harmonics is limited by multi-level topology of the inverter or by using various winding arrangements of the connecting transformers. In contrast, the emerging distribution level equipment generally uses IGBTs as the switching element with some form of Pulse Width Modulation (PWM) switching pattern. The lower order harmonics are eliminated by the PWM switching and so a simple six-pulse, two-level bridge may be used. A further advantage of the use of PWM is that the variable pulse pattern allows the magnitude of the fundamental of the output waveform of the converter to be controlled independently of the voltage of the capacitor energy store. Presently available IGBT converters for distribution use are rated at 1-2

MVA but can be connected in parallel to give overall system ratings of up to 10 MVA. CIGRE has adopted the term Custom Power for such systems.

To date, the main applications for power electronics on the distribution system have been to improve two aspects of power quality: voltage dips or sags and harmonic distortion. These applications are of commercial interest when it is possible to identify a customer who will benefit from improved power quality either through a reduction in the sags experienced or by limiting the harmonic currents injected and thus complying with the utility requirements. Then a commercial case may be made for the introduction of the power electronic equipment. It is commercially more difficult to justify equipment for general wide-area improvement of power quality throughout the distribution network as most non-industrial customers do not suffer a readily measurable financial loss caused by poor power quality. One important application has been to protect manufacturing plants of high value products (e.g. semiconductor plants or paper mills) from the effect of voltage sags which disrupt the continuous manufacturing process. A series connected converter may be used, as a so-called Dynamic Voltage Restorer (DVR), to add a voltage vector to that of the network. This solution has been adopted at a paper mill in Scotland where a 4 MVA DVR has been used to protect an 8 MVA adjustable speed drive [2]. Alternatively a shunt connected STATCOM may be used to inject reactive power. A STATCOM will provide general voltage support to the distribution network and, if this is not desired, a solid state switch can also be used to isolate the load and STATCOM for the duration of the sag.

Limited voltage sags may be compensated either by injection of reactive current or by using the small amount of energy stored in the capacitor of the converter. Single phase sags may be compensated by drawing power from the healthy phases. However, improvement of voltage during sustained 3-phase sags requires a larger energy store and for some applications it may be economic simply to increase the size of the capacitor bank. Alternative larger energy storage systems including batteries, flywheels and Superconducting Magnetic Energy Storage (SMES) are all in the early stages of commercial application or under development for this purpose.

Voltage sags are caused by faults on the power system but some loads also cause dynamic voltage changes, usually referred to as “flicker” due to their effect on incandescent lights. This is a common feature of arc-furnaces and traditionally either passive devices or conventional thyristor based Static VAr Compensators have been used to inject reactive power to stabilise the voltage. A STATCOM may also be used for this application if it is commercially attractive but rapid control is required as the human eye is most sensitive to variations in light occurring every 60 ms. A 2 MVA STATCOM has been installed at a saw mill in Canada to control flicker caused by large induction motors driving wood cutting machinery. The STATCOM was operated in a voltage control mode and had a sub-cycle response time. Test results reported are that the magnitude of the voltage variations on the 25 kV supply was halved.

Many loads draw non-sinusoidal current and this can lead to unacceptable levels of voltage distortion. Distribution utilities apply limits either to the harmonic current which may be injected or to the resulting harmonic voltages on the network. If these limits are exceeded then permission for the connection of the load is refused. If the load equipment cannot be modified economically to draw sinusoidal current then a power electronic Active Filter may be used to correct the current drawn from the network. The commercial use of active filters is well advanced in Japan and it is reported that some 500 shunt active filters have been put into service mainly for harmonic compensation although some are also used for reactive power compensation [3]. Table 14.1 summarises the shunt active filters in practical applications in Japan.

Table 14.1 Shunt active filters in commercial service in Japan [3]. ©1996 IEEE

Objective	Rating	Switching Devices	Applications
Harmonic compensation with or without reactive/negative sequence current compensation	50 kVA-2000 kVA	IGBTs	diode/thyristor rectifiers and cycloconverters for industrial loads
Voltage flicker compensation	5 MVA-50 MVA	GTO thyristors	arc furnaces
Voltage regulation	40 MVA-60 MVA	GTO thyristors	Sinkansen (the Japanese "bullet" trains)

An emerging application of power electronics on the distribution system is to interface embedded or dispersed generation plant. At present there is little generation connected to distribution networks but this is likely to increase as renewable energy, combined heat and power and distributed energy storage become more commercially significant. Many of the prime-mover systems e.g. fuel cells, photovoltaic systems, micro-turbines and variable speed wind energy converters do not directly produce 3-phase 50 Hz output and so power electronic equipment is required to interface them to the network. For most applications the network side converter is again based on a voltage source bridge. Fixed speed wind turbines must use induction generators and there has been one application of a STATCOM to provide dynamic reactive compensation for a 24 MW wind farm in Denmark [4].

One interesting recent development has been the use of voltage source converters for small-scale HVDC [5]. A 3 MW, +/- 10 kV plant has been put into operation in Sweden and further units are planned. Series connected IGBTs are used in a six-pulse, two-level bridge with a PWM switching pattern. It is reported that tests have shown that it is possible to operate into receiving end networks having no rotating machines. The use of small-scale HVDC has obvious possibilities not only for reinforcing the distribution network and supplying isolated loads but also to allow the collection of power from renewable energy sources which may be remote from centres of load and existing ac distribution circuits.

The use of a distribution STATCOM to increase the capacity of voltage limited rural distribution circuits has also been proposed [6] and is likely to be cost-effective for increasing the load or distance served by some circuits, perhaps on a temporary basis using relocatable equipment. An important limitation, noted by the study, is that the use of a STATCOM for voltage support of a feeder has little effect on losses and so for a circuit with a high load factor increasing the voltage level may be the preferable long term solution. A more comprehensive feasibility study of the use of distribution STATCOMs on a US public utility system is described in [7]. This considered a number of potential benefits:

- Improvement of power quality
- Reduction of voltage transients due to capacitor switching
- Reduction in harmonic distortion
- Increase in maximum loadability (particularly of induction motor loads)
- Improvement in long term voltage/VAr management
- Co-ordination with other control devices for voltage profile improvement

The general conclusions were that, for the case studied, there were significant potential benefits in terms of: improvement of power quality, increase in maximum loadability and improvement in long term voltage/VAr management.

14.2 Improvement of customer power quality

14.2.1 Customer power quality

In recent years, customers have become increasingly concerned over the quality of the electrical power, or more precisely the quality of the voltage waveform, which they receive. Much modern industrial load equipment uses sophisticated electronic controllers which are sensitive to poor voltage quality and will shut down if the supply voltage is depressed and may mal-operate in other ways if harmonic distortion of the supply voltage is excessive. Perversely, much of this modern load equipment itself uses electronic switching devices which then can contribute to poor network voltage quality. The introduction of competition into electrical energy supply has created greater commercial awareness of the issues of power quality while equipment is now readily available to measure the quality of the voltage waveform and so quantify the problem.

Power quality is used to describe a wide-range of deviations of the voltage wave-form from a continuous sine-wave of the appropriate magnitude [8]. These include:

1) Voltage sags or dips. The terms sag and dip are used interchangeably with sag being a term used in North America while the European standards (IEC) use the term dip. A sag or dip is a decrease to between 0.1-0.9 pu in the rms voltage (at the power frequency) of less than 1 minute. If the voltage falls to less than 0.1 pu then this is considered to be an *interruption* rather than a dip.
2) Voltage swell. A voltage swell is an increase of between 1.1-1.8 pu in the rms voltage for less than 1 minute.
3) Voltage imbalance. The extent of unbalance between phases may either be expressed as a voltage imbalance factor [9] or by the ratio of negative phase sequence voltage to positive phase sequence voltage.
4) Voltage harmonics and inter-harmonics. Harmonic distortion is measured by considering harmonic frequencies at integer multiples of the fundamental while inter-harmonics refers to the intermediate frequencies. Waveform notching due to semi-conductor device switching may be considered separately.
5) Voltage flicker refers to rapid variations in the voltage which, if applied to incandescent lights, are detected by the human eye and cause annoyance. Thus "flicker" refers to the effect rather than the cause which is rapid transient variations of voltage magnitude.

Large generators produce a close to perfect sinusoidal voltage and so the reduced power quality seen by a customer is caused either by distorting loads connected to the system or by the transmission and distribution networks. Figure 14.1 shows the origin of some of the main power quality problems. A particular load will see voltage sags and swells which are usually caused by faults on the transmission or distribution networks although some sags may also be caused by switching of large loads. Harmonic distortion and phase unbalance are created to some extent simply by the operation of transmission and distribution plant but are also caused by the cumulative effect of a very large number of small (e.g. domestic) customer loads and so appear as a network problem. In contrast, particular loads can produce harmonic or unbalanced current, draw excessive reactive power or draw rapidly varying active and reactive power to cause flicker and so are shown as imposing a power quality problem on to the network.

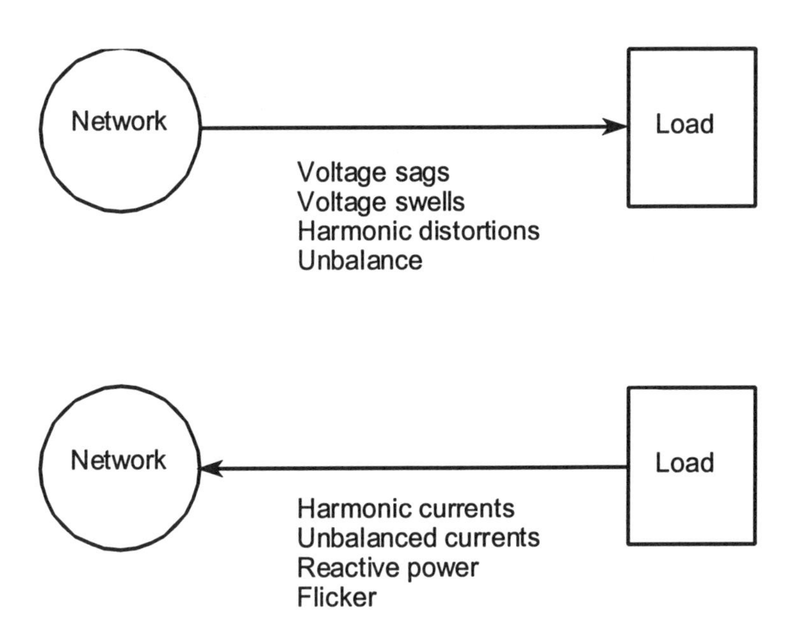

Figure 14.1 Origin of power quality problems

The effect of distorting loads on the network voltage have been known to power system engineers for many years and national standards and design codes have been developed to control them. This has generally been effective for individual large industrial loads but the cumulative effect of great numbers of small appliances, (e.g. domestic TVs or office personal computers) has been difficult to control and harmonic distortion of the network voltage has grown in many networks.

Short circuits have always caused sags in the power system but only recently has there been specific concern over them. Figure 14.2 shows a simple representation of a distribution network where it may be seen that the load is affected by sags on faults on the adjacent MV feeder as well as faults on the HV network. Faults on the interconnected higher voltage networks will also cause voltage depressions which will be experienced by a large number of customers. For most load busbars, the majority of sags experienced will have a residual magnitude above 0.8 pu and a duration of less than 200 ms. Figure 14.3 [10] shows a summary of voltage sag data gathered from four papers. It may be seen that 60-80% of the reported voltage sags last less than 200 ms and that there is a steep rise in the curve at just less than 100 ms corresponding to the minimum clearing time of the circuit breakers. There is considerable work presently in progress both on measuring voltage sags and in the use of computer based methods to predict them. In addition to changes in voltage magnitude, faults may also cause phase-jumps when the phase of the voltage suddenly changes. Phase-jumps may again lead to mal-operation of control equipment of variable speed drives.

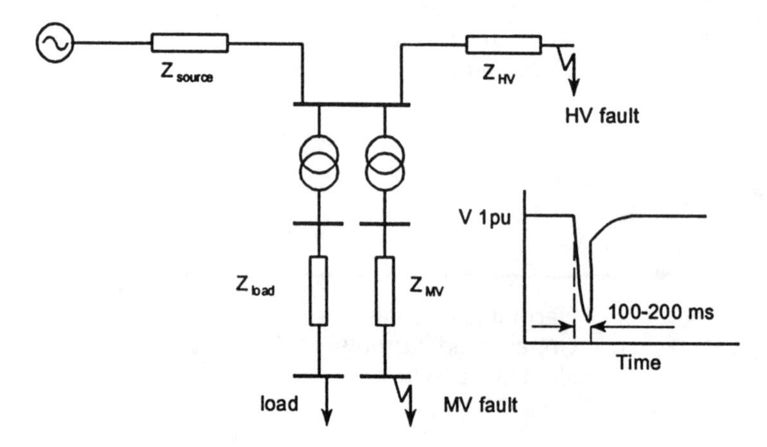

Figure 14.2 Examples of faults causing voltage sags

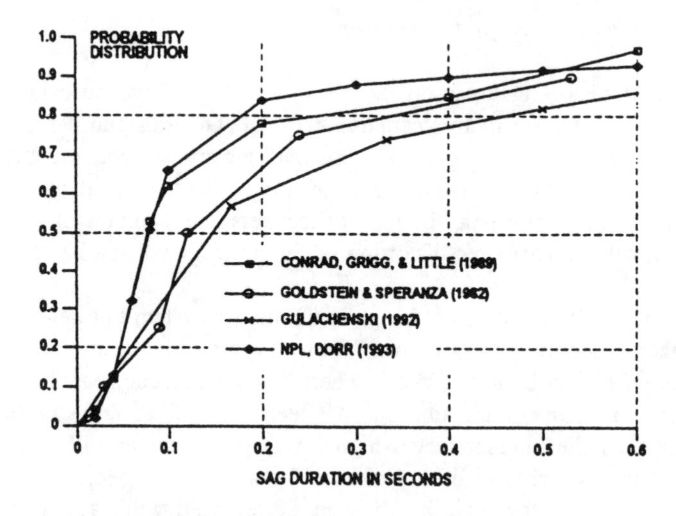

Figure 14.3 Voltage sag probability distribution [10] ©1994 IEEE

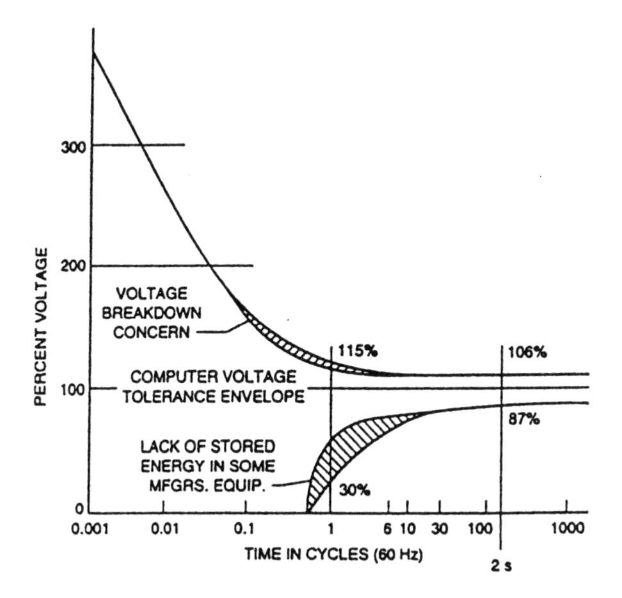

Figure 14.4 Typical design goals of power conscious computer manufacturers [11] ©1996 IEEE

Different types of electrical equipment show different tolerance to voltage sags. Early work on main-frame computers led to the development of the CBEMA (Computer Business Equipment Manufacturers Association) curve (Fig 14.4), [11] which has been widely quoted as providing some guidance for other types of equipment. This curve has recently been updated and is now known as the ITIC (Information Technology Industry Council) curve [25].

There are only a few ways in which the effects of sags may be limited. One option is to reduce the number of faults (e.g. by replacing overhead circuits with underground cables or ensuring trees under overhead lines are adequately trimmed). However, most utilities will have taken every reasonable measure to reduce the number of faults on their system and so any further measures are likely to be very expensive. An alternative approach is to change the way the distribution system is operated so that a sensitive load is connected to the shortest length of exposed network. This conflicts with the requirement to increase the reliability of the load busbar which is normally achieved by arranging duplicate feeds. Clearly, it would be desirable if the load equipment was robust enough not to be affected by the sags, but this is generally not the case. Therefore, the only remaining solution is to use some form of mitigation equipment. Although UPS systems are sometimes installed adjacent to each sensitive load this is expensive and leads to a high maintenance requirement. An alternative solution is a single larger device to provide protection for all or part of the load.

The basic component of most high power electronic systems to improve customer power quality is the conventional voltage source converter. Figure 14.5 shows a 3-phase arrangement although 3 single-phase H bridges may also be used. The IGBTs are switched at a frequency typically in the range of 2-4 kHz depending on the details of the design and the rating of the equipment. Operation at this frequency will result in switching losses, overall converter losses are quoted as some 2 % of the MVA rating, but eliminates low order harmonics. However, there will be harmonics generated at around the switching frequency and so some filtering of the output of the inverter may still be required.

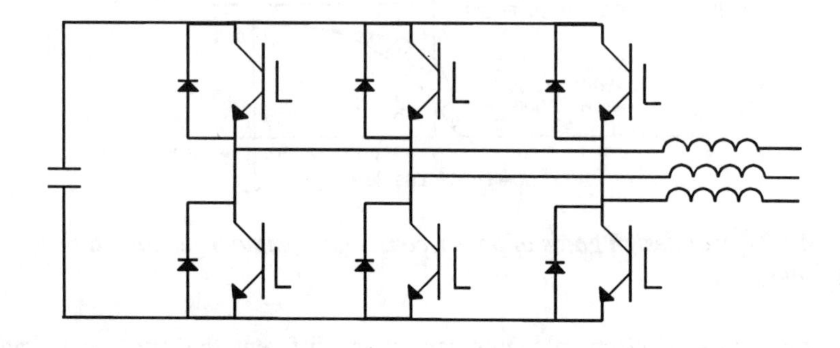

Figure 14.5 Six-pulse two-level IGBT voltage source converter

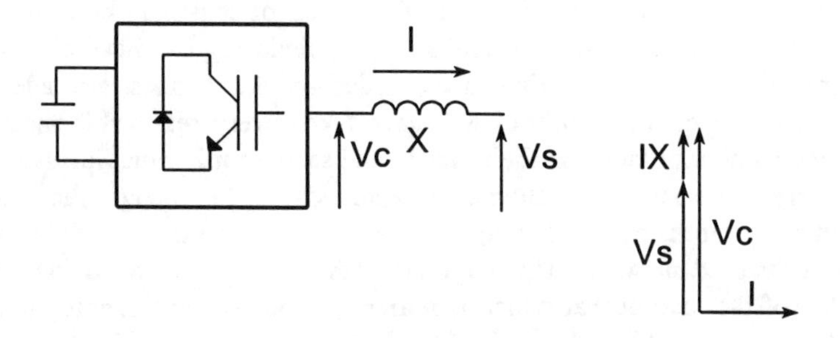

Figure 14.6 Schematic of a distribution STATCOM. Phasor diagram showing capacitive operation (generator convention)

14.2.2 Distribution STATCOM

The converter may be arranged to act as a distribution STATCOM as shown in Figure 14.6. In its simplest implementation, a sinusoidal PWM pattern may be created by combining a triangular carrier wave with a reference signal of the desired output sine wave. Hence a voltage V_c, which approximates to a sine wave is used to control the current flow through the coupling reactor X. If harmonics are ignored then the power flow across the inductor is governed by the well-known equations:

$$\boldsymbol{I} = \frac{\boldsymbol{V}_c - \boldsymbol{V}_s}{jX}$$

or in terms of active and reactive power

$$P = \frac{V_c \times V_s}{X} \sin \delta$$

$$Q = \frac{V_c \times V_s}{X} \cos \delta - \frac{V_s^2}{X}$$

where: V_c - converter output voltage
V_s - system voltage
X - coupling reactance
I - converter current
δ - angle between V_c and V_s

Thus the distribution STATCOM can operate in an identical manner to the transmission STATCOM with the real power flow controlled by the angle between the network and converter voltages and the reactive power flow controlled by the difference in voltage magnitudes. The capacitor is charged with power taken from the network and, in the steady state, the network voltage always leads the converter voltage by a small angle to provide power for converter losses. The main difference with respect to a transmission STATCOM is that the magnitude of V_c is not proportional to the capacitor voltage, as would be the case with a fixed pulse pattern, but can be varied by changing the reference of the PWM control. As with a transmission STATCOM, the capacitor acts as an energy store and its size is chosen based on control and harmonic considerations. A high value of coupling reactance, X, leads to rather slower transient performance but easier control and reduced harmonic distortion of the network.

In practice, the flexibility provided by rapid switching of the IGBTs allows more sophisticated control schemes, based on voltage space vector techniques, to be implemented to provide additional functions such as phase balancing (by injection of negative phase sequence current), active filtering (either using frequency domain or time domain techniques), flicker reduction (by very fast acting, sub-cycle, control schemes).

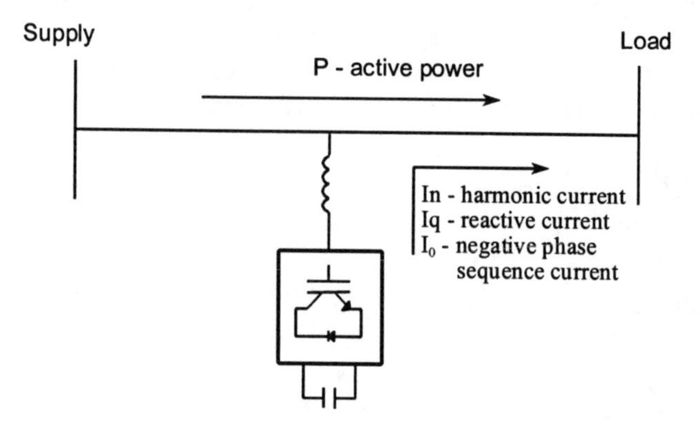

Figure 14.7 Functions of a distribution STATCOM

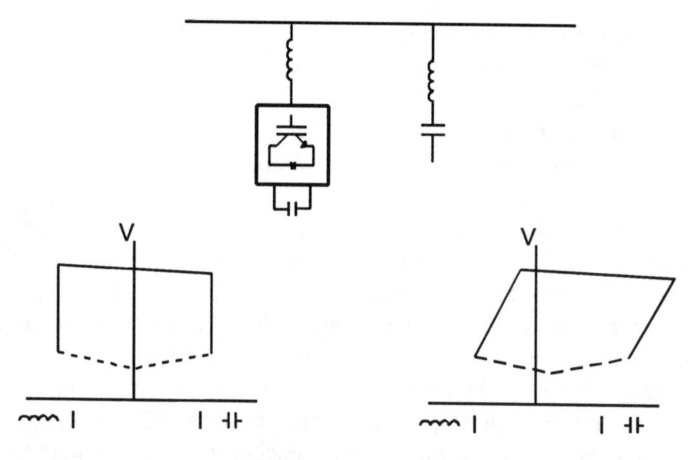

Figure 14.8 Effect of fixed capacitor/filter on STATCOM operating region

The complete STATCOM can be considered as a variable current source, injecting appropriate currents as required (Fig 14.7). For control purposes, high bandwidth measurements of the current from the supply busbar, the converter current, and the network voltage are required. The capacity of the converter is limited by the current and voltage ratings of its various component and so, for example, if the STATCOM is operating at its full capacity providing reactive power support it cannot then also provide flicker control. Hence it is necessary to budget the capacity of the converter for the particular function(s) which are desired. One way of increasing the capacitive capability is to use a fixed capacitor/filter to bias the operating region as shown in Figure 14.8. The operating region of the STATCOM is essentially rectangular as it is limited by the current and voltage ratings of the converter components. However, when combined with a fixed capacitor the operating region is skewed by the reduction in capacitor current with falling system voltage.

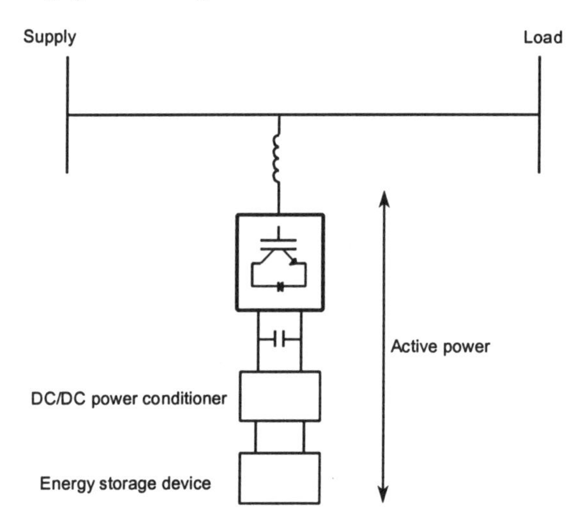

Figure 14.9 Addition of large energy store (e.g. flywheel, battery, SMES) to a STATCOM

Limited storage of energy can be achieved by increasing the size of the DC capacitor. However a number of systems have been installed recently or are under development to connect larger energy stores by an additional DC/DC converter (Figure 14.9). A 6 MJ, 750 kVA SMES system was commissioned in January 1998 in the USA for a power quality application, although the converter topology used was a conventional in-line UPS type arrangement with the battery replaced by the SMES. A 50 MW SMES based systems using shunt connected GTO thryristor converters and a super-conducting magnet operating at 1.8 K has recently been offered for commercial sale in order to provide voltage support for critical manufacturing plants for 5-10 seconds. This STATCOM based equipment will allow voltage dips to be compensated and the additional energy stored in the

SMES will provide time for standby generators to be started in the event of prolonged outages.

14.2.3 Dynamic voltage restorer (DVR)

The distribution STATCOM is a shunt connected device which may be controlled as a current source and used to compensate non-ideal loads and so reduce the distortion of current drawn from the supply. In contrast, the DVR is a series connected device which acts as a voltage source to improve the quality of the voltage supplied by the network. Thus the main function of the DVR is to eliminate or reduce voltage sags seen by sensitive loads, to reduce the phase unbalance seen by the load and to compensate any voltage harmonics existing in the supply.

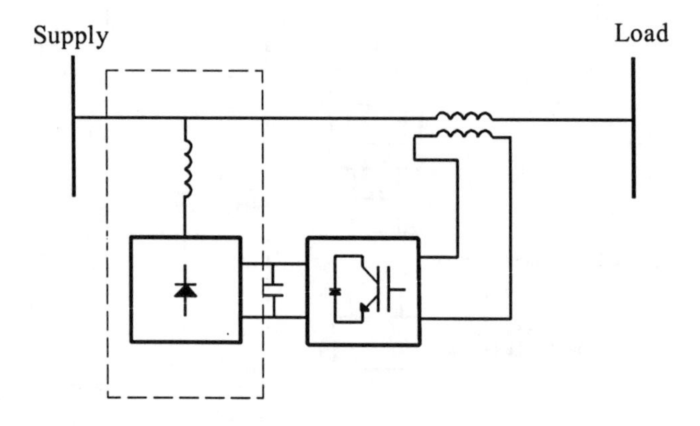

Figure 14.10 Schematic of a Dynamic Voltage Restorer (DVR). Rectifier circuit not used in some implementations

A DVR is shown schematically in Figure 14.10. A voltage source converter is used to inject voltages in series with the distribution feeder using three single phase transformers. By adjusting the angle between the injected voltages and the currents flowing in the circuit either real or reactive power may be injected into, or absorbed from, the feeder. The operation is shown in Figure 14.11. In 14.11(a) the injected voltage is in quadrature with the current flowing in the circuit. Therefore no active power is transferred. In Figure 14.11 (b) the injected voltage V_{DVR} is in phase with the load voltage and so both real and reactive power are transferred. For voltage depressions on a single phase it is possible to draw active power from the healthy phases and inject it into the phase experiencing a depressed voltage provided an overall power balance within the DVR is maintained. In most practical applications of the DVR some energy storage is desirable and, where this cannot be provided economically by capacitors, then an additional energy store can be added to the DC bus through a DC/DC converter in a manner similar to that shown in Figure 14.9 for the STATCOM.

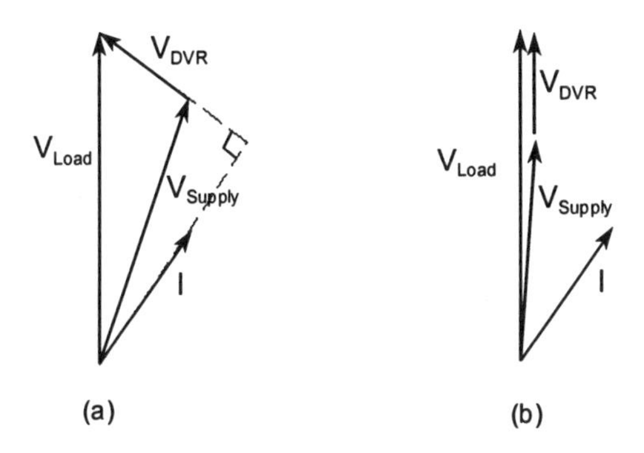

Figure 14.11 Operation of Dynamic Voltage Restorer (DVR): (a) reactive power compensation only (b) active and reactive power compensation [12]

A series connected IGBT converter has been used recently to connect a 2.2 MJ SMES operating at 4.2 K to provide a 75% voltage boost for a 1 MVA load on a paper mill in South Africa. Although the main application was to protect the paper machinery from the effect of voltage sags, it was also reported that the unprotected plant was also susceptible to phase-jumps caused by phase-phase faults on the supply circuit and that the DVR/SMES equipment successfully compensated for this type of disturbance.

As with the distribution STATCOM some filtering of the harmonics at the converter switching frequency may be necessary. The active power required to charge the capacitor and provide for switching losses may be supplied either through the series connected transformers or via the separate rectifier circuit. The converter may also be controlled to inject negative phase sequence or harmonic voltages to improve the voltage waveshape seen by the load.

The main application of the DVR to date has been to mitigate the impact of voltage dips and sags. Typical applications have used DVRs rated at 2-4 MVA, with 660-800 kJ of energy storage in capacitors to provide voltage support of 0.3-0.5 per unit for 300-500 ms. The advantage of a DVR over a STATCOM for the protection of sensitive loads from voltage sags is that a lower rated device can be used. The series transformers must, of course, be rated for full load current but the magnitude of the injected voltage is a design choice based on the assessment of the range of sags against which protection is required. Thus the DVR rating may be calculated [12] as:

$$MVA_{DVR} = MVA_{LOAD} \times \text{Injected Voltage (per unit)}$$

Table 14.2 Energy storage requirement to mitigate voltage sags using a DVR [12]

Energy storage requirement MJ per MW of load served	% of sags restored to 0.9 pu voltage
0.05	70
0.15	90
0.3	99

The paper mill described in Reference [2] has a DVR rated at approximately 50% of the sensitive load in order to restore the load voltage to 0.9 per unit for 98% of the dips seen at that point in the power system.

The other key design choice is the size of the energy store required. Calculations based on a study of power quality in New England indicate that for balanced 3-phase sags which would otherwise depress the voltage to below 0.9 per unit the energy storage requirements are as given in Table 14.2 [12].

As only a minority of faults are 3 phase then the values given in Table 14.2 may be considered to be conservative. However, faults which cause unbalanced voltages on the transmission network generate more balanced voltage sags on the distribution system due to the effect of winding arrangements of the transformers (Table 14.3).

Table 14.3 Voltages on the secondary side of a transformer due to a single-phase to ground fault (zero voltage) on the A phase of the primary side [13]

Winding Connection	Phase-phase voltages			Phase to neutral voltages		
	AB	BC	CA	A	B	C
Yn/yn, Yn/y	58%	100%	58%	0	100%	100%
Y/y, Y/yn,	58%	100%	58%	33%	88%	88%
Dyn, Dy	33%	88%	88%	58%	58%	100%

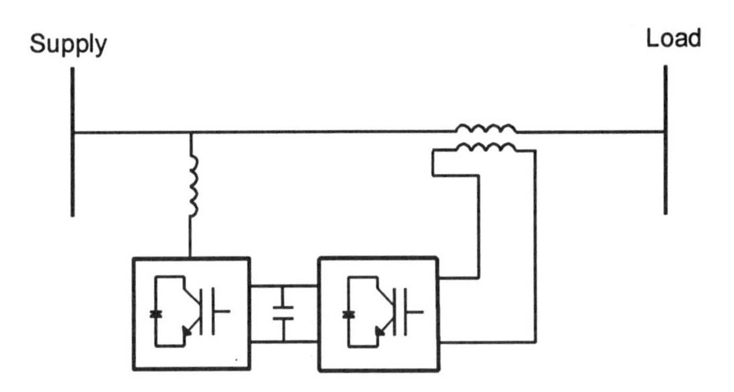

Figure 14.12 Schematic of a distribution UPFC

A further refinement is to use an IGBT bridge for both the shunt and series element. The device then resembles a Unified Power Flow Controller (Figure 14.12). Because each converter can operate with a variable pulse pattern, the operation of the two converters is completely independent and not linked by the DC capacitor voltage. Thus very flexible control is possible. Smale [14] following the work of Akagi [3] and others has investigated reversing the order of the shunt and series elements so that the series element compensates for any pre-distortions in the network voltage and the shunt element is used to compensate for any non-active current drawn by the load. The obvious drawback of using two complete IGBT converters is the cost and so far no commercial application of this topology has been reported.

14.2.4 Active filters

The origins of both the distribution STATCOM and the DVR may be traced to the ideas of transmission FACTS. The concepts were proposed by researchers active in FACTS and the early implementations have been to improve customer power quality by ameliorating sags or to improve network voltage by eliminating flicker. However, STATCOMS and DVRs which use IGBTs are extremely flexible pieces of equipment and can be controlled to carry out a number of functions including harmonic compensation. Active filters were originally proposed to compensate harmonic distortion but are now being considered as sources of fundamental frequency reactive power as well as being suitable for rapid control of network voltages. Thus the two strands of technology appear to be converging.

Active filters have been investigated for a number of years and there have been some commercial installations [3, 15, 16]. Several design variations are described in the literature including: current and voltage source converters, shunt or series connection, control in the time or frequency domain and their application to compensate either individual loads or to damp resonances in the network. Akagi [3] notes that the shunt connected voltage source PWM based inverter has been preferred for almost all the active filters put into commercial service in Japan. These active filters show better filtering characteristics than passive devices but at higher cost. Therefore, the use of hybrid filters consisting of an active and passive section may be desirable in order to reduce both capital cost and operating losses.

Control may be in the time domain where the load current or voltage is controlled to be within some tolerance of a reference signal. This may either be achieved by extracting the fundamental using a filter or, more usually, by an instantaneous orthogonal transformation [16]. The alternative control strategy is to operate in the frequency domain by carrying out a Fourier analysis of the wave form to be controlled. This has the advantage that only those harmonics which are the cause of concern may be controlled and the capacity of the filter may be employed most effectively. The disadvantage is that a fast processor is required if many harmonics are to be considered.

The main application of active filters appears still to be the compensation of particular loads or buildings with, so far little commercial interest in wide area application over the utility distribution system.

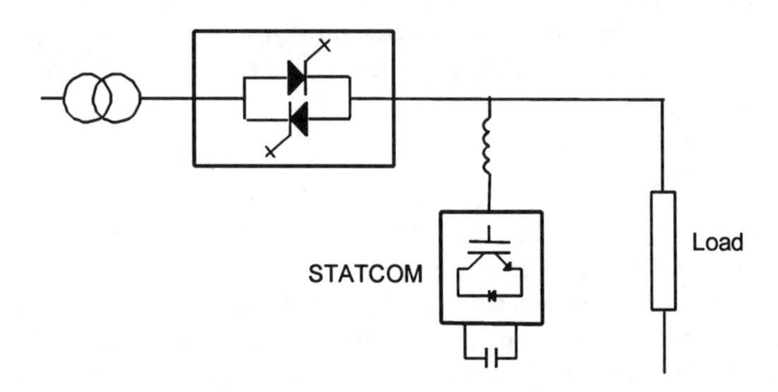

Figure 14.13 Use of a solid state switch with a STATCOM

14.2.5 Solid state switches

A number of solid state switching devices have been proposed based on anti-parallel pairs of GTO thyristors with suitable snubber circuits [17,18]. GTO thyristors can interrupt current with a negligible delay but have very limited overload capability and so it is not possible to make a simple replacement for a conventional distribution circuit breaker using only GTO thyristors. The discrimination of overcurrent protection on conventional distribution circuits is based on time-grading and so upstream circuit breakers carry fault currents for up to several seconds to allow downstream devices to open. Presently available GTO thyristors are not rated for this duty and so applications of solid state switches which do not require an overload capability have been investigated.

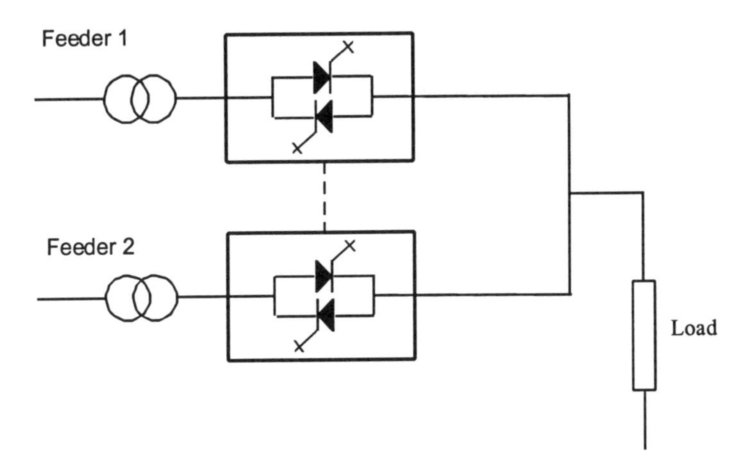

Figure 14.14 Solid state transfer switch

Figure 14.13 shows on application of a solid state switch which is used to isolate a distribution STATCOM and a sensitive load. When a voltage sag is detected the solid state switch opens and both active and reactive power is supplied to the load from the STATCOM. When the network voltage is restored the load is reconnected by the solid state switch. Figure 14.14 shows an alternative approach to improving power quality using a solid state transfer switch. The sensitive load is fed from two alternative independent feeders. If a voltage sag or interruption is detected on the feeder which is supplying the load then that switch is opened and the load fed from the alternative supply. The use of the solid state transfer switch allows the reliability of a load to be improved, by using duplicate supplies, without the degradation in power quality which occurs if additional lengths of circuit are connected to a load busbar.

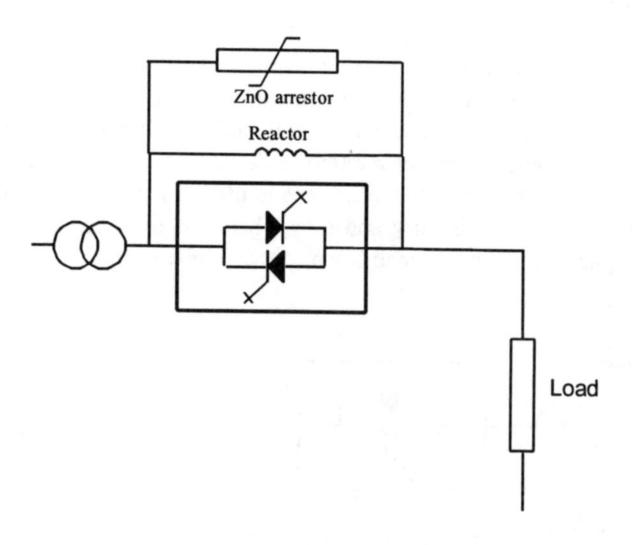

Figure 14.15 Solid state current limiter

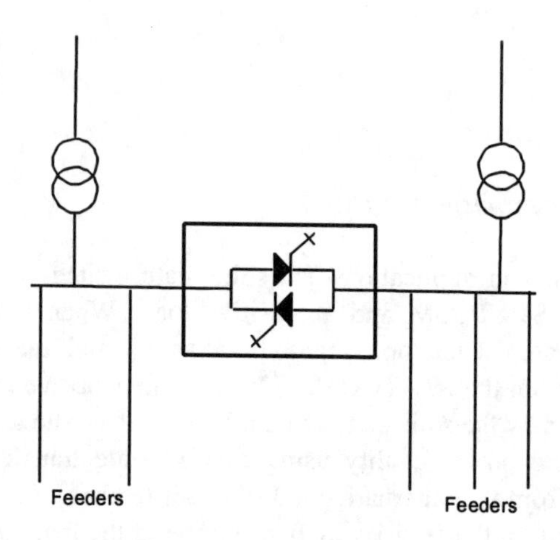

Figure 14.16 Solid state bus section switch

Figure 14.15 shows a solid state fault current limiter where, in the event of a fault downstream of the limiter, the solid state switch opens and inserts a reactor in the circuit. A Zinc Oxide arrestor is required to control the voltage across the limiter. Solid state switches have also been proposed for use as bus-section devices in distribution switchboards. This is shown in Figure 14.16. Use as a bus-section switch is attractive as the losses are low, there is no requirement for discrimination and the device can operate to rapidly limit fault current.

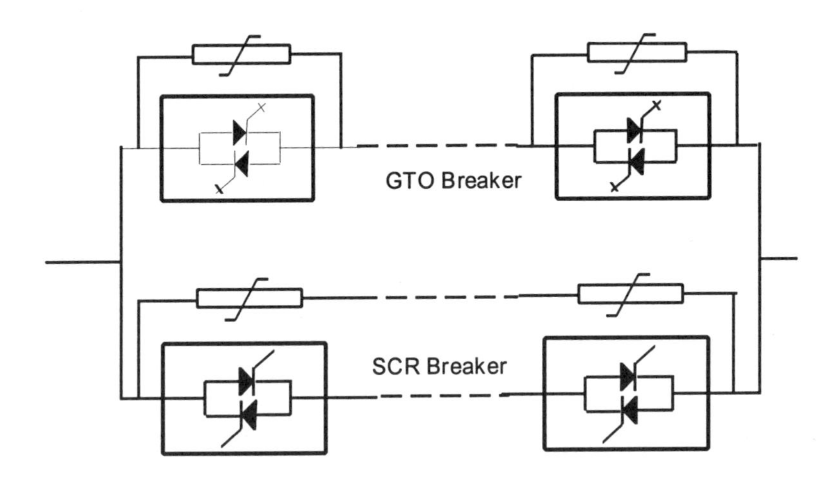

Figure 14.17 Solid state circuit breaker – using both GTO thyristors and SCRs

A complete solid state circuit breaker has been developed using a combination of both GTO thyristors and SCRs (Silicon Controlled Rectifiers or thyristors) [19]. The GTO thyristors provide rapid interruption of the fault current but the parallel path through the SCRs is used to provide sustained fault current to allow downstream devices time to operate, Figure 14.17. In normal operation the feeder current flows through the GTO thyristors with the SCRs not conducting. When either the current or di/dt exceeds a threshold the GTO switch is turned off and, simultaneously, the SCR switch is turned on. The ZnO arrestors are to protect the switches against over-voltages. The snubber circuits are not shown. After a specified time delay, to give down stream protection a chance to operate, the gate drive to the SCR switch is removed and current is interrupted at the next zero-crossing. Thus the circuit breaker uses the fast turn-off characteristic of the GTO switch and the higher sustained rating of the SCR switch. The requirement for both types of semi-conductor switch are due to limitations in the devices and, if improved semiconductors become available, then simpler and more cost effective solid state circuit breakers will be possible.

14.3 Power electronic applications for renewable energy

14.3.1 Generation from new renewable energy sources

New renewable energy sources are generally taken to include: wind, solar, biomass, geothermal and small hydro but excluding large scale hydro-generation schemes [20]. Over the last 5 years, there has been a dramatic change in the perception of these technologies from interesting research topics to potential major sources of electrical energy. A significant driving force behind the implementation of renewable energy generation schemes has been concern with the environment and the need to reduce the environmental impact of electrical power generation. Thus, in many countries the exploitation of new renewable energy sources is receiving considerable government support as one of the ways of reducing CO_2 emissions.

Land based wind turbines have now emerged as a maturing technology with commercially available machines having rotor diameters up to 60 m and ratings of 1.5 MW. The turbines may either be distributed individually or grouped into wind farms of up to 50 machines giving combined ratings of up to 75 MW. Even larger turbines are under development and some 8 GW of wind energy capacity has now been installed throughout the world. The next major development in wind energy is likely to be large installations offshore. There are already three "offshore" wind farms located in rather shallow waters close to land off the coasts of Denmark and Holland but future developments will be as much as 40 km offshore. The Danish government has a commitment to install 4000 MW of offshore wind turbine capacity by 2030 and is commencing with a programme of 750 MW. The UK government has yet to announce its intentions but they are likely to be some 1000-2000 MW by 2010 if the target to provide 10% of UK electricity from renewable energy sources by that date is to be achieved. Initially, the turbines used offshore will be similar to those used on land, however, as the production volumes increase, it is likely that the designs will become tailored for offshore. HVDC using voltage source converters [5] is being investigated as one way of transporting the power from the offshore wind farms to the shore.

Direct conversion of solar energy using photovoltaics (PV) devices is also attracting very significant interest with several multi-national energy companies making large commercial investments. Previously PV was seen as commercially viable only for off-grid power supplies in remote locations. However, the main anticipated market for PV is now in grid-connected applications to provide electrical power into utility distribution networks. One major proposed application is the integration of PV cells into the roofs and facades of commercial buildings. Integration of the cells into the building fabric serves to reduce installation costs as well as ensuring generation close to the loads, during daylight hours, when the buildings are occupied.

Table14.4 Estimated contribution to EU energy supply (source European Commission White Paper on Renewable Energy, November 1997). 1Mtoe is equal to approximately 12 TWh at 100% conversion efficiency.

Energy source	1995	2010
Wind	2.5 GW	40 GW
Hydro	92 GW	105 GW
- Large	82.5 GW	91 GW
- Small	9.5 GW	14 GW
PV	0.03 GW	3 GW
Biomass	44.8 Mtoe	135 Mtoe
Geothermal - Electricity	0.5 GW	1 GW

The biomass resource takes a number of forms including: gas from landfill waste, municipal solid waste and agricultural and forestry crops or residues. The resource can either be used to supply heat loads directly or passed through some form of more or less conventional thermal generating plant. Thus landfill gas is used to fuel reciprocating engines while agricultural and forestry products are burnt in steam raising boilers. The generators used are conventional synchronous rotating machines.

The Commission of the European Union has ambitious plans to encourage the introduction of new renewable energy schemes and Table 14.4 is taken from a White Paper issued for discussion in November 1997.

Campaign targets proposed in the White Paper included:

- Introducing 1 million new PV systems. 500,000 of these would be in the EU on roofs or building facades, the other 500,000 would be exported for decentralised electrification systems in other regions of the world.
- Installation of 10,000 MW of large wind farms
- Development of 10,000 MW of biomass installations

Although not official policy and still the subject of debate, the White Paper gives useful insights into where a significant contribution to bulk energy production might be made. It may be seen that the main contenders are wind energy and biomass. Future hydro capacity is limited by the availability of sites and PV is likely to remain expensive and with a relatively small total capacity during the time-scales envisaged although the rate of growth projected for PV installations is remarkable.

Generation from these new renewable sources differs from conventional central generation in that the sources (e.g. wind, solar and biomass) are diffuse and need to be collected over a wide geographical area. Thus large number of renewable generation plants, ranging from 200 W domestic PV schemes to 50 MW wind farms or biomass plants, are embedded into distribution networks giving rise to the description *dispersed or embedded generation* [21]. Unlike fossil fuel it is not possible to store renewable energy economically and so the generation plant must operate in response to the available resource rather than the required load. As always, storage of electrical energy would be highly desirable and would increase the value of the energy generated by the fluctuating renewable resources. However, bulk storage of electrical energy remains prohibitively expensive. Some renewable generation technologies (e.g. PV and variable speed wind turbines) do not produce 3-phase 50 Hz power from conventional synchronous generators and so require electronic power conditioners.

The main application of power electronics for renewable energy is to interface the renewable energy conversion system with the utility distribution network and to control the distribution network so that it can accept the generation, for which it was not originally designed. Although a variety of innovative equipment has been proposed, the most popular converter remains the voltage source, PWM, IGBT converter described earlier and a number of manufacturers offer the same basic equipment for power quality applications as for interfacing renewable energy conversion plant.

14.3.2 Wind energy

The simplest design of wind turbine operates at fixed rotational speed and consists of an aerodynamic rotor, a gearbox and an induction or asynchronous generator.

It is not possible to use a synchronous generator as this type of electrical machine does not provide sufficient damping in the drive train. Considerable damping is required to control the periodic torque pulsations caused by the rotor blades as they pass through the different wind velocities which increase with height but are reduced in front of the tower. Induction generators are not usually used on large conventional generating plant as they suffer from a number of disadvantages compared to synchronous machines.

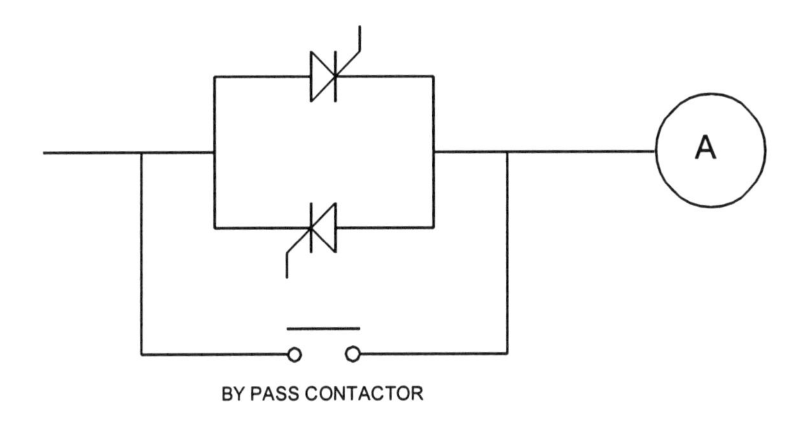

Figure 14.18 Wind turbine soft start

An induction generator draws its magnetising current from the network and so, on direct connection, will draw a large transient reactive current. Also if the rotational speed of the generator is not matched to the network frequency then the connection transient will include a slower component of active power which accelerates or decelerates the drive train. The magnitude of the connection transient is controlled using an anti-parallel thyristor soft-start Figure 14.18. The firing angle of the thyristors is controlled to apply the network voltage to the generator at a defined rate. Once the generator is fully fluxed and the appropriate slip speed achieved, the by-pass contactor is closed. The by-pass contactor is used to reduce operating losses but also because local shunt capacitors are then connected to improve the power factor of the generator and these can be excited into resonance by harmonic currents produced by a soft-start unit. The performance of a soft-start in controlling magnetising inrush can be very good and these currents can be maintained to less than full load values. The size of the active power component depends on how well the rotational speed of the generator is matched to the network frequency and the acceleration of the drive train.

Once connected to the network, an induction generator will operate at a power factor determined by its active power output. It is usual to improve the power factor of each turbine by connecting local fixed shunt capacitors. The disadvantage of this approach is that the degree of compensation is fixed and there is also the possibility of large overvoltages if the connection to the network is lost and the induction generator/shunt capacitor combination resonate to give self-excitation. In order to investigate how a STATCOM might be used to improve the power quality of a large fixed speed wind farm an 8 MVAr, GTO, 12 pulse, 3 level unit was installed at the 24 MW (40 x 600 kW turbines) Rejsby Hede wind farm in Denmark [4, 22]. The objective of the installation was to allow the wind farm to operate at unity power factor without the danger of self-excitation. Hence the control strategy adopted was to minimise the reactive power flowing from the 60 kV network and to control any overvoltage in the event of the 60 kV system becoming isolated from the transmission network.

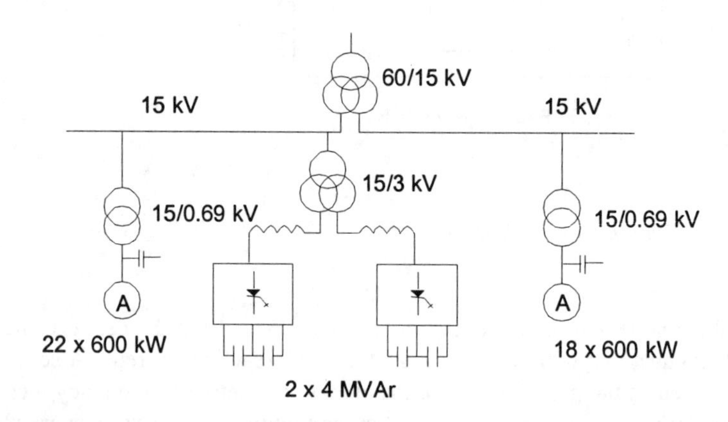

Figure 14.19 Schematic of the Rejsby Hede wind farm and STATCOM

A single line diagram of the installation is shown in Figure 14.19. The converters used were based on existing designs with GTO thyristors. In order to limit network harmonic distortion a three level topology was used with a selective harmonic elimination pulse pattern arranged to eliminate the 11th and 13th harmonics and minimise the 5th and 7th harmonics. The three winding transformer used to connect the STATCOM to the 15 kV busbar also served to reduce the 5th and 7th harmonic distortion. The 60/15 kV transformer was also a three winding type although, in this case, the secondary was divided in order to limit the fault level on the 15 kV busbars. The STATCOM was put into commercial service in early 1998.

A series of studies was carried out to investigate other potential control strategies for a STATCOM on a wind farm of fixed speed wind turbines. This indicated such equipment could be used for voltage control and so increase the rating of a wind farm which might be connected to a Distribution circuit without violating voltage limits. Simulations were used to show that it was possible to increase voltage stability as well as reduce flicker using such equipment.

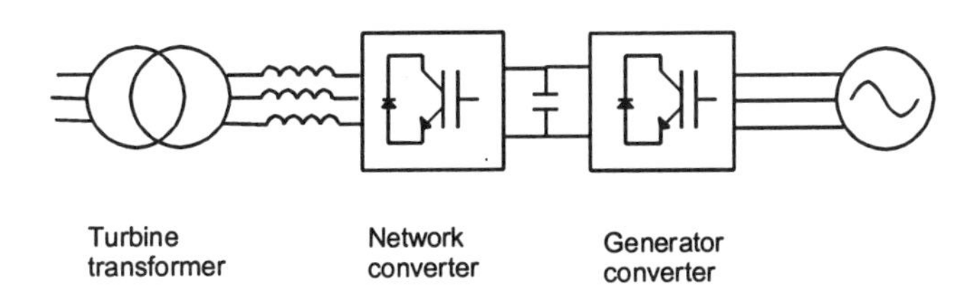

Figure 14.20 Wide range variable speed wind turbine

Although some 1.5 MW wind turbines operate at fixed speed using induction generators other manufacturers prefer to allow the rotor speed to vary. This reduces mechanical loads on the drive train and so allows a lighter design as well as improving output power quality. In principal, variable speed operation also leads to higher energy capture although the additional losses in the converter must be set against this. Operation over a wide speed range may be obtained by passing all the wind turbine current through an AC/DC/AC system. (Figure 14.20), [23]. The generator converter is used to control the voltage on the DC bus while the network converter is used to control the power flow to the network. Usually the wind turbine will also have pitch control of the blades so that once rated power is reached the aerodynamic input power may be reduced. This concept can be combined with the use of large diameter direct drive synchronous generators to eliminate the wind turbine gearbox. The obvious disadvantage of passing all the power through the converters is that the rating of the equipment and the losses will be high (typically 2% per converter). An alternative approach, which was first applied many years ago in very large experimental wind turbines, is to use a wound rotor induction machine and pass only the rotor current through the frequency converter (Figure 14.21). The available speed range is smaller and a slip ring is required but both the losses and the cost of the converters are reduced. For a speed variation of up to plus 10% only, one manufacturer has mounted a controlled resistor on the rotor of a wound induction generator. The rotor resistance is then varied to allow small speed variations and hence reduce torque fluctuations without the expense of a full frequency converter.

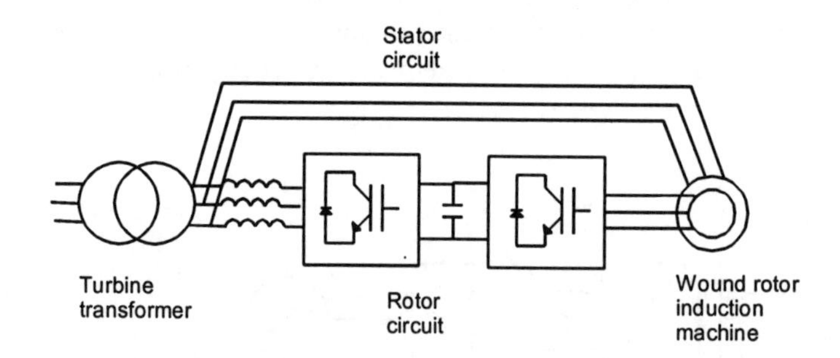

Figure 14.21 Narrow range variable speed wind turbine

If a fully rated network converter is used for wide range variable speed operation then it is also possible, in principal, to use it for reactive power control, network voltage control or even for active harmonic injection. The main power stage is the same voltage source converter bridge discussed earlier. However, injection of reactive power into the network requires a higher DC voltage and the impact of any harmonic currents on the components must be considered. So far little use has been made of the power converters of variable speed wind turbines to control actively the voltage of the distribution network as there is no commercial incentive for the wind farm operator to do this. However, it may be seen that the combined converters of a 50 MW wind farm of variable speed wind turbines could be very effective in improving the distribution network voltage or power quality.

14.3.3 Solar photovoltaic generation

Some large (MW) scale photovoltaic installations have been constructed as demonstration projects, including one installation of 6.5 MW in the US [20] but attention is now focused on the integration of PV equipment into the fabric of buildings. If this becomes commercially attractive it will result in an extremely large number of small generators connected to the distribution network. An individual mono-crystalline or polycrystalline silicon PV cell will develop approximately 0.6 V with its current proportional to the incident radiation up to a maximum of some 30-40 mA/cm^2 and so both series and parallel connection of cells are made to form modules or panels rated at up to 100-200 W. Typical ratings of complete PV installations on commercial buildings may be up to 50-100 kW although they are likely to be much smaller (< 1kW) for domestic homes.

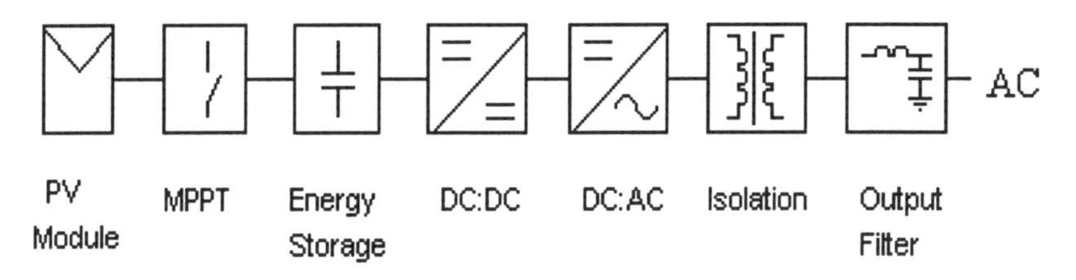

Figure 14.22 Functional block diagram of a PV inverter

Figure 14.22 shows the functional block diagram of a small PV inverter to convert the DC output of one or more PV modules to AC for connection to the network. A Maximum Power Point Tracker (MPPT) is required to draw the maximum available DC power from the PV module. An energy storage element is required to decouple the DC power input from the AC power output. The low voltage DC module output must then be converted to a higher DC voltage and inverted to 230/415 V ac output. Isolation between the PV module and the distribution network is often included and an output filter is used to remove power electronic switching harmonics. The reactance of the output filter is also used to provide the phase shift between the inverter synthesised wave form and the distribution network. There are many designs of such converters, some operating at very high frequency, and opinion is divided on whether, for a large installation, it is better to use a single large inverter, an inverter for each module "string" or even an inverter for each module, the so-called AC module concept [24]. However, the output stage of most converters is of the voltage source type and so will operate in the same basic manner as the very much larger voltage source converters used for variable speed wind turbines. It remains a matter of speculation as to whether large numbers of small PV systems will, in fact, be installed but if commercial conditions are such as to encourage the widespread adoption of PV, then this will result in a very large number of power electronic converters being used to inject power into the distribution network.

14.4 Summary

At present very little power electronics is applied to the distribution system although there are various strands of technology and commercial/administrative changes which are likely to encourage its use. These include the development of improved power semiconductors, increased use of high value and sensitive load equipment as well as the opening of the distribution network to third party access for embedded generation. As FACTS equipment becomes more common on transmission networks then its derivatives on the distribution network (e.g. the distribution STATCOM and the DVR) are also likely to gain acceptance. The initial applications seen to date have been to improve the Power Quality of customers with a high-value load and there is no reason to believe that this market will not develop. The technology of active filters has been demonstrated and been shown to be technically successful, although solid state switchgear appears to be limited by the semiconductor devices presently available. More speculatively, one may consider that the wide-spread development of embedded or dispersed generation will require a major application of power electronics. This can already be seen in the application of power electronic interfaces to the network, in the case of variable speed wind turbines, but there has been less progress in the active control of the distribution network itself. In summary, it would appear that a number of technically interesting power electronic devices and systems have been investigated and demonstrated on the distribution system but that the large scale market has yet to develop.

14.5 Acknowledgments

Thanks are due to the following for providing invaluable sources of information for this chapter: Dr M. Weinhold, Dr J. Hill, Mr A. Campbell and Dr M. Barnes.

14.6 References

1 Michael, F., Weinhold, M., and Zurowski, R., "Application of drive converters for improvement of power quality in distribution systems", *Power Quality*, June 1997 Proceedings, pp. 199-208, 1997.

2 Taylor, G.A., Hill, J.E., Burdon, A.B., and Mattern, K., "Responding to the changing demands of lower voltage networks – the utilisation of custom power systems", CIGRE session 1998, paper 14-303.

3 Akagi, H., "New trends in active filters for power conditioning", *IEEE Transactions on Industry Applications*, 32, (6), pp. 1312-22, 1996.

4 Sobrink, K.H., Jenkins, N., Schettler, F., Pedersen, J.K., Pedersen, K.O.H., and Bergmann, K., "Reactive power compensation of a 24 MW wind farm using a 12-pulse voltage source converter", CIGRE session 1998, paper 14-301.

5 Asplund, G., Lindberg, J., Eriksson, K., Palsson, R., Jiang, H., and Svensson, K., "DC Transmission based on voltage source converters", CIGRE session 1998 paper 14-302.

6 Ramsey, S.M., Cronin, P.E., Nelson, R.J., Bian. J., and Menedez, F.E., "Using distribution static compensators (D-STATCOMS) to extend the capability of voltage limited distribution feeders", *IEEE Rural Electric Power Conference*, April 1996, pp. A4/18-24.

7 Paserba, J., Larsen, E., Lauby, M., et al, "Feasibility study for a Distribution level STATCON" in FACTS Applications, IEEE Power Engineering Society, 1996, IEEE Catalog Number 96TP 116-0.

8 Dugan, R.C., McGranaghan, M.F., and Beaty, H.W., *Electrical power systems quality,* McGraw Hill, New York, 1996.

9 ANSI/IEEE, "IEEE recommended practice for electric power distribution in industrial plants", IEEE Std 141-1986.

10 Becker, C., et al, "Proposed chapter 9 for predicting voltage sags (dips) in revision to IEEE Std 493, the Gold Book", *IEEE Transactions on Industry Applications*, 30, (3), pp 805-21, 1994.

11 ANSI/IEEE, "IEEE recommended practice for powering and grounding sensitive electronic equipment", IEEE Std 1100-1992.

12 Nelson, R.J., Legro, J.R., Gurlaskie, G.T., Woodley, N.H., Sarkozi, M., and Sundaram, A., "Voltage sag relief: Guidelines to estimate DVR equipment ratings", Proceedings of American Power Conference April, 1996.

13 Ping Wang, "The use of FACTS devices to mitigate voltage sags", PhD thesis UMIST, 1997.

14 Smale, M., Jenkins, N., and Collinson, A., "The unified power flow controller as an energy storage interface to the power system", Proceedings of the International Conference on Electrical Energy Storage Systems Applications and Technologies, Chester 16-18 June 1998, pp. 195-200.

15 Akagi, H., "Control strategy and site selection of a shunt active filter for damping harmonic propagation in power distribution systems", *IEEE Transactions on Power Delivery*, 12, (1), pp. 354-63, 1997.

16 Grady, W.M., Samotyj, M.J., and Noyola, A.H., "Survey of active power line conditioning methodologies", *IEEE Transactions on Power Delivery*, 5, (3), pp. 1536-1542, July 1990.

17 Smith, R.K., Slade, P.G., Sarkozi, M., Stacey, E.J., Bank, J.J., and Mehta, H., "Solid state distribution current limiter and circuit breaker: application requirements and control strategies", Paper 92 SM 572-8 PWRD presented at IEEE Summer Power Meeting, 1992.

18 Woodley, N., Sarkhozi, M., Lopez, F., Tahiliani, V., and Malkin, P., "Solid-state 13 kV distribution class circuit breaker: planning, development and demonstration", *Proceedings of 4th International Conference on "Trends in Distribution Switchgear"*, IEE Publication 400 pp 163-7.

19 Abi-Samra, N.C., Johnson, F.O., Labos, and Richardson, R.D., "Applications of the new distribution class solid state breaker", Westinghouse Electric Corporation, Custom Power Technical Papers, March 1996.

20 Boyle, G. (Ed), *Renewable energy, power for a sustainable future*, Oxford University Press, 1996.
21 Jenkins, N., "Tutorial on Embedded generation Parts 1 & 2", *IEE Power Engineering Journal*, June 1995, pp.145-50 and October 1996, pp. 233-39.
22 Bergmann, K., Renz, K., Schettler, F., Stober, R., Tyll, H., Weinhold, M., Sobrink, K., Jenkins, N., and Pederson, K., "Application of GTO-based SVC's for improved performance of the Rejsby Hede windfarm", Proceedings of the Ninth National Power Systems Conference, Kanpur India, 19-21 December 1996.
23 Jones, R., "Power electronic converters for variable speed wind turbines", IEE Colloquium on Power Electronics for Renewable Energy, 16 June 1997, IEE Digest No 1997/170, pp 1/1-8.
24 de Haan, S.W.H., Oldenkamp, H., Frumau, C.F.A., and Bonim, W., "Development of a 100W resonant inverter for ac modules", Proceedings 12th European Photovoltaic Solar Energy Conference, Amsterdam 11-15 April 1994, pp. 395-8.
25 http://www.itic.org/iss_pol/techdocs/iticurv.pdf

Index